# MS&A

Volume 9



For further volumes:
http://www.springer.com/series/8377

Alfio Quarteroni · Gianluigi Rozza
Editors

# Reduced Order Methods for Modeling and Computational Reduction

Springer

Alfio Quarteroni
CMCS-MATHICSE
Ecole Polytechnique Fédérale de Lausanne
Lausanne, Switzerland
and
MOX, Department of Mathematics
"F. Brioschi"
Politecnico di Milano
Milan, Italy

Gianluigi Rozza
SISSA mathLab
International School for Advanced Studies
Trieste, Italy

ISSN: 2037-5255 ISSN: 2037-5263 (electronic)
MS&A – Modeling, Simulation & Applications
ISBN 978-3-319-02089-1 ISBN 978-3-319-02090-7 (eBook)
DOI 10.1007/978-3-319-02090-7
Springer Cham Heidelberg New York Dordrecht London

Library of Congress Control Number: 2013945788



Cover Design: Beatrice ᗺ, Milano
Cover figure: from the top-left to bottom-right. Heuristic representation of a smooth manifold of parametrized solutions and its approximation by selected snapshots used as basis functions; parametrized airfoils (thickness and angle of attack) in a potential flow: visualization of pressure fields in four different configurations; aero-elastic deformations applied on a commercial aircraft model by free-form-deformation technique acting on few geometrical control points. Courtesy of D. Forti and A. Manzoni
Typesetting with LaTeX: PTP-Berlin, Protago TeX-Production GmbH, Germany (www.ptp-berlin.de)
Printing and Binding: Grafiche Porpora, Segrate (MI)

Springer is a part of Springer Science+Business Media (www.springer.com)

# Preface

This book contains selected peer-reviewed contributions submitted by the plenary speakers to the workshop "Reduced Basis, POD and Reduced Order Methods for model and computational reduction: towards real-time computing and visualization?" funded by CECAM (European Center for Atomistic and Molecular Computing) hosted at Ecole Polytechnique Fédérale de Lausanne, Switzerland on 14–16 May 2012 (More info: http://www.cecam.org/workshop-681.html).

This book addresses a wide range of model reduction strategies with applications in various fields.

The increasing complexity of mathematical models used to predict real-world systems, such as climate or the human cardiovascular system, calls for the development of model reduction strategies, that is computationally cheaper algorithms that however still accurately capture the most important features of the phenomena being modelled. Model reduction strategies can be classified according to two main approaches: "reduce-then-model" and "discretize-then-reduce". In the former approach the continuous equations representing the underlying physics are first reduced, e.g. by symmetry assumptions that allow us to consider 1D or 2D equations instead of the full 3D equations, before a computational model is derived. In the latter approach a computational model is obtained by discretizing the continuous equations and only then a reduced model is sought. Some subtopics include spatial dimensionality reduction and multiscale modelling frameworks in the "reduce-then-model" category; state space and parameter space reduction – with a special accent on reduced basis and proper orthogonal decomposition – in the "discretize-then-reduce" category. This monograph focuses more on this second aspect.

It can be regarded as a state of the art survey and integration of several contributions on model order reduction developed in the last few years in different fields and with different purposes, in order to:

1. facilitate a stronger interaction between scientists doing model order reduction on ordinary and partial differential equations (both theory and applications);
2. enhance the state of the art in model reduction making it possible to perform real-time computing for complex systems;

3. address the reliability of reduced order models when compared to more classical high fidelity discretization techniques (and discuss the trade-off between accuracy and costs);
4. improve and generalize parametrization techniques from both a physical and a geometrical point of view, in order to better deal with realistic parametrized geometries and complex parametrized systems;
5. propose and analyze novel sampling and parameter space exploration techniques;
6. certify reduced order modelling for time-dependent problems by simulating long-time phenomena;
7. explore several possible combinations of reduction strategies (like POD, RB, PGD).

The monograph emphasizes model reduction topics in several areas:

1. design, optimization, and control theory in real-time with applications in engineering;
2. data assimilation, geometry registration, and parameter estimation with a special attention to real-time computing in biomedical engineering and computational physics;
3. the treatment of high-dimensional problems in state space, physical space or parameter space;
4. the interactions between different model reduction and dimensionality reduction approaches;
5. the development of general error estimation frameworks which accommodate both model and discretization effects.

The book deals with mathematical models based on both ordinary and partial differential equations with emphasis on engineering and life-sciences applications, including continuum mechanics, fluid dynamics, and transport problems with a methodological focus.

We anticipate a wide range of both academic and industrial problems of high complexity to motivate, stimulate, and ultimately demonstrate the meaningfulness and efficiency of the selected approaches.

The proposed topics open new perspectives in the development of efficient methodologies related with new frontiers in computational science and engineering in order to assist scientists and engineers during design, construction, manufacturing or production phases, and even medical doctors during surgery or diagnosis.

The methodologies we consider are motivated by, optimized for, and applied within two particular contexts: real-time (e.g., parameter estimation or control) and many query (e.g., design, optimization or multimodel/scale simulation). Both contexts are crucial to computational science and engineering and to more widespread adoption and application of numerical methods for partial and/or ordinary differential equations in engineering practice and education.

The real-time context can be found in engineering situations dealing with in-the-field robust parameter estimation (or inverse problems, or nondestructive evaluation), design and optimization, and control. On the other side the many-query context involves multiscale (temporal or spatial) or multiphysics models in which behavior at a larger scale must "invoke" many spatial or temporal realizations of parametrized behavior at a smaller scale.

Both the real-time and many-query contexts present a significant and often unsurmountable challenge to "classical" numerical techniques such as the finite element method (or finite difference, finite volume, spectral methods). These contexts are often much better served by the reduced order modelling techniques (even associated with a posteriori error estimation techniques).

The development of reduced order modelling can perhaps be viewed as a response to the considerations and imperatives described above. In particular, the parametric real-time and many query contexts represent not only computational challenges, but also computational opportunities.

The state of the art is currently moving towards interaction between different reduced order modelling techniques. For example, reduced basis method and proper orthogonal decomposition are combined to solve time-dependent parametrized diffusion-reaction problems with certification of accuracy for the reduced model provided by a posteriori error bounds.

Theoretical studies are being carried out to ensure a better understanding of model order reduction and the reliability and the applicability of the methodologies proposed. Parametrization of systems is advancing by proposing new techniques to deal with more complex configurations and more parameters. Techniques to improve the exploration of parameter space (sampling procedures, greedy algorithms) have been refined, combined, and specialized.

Advances made in computer graphics and physics-based simulation communities can be adapted to produce new methodologies satisfying the real-time needs of applications.

The book is organized as it follows. Chapter 1 deals with model order reduction techniques for coupled multiphysics problems, then Chap. 2 introduces a case study to compare reduced basis method in a time domain and the Loewner rational interpolation in a frequency domain. Chapter 3 focuses on the comparison with some reduced representation approximations by showing different features, in Chap. 4 the emphasis is on reduction techniques for nonlinear parametrized problems. Then Chap. 5 deals with efficient sampling techniques using nonlinear optimization; Chapter 6 introduces parametrized model order reduction by implicit moment matching. In Chap. 7 the focus is on reduced basis method for *parareal* time integration. Stability of reduced order linearized models in computational fluid dynamics is discussed in Chap. 8, followed by Chap. 9 with some more challenges and perspectives for model order reduction in fluid dynamics; window proper orthogonal decomposition and applications is the content of Chap. 10, followed by Chap. 11 with applications of reduced order modeling in aeronautics and medicine.

We would like to thank the reviewers of each chapter for their remarks, criticism and insights that have allowed a significant improvement of the book's content.

We acknowledge the support provided for the workshop also by the MATHICSE Institute of EPFL and by CADMOS (Center for Advanced Modeling Science), a joint-initiative by Ecole Polytechnique Fédérale de Lausanne, University of Lausanne and University of Geneva.

Last, but not least, special thanks to Francesca Bonadei and Francesca Ferrari of Springer Milano for their invaluable help and care.

Lausanne, Milano and Trieste
September 2013

Alfio Quarteroni
Gianluigi Rozza

# Contents

1

# A Novel Approach to Model Order Reduction for Coupled Multiphysics Problems

**Wil H.A. Schilders and Agnieszka Lutowska**

**Abstract** Model order reduction (MOR) has become an important tool in the design of complex high-tech systems. It can be used to find a low-order model that approximates the behavior of the original high-order model, where this low-order approximation facilitates both the computationally efficient analysis and controller design for the system to induce desired behavior. This chapter introduces MOR techniques that are designed especially for coupled problems, meaning that different physical phenomena are simulated in conjunction with each other. The method developed makes use of the reduction of the individual systems, and low rank approximations of the coupling blocks. This is done in such a way that existing software for industrial problems can be adapted in a straightforwward way. An industrial test case is described in detail, so as to demonstrate the effectiveness of the reduction technique.

## 1.1 Introduction

This chapter focuses on the development of a model reduction methodology for coupled multi-physical models to serve the efficient simulation-based design of the underlying coupled systems. Examples of coupled systems are larger systems such as magnetic resonance imaging (*MRI*) scanners, printers/copiers, precision motion stages, foldable solar panels of a space-telescope, down to very small systems such as very large scale integrated (*VLSI*) systems (see for instance [12, 21]) and microelectromechanical systems (*MEMS*) (see for instance [15]). Figure 1.1 shows such examples.

W.H.A. Schilders (✉)
TU Eindhoven, Centre for Analysis, Scientific Computing and Applications
e-mail: w.h.a.schilders@tue.nl

A. Lutowska
TU Eindhoven, Centre for Analysis, Scientific Computing and Applications

A. Quarteroni, G. Rozza (eds.): *Reduced Order Methods for Modeling and Computational Reduction*, MS&A 9. DOI 10.1007/978-3-319-02090-7_1, 

(a)

(b)

**Fig. 1.1.** Coupled systems. (a) Foldable solar panels (courtesy ESA); (b) A MEMS comb drive

The word *system*, which originates from the Greek word s'ustema and the Latin word *sustema*, stands for "a set of interacting or interdependent components forming an integrated whole". In this chapter, the integrated whole is called the *system* or *coupled system* and its individual components are called *sub-systems*. The word *model* as in "physical model" stands for a "representation" for the system under consideration, usually in terms of a set of physical quantities and relations. A coupled system's model consists of the coupled sub-systems' models. A *multi-physical model* is a model which is represented by multiple physical quantities such as temperature, structural mechanical displacements [7], electro-magnetic fields, and so forth. Simple systems in an insulated environment can often be described with few physical quantities and relations, while interacting systems frequently require more of such quantities and relations.

This chapter is about sub-systems which *interact*. When the interaction takes place inside a domain of interest or through the boundary which separates, such a domain of interest from the outside world such a system is called a *coupled system*. If the physical quantities interact through a discrete amount of *inputs* and *outputs* in space, then the system is said to be an *interconnected system* (see for instance [24]) rather than a *coupled system*.

To explain the envisioned reduction, first note that most physical models cannot be solved exactly with contemporary computers. To calculate an approximate solution, the involved physical quantities such as an electromagnetic field are first discretized, i.e., represented by a finite number of *degrees of freedom*, after which the physical equations are reformulated for the discretized physical quantities, leading to a *discrete system* of equations. This process is called *discretization* of the model. An accurate representation of physical quantities such as an electromagnetic field can require millions of degrees of freedom and consume a considerable amount of data storage and computation time. Therefore, an analysis of a coupled system's dynamic behavior can require excessive amounts of data storage and computation time.

We focus on state-of-the-art model order reduction techniques which *reduce the system as a whole based on available reduction techniques for the individual sub-systems*. Such methods are scarcely available and mostly in development. They have an advantage that the individual sub-systems can be reduced in parallel (see [3]) with the method best suited for each of them. This can save a considerable amount of data storage and computational time since these systems are also smaller than the system as a whole. On the other hand, one must figure out how to couple the individually reduced models to a reduced model for the whole, i.e., need to figure out how to effectively deal with interior couplings/interconnections.

Our reduction methods are primarily for coupled time-invariant linear models. Time-dependent linear models, *affine models* (such as presented in [4]) and *non-linear models* (see for instance [14, 23]) require other than the presented reduction techniques. Furthermore, we restrict ourselves to Krylov subspace projection techniques (see [11]).

In more detail, without loss of generality, we focus at systems which consist of two coupled subsystems. We suggest a method for the parallel reduction of the individual sub-systems, call it the *Separate Bases Reduction* algorithm (SBR), and show how to create a reduced model for the whole system based on the reduced parts. Furthermore, we show that this algorithm applied to coupled systems matches at least the same amount of moments as a standard method applied to the whole system would (see [24] for interconnected systems). We establish that a large amount of internal couplings leads to large and hence undesirable reduced models and show that this can be overcome with the use of a generalized singular value decomposition (GSVD) based reduction of the coupling blocks. However, the use of a GSVD-based approximation leads to an approximation of the moments – which as benchmark examples show can still be quite accurate.

The remainder of this chapter is focused on the presentation of the SBR algorithm and the GSVD reduction of the internal couplings. It is organized as follows. Section 1.2 describes Krylov subspace techniques, focusing on *coupled* and *interconnected* time-invariant linear systems. First, it shows what happens if standard techniques are applied to the coupled system as a whole – it shows that the *block structure* is lost. Next, it introduces existing techniques from the literature such as [1,6,9], still based on Krylov subspace methods for the coupled system as a whole, which preserve the block-structure and the number of matched moments. At the end of this chapter, we show an alternative method to efficiently calculate the second Krylov projector and extend the proof of [6] to a more general case, under assumptions.

In Sect. 1.3 we assume that Krylov subspace reduction methods are already available for the individual sub-systems and based thereon, we focus on the construction of a reduced-order model for the system as a whole. We show that this is possible (and also that moments are matched) in Theorem 1.2 and call the approach the Separate Bases Reduction algorithm (SBR). In Subsection 1.3.6 we show that the SBR algorithm also matches the standard double amount of moments if one uses two Krylov subspace projectors instead of one.

In Sect. 1.4 we show that the replacement of the coupling blocks by an explicitly rank-revealing GSVD based components leads to the same Krylov subspaces and hence matched moments. Approximations based on a few of the dominant modes lead to quite accurate moment approximation.

Finally, in Sect. 1.5 we apply the SBR algorithm to a benchmark system. The system under consideration is scaled in a specific manner such that it is numerically better conditioned. We conclude with some remarks and recommendations for further research in Sect. 1.6.

## 1.2 Block-Structure Preserving Model Order Reduction

Model order reduction is frequently based on Krylov subspace projections. The starting point is a linear time-invariant system, that in the Laplace domain is given by (later we will also use small letters $\mathbf{x}$, $\mathbf{y}$ for unknowns in the Laplace domain)

$$\begin{aligned} s\mathbf{E}\mathbf{X}(s) &= \mathbf{A}\mathbf{X}(s) + \mathbf{B}\mathbf{U}(s) \\ \mathbf{Y}(s) &= \mathbf{C}^T\mathbf{X}(s). \end{aligned} \tag{1.1}$$

The left side of Fig. 1.2 represents a schematic model of an *interconnected system* which consists of four sub-systems and a number of interconnections. These interconnections can be realized in different ways, which will be focused on in Sect. 1.3. The right side of Fig. 1.2 shows the system matrix $\mathbf{A}$ which corresponds to the graph on the left. The matrix $\mathbf{A}$ has a visible block-structure. Each of the gray diagonal blocks corresponds to one *sub-system*. The off-diagonal blocks are related to the *interconnections*. The blue dots in the off-diagonal blocks show that the two corresponding sub-systems are *interconnected*. The empty off-diagonal blocks show that there is no coupling between the corresponding two sub-systems.

In general, a system of $k$ components, can be described by a linear system

$$\begin{aligned} s\begin{bmatrix} \mathbf{E}_{11} & \cdots & \mathbf{E}_{1k} \\ \vdots & \ddots & \vdots \\ \mathbf{E}_{k1} & \cdots & \mathbf{E}_{kk} \end{bmatrix} \begin{bmatrix} \mathbf{X}_1 \\ \vdots \\ \mathbf{X}_k \end{bmatrix} &= \begin{bmatrix} \mathbf{A}_{11} & \cdots & \mathbf{A}_{1k} \\ \vdots & \ddots & \vdots \\ \mathbf{A}_{k1} & \cdots & \mathbf{A}_{kk} \end{bmatrix} \begin{bmatrix} \mathbf{X}_1 \\ \vdots \\ \mathbf{X}_k \end{bmatrix} + \begin{bmatrix} \mathbf{B}_1 \\ \vdots \\ \mathbf{B}_k \end{bmatrix} \mathbf{U} \\ \mathbf{Y} &= [\mathbf{C}_1^T, \cdots, \mathbf{C}_k^T] \begin{bmatrix} \mathbf{X}_1 \\ \vdots \\ \mathbf{X}_k \end{bmatrix}, \end{aligned} \tag{1.2}$$

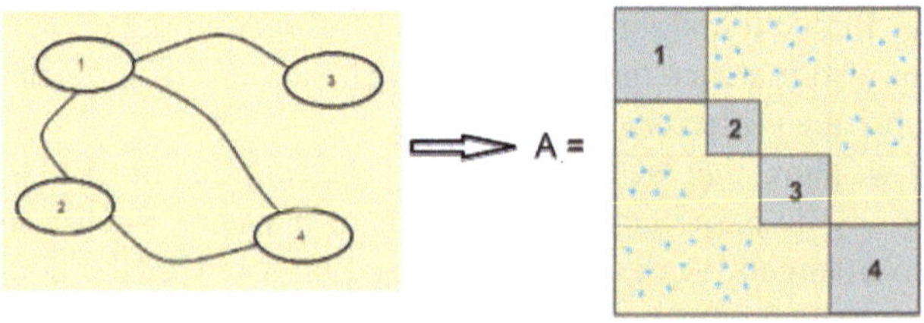


**Fig. 1.2.** Modeling of a coupled system

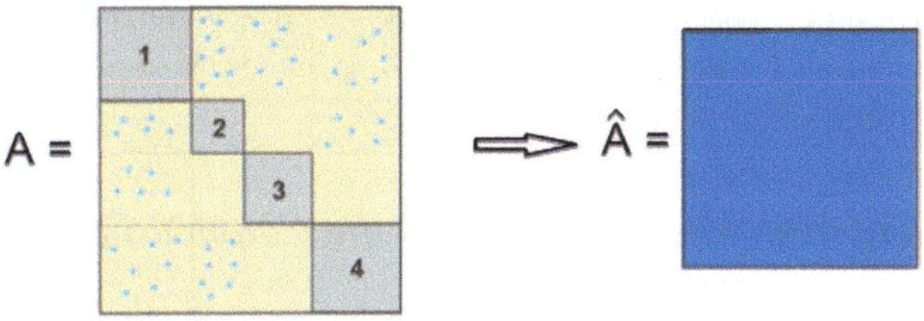


**Fig. 1.3.** Loosing of the structure in the reduced-order matrix $\hat{\mathbf{A}}$

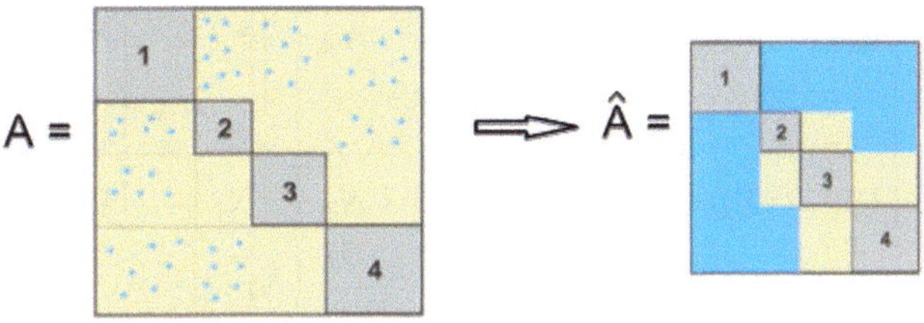


**Fig. 1.4.** Block structure preservation in the reduced-order matrix $\hat{\mathbf{A}}$

where the $\mathbf{X}_i \in \mathbb{R}^{N_i}$, $\mathbf{N}_i \in \mathbb{N}$, $i = 1, \ldots, k$, and the corresponding sub-blocks have compatible dimensions, where typically the off-diagonal blocks are not square. Naturally, we would like to still be able to recognize this type of *block-structure* in a reduced-order system matrix $\hat{\mathbf{A}}$. Unfortunately, if we apply a standard Krylov subspace reduction technique to the matrix $\mathbf{A}$ we unavoidably lose the block-structure and obtain a non-structured *dense* reduced-order matrix $\hat{\mathbf{A}}$ as shown in Fig. 1.3. In the next two subsections, we present a brief overview of Krylov-subspace based block-structure preserving reduction techniques. Such techniques applied to a structured matrix $\mathbf{A}$ result in a reduced-order matrix $\hat{\mathbf{A}}$ like the one shown in Fig. 1.4. Although the potential sparse nature of the interconnection off-diagonal blocks is lost, one can still recognize the system's general block-structure. The diagonal blocks still correspond to the reduced-order sub-systems and the zero blocks related to uncoupled sub-systems are preserved. The reduction techniques of this type are called block-structure preserving (BSP) methods (see for instance [9]). For more information about this type of technique the reader can consult for instance [18].

For the sake of simplicity assume that there are two coupled sub-systems ($k = 2$ in (1.2)). Then the system matrix has the block structure

$$\begin{bmatrix} \mathbf{A}_{11} & \mathbf{A}_{12} \\ \mathbf{A}_{21} & \mathbf{A}_{22} \end{bmatrix}$$

We call such a system an *interconnected system* if $\mathbf{A}_{12}$ and $\mathbf{A}_{21}$ are explicitly defined by means of their inputs and outputs, i.e., if for instance $\mathbf{A}_{12} = \mathbf{B}_3\mathbf{C}_4^{\mathrm{T}}$. Otherwise, if $\mathbf{A}_{12}$ and $\mathbf{A}_{21}$ are specified in unfactored form, we call the system a *coupled system*. However, it is reasonable to assume that even for the blocks specified in unfactored form there might be defined related input and output operators, i.e., that there can be constructed $\mathbf{B}_3$ and $\mathbf{C}_4$ such that for instance $\mathbf{A}_{12} = \mathbf{B}_3\mathbf{C}_4^{\mathrm{T}}$. [13] considers possible construction methods for the input and output maps when $\mathbf{A}_{12}$ and $\mathbf{A}_{21}$ are specified in unfactored form.

### 1.2.1 Moment Matching Methods for the Coupled Formulations

We will begin with BSP methods that are directly applicable to coupled systems of the form (1.2)

$$s\begin{bmatrix}\mathbf{E}_{11} & \cdots & \mathbf{E}_{1k}\\ \vdots & \ddots & \vdots\\ \mathbf{E}_{k1} & \cdots & \mathbf{E}_{kk}\end{bmatrix}\begin{bmatrix}\mathbf{X}_1\\ \vdots\\ \mathbf{X}_k\end{bmatrix} = \begin{bmatrix}\mathbf{A}_{11} & \cdots & \mathbf{A}_{1k}\\ \vdots & \ddots & \vdots\\ \mathbf{A}_{k1} & \cdots & \mathbf{A}_{kk}\end{bmatrix}\begin{bmatrix}\mathbf{X}_1\\ \vdots\\ \mathbf{X}_k\end{bmatrix} + \begin{bmatrix}\mathbf{B}_1\\ \vdots\\ \mathbf{B}_k\end{bmatrix}\mathbf{U}$$

$$\mathbf{Y} = [\mathbf{C}_1^T, \cdots, \mathbf{C}_k^T]\begin{bmatrix}\mathbf{X}_1\\ \vdots\\ \mathbf{X}_k\end{bmatrix}.$$

This type of methods is studied in more detail in for instance [2,6,9]. These methods aim at the creation of a reduced-order model whose matrices exhibit the original block-structure and whose transfer function matches a number of moments of the transfer function of the original system. As for standard Krylov methods, the moment matching property is realized by projecting the original system matrices onto the appropriate input- and/or output-based Krylov subspaces by using the matrices $\tilde{V}$ and $\tilde{W}$ for a chosen expansion point $s_0 \in \mathbb{C}$. However, to preserve the block structure of the original system, the reduction bases also need to have a special shape. They are created by partitioning the matrices $\tilde{\mathbf{V}}$ and $\tilde{\mathbf{W}}$ into $k$ sub-blocks (with $k$ being the number of sub-systems)

$$\tilde{\mathbf{V}} = \begin{bmatrix}\mathbf{V}_1\\ \vdots\\ \mathbf{V}_k\end{bmatrix} \quad \text{and} \quad \tilde{\mathbf{W}} = \begin{bmatrix}\mathbf{W}_1\\ \vdots\\ \mathbf{W}_k\end{bmatrix},$$

where the number of rows in the blocks $\mathbf{V}_i,\ \mathbf{W}_i,\ i = 1,\ldots,k$ corresponds to the number of rows of the diagonal blocks $\mathbf{A}_{ii}$. Next, the blocks $\mathbf{V}_i$ and $\mathbf{W}_i$ are used to build block-diagonal reduction matrices $\mathbf{V}$ and $\mathbf{W}$

$$\mathbf{V} = \begin{bmatrix}\mathbf{V}_1 & & \\ & \ddots & \\ & & \mathbf{V}_k\end{bmatrix} \quad \text{and} \quad \mathbf{W} = \begin{bmatrix}\mathbf{W}_1 & & \\ & \ddots & \\ & & \mathbf{W}_k\end{bmatrix} \qquad (1.3)$$

and the reduced-order system is obtained by projecting the original matrices

$$\hat{\mathbf{A}} = \mathbf{W}^T \mathbf{A}\mathbf{V}, \quad \hat{\mathbf{E}} = \mathbf{W}^T \mathbf{E}\mathbf{V}, \quad \hat{\mathbf{B}} = \mathbf{W}^T \mathbf{B}, \quad \hat{\mathbf{C}} = \mathbf{V}^T \mathbf{C}. \tag{1.4}$$

Note that since the splitting of the matrices $\tilde{\mathbf{V}}$ and $\tilde{\mathbf{W}}$ into sub-blocks may create linearly dependent columns, one needs to apply a *re-orthogonalization* of the matrices $\mathbf{V}$ and $\mathbf{W}$ to remove every possible linear dependence. Moreover, after re-orthogonalization, one has to assure, that the matrices $\mathbf{V}$ and $\mathbf{W}$ have the same number of columns. This can be done by adding the necessary number of random orthogonal columns to the matrix with the smallest amount of columns.

For the reduction bases created in the way described above, the following theorem holds.

**Theorem 1.1** *Let $\tilde{\mathbf{V}}$ and $\tilde{\mathbf{W}}$ span the input- and output-based Krylov subspaces of the rth order around the expansion point $s \in \mathbb{C}$ for the system* (1.2). *If*

$$colspan\tilde{\mathbf{V}} \subseteq colspan\mathbf{V} \qquad and \qquad colspan\tilde{\mathbf{W}} \subseteq colspan\mathbf{W},$$

*then a reduced-order system computed as in* (1.4) *has the transfer function that matches $2p$ moments of the transfer function of the original system* (1.2).

There are several examples of methods that satisfy the foregoing. Paper [6] presents *SPRIM*, a structure preserving reduced order method for interconnect macromodeling. It focuses on an *RLC circuit* application, as model order reduction methods are of importance to *microchip* manufacturers since complex microchips such as processors contain many interconnected substructures. The relevant equations are (notation as in [6])

$$\mathscr{G}\mathbf{x} + \mathscr{C}\mathbf{x}' = \mathscr{B}u \tag{1.5}$$

with

$$\mathscr{G} = \begin{bmatrix} \mathbf{E}_g^{\mathrm{T}}\mathbf{G}\mathbf{E}_g & \mathbf{E}_l^{\mathrm{T}} \\ -\mathbf{E}_l & \mathbf{0} \end{bmatrix}, \quad \mathscr{C} = \begin{bmatrix} \mathbf{E}_c^{\mathrm{T}}\mathbf{C}\mathbf{E}_c & \mathbf{0} \\ \mathbf{0} & \mathbf{L} \end{bmatrix}, \quad \mathscr{B} = \begin{bmatrix} \mathbf{E}_i^{\mathrm{T}} \\ \mathbf{0} \end{bmatrix},$$

where $\mathbf{G}$, $\mathbf{C}$, and $\mathbf{L}$ are *symmetric positive definite* (square) matrices. The matrices $\mathbf{E}_g$, $\mathbf{E}_c$, $\mathbf{E}_l$ and $\mathbf{E}_i$ are parts of an adjacency matrix $\mathbf{E}$ which describes the connectivity of the electronic circuit, the subscripts $g, c, l, i$ stand for branches containing resistors, capacitors, inductors and current sources. The SPRIM related Laplace domain transfer function $\mathbf{H}_{\text{SPRIM}}$ is

$$\mathbf{H}_{\text{SPRIM}}(s) = \mathscr{B}^{\mathrm{T}} \left(\mathscr{G} + s\mathscr{C}\right)^{-1} \mathscr{B}$$

where $\mathscr{B}$, $\mathscr{C}$ and $\mathscr{G}$ are re-written

$$\mathscr{C} = \begin{bmatrix} \mathbf{C}_1 & \mathbf{0} \\ \mathbf{0} & \mathbf{C}_2 \end{bmatrix}, \quad \mathscr{G} = \begin{bmatrix} \mathbf{G}_1 & \mathbf{G}_2^{\mathrm{T}} \\ -\mathbf{G}_2 & \mathbf{0} \end{bmatrix}, \quad \mathscr{B} = \begin{bmatrix} \mathbf{B}_1 \\ \mathbf{0} \end{bmatrix}.$$

The paper presents a reduction basis $\mathbf{V}$ of the type (1.3) in [6, (21)] and proves in [6, Theorem 3] that it ($\mathbf{W} = \mathbf{V}$) preserves $2p$ moments, double the amount preserved by *PRIMA*.

The technique proposed in [5] is motivated by the fact, that for some applications the single-point expansion does not give a sufficient approximation accuracy in the frequency range. On the other hand, using a multi-point expansion can result in excessively large models, especially for systems with many external inputs and outputs. The method proposed in the paper mentioned above, is based on creating a reduction space that consists of a number of sampling matrices $\mathbf{Z}_j,\ j = 1,\dots,p$, computed for the system (1.2) for $p$ sampling points $s_j$ as follows

$$\mathbf{Z}_j = (s_j\mathbf{E} - \mathbf{A})^{-1}\mathbf{B}.$$

In other words, $\mathbf{Z}_j,\ j = 1,\dots,p$ is a vector (or a matrix) that, after projecting the system (1.2) onto, will match the 0th moment around the point $s_j$ of the original transfer function, since it consists of the input based starting matrix for the Krylov subspace for $s_j$. After computing $p$ samples, the total sampling matrix $\mathbf{Z}$ is defined as

$$\mathbf{Z} = [\mathbf{Z}_1,\dots,\mathbf{Z}_p].$$

Next, following the block-structure presented by the system matrices, matrix $\mathbf{Z}$ is split row-wise into $k$ blocks $\tilde{\mathbf{V}}_i,\ i = 1,\dots,k$

$$\mathbf{Z} = \begin{bmatrix} \tilde{\mathbf{V}}_1 \\ \vdots \\ \tilde{\mathbf{V}}_k \end{bmatrix}$$

and a block-diagonal projector is created

$$\tilde{\mathbf{V}} = \begin{bmatrix} \tilde{\mathbf{V}}_1 & & \\ & \ddots & \\ & & \tilde{\mathbf{V}}_k \end{bmatrix}.$$

Finally, the singular value decomposition (SVD) is performed on each of the blocks separately, to produce the orthogonal matrix $\mathbf{V}$

$$\mathbf{V} = \begin{bmatrix} \mathbf{V}_1 & & \\ & \ddots & \\ & & \mathbf{V}_k \end{bmatrix}$$

where $\mathbf{V}_i, i = 1,\dots,k$ is an orthogonal basis for $\tilde{\mathbf{V}}_i$. At this point, further reduction in size is possible, by removing from the bases $\mathbf{V}_i, i = 1,\dots,k$ the columns that correspond to to small singular values. Having the reduction bases $\mathbf{V}$, one can project the original system in the way defined in (1.4).

A noticeable advantage of the technique described above is, next to the block-structure preservation, the possibility of reducing different sub-systems with different reduction ratio, determined for each sub-system separately, based on the singular values related to this sub-block as well as the importance of the considered sub-system in the total coupled system.

### *1.2.2 Two-Sided Structure Preserving Methods*

In this section we will explain how the two-sided projection idea can be implemented in case of the block-structure preserving methods. A detailed explanation of the two-sided methods one can find for instance in [8]. Generally speaking, the use of a two-sided reduction method means, that the system is projected onto two subspaces, **V** and **W**, based on input and output matrices, respectively. In case of the coupled system (1.10) (defined somewhat later), the reduction matrices **V** and **W**, for an expansion point $s_0 \in \mathbb{C}$, are built according to the following algorithm:

1. Create matrix $\tilde{\mathbf{V}}$, whose columns span the $n$th Krylov subspace around $s_0 \in \mathbb{C}$

$$\tilde{\mathbf{V}} = \mathscr{K}_n(\mathbf{P}(s_0), \mathbf{R}(s_0)),$$

   where $\mathbf{P}(s_0)$ and $\mathbf{R}(s_0)$ are

$$\mathbf{P}(s_0) = (s_0\mathbf{E} - \mathbf{A})^{-1}\mathbf{E} \quad \text{and} \quad \mathbf{R}(s_0) = (s_0\mathbf{E} - \mathbf{A})^{-1}\mathbf{B}.$$

2. Create matrix $\tilde{\mathbf{W}}$, whose columns span the $n$th Krylov subspace around $s_0 \in \mathbb{C}$

$$\tilde{\mathbf{W}} = \mathscr{K}_n(\mathbf{S}(s_0), \mathbf{T}(s_0)),$$

   where $\mathbf{S}(s_0)$ and $\mathbf{T}(s_0)$ are

$$\mathbf{S}(s_0) = (s_0\mathbf{E} - \mathbf{A})^{-T}\mathbf{E}^T \quad \text{and} \quad \mathbf{T}(s_0) = (s_0\mathbf{E} - \mathbf{A})^{-T}\mathbf{C}.$$

3. Build the block-diagonal reduction matrix **V** with $N_1 + N_2 = N$ rows

$$\mathbf{V} = \begin{bmatrix} \mathbf{V}_1 & \mathbf{0} \\ \mathbf{0} & \mathbf{V}_2 \end{bmatrix},$$

   where $\mathbf{V}_1$ and $\mathbf{V}_2$ contain the first $N_1$ respectively last $N_2$ rows of the matrix $\tilde{\mathbf{V}}$.
4. Build the block-diagonal reduction matrix **W** with $N_1 + N_2 = N$ rows

$$\mathbf{W} = \begin{bmatrix} \mathbf{W}_1 & \mathbf{0} \\ \mathbf{0} & \mathbf{W}_2 \end{bmatrix},$$

   where $\mathbf{W}_1$ and $\mathbf{W}_2$ contain the first $N_1$ respectively last $N_2$ rows of the matrix $\tilde{\mathbf{W}}$.

Different algorithms lead to $\tilde{\mathbf{V}}$ and $\tilde{\mathbf{W}}$ (and hence **V** and **W**) with different specific properties (such as orthogonality or bi-orthogonality). Some properties and their advantages and disadvantages are discussed in [17].

The described BSP algorithm results in a block-structured reduced order system and uses both inputs *and* outputs. Consequently, the BSP-based reduced order system's transfer function matches twice as many moments of the original system's transfer function.

## 1.3 Separate Bases Reduction Algorithm

Model order reduction techniques, designed especially for coupled or interconnected systems, became a new field of research in recent years. The common feature of this type of methods is the use of a special block-diagonal form reduction basis $\mathbf{V}$

$$\mathbf{V} = \begin{bmatrix} \mathbf{V}_1 & & \\ & \ddots & \\ & & \mathbf{V}_k \end{bmatrix} \tag{1.6}$$

that results from the splitting a matrix $\tilde{\mathbf{V}}$ created by a Krylov method applied directly to the coupled system. This approach allows for preservation of the zero-blocks in the coupled system's coefficient matrix. Such blocks appear when two of the sub-systems are not coupled (interconnected) or the coupling holds only in one direction. An example of uni-directional coupling can be a case of a vibrating structure, where the movement of the structure causes acoustic noise, but there is no influence (feed-back) of the acoustic behavior of the system on it's dynamics.

Due to the fact that the zero-blocks are preserved in the reduced system, such MOR techniques are called *block structure preserving* (*BSP*) model reduction methods. Their application usually results in a good approximation of the original model. For most of them one can prove the moment matching property. However, this type of methods also has three important drawbacks:

- Though $\mathbf{V}$ in (1.6) (possibly) matches the same (number of) moments as $\tilde{\mathbf{V}}$, it has $k$ times more column vectors and therefore leads to a $k$ times larger reduced system.
- The calculation of $\tilde{\mathbf{V}}$ requires (repeatedly) solving systems with the entire coupled system's coefficient matrix which can be computationally (time- and memory-wise) expensive.
- In practice, the reduction techniques based on an uncoupled formulation of the system (see e.g. [24]) are restricted to the case of interconnected systems with a limited number of interconnections. Otherwise, the reduction procedure is not very efficient, since the dimension of the reduction basis (hence, the reduced-order model) grows very fast. Moreover, such techniques assume that the inputs $\mathbf{B}$ and outputs $\mathbf{C}$ of the sub-systems are both explicitly available. In case of a coupled system these are not explicitly available, only their product $\mathbf{BC}$ is.

In the remainder of this chapter, we will focus on the second and third issue. We present a reduction algorithm suitable for systems, coupled through a large number of couplings. We introduce a reduction technique based on an uncoupled formulation of a coupled system, called *Separate Bases Reduction* (*SBR*) algorithm.

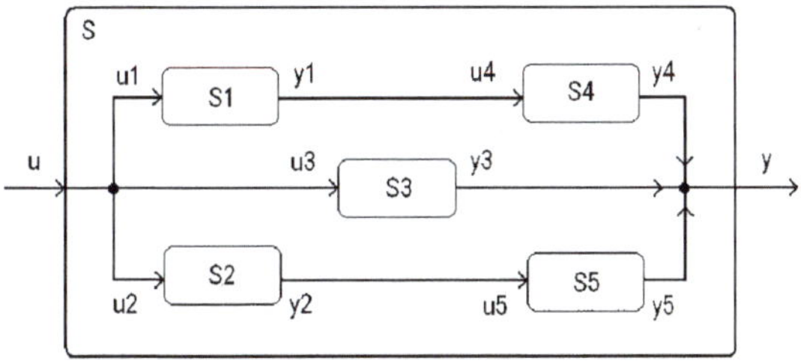


**Fig. 1.5.** Schematic representation of the interconnected system $S$

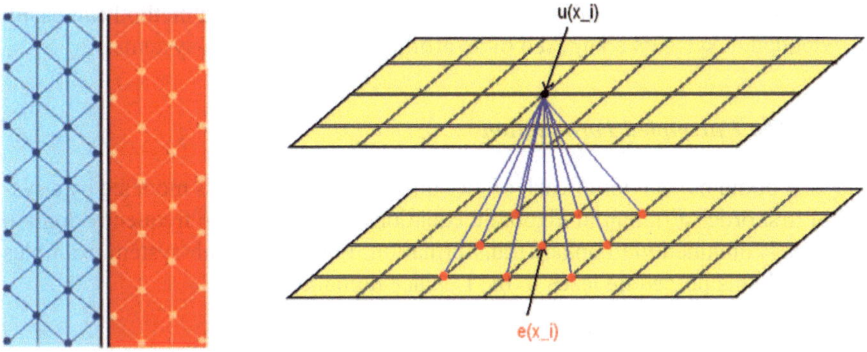


(a) (b)

**Fig. 1.6.** Different types of strong coupling. (a) an interface coupling; (b) a strong coupling between different physical domains

It creates a reduction basis for each sub-system separately, hence is computationally cheaper compared to the reduction techniques that use a coupled formulation such as the BSP methods discussed in Sect. 1.2. However, the algorithm still suffers from the third point of the drawback list presented above. They can be easily applied to interconnected systems of a form shown in Fig. 1.5, where the sub-systems are not strongly interconnected (i.e. each sub-system exchanges information only with a small number of other sub-systems). We suggest a way to relax this limitation, and will also show how to apply the SBR algorithm to strongly coupled systems, i.e. to the systems, where many degrees of freedom of one sub-system are coupled to many degrees of freedom of other sub-systems and where the internal input and output matrices are not explicitly given in the system formulation. Examples of these types of coupled problems are shown in Fig. 1.6. Figure 6(a) presents a coupled system that consists of two sub-structures, for instance a solid body and a fluid. The coupling occurs at the interface, where all degrees of freedom of one sub-domain which are sufficiently close to the interface influence similar degrees of freedom of the second sub-domain and vice versa. A different type of *strong coupling* is shown in Fig. 6(b). This picture shows a situation, where all degrees of freedom related to both physical quantities $u$ and $e$ are located inside the same domain. Such situations appear for instance in case of modeling of systems, where the dynamics of the structure is influenced by an electromagnetic field (and vice versa). In the depicted case the change of the velocity of the node $u(x_i)$ influences the electromagnetic field $x \mapsto e(x)$ at the node $x_i$, and at many nodes in the neighborhood of $x_i$.

## 1.3.1 Interconnected System – System Definition

In this subsection we introduce the family of linear *interconnected systems* to which the reduction algorithm is to be applied to. For the sake of simplicity, we focus on a system of two-subsystems where one sub-system's output is used as a part of the other sub-system's input and vice versa. However, the proposed method can easily be extended to systems composed of an arbitrary number of sub-systems.

### 1.3.1.1 The Uncoupled Formulation

The time domain behavior of each of the sub-systems $S_1$ and $S_2$ is modeled by a system of first order differential-algebraic equations after which the frequency domain behavior is obtained via Laplace transformation. For the two sub-system examples in Fig. 1.7, this procedure leads to the Laplace domain systems

$$S_1 : \begin{cases} s\mathbf{E}_{11}\mathbf{x}_1 = \mathbf{A}_{11}\mathbf{x}_1 + \mathbf{B}_1\mathbf{u}_1 + \mathbf{B}_3\mathbf{u}_3, \\ \mathbf{y}_1 = \mathbf{C}_1^T\mathbf{x}_1, \\ \mathbf{y}_3 = \mathbf{C}_3^T\mathbf{x}_1, \end{cases}$$

$$S_2 : \begin{cases} s\mathbf{E}_{22}\mathbf{x}_2 = \mathbf{A}_{22}\mathbf{x}_2 + \mathbf{B}_2\mathbf{u}_2 + \mathbf{B}_4\mathbf{u}_4, \\ \mathbf{y}_2 = \mathbf{C}_2^T\mathbf{x}_2, \\ \mathbf{y}_4 = \mathbf{C}_4^T\mathbf{x}_2. \end{cases}$$

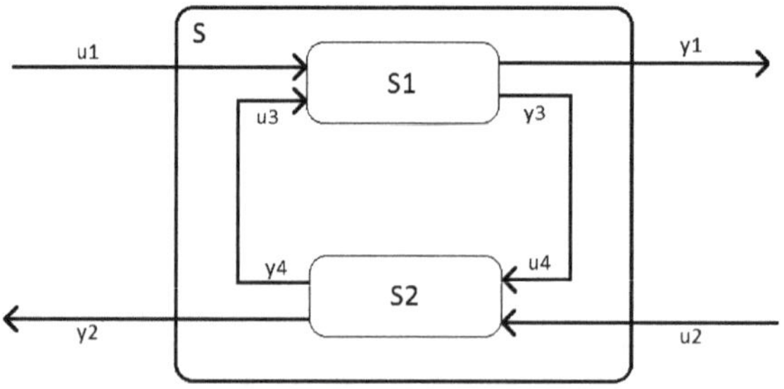


**Fig. 1.7.** Schematic representation of the interconnected system

Using matrix notation, the *system* $S_1$ and *system* $S_2$ can be described as

$$S_1 : \begin{cases} s\mathbf{E}_{11}\mathbf{x}_1 = \mathbf{A}_{11}\mathbf{x}_1 + [\mathbf{B}_1\ \mathbf{B}_3]\begin{bmatrix}\mathbf{u}_1\\ \mathbf{u}_3\end{bmatrix} \\ \begin{bmatrix}\mathbf{y}_1\\ \mathbf{y}_3\end{bmatrix} = \begin{bmatrix}\mathbf{C}_1^T\\ \mathbf{C}_3^T\end{bmatrix}\mathbf{x}_1, \end{cases} \tag{1.7}$$

$$S_2 : \begin{cases} s\mathbf{E}_{22}\mathbf{x}_2 = \mathbf{A}_{22}\mathbf{x}_2 + [\mathbf{B}_2\ \mathbf{B}_4]\begin{bmatrix}\mathbf{u}_2\\ \mathbf{u}_4\end{bmatrix} \\ \begin{bmatrix}\mathbf{y}_2\\ \mathbf{y}_4\end{bmatrix} = \begin{bmatrix}\mathbf{C}_2^T\\ \mathbf{C}_4^T\end{bmatrix}\mathbf{x}_2. \end{cases} \tag{1.8}$$

#### 1.3.1.2 The Coupled System

When the output of $S_1$ is used as an input of $S_2$ and the output of $S_2$ is used as an input of $S_1$, equations (1.7) and (1.8) reduce to an interconnected Laplace domain system. Due to the design of the system depicted in Fig. 1.7 one has

$$\begin{cases} \mathbf{u}_3 = \mathbf{y}_4 = \mathbf{C}_4^T\mathbf{x}_2 \\ \mathbf{u}_4 = \mathbf{y}_3 = \mathbf{C}_3^T\mathbf{x}_1, \end{cases} \tag{1.9}$$

which in addition implies

$$\begin{cases} m_3 = p_4 \\ m_4 = p_3. \end{cases}$$

Using relation (1.9), the interconnected system (1.7) can be represented as a single *coupled system* **S** of equations

$$S : \begin{cases} s\mathbf{E}_{11}\mathbf{x}_1 = \mathbf{A}_{11}\mathbf{x}_1 + \mathbf{B}_1\mathbf{u}_1 + \mathbf{B}_3\mathbf{C}_4^T\mathbf{x}_2, \\ s\mathbf{E}_{22}\mathbf{x}_2 = \mathbf{A}_{22}\mathbf{x}_2 + \mathbf{B}_2\mathbf{u}_2 + \mathbf{B}_4\mathbf{C}_3^T\mathbf{x}_1, \\ \mathbf{y}_1 = \mathbf{C}_1^T\mathbf{x}_1, \\ \mathbf{y}_2 = \mathbf{C}_2^T\mathbf{x}_2 \end{cases}$$

and in matrix form

$$S : \begin{cases} s\begin{bmatrix}\mathbf{E}_{11} & \mathbf{0}\\ \mathbf{0} & \mathbf{E}_{22}\end{bmatrix}\begin{bmatrix}\mathbf{x}_1\\ \mathbf{x}_2\end{bmatrix} = \begin{bmatrix}\mathbf{A}_{11} & \mathbf{B}_3\mathbf{C}_4^T\\ \mathbf{B}_4\mathbf{C}_3^T & \mathbf{A}_{22}\end{bmatrix}\begin{bmatrix}\mathbf{x}_1\\ \mathbf{x}_2\end{bmatrix} + \begin{bmatrix}\mathbf{B}_1 & \mathbf{0}\\ \mathbf{0} & \mathbf{B}_2\end{bmatrix}\begin{bmatrix}\mathbf{u}_1\\ \mathbf{u}_2\end{bmatrix} \\ \begin{bmatrix}\mathbf{y}_1\\ \mathbf{y}_2\end{bmatrix} = \begin{bmatrix}\mathbf{C}_1^T & \mathbf{0}\\ \mathbf{0} & \mathbf{C}_2^T\end{bmatrix}\begin{bmatrix}\mathbf{x}_1\\ \mathbf{x}_2\end{bmatrix}. \end{cases} \tag{1.10}$$

Let $N = N_1 + N_2,\ m = m_1 + m_2,\ p = p_1 + p_2$ and define

$$\mathbf{E} = \begin{bmatrix} \mathbf{E}_{11} & \mathbf{0} \\ \mathbf{0} & \mathbf{E}_{22} \end{bmatrix},\quad \mathbf{A} = \begin{bmatrix} \mathbf{A}_{11} & \mathbf{B}_3\mathbf{C}_4^T \\ \mathbf{B}_4\mathbf{C}_3^T & \mathbf{A}_{22} \end{bmatrix},\quad \mathbf{B} = \begin{bmatrix} \mathbf{B}_1 & \mathbf{0} \\ \mathbf{0} & \mathbf{B}_2 \end{bmatrix},\quad \mathbf{C} = \begin{bmatrix} \mathbf{C}_1 & \mathbf{0} \\ \mathbf{0} & \mathbf{C}_2 \end{bmatrix},\tag{1.11}$$

where $\mathbf{A},\mathbf{E} \in \mathbb{R}^{N\times N}$, $\mathbf{B} \in \mathbb{R}^{N\times m}$, $\mathbf{C} \in \mathbb{R}^{N\times p}$. The matrices defined in (1.11) show a special block structure. The sub-systems' matrices $\mathbf{A}_{11}$ and $\mathbf{A}_{22}$ form the diagonal blocks of the system matrix $\mathbf{A}$ of $S$. The off-diagonal blocks are the products $\mathbf{B}_3\mathbf{C}_4^T$ and $\mathbf{B}_4\mathbf{C}_3^T$ of the internal input and output matrices of the sub-system. The input and output matrices $\mathbf{B}$ and $\mathbf{C}$ are block structured, as well as the matrix $\mathbf{E}$.

### *1.3.2 Transfer Functions of the Uncoupled and Coupled Systems*

One of the questions arising at this point is the relation between the transfer functions of the sub-systems S1 and S2, and the transfer function of the coupled system. In this subsection we will study this issue. Let us begin with the uncoupled sub-systems. At $s \in \mathbb{C}$ the transfer function of sub-system S1 defined in (1.7) is given by

$$\begin{aligned}\mathbf{H}(s) &= \begin{bmatrix} \mathbf{C}_1^T \\ \mathbf{C}_3^T \end{bmatrix} (s\mathbf{E}_{11} - \mathbf{A}_{11})^{-1}[\mathbf{B}_1\ \mathbf{B}_3] \\ &= \begin{bmatrix} \mathbf{C}_1^T(s\mathbf{E}_{11} - \mathbf{A}_{11})^{-1}\mathbf{B}_1 & \mathbf{C}_1^T(s\mathbf{E}_{11} - \mathbf{A}_{11})^{-1}\mathbf{B}_3 \\ \mathbf{C}_3^T(s\mathbf{E}_{11} - \mathbf{A}_{11})^{-1}\mathbf{B}_1 & \mathbf{C}_3^T(s\mathbf{E}_{11} - \mathbf{A}_{11})^{-1}\mathbf{B}_3 \end{bmatrix} = \begin{bmatrix} \mathbf{H}_{11}(s) & \mathbf{H}_{12}(s) \\ \mathbf{H}_{21}(s) & \mathbf{H}_{22}(s) \end{bmatrix}.\end{aligned}\tag{1.12}$$

For the sub-system S2 defined in (1.8), similarly

$$\begin{aligned}\mathbf{G}(s) &= \begin{bmatrix} \mathbf{C}_2^T \\ \mathbf{C}_4^T \end{bmatrix} (s\mathbf{E}_{22} - \mathbf{A}_{22})^{-1}[\mathbf{B}_2\ \mathbf{B}_4] \\ &= \begin{bmatrix} \mathbf{C}_2^T(s\mathbf{E}_{22} - \mathbf{A}_{22})^{-1}\mathbf{B}_2 & \mathbf{C}_2^T(s\mathbf{E}_{22} - \mathbf{A}_{22})^{-1}\mathbf{B}_4 \\ \mathbf{C}_4^T(s\mathbf{E}_{22} - \mathbf{A}_{22})^{-1}\mathbf{B}_2 & \mathbf{C}_4^T(s\mathbf{E}_{22} - \mathbf{A}_{22})^{-1}\mathbf{B}_4 \end{bmatrix} = \begin{bmatrix} \mathbf{G}_{11}(s) & \mathbf{G}_{12}(s) \\ \mathbf{G}_{21}(s) & \mathbf{G}_{22}(s) \end{bmatrix}.\end{aligned}\tag{1.13}$$

At $s \in \mathbb{C}$ the transfer function of the coupled system (1.10) is

$$\begin{aligned}\mathbf{Z}(s) = \mathbf{C}^T(s\mathbf{E} - \mathbf{A})^{-1}\mathbf{B} &= \begin{bmatrix} \mathbf{C}_1^T & \mathbf{0} \\ \mathbf{0} & \mathbf{C}_2^T \end{bmatrix} \left( s \begin{bmatrix} \mathbf{E}_{11} & \mathbf{0} \\ \mathbf{0} & \mathbf{E}_{22} \end{bmatrix} - \begin{bmatrix} \mathbf{A}_{11} & \mathbf{B}_3\mathbf{C}_4^T \\ \mathbf{B}_4\mathbf{C}_3^T & \mathbf{A}_{22} \end{bmatrix} \right)^{-1} \begin{bmatrix} \mathbf{B}_1 & \mathbf{0} \\ \mathbf{0} & \mathbf{B}_2 \end{bmatrix} \\ &= \begin{bmatrix} \mathbf{Z}_{11}(s) & \mathbf{Z}_{12}(s) \\ \mathbf{Z}_{21}(s) & \mathbf{Z}_{22}(s) \end{bmatrix}.\end{aligned}\tag{1.14}$$

Based on definitions Eqs. (1.12) to (1.14) we will express the components of the transfer function $\mathbf{Z}(s)$ in terms of the components of the transfer functions $\mathbf{H}(s)$ and $\mathbf{G}(s)$ in two manners. First we follow the typical approach used in the field of systems and control (more details can be found in for instance [19]). Secondly we use the Sherman-Morrison-Woodbury formula.

### *The Systems and Control Approach*

The starting point of this approach are two transfer functions $\mathbf{H}(s)$ and $\mathbf{G}(s)$ of the sub-systems 1 and 2, respectively. For each sub-system, its transfer function relates its inputs to outputs:

$$\begin{bmatrix}\mathbf{y}_1\\ \mathbf{y}_3\end{bmatrix} = \begin{bmatrix}\mathbf{H}_{11}(s) & \mathbf{H}_{12}(s)\\ \mathbf{H}_{21}(s) & \mathbf{H}_{22}(s)\end{bmatrix}\begin{bmatrix}\mathbf{u}_1\\ \mathbf{u}_3\end{bmatrix}, \quad \begin{bmatrix}\mathbf{y}_2\\ \mathbf{y}_4\end{bmatrix} = \begin{bmatrix}\mathbf{G}_{11}(s) & \mathbf{G}_{12}(s)\\ \mathbf{G}_{21}(s) & \mathbf{G}_{22}(s)\end{bmatrix}\begin{bmatrix}\mathbf{u}_2\\ \mathbf{u}_4\end{bmatrix}$$

and

$$\begin{bmatrix}\mathbf{y}_1\\ \mathbf{y}_2\end{bmatrix} = \begin{bmatrix}\mathbf{Z}_{11}(s) & \mathbf{Z}_{12}(s)\\ \mathbf{Z}_{21}(s) & \mathbf{Z}_{22}(s)\end{bmatrix}\begin{bmatrix}\mathbf{u}_1\\ \mathbf{u}_2\end{bmatrix}. \tag{1.15}$$

Systems (1.12) and (1.13) in combination with relation (1.9) lead to

$$\mathbf{y}_1 = \mathbf{H}_{11}(s)\mathbf{u}_1 + \mathbf{H}_{12}(s)\mathbf{y}_4 \tag{1.16}$$
$$\mathbf{y}_3 = \mathbf{H}_{21}(s)\mathbf{u}_1 + \mathbf{H}_{22}(s)\mathbf{y}_4 \tag{1.17}$$
$$\mathbf{y}_2 = \mathbf{G}_{11}(s)\mathbf{u}_2 + \mathbf{G}_{12}(s)\mathbf{y}_3 \tag{1.18}$$
$$\mathbf{y}_4 = \mathbf{G}_{21}(s)\mathbf{u}_2 + \mathbf{G}_{22}(s)\mathbf{y}_3. \tag{1.19}$$

Substituting $\mathbf{y}_4$ of (1.19) for $\mathbf{y}_4$ in (1.17) we obtain

$$\mathbf{y}_3 = \mathbf{H}_{21}(s)\mathbf{u}_1 + \mathbf{H}_{22}(s)\mathbf{y}_4 = \mathbf{H}_{21}(s)\mathbf{u}_1 + \mathbf{H}_{22}(s)[\mathbf{G}_{21}(s)\mathbf{u}_2 + \mathbf{G}_{22}(s)\mathbf{y}_3]$$

and hence

$$\mathbf{y}_3 = [\mathbf{I} - \mathbf{H}_{22}(s)\mathbf{G}_{22}(s)]^{-1}[\mathbf{H}_{21}(s)\mathbf{u}_1 + \mathbf{H}_{22}(s)\mathbf{G}_{21}(s)\mathbf{u}_2]. \tag{1.20}$$

With this result and (1.19), we can also express $\mathbf{y}_4$ in terms of $\mathbf{u}_1$ and $\mathbf{u}_2$

$$\begin{aligned}\mathbf{y}_4 &= \mathbf{G}_{21}(s)\mathbf{u}_2 + \mathbf{G}_{22}(s)\mathbf{y}_3 = \mathbf{G}_{21}(s)\mathbf{u}_2\\ &\quad + \mathbf{G}_{22}(s)[\mathbf{I} - \mathbf{H}_{22}(s)\mathbf{G}_{22}(s)]^{-1}[\mathbf{H}_{21}(s)\mathbf{u}_1 + \mathbf{H}_{22}(s)\mathbf{G}_{21}(s)\mathbf{u}_2].\end{aligned} \tag{1.21}$$

Using (1.20) and (1.21) in (1.16) and (1.18), we arrive at

$$\begin{aligned}\mathbf{y}_1 &= \mathbf{H}_{11}(s)\mathbf{u}_1 + \mathbf{H}_{12}(s)\mathbf{y}_4 = \mathbf{H}_{11}(s)\mathbf{u}_1\\ &\quad + \mathbf{H}_{12}(s)\Big(\mathbf{G}_{21}(s)\mathbf{u}_2 + \mathbf{G}_{22}(s)[\mathbf{I} - \mathbf{H}_{22}(s)\mathbf{G}_{22}(s)]^{-1}\\ &\qquad \times [\mathbf{H}_{21}(s)\mathbf{u}_1 + \mathbf{H}_{22}(s)\mathbf{G}_{21}(s)\mathbf{u}_2]\Big)\\ &= \Big(\mathbf{H}_{11}(s) + \mathbf{H}_{12}(s)\mathbf{G}_{22}(s)[\mathbf{I} - \mathbf{H}_{22}(s)\mathbf{G}_{22}(s)]^{-1}\mathbf{H}_{21}(s)\Big)\mathbf{u}_1\\ &\quad + \Big(\mathbf{H}_{12}(s)\mathbf{G}_{21}(s) + \mathbf{H}_{12}(s)\mathbf{G}_{22}(s)[\mathbf{I} - \mathbf{H}_{22}(s)\mathbf{G}_{22}(s)]^{-1}\mathbf{H}_{22}(s)\mathbf{G}_{21}\Big)\mathbf{u}_2\end{aligned}$$

and

$$\begin{aligned}\mathbf{y}_2 &= \mathbf{G}_{11}(s)\mathbf{u}_2 + \mathbf{G}_{12}(s)\mathbf{y}_3 = \mathbf{G}_{11}(s)\mathbf{u}_2 + \mathbf{G}_{12}(s)[\mathbf{I} - \mathbf{H}_{22}(s)\mathbf{G}_{22}(s)]^{-1}[\mathbf{H}_{21}(s)\mathbf{u}_1 \\ &\quad + \mathbf{H}_{22}(s)\mathbf{G}_{21}(s)\mathbf{u}_2] = \mathbf{G}_{12}(s)[\mathbf{I} - \mathbf{H}_{22}(s)\mathbf{G}_{22}(s)]^{-1}\mathbf{H}_{21}(s)\mathbf{u}_1 \\ &\quad + \Big(\mathbf{G}_{11}(s) + \mathbf{G}_{12}(s)[\mathbf{I} - \mathbf{H}_{22}(s)\mathbf{G}_{22}(s)]^{-1}\mathbf{H}_{22}(s)\mathbf{G}_{21}(s)\Big)\mathbf{u}_2.\end{aligned}$$

This shows that the components of $\mathbf{Z}(s)$, as defined in (1.15), are

$$\mathbf{Z}_{11}(s) = \mathbf{H}_{11}(s) + \mathbf{H}_{12}(s)\mathbf{G}_{22}(s)[\mathbf{I} - \mathbf{H}_{22}(s)\mathbf{G}_{22}(s)]^{-1}\mathbf{H}_{21}(s) \tag{1.22}$$

$$\mathbf{Z}_{12}(s) = \mathbf{H}_{12}(s)\mathbf{G}_{21}(s) + \mathbf{H}_{12}(s)\mathbf{G}_{22}(s)[\mathbf{I} - \mathbf{H}_{22}(s)\mathbf{G}_{22}(s)]^{-1}\mathbf{H}_{22}(s)\mathbf{G}_{21}(s) \tag{1.23}$$

$$\mathbf{Z}_{21}(s) = \mathbf{G}_{12}(s)[\mathbf{I} - \mathbf{H}_{22}(s)\mathbf{G}_{22}(s)]^{-1}\mathbf{H}_{21}(s) \tag{1.24}$$

$$\mathbf{Z}_{22}(s) = \mathbf{G}_{11}(s) + \mathbf{G}_{12}(s)[\mathbf{I} - \mathbf{H}_{22}(s)\mathbf{G}_{22}(s)]^{-1}\mathbf{H}_{22}(s)\mathbf{G}_{21}(s). \tag{1.25}$$

### *Computing the Transfer Function of the Coupled System Using the Sherman–Morrison–Woodbury Formula*

The evaluation of the transfer function of the coupled system, as defined in (1.14), requires a computation of an inverse of a block matrix. For a system consisting of an arbitrary number of sub-systems, a suitable tool towards this end is the Sherman–Morrison–Woodbury formula (see for instance [10] and references therein). This formula allows for a computationally cheap matrix inversion, as long as the considered matrix can be easily expressed as a sum of a matrix for which an inverse is known (or easy to compute) and a (low rank) correction. Let $\mathbf{L}$ be non-singular and let matrices $\mathbf{J}$, $\mathbf{M}$, $\mathbf{N}$ be of compatible size. Then the formula of $\mathbf{K} = \mathbf{L} + \mathbf{MJN}^T$ is (after [10])

$$\mathbf{K}^{-1} = (\mathbf{L} + \mathbf{MJN}^T)^{-1} = \mathbf{L}^{-1} - \mathbf{L}^{-1}\mathbf{M}(\mathbf{J}^{-1} + \mathbf{N}^T\mathbf{L}^{-1}\mathbf{M})^{-1}\mathbf{N}^T\mathbf{L}^{-1}, \tag{1.26}$$

In our case, the matrix to be inverted can be decomposed into

$$\begin{aligned}\mathbf{K} &= s\begin{bmatrix}\mathbf{E}_{11} & \mathbf{0}\\ \mathbf{0} & \mathbf{E}_{22}\end{bmatrix} - \begin{bmatrix}\mathbf{A}_{11} & \mathbf{B}_3\mathbf{C}_4^T\\ \mathbf{B}_4\mathbf{C}_3^T & \mathbf{A}_{22}\end{bmatrix} = \begin{bmatrix}s\mathbf{E}_{11} - \mathbf{A}_{11} & -\mathbf{B}_3\mathbf{C}_4^T\\ -\mathbf{B}_3\mathbf{C}_3^T & s\mathbf{E}_{22} - \mathbf{A}_{22}\end{bmatrix}\\ &= \begin{bmatrix}s\mathbf{E}_{11} - \mathbf{A}_{11} & \mathbf{0}\\ \mathbf{0} & s\mathbf{E}_{22} - \mathbf{A}_{22}\end{bmatrix} - \begin{bmatrix}\mathbf{0} & \mathbf{B}_3\mathbf{C}_4^T\\ \mathbf{B}_4\mathbf{C}_3^T & \mathbf{0}\end{bmatrix}\\ &= \mathbf{L} - \begin{bmatrix}\mathbf{0} & \mathbf{B}_3\mathbf{C}_4^T\\ \mathbf{B}_4\mathbf{C}_3^T & \mathbf{0}\end{bmatrix}\end{aligned}$$

where $\mathbf{L}$ is a block-diagonal matrix, whose inverse can be calculated by computing the inverses of each sub-block separately and the correction matrix can be factored

$$\begin{bmatrix}\mathbf{0} & \mathbf{B}_3\mathbf{C}_4^T\\ \mathbf{B}_4\mathbf{C}_3^T & \mathbf{0}\end{bmatrix} = \begin{bmatrix}\mathbf{0} & \mathbf{B}_3\\ \mathbf{B}_4 & \mathbf{0}\end{bmatrix}\mathbf{I}\begin{bmatrix}\mathbf{C}_3^T & \mathbf{0}\\ \mathbf{0} & \mathbf{C}_4^T\end{bmatrix} = \mathbf{MJN}^T.$$

Abbreviate $\mathbf{G}_i(s) = (s\mathbf{E}_{ii} - \mathbf{A}_{ii})^{-1}$, $\mathbf{P}_i(s) = \mathbf{G}_i(s)\mathbf{E}_{ii}$ and $\mathbf{R}_i(s) = \mathbf{G}_i(s)[\mathbf{B}_i\ \mathbf{B}_{2+i}]$, $i = 1,2$ and omit the argument $s$ when possible. Note that $\mathbf{R}_i = [\mathbf{R}_{i1}, \mathbf{R}_{i2}] = [\mathbf{G}_i\mathbf{B}_i, \mathbf{G}_i\mathbf{B}_{2+i}]$ consists of two blocks. Substituting the formulas for $\mathbf{L}, \mathbf{M}, \mathbf{J}$ and $\mathbf{N}$ into the the Sherman–Morrison–Woodbury formula, we get

$$
\begin{aligned}
&\left(s\begin{bmatrix}\mathbf{E}_{11} & \mathbf{0}\\ \mathbf{0} & \mathbf{E}_{22}\end{bmatrix} - \begin{bmatrix}\mathbf{A}_{11} & \mathbf{B}_3\mathbf{C}_4^T\\ \mathbf{B}_4\mathbf{C}_3^T & \mathbf{A}_{22}\end{bmatrix}\right)^{-1}\\
&= \left(\begin{bmatrix}s\mathbf{E}_{11} - \mathbf{A}_{11} & \mathbf{0}\\ \mathbf{0} & s\mathbf{E}_{22} - \mathbf{A}_{22}\end{bmatrix} - \begin{bmatrix}\mathbf{0} & \mathbf{B}_3\\ \mathbf{B}_4 & \mathbf{0}\end{bmatrix}\begin{bmatrix}\mathbf{C}_3^T & \mathbf{0}\\ \mathbf{0} & \mathbf{C}_4^T\end{bmatrix}\right)^{-1}\\
&= \begin{bmatrix}\mathbf{G}_1 & \mathbf{0}\\ \mathbf{0} & \mathbf{G}_2\end{bmatrix} + \begin{bmatrix}\mathbf{G}_1 & \mathbf{0}\\ \mathbf{0} & \mathbf{G}_2\end{bmatrix}\begin{bmatrix}\mathbf{0} & \mathbf{B}_3\\ \mathbf{B}_4 & \mathbf{0}\end{bmatrix} \circ\\
&\quad \left(\mathbf{I} - \begin{bmatrix}\mathbf{C}_3^T & \mathbf{0}\\ \mathbf{0} & \mathbf{C}_4^T\end{bmatrix}\begin{bmatrix}\mathbf{G}_1 & \mathbf{0}\\ \mathbf{0} & \mathbf{G}_2\end{bmatrix}\begin{bmatrix}\mathbf{0} & \mathbf{B}_3\\ \mathbf{B}_4 & \mathbf{0}\end{bmatrix}\right)^{-1}\begin{bmatrix}\mathbf{C}_3^T & \mathbf{0}\\ \mathbf{0} & \mathbf{C}_4^T\end{bmatrix}\begin{bmatrix}\mathbf{G}_1 & \mathbf{0}\\ \mathbf{0} & \mathbf{G}_2\end{bmatrix}\\
&= \begin{bmatrix}\mathbf{G}_1 & \mathbf{0}\\ \mathbf{0} & \mathbf{G}_2\end{bmatrix} + \begin{bmatrix}\mathbf{0} & \mathbf{R}_{12}\\ \mathbf{R}_{22} & \mathbf{0}\end{bmatrix} \circ\\
&\quad \left(\mathbf{I} - \begin{bmatrix}\mathbf{0} & \mathbf{C}_3^T\mathbf{R}_{12}\\ \mathbf{C}_4^T\mathbf{R}_{22} & \mathbf{0}\end{bmatrix}\right)^{-1}\begin{bmatrix}\mathbf{C}_3^T\mathbf{G}_1 & \mathbf{0}\\ \mathbf{0} & \mathbf{C}_4^T\mathbf{G}_2\end{bmatrix},
\end{aligned}
\tag{1.27}
$$

where the entries of $\mathbf{G}$ and $\mathbf{R}$ depend on $s$. Using this result and Eqs. (1.12), (1.13), (1.14), one can find the formula for the transfer function of the coupled system

$$
\begin{aligned}
\mathbf{Z}(s) &= \begin{bmatrix}\mathbf{C}_1^T & \mathbf{0}\\ \mathbf{0} & \mathbf{C}_2^T\end{bmatrix}\left(\begin{bmatrix}\mathbf{G}_1 & \mathbf{0}\\ \mathbf{0} & \mathbf{G}_2\end{bmatrix} + \begin{bmatrix}\mathbf{0} & \mathbf{R}_{12}\\ \mathbf{R}_{22} & \mathbf{0}\end{bmatrix} \circ\right.\\
&\quad \left.\left(\mathbf{I} - \begin{bmatrix}\mathbf{0} & \mathbf{C}_3^T\mathbf{R}_{12}\\ \mathbf{C}_4^T\mathbf{R}_{22} & \mathbf{0}\end{bmatrix}\right)^{-1}\begin{bmatrix}\mathbf{C}_3^T\mathbf{G}_1 & \mathbf{0}\\ \mathbf{0} & \mathbf{C}_4^T\mathbf{G}_2\end{bmatrix}\right)\begin{bmatrix}\mathbf{B}_1 & \mathbf{0}\\ \mathbf{0} & \mathbf{B}_2\end{bmatrix}\\
&= \begin{bmatrix}\mathbf{C}_1^T\mathbf{R}_{11} & \mathbf{0}\\ \mathbf{0} & \mathbf{C}_2^T\mathbf{R}_{21}\end{bmatrix} + \begin{bmatrix}\mathbf{0} & \mathbf{C}_1^T\mathbf{R}_{12}\\ \mathbf{C}_2^T\mathbf{R}_{22} & \mathbf{0}\end{bmatrix} \circ\\
&\quad \left(\mathbf{I} - \begin{bmatrix}\mathbf{0} & \mathbf{C}_3^T\mathbf{R}_{12}\\ \mathbf{C}_4^T\mathbf{R}_{22} & \mathbf{0}\end{bmatrix}\right)^{-1}\begin{bmatrix}\mathbf{C}_3^T\mathbf{R}_{11} & \mathbf{0}\\ \mathbf{0} & \mathbf{C}_4^T\mathbf{R}_{21}\end{bmatrix}\\
&= \begin{bmatrix}\mathbf{H}_{11}(s) & \mathbf{0}\\ \mathbf{0} & \mathbf{G}_{11}(s)\end{bmatrix} + \begin{bmatrix}\mathbf{0} & \mathbf{H}_{12}(s)\\ \mathbf{G}_{12}(s) & \mathbf{0}\end{bmatrix} \circ\\
&\quad \left(\mathbf{I} - \begin{bmatrix}\mathbf{0} & \mathbf{H}_{22}(s)\\ \mathbf{G}_{22}(s) & \mathbf{0}\end{bmatrix}\right)^{-1}\begin{bmatrix}\mathbf{H}_{21}(s) & \mathbf{0}\\ \mathbf{0} & \mathbf{G}_{21}(s)\end{bmatrix}.
\end{aligned}
\tag{1.28}
$$

It is easy to show, that the formulation (1.28) is equivalent to the formulation given by Eqs. (1.22) to (1.25). Moreover, (1.28) provides an elegant relationship between the components of the transfer functions of the sub-systems and the coupled system, that reveals the symmetry and the structure of the coupled system. In addition it shows that the relation between the transfer functions is not straightforward. Since several sub-expressions such as $(s\mathbf{E}_{ii} - A_{ii})^{-1}$ reoccur frequently, we will introduce abbreviations in the upcoming sections.

Formula (1.28) reveals a structure which is more difficult to find in (1.22)–(1.25) and can be used to calculate the transfer function of the coupled system if the transfer functions of the individual sub-systems are available. The involved inverse is of a small matrix which means that calculation of the transfer function of the coupled system is relatively cheap.

### *1.3.3 Standard Block Structure Preserving Reduction*

In this section we will recall the general ideas of the standard block-structure preserving methods.

A typical block structure preserving (*BSP*) model reduction method applied to the system (1.10) consists of the following three steps:

1. Create the matrix $\tilde{\mathbf{V}}$ whose columns span the $n$th Krylov subspace around $s_0 \in \mathbb{C}$

$$\tilde{\mathbf{V}} = \mathscr{K}_n(\mathbf{P}(s_0), \mathbf{R}(s_0)),$$

   where $\mathbf{P}(s_0)$ and $\mathbf{R}(s_0)$ are

$$\mathbf{P}(s_0) = (s_0\mathbf{E} - \mathbf{A})^{-1}\mathbf{E} \in \mathbb{R}^{N\times N} \quad \text{and} \quad \mathbf{R}(s_0) = (s_0\mathbf{E} - \mathbf{A})^{-1}\mathbf{B} \in \mathbb{R}^{N}.$$

2. Build a the block-diagonal reduction matrix $\mathbf{V}$ with $N_1 + N_2 = N$ rows

$$\mathbf{V} = \begin{bmatrix} \mathbf{V}_1 & \mathbf{0} \\ \mathbf{0} & \mathbf{V}_2 \end{bmatrix},$$

   where $\mathbf{V}_1$ and $\mathbf{V}_2$ contain the first $N_1$ respectively last $N_2$ rows of the matrix $\tilde{\mathbf{V}}$.
3. Project the original system onto a lower-dimensional space

$$\hat{\mathbf{E}} = \mathbf{V}^T\mathbf{E}\mathbf{V}, \quad \hat{\mathbf{A}} = \mathbf{V}^T\mathbf{A}\mathbf{V}, \quad \hat{\mathbf{B}} = \mathbf{V}^T\mathbf{B}, \quad \hat{\mathbf{C}} = \mathbf{V}^T\mathbf{C}.$$

When possible we write $\mathbf{P}$ and $\mathbf{R}$ rather than $\mathbf{P}(s_0)$ respectively $\mathbf{R}(s_0)$. The model reduction methods based on this idea are widely applied and popular due to a good accuracy of the reduced-order systems that they deliver. However, they have a few drawbacks, one of them being the high cost of the construction of the reduction basis. The main computational cost of this type of methods is related to evaluation of $\mathbf{x} \mapsto (s_0\mathbf{E} - \mathbf{A})^{-1}\mathbf{x}$, which involves solving a system of equations with a large coefficient matrix. In the next section we introduce an alternative structure preserving method which for some cases can significantly reduce the computational costs.

### *1.3.4 Separate Bases Reduction Algorithm*

In the classical case, the reduction basis is built using the coupled formulation of the system (1.10). The construction of this basis requires repeated evaluations of

$\mathbf{x} \mapsto (s_0\mathbf{E} - \mathbf{A})^{-1}\mathbf{x}$ where $s_0\mathbf{E} - \mathbf{A}$ is an $N \times N$ matrix. For large $N$ this procedure can be computationally very expensive or even unfeasible. In such cases one can try to make use of a natural block structure of the coupled system and for instance replace the evaluations involving $(s_0\mathbf{E} - \mathbf{A})^{-1}$ by evaluations involving $(s_0\mathbf{E}_{11} - \mathbf{A}_{11})^{-1}$ and $(s_0\mathbf{E}_{22} - \mathbf{A}_{22})^{-1}$, i.e., by evaluations involving only the coefficient matrices of both sub-systems. If $N$ is large and for instance $N_1 = N_2 = N/2$ then the serial computation of $(s_0\mathbf{E}_{11} - \mathbf{A}_{11})^{-1}$ and $(s_0\mathbf{E}_{22} - \mathbf{A}_{22})^{-1}$ may be much faster that of $(s_0\mathbf{E} - \mathbf{A})^{-1}$. Further acceleration can be achieved through parallelism.

Following this idea, we introduce a new model reduction algorithm, called Separate Bases Reduction (*SBR*) algorithm. Here the Krylov subspaces that create the reduction bases correspond to the uncoupled sub-systems (as defined in (1.7) and (1.8)) rather than to the coupled system (1.10). The procedure is as follows:

1. Create two matrices $\mathbf{V}_1$ and $\mathbf{V}_2$, one for each sub-system:
   - For the sub-system $S_1$, build a matrix $\mathbf{V}_1$, whose columns span the $n_1$th Krylov subspace around $s_0 \in \mathbb{C}$

     $$\mathbf{V}_1 = \mathscr{K}_{n_1}(\mathbf{P}_1(s_0), \mathbf{R}_1(s_0)),$$

     where $\mathbf{P}_1(s_0)$ and $\mathbf{R}_1(s_0)$ are

     $$\mathbf{P}_1(s_0) = (s_0\mathbf{E}_{11} - \mathbf{A}_{11})^{-1}\mathbf{E}_{11} \quad \text{and} \quad \mathbf{R}_1(s_0) = (s_0\mathbf{E}_{11} - \mathbf{A}_{11})^{-1}[\mathbf{B}_1\ \mathbf{B}_3].$$

     Matrix $\mathbf{V}_1$ has $N_1$ rows.
   - For the sub-system $S_2$, build a matrix $\mathbf{V}_2$, whose columns span the $n_2$th Krylov subspace around $s_0 \in \mathbb{C}$

     $$\mathbf{V}_2 = \mathscr{K}_{n_2}(\mathbf{P}_2(s_0), \mathbf{R}_2(s_0)),$$

     where $\mathbf{P}_2(s_0)$ and $\mathbf{R}_2(s_0)$ are

     $$\mathbf{P}_2(s_0) = (s_0\mathbf{E}_{22} - \mathbf{A}_{22})^{-1}\mathbf{E}_{22} \quad \text{and} \quad \mathbf{R}_2(s_0) = (s_0\mathbf{E}_{22} - \mathbf{A}_{22})^{-1}[\mathbf{B}_2\ \mathbf{B}_4].$$

     Matrix $\mathbf{V}_2$ has $N_2$ rows.
2. Build the block-diagonal reduction matrix $\mathbf{V}$ with $N_1 + N_2 = N$ rows

   $$\mathbf{V} = \begin{bmatrix} \mathbf{V}_1 & \mathbf{0} \\ \mathbf{0} & \mathbf{V}_2 \end{bmatrix}.$$

3. Project the original system onto a lower-dimensional space

   $$\hat{\mathbf{E}} = \mathbf{V}^T\mathbf{E}\mathbf{V}, \quad \hat{\mathbf{A}} = \mathbf{V}^T\mathbf{A}\mathbf{V}, \quad \hat{\mathbf{B}} = \mathbf{V}^T\mathbf{B}, \quad \hat{\mathbf{C}} = \mathbf{V}^T\mathbf{C}.$$

In the sequel, when possible without causing confusion, we omit the argument $s_0$ of $\mathbf{P}_i$ and $\mathbf{R}_i$, $i = 1, 2$. In the next subsection, we will compare the SBR algorithm with a standard BSP reduction method, by examining their most important properties.

## *1.3.5 Separate Bases Reduction Algorithm – Properties*

In this subsection we will discuss the differences and similarities between Separate Bases Reduction algorithm and standard block structure preserving model reduction methods.

### *Block-Structure Preservation*

As described in subsection 1.3.4, the SBR algorithm uses reduction matrices of the block-diagonal form

$$\mathbf{V} = \begin{bmatrix} \mathbf{V}_1 & \mathbf{0} \\ \mathbf{0} & \mathbf{V}_2 \end{bmatrix}.$$

Therefore, its application preserves the block structure of the coupled system matrices.

### *Rank and Orthogonality*

The sub-blocks $\mathbf{V}_1$ and $\mathbf{V}_2$ of the projector $\mathbf{V}$ are constructed separately, using one of the Krylov basis building algorithms. Hence, both of them have a full column rank and, as a result, the matrix $\mathbf{V}$ also has a full column rank. If the sub-blocks $\mathbf{V}_1$ and $\mathbf{V}_2$ have orthogonal columns then also matrix $\mathbf{V}$ has (automatically) orthogonal columns, i.e., no explicit orthogonalization has to be applied.

### *Computational Cost*

The difference between the computational costs for a standard block structure preserving method and the Separate Bases Reduction algorithm comes from the fact, that the SBR algorithm computes the reduction bases for the set of uncoupled systems instead of using the coupled formulation of the system. This approach can significantly reduce the computational time and storage requirements needed during the model reduction process.

The main cost of the Krylov basis construction lies in the evaluation of the matrix pencil inverse function $\mathbf{x} \mapsto (s_0\mathbf{E} - \mathbf{A})^{-1}\mathbf{x}$. For coupled models with many degrees $n$ of freedom this evaluation may be unfeasible. But for sub-problems of smaller size evaluation may be possible. The amount of computational work required for the solution of $(s_0\mathbf{E} - \mathbf{A})\mathbf{x} = \mathbf{d}$ depends on the employed solution method which at its turn relies on specific properties of the matrix $s_0\mathbf{E} - \mathbf{A}$ (symmetry, monotone, positive definite, etc.). Different methods lead to different amounts of computational work: The minimal amount of work of $O(n)$ operations is usually achieved by multigrid methods (see [25]), other methods such as GMRES, PCG, CGS and BiCGstab(l) (see [16, 20, 22]) are more expensive. Classical fixed point methods such as Jacobi, Gauss-Seidel and matrix-splitting based methods are usually even slower.

***Size of the Reduction Space***

Another difference with respect to the standard BSP reduction methods is the size of the reduction matrix $\mathbf{V}$ and, as a result, dimension of the reduced order model.

Let us consider the coupled system (1.10) and assume, for simplicity, that there is no need for deflation (all columns turn out to be linearly independent) while building the matrix $\mathbf{V}$. We will apply a typical reduction procedure like described in subsection 1.3.3 and the SBR algorithm. In both cases, we will build a Krylov subspace of order $n$ and estimate the size of the reduction space and reduced order model.

We begin with the analysis of the standard structure preserving algorithm. The $n$th Krylov subspace built for the coupled system for the starting matrices as defined in subsection 1.3.3 will be of the form

$$\tilde{\mathbf{V}} = \mathscr{K}_n\Big(\mathbf{P},\mathbf{R}\Big) = \operatorname{colspan}\{\mathbf{R},\ldots,\ \mathbf{P}^{n-1}\mathbf{R}\}$$

where $\mathbf{P} = (s_0\mathbf{E} - \mathbf{A})^{-1}\mathbf{E}$ and $\mathbf{R} = (s_0\mathbf{E} - \mathbf{A})^{-1}\mathbf{B}$. Since $\mathbf{B} \in \mathbb{R}^{N\times m}$, each of the components $\mathbf{P}^j\mathbf{R}$ of the matrix $\tilde{\mathbf{V}}$ has $m$ columns. Thus, for a degree $n$ Krylov space, assuming no deflation, the size of $\tilde{\mathbf{V}}$ is $N\times(nm)$. Next, the block-diagonal reduction matrix $\mathbf{V}$ is created by splitting the rows of $\tilde{\mathbf{V}}$ according to the dimensions of the sub-problems. In our case, the coupled system consists of two sub-systems, so the final size of the reduction matrix $\mathbf{V}$ is $N\times(2nm)$. This leads to a reduced model of order $2nm$.

Next, we will focus on the *SBR* algorithm. In this case two matrices $\mathbf{V}_1$ and $\mathbf{V}_2$, are built separately and we assume that each of them corresponds to an $n$th degree Krylov subspace based on the appropriate matrices (for $i = 1,2$ define $\mathbf{G}_i(s_0) = (s_0\mathbf{E}_{ii} - \mathbf{A}_{ii})^{-1}$, $\mathbf{P}_i(s_0) = \mathbf{G}_i\mathbf{E}_{ii}$ and $\mathbf{R}_i(s_0) = \mathbf{G}_i[\mathbf{B}_i\ \ \mathbf{B}_{2+i}]$ and observe that $\mathbf{R}_i = [\mathbf{R}_{i1},\mathbf{R}_{i2}]$ where $\mathbf{R}_{i1}$ and $\mathbf{R}_{i2}$ are $\mathbf{G}_i\mathbf{B}_i$, respectively $\mathbf{G}_i\mathbf{B}_{2+i}$). For the sub-system $S_1$, we create the matrix $\mathbf{V}_1$

$$\mathbf{V}_1 = \mathscr{K}_n\Big(\mathbf{P}_1,\ \mathbf{R}_1\Big).$$

Here, $\mathbf{R}_1,[\mathbf{B}_1\ \ \mathbf{B}_3] \in \mathbb{R}^{N_1\times(m_1+m_3)}$, so each component $\mathbf{P}_1^j\mathbf{R}_1$ of the matrix $\mathbf{V}_1$ has $(m_1+m_3)$ columns whence $\mathbf{V}_1$ has $n\times(m_1+m_3)$ columns.
For the sub-system $S_2$, we create

$$\mathbf{V}_2 = \mathscr{K}_n\Big(\mathbf{P}_2,\ \mathbf{R}_2\Big).$$

Similarly, since $\mathbf{R}_2,[\mathbf{B}_2\ \ \mathbf{B}_4] \in \mathbb{R}^{N_2\times(m_2+m_4)}$, every component $\mathbf{P}_2^j\mathbf{R}_2$ of the matrix $\mathbf{V}_2$ has $(m_2+m_4)$ columns, and matrix $\mathbf{V}_2$ has $n\times(m_2+m_4)$ columns.

Next, matrices $\mathbf{V}_1$ and $\mathbf{V}_2$ are used as diagonal blocks of the reduction matrix $\mathbf{V}$, resulting in a reduced model of order

$$n\times(m_1+m_3)+n\times(m_2+m_4) = n\times(m+m_3+m_4).$$

This result shows that the SBR algorithm creates a smaller reduced order model than standard BSP methods if $(m_3+m_4) < m$. This is for instance the case for coupled systems for which the number of internal inputs is not larger than the number of external inputs. If there are many more internal inputs than external ones, the size of the SBR algorithm based reduction matrix will grow very fast compared to the size of the BSP reduction matrix. However, this problem can be avoided for the category of systems for which the internal input matrices $\mathbf{B}_2$ and $\mathbf{B}_4$ can be approximated by only a small number of dominant components. This approach will be explained in more detail in the next section.

### *The Moment Matching Property*

In order to assess the *SBR moment matching* properties we compare the column-spaces of the BPS and SBR reduction matrices. For simplicity, without loss of generalization, we focus on the *SISO* case (the coupled system is SISO) where in addition $\mathbf{B}_i,\ \mathbf{C}_i,\ i=1,\ldots,4$ related to the sub-systems are column-vectors which implies that all products $\mathbf{C}_i^T(\ldots)\mathbf{B}_j,\ i,j=1,\ldots,4$, are scalars. A similar analysis is possible for the *MIMO* case (a MIMO coupled system with sub-system matrices $\mathbf{B}_i,\ \mathbf{C}_i$).

**Theorem 1.2** *Let the coupled system be as in Fig. 1.7, described by* (1.7) *and* (1.8). *Assume that all inputs and outputs are column-vectors, i.e., $m_i=p_i=1,\ i=1,2,3,4$. Then the SBR reduced-order model transfer function matches at least the same (number of) moments as the BSP reduced-order model transfer function.*

*Proof* First, we examine the reduction space built by a standard BSP method. To match the first $k$ moments at $s_0\in\mathbb{C}$, of the coupled system of the form (1.10), one has to construct the Krylov space

$$\tilde{\mathbf{V}}=\mathscr{K}_k(\mathbf{P},\mathbf{R}),$$

where

$$\mathbf{P}=(s\mathbf{E}-\mathbf{A})^{-1}\mathbf{E}\quad\text{and}\quad\mathbf{R}=(s\mathbf{E}-\mathbf{A})^{-1}\mathbf{B}.$$

The $i^{th}$ Krylov step for the BSP method adds to the reduction basis the column span of the following matrix $\mathbf{V}_{\mathrm{BSP}}^{(i)}$

$$\mathbf{V}_{\mathrm{BSP}}^{(i)}=\begin{bmatrix}\mathbf{V}_{11}^{(i)} & \mathbf{V}_{12}^{(i)} & 0 & 0\\ 0 & 0 & \mathbf{V}_{21}^{(i)} & \mathbf{V}_{22}^{(i)}\end{bmatrix}\tag{1.29}$$

with blocks of the form

$$\begin{aligned}\tilde{\mathbf{V}}_{\mathrm{BSP}}^{(i)}&=\begin{bmatrix}\mathbf{V}_{11}^{(i)} & \mathbf{V}_{12}^{(i)}\\ \mathbf{V}_{21}^{(i)} & \mathbf{V}_{22}^{(i)}\end{bmatrix}\\ &=\begin{bmatrix}\mathbf{P}_1^{i-1}\mathbf{R}_{11}+\sum_{j=0}^{i-1}\alpha_j\mathbf{P}_1^j\mathbf{R}_{12} & \sum_{j=0}^{i-1}\gamma_j\mathbf{P}_1^j\mathbf{R}_{12}\\ \sum_{j=0}^{i-1}\beta_j\mathbf{P}_2^j\mathbf{R}_{22} & \mathbf{P}_2^{i-1}\mathbf{R}_{21}+\sum_{j=0}^{i-1}\delta_j\mathbf{P}_2^j\mathbf{R}_{22}\end{bmatrix}.\end{aligned}\tag{1.30}$$

By (1.27) there exist scalars $a,\ b,\ c,\ d$ and by construction (induction) there exist coefficient vectors $\alpha = [\alpha_1, \ldots, \alpha_j] \in \mathbb{R}^{i-2}$, $\beta, \gamma, \delta \in \mathbb{R}^{i-2}$ such that

$$
\begin{aligned}
\tilde{\mathbf{V}}^{(i)}_{\mathrm{BSP}} &= \mathbf{P}\tilde{\mathbf{V}}^{(i-1)}_{\mathrm{BSP}} \\
&= \left( s \begin{bmatrix} \mathbf{E}_{11} & \mathbf{0} \\ \mathbf{0} & \mathbf{E}_{22} \end{bmatrix} - \begin{bmatrix} \mathbf{A}_{11} & \mathbf{B}_3\mathbf{C}_4^T \\ \mathbf{B}_4\mathbf{C}_3^T & \mathbf{A}_{22} \end{bmatrix} \right)^{-1} \begin{bmatrix} \mathbf{E}_{11} & \mathbf{0} \\ \mathbf{0} & \mathbf{E}_{22} \end{bmatrix} \tilde{\mathbf{V}}^{(i-1)}_{\mathrm{BSP}} \\
&\overset{(1.27)}{\underset{\text{induction}}{=}} \left( \begin{bmatrix} \mathbf{G}_1 & \mathbf{0} \\ \mathbf{0} & \mathbf{G}_2 \end{bmatrix} + \begin{bmatrix} \mathbf{0} & \mathbf{R}_{12} \\ \mathbf{R}_{22} & \mathbf{0} \end{bmatrix} \begin{bmatrix} a & b \\ c & d \end{bmatrix} \begin{bmatrix} \mathbf{C}_3^T\mathbf{G}_1 & \mathbf{0} \\ \mathbf{0} & \mathbf{C}_4^T\mathbf{G}_2 \end{bmatrix} \right) \begin{bmatrix} \mathbf{E}_{11} & \mathbf{0} \\ \mathbf{0} & \mathbf{E}_{22} \end{bmatrix} \tilde{\mathbf{V}}^{(i-1)}_{\mathrm{BSP}} \\
&= \left( \begin{bmatrix} \mathbf{P}_1 & \mathbf{0} \\ \mathbf{0} & \mathbf{P}_2 \end{bmatrix} + \begin{bmatrix} \mathbf{0} & \mathbf{R}_{12} \\ \mathbf{R}_{22} & \mathbf{0} \end{bmatrix} \begin{bmatrix} a & b \\ c & d \end{bmatrix} \begin{bmatrix} \mathbf{C}_3^T\mathbf{P}_1 & \mathbf{0} \\ \mathbf{0} & \mathbf{C}_4^T\mathbf{P}_2 \end{bmatrix} \right) \\
&\quad \begin{bmatrix} \mathbf{P}_1^{i-2}\mathbf{R}_{11} + \sum_{j=0}^{i-2}\alpha_j\mathbf{P}_1^j\mathbf{R}_{12} & \sum_{j=0}^{i-2}\gamma_j\mathbf{P}_1^j\mathbf{R}_{12} \\ \sum_{j=0}^{i-2}\beta_j\mathbf{P}_2^j\mathbf{R}_{22} & \mathbf{P}_2^{i-2}\mathbf{R}_{21} + \sum_{j=0}^{i-2}\delta_j\mathbf{P}_2^j\mathbf{R}_{22} \end{bmatrix} \\
&= \begin{bmatrix} \mathbf{P}_1^{i-1}\mathbf{R}_{11} + \sum_{j=1}^{i-1}\alpha_j\mathbf{P}_1^j\mathbf{R}_{12} & \sum_{j=1}^{i-1}\gamma_j\mathbf{P}_1^j\mathbf{R}_{12} \\ \sum_{j=1}^{i-1}\beta_j\mathbf{P}_2^j\mathbf{R}_{22} & \mathbf{P}_2^{i-1}\mathbf{R}_{21} + \sum_{j=1}^{i-1}\delta_j\mathbf{P}_2^j\mathbf{R}_{22} \end{bmatrix} + \\
&\quad \begin{bmatrix} \mathbf{0} & \mathbf{R}_{12} \\ \mathbf{R}_{22} & \mathbf{0} \end{bmatrix} \begin{bmatrix} a & b \\ c & d \end{bmatrix} \begin{bmatrix} \mathbf{C}_3^T\mathbf{P}_1 & \mathbf{0} \\ \mathbf{0} & \mathbf{C}_4^T\mathbf{P}_2 \end{bmatrix} \\
&\quad \begin{bmatrix} \mathbf{P}_1^{i-2}\mathbf{R}_{11} + \sum_{j=0}^{i-2}\alpha_j\mathbf{P}_1^j\mathbf{R}_{12} & \sum_{j=0}^{i-2}\gamma_j\mathbf{P}_1^j\mathbf{R}_{12} \\ \sum_{j=0}^{i-2}\beta_j\mathbf{P}_2^j\mathbf{R}_{22} & \mathbf{P}_2^{i-2}\mathbf{R}_{21} + \sum_{j=0}^{i-2}\delta_j\mathbf{P}_2^j\mathbf{R}_{22} \end{bmatrix} \\
&\underset{\mathbf{C}_i^T(\ldots)\mathbf{B}_j\in\mathbb{C}}{=} \begin{bmatrix} \mathbf{P}_1^{i-1}\mathbf{R}_{11} + \sum_{j=1}^{i-1}\alpha_j\mathbf{P}_1^j\mathbf{R}_{12} & \sum_{j=1}^{i-1}\gamma_j\mathbf{P}_1^j\mathbf{R}_{12} \\ \sum_{j=1}^{i-1}\beta_j\mathbf{P}_2^j\mathbf{R}_{22} & \mathbf{P}_2^{i-1}\mathbf{R}_{21} + \sum_{j=1}^{i-1}\delta_j\mathbf{P}_2^j\mathbf{R}_{22} \end{bmatrix} + \\
&\quad \begin{bmatrix} \mathbf{0} & \mathbf{R}_{12} \\ \mathbf{R}_{22} & \mathbf{0} \end{bmatrix} \begin{bmatrix} a & b \\ c & d \end{bmatrix} \begin{bmatrix} * & * \\ * & * \end{bmatrix} \\
&= \begin{bmatrix} \mathbf{P}_1^{i-1}\mathbf{R}_{11} + \sum_{j=1}^{i-1}\alpha_j\mathbf{P}_1^j\mathbf{R}_{12} & \sum_{j=1}^{i-1}\gamma_j\mathbf{P}_1^j\mathbf{R}_{12} \\ \sum_{j=1}^{i-1}\beta_j\mathbf{P}_2^j\mathbf{R}_{22} & \mathbf{P}_2^{i-1}\mathbf{R}_{21} + \sum_{j=1}^{i-1}\delta_j\mathbf{P}_2^j\mathbf{R}_{22} \end{bmatrix} + \\
&\quad \begin{bmatrix} \mathbf{0} & \mathbf{R}_{12} \\ \mathbf{R}_{22} & \mathbf{0} \end{bmatrix} \begin{bmatrix} \mu_1 & \mu_2 \\ \mu_3 & \mu_4 \end{bmatrix} \\
&= \begin{bmatrix} \mathbf{P}_1^{i-1}\mathbf{R}_{11} + \sum_{j=1}^{i-1}\alpha_j\mathbf{P}_1^j\mathbf{R}_{12} & \sum_{j=1}^{i-1}\gamma_j\mathbf{P}_1^j\mathbf{R}_{12} \\ \sum_{j=1}^{i-1}\beta_j\mathbf{P}_2^j\mathbf{R}_{22} & \mathbf{P}_2^{i-1}\mathbf{R}_{21} + \sum_{j=1}^{i-1}\delta_j\mathbf{P}_2^j\mathbf{R}_{22} \end{bmatrix} + \\
&\quad \begin{bmatrix} \mu_1\mathbf{R}_{12} & \mu_2\mathbf{R}_{12} \\ \mu_3\mathbf{R}_{22} & \mu_4\mathbf{R}_{22} \end{bmatrix} \\
&= \begin{bmatrix} \mathbf{P}_1^{i-1}\mathbf{R}_{11} + \sum_{j=0}^{i-1}\hat{\alpha}_j\mathbf{P}_1^j\mathbf{R}_{12} & \sum_{j=0}^{i-1}\hat{\gamma}_j\mathbf{P}_1^j\mathbf{R}_{12} \\ \sum_{j=0}^{i-1}\hat{\beta}_j\mathbf{P}_2^j\mathbf{R}_{22} & \mathbf{P}_2^{i-1}\mathbf{R}_{21} + \sum_{j=0}^{i-1}\hat{\delta}_j\mathbf{P}_2^j\mathbf{R}_{22} \end{bmatrix} \qquad (1.31)
\end{aligned}
$$

where $\hat{\alpha} = [\mu_1, \alpha]$, $\hat{\beta} = [\mu_3, \beta]$, $\hat{\gamma} = [\mu_2, \gamma]$, $\hat{\delta} = [\mu_4, \delta]$, and the matrix with '*' is a full matrix. Now it is easy to see that the column span of the matrix constructed from the matrix $\tilde{\mathbf{V}}_{\text{BSP}}^{(i)}$ by splitting its rows, has the same column span as the matrix defined in (1.29). Finally, the reduction basis $\mathbf{V}_{\text{BSP}}$ after $k$ steps of the BSP algorithm has the following form

$$\mathbf{V}_{\text{BSP}} = [\mathbf{V}_{\text{BSP}}^{(1)}, \ldots, \mathbf{V}_{\text{BSP}}^{(k)}]. \tag{1.32}$$

Now we will examine the SBR reduction space algorithm. Let $\mathbf{P}_i, \mathbf{R}_i = [\mathbf{R}_{i1}, \mathbf{R}_{i2}]$, $i = 1,2$ be as defined before. For $s \in \mathbb{C}$ SBR builds two Krylov subspaces

$$\mathbf{V}_1 = \mathscr{K}_k(\mathbf{P}_1, \mathbf{R}_1), \quad \text{and} \quad \mathbf{V}_2 = \mathscr{K}_k(\mathbf{P}_2, \mathbf{R}_2).$$

One can easily prove, that the $i^{th}$ step of the Krylov iteration within the SBR algorithm adds to the reduction basis the column span of the following matrix $\mathbf{V}_{\text{SBR}}^{(i)}$

$$\mathbf{V}_{\text{SBR}}^{(i)} = \begin{bmatrix} \mathbf{V}_1^{(i)} & 0 \\ 0 & \mathbf{V}_2^{(i)} \end{bmatrix}, \tag{1.33}$$

where

$$\mathbf{V}_1^{(i)} = \left[\mathbf{P}_1^{i-1}\mathbf{R}_{11}, \mathbf{P}_1^{i-1}\mathbf{R}_{12}\right]$$

and

$$\mathbf{V}_2^{(i)} = \left[\mathbf{P}_2^{i-1}\mathbf{R}_{21}, \mathbf{P}_2^{i-1}\mathbf{R}_{22}\right].$$

Finally, the reduction basis $\mathbf{V}_{\text{SBR}}$ after $k$ steps of the SBR algorithm has the following form

$$\mathbf{V}_{\text{SBR}} = [\mathbf{V}_{\text{SBR}}^{(1)}, \ldots, \mathbf{V}_{\text{SBR}}^{(k)}]. \tag{1.34}$$

Comparing (1.30) and (1.33), we observe that

$$\text{colspan}\,\mathbf{V}_{\text{BSP}} \subset \text{colspan}\,\mathbf{V}_{\text{SBR}}.$$

Because the dimensions of the spaces are equal for our case (SISO external and column-vectors $\mathbf{B}_i, \mathbf{C}_i$ for the sub-systems) one finds that in addition

$$\text{colspan}\,\mathbf{V}_{\text{BSP}} = \text{colspan}\,\mathbf{V}_{\text{SBR}}. \tag{1.35}$$

Because $\text{colspan}\,\mathbf{V}_{\text{BSP}} \subset \text{colspan}\,\mathbf{V}_{\text{SBR}}$ the SBR reduced-order model transfer function matches (at least) the same (number of) moments as the BSP reduced-order model transfer function which at its turn (Theorem 2, [6]) matches the same (number of) moments as the original coupled system's transfer function. For the more general case where $\mathbf{B}_i$, $\mathbf{C}_i$, $i = 1, \ldots, 4$ are matrices one should also obtain

$$\text{colspan}\,\mathbf{V}_{\text{BSP}} \subseteq \text{colspan}\,\mathbf{V}_{\text{SBR}} \tag{1.36}$$

which is sufficient to prove the moment matching property of the SBR reduced-order system.

### 1.3.6 Two-Sided Separate Bases Reduction Algorithm

The two-sided projection technique introduced in the previous section can be adapted to similarly improve the moment matching properties of the SBR algorithm, where we assume, as in the previous section, thath the $B_i$ and $C_i$ are column vectors. With the uncoupled formulation (1.7) and (1.8) in mind we define the reduction algorithm as follows.

1. For the sub-system $S_1$, create two matrices:
   - Matrix $\mathbf{V}_1$, whose columns span the $n_1$th Krylov subspace around $s_0 \in \mathbb{C}$
   $$\mathbf{V}_1 = \mathscr{K}_{n_1}(\mathbf{P}_1(s_0), \mathbf{R}_1(s_0)),$$
   where $\mathbf{P}_1(s_0)$ and $\mathbf{R}_1(s_0)$ are
   $$\mathbf{P}_1(s_0) = (s_0\mathbf{E}_{11} - \mathbf{A}_{11})^{-1}\mathbf{E}_{11} \quad \text{and} \quad \mathbf{R}_1(s_0) = (s_0\mathbf{E}_{11} - \mathbf{A}_{11})^{-1}[\mathbf{B}_1\ \mathbf{B}_3].$$
   Matrix $\mathbf{V}_1$ has $N_1$ rows.
   - Matrix $\mathbf{W}_1$, whose columns span the $n_1$th Krylov subspace around $s_0 \in \mathbb{C}$
   $$\mathbf{W}_1 = \mathscr{K}_{n_1}(\mathbf{S}_1(s_0), \mathbf{T}_1(s_0)),$$
   where $\mathbf{S}_1(s_0)$ and $\mathbf{T}_1(s_0)$ are
   $$\mathbf{S}_1(s_0) = (s_0\mathbf{E}_{11} - \mathbf{A}_{11})^{-T}\mathbf{E}_{11}^T \quad \text{and} \quad \mathbf{T}_1(s_0) = (s_0\mathbf{E}_{11} - \mathbf{A}_{11})^{-T}[\mathbf{C}_1\ \mathbf{C}_3].$$
   Matrix $\mathbf{W}_1$ has $N_1$ rows.
2. For the sub-system $S_2$, create two matrices:
   - Matrix $\mathbf{V}_2$, whose columns span the $n_2$th Krylov subspace around $s_0 \in \mathbb{C}$
   $$\mathbf{V}_2 = \mathscr{K}_{n_2}(\mathbf{P}_2(s_0), \mathbf{R}_2(s_0)),$$
   where $\mathbf{P}_2(s_0)$ and $\mathbf{R}_2(s_0)$ are
   $$\mathbf{P}_2(s_0) = (s_0\mathbf{E}_{22} - \mathbf{A}_{22})^{-1}\mathbf{E}_{22} \quad \text{and} \quad \mathbf{R}_2(s_0) = (s_0\mathbf{E}_{22} - \mathbf{A}_{22})^{-1}[\mathbf{B}_2\ \mathbf{B}_4].$$
   Matrix $\mathbf{V}_2$ has $N_2$ rows.
   - Matrix $\mathbf{W}_2$, whose columns span the $n_2$th Krylov subspace around $s_0 \in \mathbb{C}$
   $$\mathbf{W}_2 = \mathscr{K}_{n_2}(\mathbf{S}_2(s_0), \mathbf{T}_2(s_0)),$$
   where $\mathbf{S}_2(s_0)$ and $\mathbf{T}_2(s_0)$ are
   $$\mathbf{S}_2(s_0) = (s_0\mathbf{E}_{22} - \mathbf{A}_{22})^{-T}\mathbf{E}_{22}^T \quad \text{and} \quad \mathbf{T}_2(s_0) = (s_0\mathbf{E}_{22} - \mathbf{A}_{22})^{-T}[\mathbf{C}_2\ \mathbf{C}_4].$$
   Matrix $\mathbf{W}_2$ has $N_2$ rows.

3. Build two block-diagonal reduction matrices $\mathbf{V}$ and $\mathbf{W}$ with $N_1 + N_2 = N$ rows

$$\mathbf{V} = \begin{bmatrix} \mathbf{V}_1 & \mathbf{0} \\ \mathbf{0} & \mathbf{V}_2 \end{bmatrix} \quad \text{and} \quad \mathbf{W} = \begin{bmatrix} \mathbf{W}_1 & \mathbf{0} \\ \mathbf{0} & \mathbf{W}_2 \end{bmatrix}.$$

4. Project the original system onto the lower-dimensional space

$$\hat{\mathbf{E}}_{\mathrm{SBR}} = \mathbf{W}^T \mathbf{E} \mathbf{V}, \quad \hat{\mathbf{A}}_{\mathrm{SBR}} = \mathbf{W}^T \mathbf{A} \mathbf{V}, \quad \hat{\mathbf{B}}_{\mathrm{SBR}} = \mathbf{W}^T \mathbf{B}, \quad \hat{\mathbf{C}}_{\mathrm{SBR}} = \mathbf{V}^T \mathbf{C}.$$

Again, different algorithms lead to $\mathbf{V}_1$, $\mathbf{V}_2$ and $\mathbf{W}_1$, $\mathbf{W}_2$ with different properties. Also the above SBR algorithm results in a block-structured reduced order system and uses all of the inputs *and* outputs. Consequently, also the above SBR-based reduced order system's transfer function matches twice as many moments of the original system's transfer function as the only inputs based one in Sect. 1.3.3 (the moment matching property follows from the BSP algorithm, Theorem 1 and Theorem 2).

## 1.4 Low-Rank Approximations Based SBR Algorithm

In Sect. 1.3 we presented the Separate Bases Reduction algorithm – a block-structure preserving model reduction method for coupled systems. As discussed in that section, one of the SBR method's disadvantages is that the sizes of the its Krylov subspaces increase very fast for systems with a large number of internal inputs and outputs. Hence, the use of the SBR algorithm was recommended for the cases, in which the number of internal inputs and outputs was considerably smaller than the dimension of the system or comparable to the number of the external inputs and outputs. In this section, we approximate the internal inputs (outputs) by their GSVD-based dominant parts. This improves the efficiency of the SBR method. In addition we will prove that both the SBR algorithm and its low-rank based variant can be applied to coupled systems for which the internal input and output operators $\mathbf{B}$ and $\mathbf{C}$ are not explicitly availabe.

### *1.4.1 Implicitly Defined Couplings*

In Sect. 1.3, we introduced the interconnected system (1.10) as a result of the coupling of the two sub-systems, (1.7) and (1.8). Here, the coupling blocks are given by the explicit products of the internal inputs and outputs of the two sub-systems, namely $\mathbf{B}_3\mathbf{C}_4^T$ and $\mathbf{B}_4\mathbf{C}_3^T$. Having such a formulation at our disposal, we can apply the SBR algorithm in a straightforward way. However, for some applications it may be impossible to obtain matrices $\mathbf{B}_3$, $\mathbf{B}_4$, $\mathbf{C}_3$ and $\mathbf{C}_4$. In the following sections we propose a way of transforming an interconnected system with *implicitly defined couplings* of a form

$$S: \begin{cases} s\begin{bmatrix}\mathbf{E}_{11} & \mathbf{0}\\ \mathbf{0} & \mathbf{E}_{22}\end{bmatrix}\begin{bmatrix}\mathbf{x}_1\\ \mathbf{x}_2\end{bmatrix} = \begin{bmatrix}\mathbf{A}_{11} & \mathbf{A}_{12}\\ \mathbf{A}_{21} & \mathbf{A}_{22}\end{bmatrix}\begin{bmatrix}\mathbf{x}_1\\ \mathbf{x}_2\end{bmatrix} + \begin{bmatrix}\mathbf{B}_1 & \mathbf{0}\\ \mathbf{0} & \mathbf{B}_2\end{bmatrix}\begin{bmatrix}\mathbf{u}_1\\ \mathbf{u}_2\end{bmatrix} \\ \begin{bmatrix}\mathbf{y}_1\\ \mathbf{y}_2\end{bmatrix} = \begin{bmatrix}\mathbf{C}_1^T & \mathbf{0}\\ \mathbf{0} & \mathbf{C}_2^T\end{bmatrix}\begin{bmatrix}\mathbf{x}_1\\ \mathbf{x}_2\end{bmatrix} \end{cases}, \tag{1.37}$$

with

$$\mathbf{E} = \begin{bmatrix}\mathbf{E}_{11} & \mathbf{0}\\ \mathbf{0} & \mathbf{E}_{22}\end{bmatrix}, \quad \mathbf{A} = \begin{bmatrix}\mathbf{A}_{11} & \mathbf{A}_{12}\\ \mathbf{A}_{21} & \mathbf{A}_{22}\end{bmatrix}, \quad \mathbf{B} = \begin{bmatrix}\mathbf{B}_1 & \mathbf{0}\\ \mathbf{0} & \mathbf{B}_2\end{bmatrix}, \quad \mathbf{C} = \begin{bmatrix}\mathbf{C}_1 & \mathbf{0}\\ \mathbf{0} & \mathbf{C}_2\end{bmatrix} \tag{1.38}$$

into a form that can be reduced using the SBR algorithm. Our goal is to find decompositions (factorizations) of the two coupling blocks

$$\mathbf{A}_{12} = \tilde{\mathbf{B}}_3\tilde{\mathbf{C}}_4^T \qquad \text{and} \qquad \mathbf{A}_{21} = \tilde{\mathbf{B}}_4\tilde{\mathbf{C}}_3^T \tag{1.39}$$

that provide a good (with respect to the corresponding Krylov subspaces) approximation of the original internal inputs and outputs of the coupled system (1.37). A factorization of the type $\mathbf{A} = \mathbf{BC}$ is not be unique. The next section shows how to deal with this.

## 1.4.2 Decomposition Theorem

In this section, related to (1.37), first we show that a factorization $\mathbf{A} = \mathbf{BC}$ is not unique and next we prove that if $\mathbf{A}_{12} = \mathbf{B}_1\mathbf{C}_1$ and simultaneously $\mathbf{A}_{12} = \mathbf{B}_2\mathbf{C}_2$ then $\mathscr{K}_p(\mathbf{A}_{11}, \mathbf{B}_1) = \mathscr{K}_p(\mathbf{A}_{11}, \mathbf{B}_2)$ if $\mathbf{C}_1$ and $\mathbf{C}_2$ are of *full column rank*. The proofs will be for the input-based Krylov subspaces. Similar theory applies to the output-based Krylov subspaces.

First, a factorization of the type $\mathbf{A} = \mathbf{BC}$ is not unique since $\mathbf{A} = \mathbf{IA}$ and $\mathbf{A} = \mathbf{AI}$ are two different factorizations. Even a QR factorization $\mathbf{A} = \mathbf{QR}$ is not unique since if $\mathbf{A} = \mathbf{QR}$ then $\mathbf{A} = (\mathbf{Q}\overline{\mathbf{S}})(\mathbf{SR})$ for all complex valued diagonal matrices $\mathbf{S}$ with unit-length diagonal elements ($\overline{\mathbf{S}}$ denotes the complex conjugate of $\mathbf{S}$). Also other factorizations such as Gaussian-elimination based $\mathbf{A} = \mathbf{LU}$ exist.

Since we aim at the use of $\mathbf{B}$ for the generation of a Krylov subspace $\mathscr{K}_p(\mathbf{A}_{11}, \mathbf{B}_1)$ we will next show that the non-uniqueness does not need to be an issue. To this end we prove the following Lemma 1.1 and Theorem 1.3.

**Lemma 1.1** *Let* $\mathbf{B} \in \mathbb{R}^{n\times p}$, $\mathbf{C} \in \mathbb{R}^{p\times m}$, $m,n,p \in \mathbb{N}$. *Then*

$$\operatorname{rank}(\mathbf{C}) = p \implies \textit{colspan}\,\mathbf{BC} = \textit{colspan}\mathbf{B}.$$

*Proof* Matrix $\mathbf{C}$ has rank $p$ which implies $p \le m$ and that $\mathbf{C}$ has $p$ linearly independent columns of length $p$. Thus based on

$$\text{colspan}\mathbf{C} = \{\mathbf{Cx}\colon \mathbf{x} \in \mathbb{R}^m\} \tag{1.40}$$

one finds

$$\text{colspan}\mathbf{C} \underset{(1.40)}{=} \{\mathbf{C}\mathbf{x}\colon \mathbf{x} \in \mathbb{R}^m\} \underset{p \le m}{=} \mathbb{R}^p \tag{1.41}$$

whence

$$\text{colspan}\mathbf{BC} \underset{(1.40)}{=} \{\mathbf{BC}\mathbf{x}\colon \mathbf{x} \in \mathbb{R}^m\} \underset{(1.41)}{=} \{\mathbf{B}\mathbf{y}\colon \mathbf{y} \in \mathbb{R}^p\} \underset{(1.40)}{=} \text{colspan}\mathbf{B}.$$

Note: The condition that $\mathbf{C}$ has full column rank is sufficient but not necessary. It can be relaxed: If for instance $\mathbf{B}$ has only $2 \le p$ linearly independent columns, e.g. the $i$th and the $j$th column, then a sufficient condition is $\text{colspan}\mathbf{C} = \text{colspan}\,\{\mathbf{e}_i, \mathbf{e}_j\} \subset \mathbb{R}^p$.

**Theorem 1.3** *Let* $\mathbf{B}_1, \mathbf{B}_2 \in \mathbb{R}^{n \times p}$, $\mathbf{C}_1, \mathbf{C}_2 \in \mathbb{R}^{p \times m}$ *and* $m, n, p \in \mathbb{N}$.
*If*

$$\text{rank}(\mathbf{C}_1) = \text{rank}(\mathbf{C}_2) = p \quad \textit{and} \quad \mathbf{B}_1\mathbf{C}_1 = \mathbf{B}_2\mathbf{C}_2$$

*then*

$$\textit{colspan}\mathbf{B}_1 = \textit{colspan}\mathbf{B}_2.$$

*Proof* Observe that

$$\text{colspan}\mathbf{B}_1 \underset{\text{Lem.1.1}}{=} \text{colspan}\mathbf{B}_1\mathbf{C}_1 = \text{colspan}\mathbf{B}_2\mathbf{C}_2 \underset{\text{Lem.1.1}}{=} \text{colspan}\mathbf{B}_2.$$

Next we prove that certain Krylov subspaces are identical.

**Theorem 1.4** *Let* $\mathbf{A} \in \mathbb{R}^{n \times n}$ *is non-singular and* $\mathbf{B}_1, \mathbf{B}_2 \in \mathbb{R}^{n \times m}$, $n, m \in \mathbb{N}$. *Then*

$$\textit{colspan}\mathbf{B}_1 = \textit{colspan}\mathbf{B}_2 \implies \mathscr{K}_p(\mathbf{A}, \mathbf{B}_1) = \mathscr{K}_p(\mathbf{A}, \mathbf{B}_2).$$

*Proof* Note that

$$\begin{aligned} \text{colspan}\mathbf{B}_1 &= \text{colspan}\mathbf{B}_2 \iff \\ \{\mathbf{B}_1\mathbf{x}\colon \mathbf{x} \in \mathbb{R}^m\} &= \{\mathbf{B}_2\mathbf{x}\colon \mathbf{x} \in \mathbb{R}^m\} \iff \\ \{\mathbf{AB}_1\mathbf{x}\colon \mathbf{x} \in \mathbb{R}^m\} &= \{\mathbf{AB}_2\mathbf{x}\colon \mathbf{x} \in \mathbb{R}^m\} \iff \\ \text{colspan }\mathbf{AB}_1 &= \text{colspan }\mathbf{AB}_2, \end{aligned}$$

which, repeatedly applied, shows that $\text{colspan}\mathbf{A}^k\mathbf{B}_1 = \text{colspan}\mathbf{A}^k\mathbf{B}_2$ for all $k \ge 0$ whence $\mathscr{K}_p(\mathbf{A}, \mathbf{B}_1) = \mathscr{K}_p(\mathbf{A}, \mathbf{B}_2)$.

Theorem 1.3 in combination with Theorem 1.4 show that every factorization of an off-diagonal block of the form $\mathbf{A}_{12} = \mathbf{BC}^{\mathrm{T}}$ with $\mathbf{C}$ of *full column rank* leads to the same krylov space $\mathscr{K}_p(\mathbf{A}_{11}, \mathbf{B})$. The following sections show how to use this property for the application of the SBR method to an arbitrary coupled system (1.37).

### *1.4.3 Decomposition Theorem – Numerical Example*

In Sect. 1.4.2 we showed that the Krylov space does not depend on the factors of the decomposition $\mathbf{A}_{12} = \mathbf{BC}^{\mathrm{T}}$ when these factors are of maximal column rank. To illustrate this numerically, we calculate these factors of $\mathbf{A}_{12}$ with different factorization techniques, based on a *QR factoration* and *LU factorization*. For simplicity, we use a one-sided variant of the SBR method. The system used for the test is a linear beam coupled to a controller. Only the beam system has an external input and external output. Hence, the considered system is of a form

$$S: \begin{cases} s\begin{bmatrix} \mathbf{I}_{11} & \mathbf{0} \\ \mathbf{0} & \mathbf{I}_{22} \end{bmatrix}\begin{bmatrix} \mathbf{x}_1 \\ \mathbf{x}_2 \end{bmatrix} = \begin{bmatrix} \mathbf{A}_{11} & \mathbf{B}_3\mathbf{C}_4^T \\ \mathbf{B}_4\mathbf{C}_3^T & \mathbf{A}_{22} \end{bmatrix}\begin{bmatrix} \mathbf{x}_1 \\ \mathbf{x}_2 \end{bmatrix} + \begin{bmatrix} \mathbf{B}_1 \\ \mathbf{0} \end{bmatrix}\mathbf{u}_1 \\ \mathbf{y}_1 = \begin{bmatrix} \mathbf{C}_1^T & \mathbf{0} \end{bmatrix}\begin{bmatrix} \mathbf{x}_1 \\ \mathbf{x}_2 \end{bmatrix}. \end{cases} \tag{1.42}$$

Let $\mathbf{A}_{12} = \mathbf{B}_3\mathbf{C}_4^T$ and $\mathbf{A}_{21} = \mathbf{B}_4\mathbf{C}_3^T$. Here, the full coupled system has 80 degrees of freedom, 40 for each sub-system. Both of the sub-systems have 5 internal inputs and 5 internal outputs. It means, that the coupling blocks $\mathbf{A}_{12}$ and $\mathbf{A}_{21}$ are of rank 5. For all cases, the same number of Krylov iterations is performed and the reduced-order systems are of the order 55 (originally 80). The first sub-system was reduced from order 40 down to 30 and the second from order 40 down to 25.

To reduce the original system, we will build three reduction matrices involving an $n$th-order Krylov sub-space as follows:

- **Reduction matrix based on the original internal input blocks**
  The diagonal sub-blocks of the reduction matrix span the Krylov subspaces

$$\mathbf{V}_1 = \mathscr{K}_n(\mathbf{P}_1, \mathbf{R}_1),$$

  where

$$\mathbf{P}_1 = (s\mathbf{I}_{11} - \mathbf{A}_{11})^{-1} \qquad \text{and} \qquad \mathbf{R}_1 = (s\mathbf{I}_{11} - \mathbf{A}_{11})^{-1}[\mathbf{B}_1\ \mathbf{B}_3]$$

  and

$$\mathbf{V}_2 = \mathscr{K}_n(\mathbf{P}_2, \mathbf{R}_2),$$

  where

$$\mathbf{P}_2 = (s\mathbf{I}_{22} - \mathbf{A}_{22})^{-1} \qquad \text{and} \qquad \mathbf{R}_2 = (s\mathbf{I}_{22} - \mathbf{A}_{22})^{-1}\mathbf{B}_4.$$

  The block-diagonal reduction matrix $\mathbf{V}$ is of the form

$$\mathbf{V} = \begin{bmatrix} \mathbf{V}_1 & \mathbf{0} \\ \mathbf{0} & \mathbf{V}_2 \end{bmatrix}.$$

- **Reduction matrix based on a QR decomposition of the coupling blocks**
  Based on a QR decomposition of the coupling matrices $\mathbf{A}_{12}$ and $\mathbf{A}_{21}$, we get

$$\mathbf{A}_{12} = \mathscr{Q}_1\mathscr{R}_1 \qquad \text{and} \qquad \mathbf{A}_{21} = \mathscr{Q}_2\mathscr{R}_2.$$

  We use an *rank-revealing* version of the QR algorithm, i.e., $\mathscr{Q}_1$, $\mathscr{Q}_2$, $\mathscr{R}_1^{\mathrm{T}}$, $\mathscr{R}_2^{\mathrm{T}}$ are of full column rank. Hence, the matrices $\mathscr{Q}_1$ and $\mathscr{Q}_2$ used to build the Krylov subspaces have the same rank (and most likely amount of columns) as $\mathbf{B}_3$ and $\mathbf{B}_4$. Next, the reduction sub-blocks are created

$$\mathbf{V}_1^{QR} = \mathscr{K}_n(\mathbf{P}_1^{QR}, \mathbf{R}_1^{QR}),$$

where

$$\mathbf{P}_1^{QR} = (s\mathbf{I}_{11} - \mathbf{A}_{11})^{-1} \qquad \text{and} \qquad \mathbf{R}_1^{QR} = (s\mathbf{I}_{11} - \mathbf{A}_{11})^{-1}[\mathbf{B}_1 \ \mathscr{Q}_1]$$

and

$$\mathbf{V}_2^{QR} = \mathscr{K}_n(\mathbf{P}_2^{QR}, \mathbf{R}_2^{QR}),$$

where

$$\mathbf{P}_2^{QR} = (s\mathbf{I}_{22} - \mathbf{A}_{22})^{-1} \qquad \text{and} \qquad \mathbf{R}_2^{QR} = (s\mathbf{I}_{22} - \mathbf{A}_{22})^{-1}\mathscr{Q}_2.$$

The block-diagonal reduction matrix $\mathbf{V}^{QR}$ is of the form

$$\mathbf{V}^{QR} = \begin{bmatrix} \mathbf{V}_1^{QR} & \mathbf{0} \\ \mathbf{0} & \mathbf{V}_2^{QR} \end{bmatrix}.$$

- **Reduction matrix based on the LU decomposition of the coupling blocks** Based on the LU decomposition of the coupling matrices $\mathbf{A}_{12}$ and $\mathbf{A}_{21}$, we get

$$\mathbf{A}_{12} = \mathscr{L}_1\mathscr{U}_1 \qquad \text{and} \qquad \mathbf{A}_{21} = \mathscr{L}_2\mathscr{U}_2.$$

We use a *rank-revealing* version of the LU algorithm, i.e., $\mathscr{L}_1$, $\mathscr{L}_2$, $\mathscr{U}_1^{\mathrm{T}}$ and $\mathscr{U}_2^{\mathrm{T}}$ are of full column rank. Hence, the matrices $\mathscr{L}_1$ and $\mathscr{L}_2$ used to build the Krylov subspaces have the same rank (and most likely amount of columns) as $\mathbf{B}_3$ and $\mathbf{B}_4$. Next, the reduction sub-blocks are created

$$\mathbf{V}_1^{LU} = \mathscr{K}_n(\mathbf{P}_1^{LU}, \mathbf{R}_1^{LU}),$$

where

$$\mathbf{P}_1^{LU} = (s\mathbf{I}_{11} - \mathbf{A}_{11})^{-1} \qquad \text{and} \qquad \mathbf{R}_1^{LU} = (s\mathbf{I}_{11} - \mathbf{A}_{11})^{-1}[\mathbf{B}_1 \ \mathscr{L}_1]$$

and

$$\mathbf{V}_2^{LU} = \mathscr{K}_n(\mathbf{P}_2^{LU}, \mathbf{R}_2^{LU}),$$

where

$$\mathbf{P}_2^{LU} = (s\mathbf{I}_{22} - \mathbf{A}_{22})^{-1} \qquad \text{and} \qquad \mathbf{R}_2^{LU} = (s\mathbf{I}_{22} - \mathbf{A}_{22})^{-1}\mathscr{L}_2.$$

The block-diagonal reduction matrix $\mathbf{V}^{LU}$ is of the form

$$\mathbf{V}^{LU} = \begin{bmatrix} \mathbf{V}_1^{LU} & \mathbf{0} \\ \mathbf{0} & \mathbf{V}_2^{LU} \end{bmatrix}.$$

Figure 1.8 shows the magnitude plots with respect to the frequency of the frequency response functions of the three reduced-order systems, created using original, QR-,

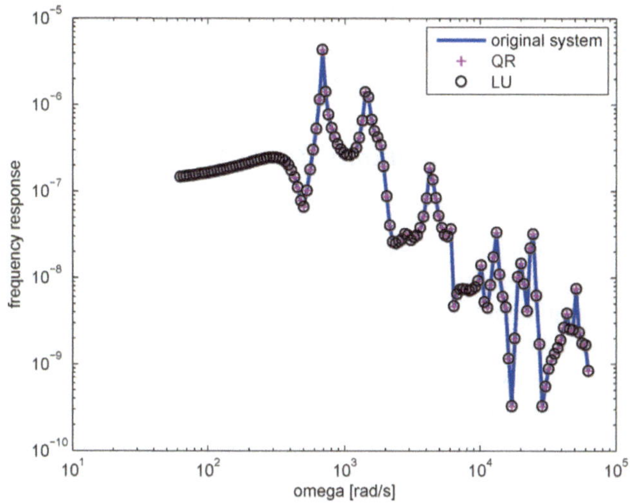


**Fig. 1.8.** Magnitude plots of the frequency response functions of the reduced-order systems based on different decompositions of the coupling blocks

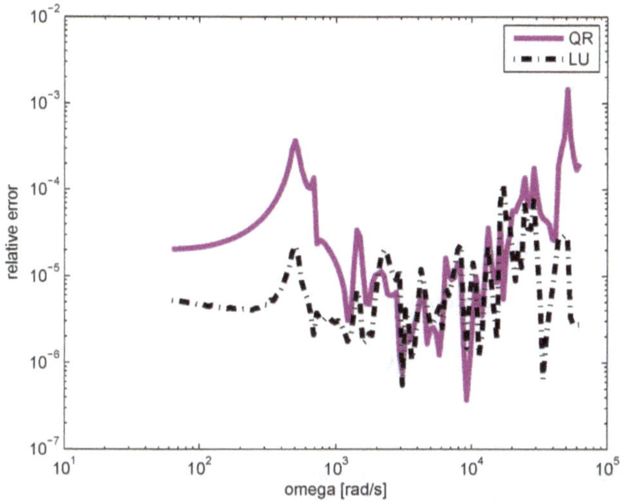


**Fig. 1.9.** Magnitude plots of the relative errors of the reduced-order frequency response functions based on different decompositions of the coupling blocks with respect to the reduced-order frequency response function based on the original input and output matrices

and LU-decomposition based input matrices. The plots are almost identical, which is confirmed in Fig. 1.9, that shows the relative errors between the reduced-order frequency response function of the original system and the frequency response functions computed based on both decompositions. The small differences between the three frequency response functions should be caused by round-off errors.

The next section shows how Theorems 1.3 and 1.4 in combination with GSVD can be used to improve the performance of the SBR algorithm applied to coupled systems with a high number of couplings (or interconnections).

### 1.4.4 Low-Rank Approximations Based SBR Algorithm

For coupled systems it is not always necessary to take into account all of the coupling components. Sometimes only a small number of them determines the behavior of the system and the rest can be neglected without much loss of accuracy. This section extends the application of the SBR algorithm to coupled (or interconnected) systems characterized by a high number of couplings of which only a small percentage is relevant to obtain an accurate solution.

Section 1.3 pointed out that the standard SBR method should be applied only for the systems with a relatively small number of internal inputs and outputs. That is, only for coupled systems where few degrees of freedom of one sub-system (related to one physical domain or to a physcial quantity) are coupled/connected to the other sub-system, which implies that the coupling blocks $\mathbf{A}_{12}$ and $\mathbf{A}_{21}$ of the system (1.37) are of low rank. Otherwise, the SBR method produces reduction bases which increase in size too fast with respect to the number of Krylov iterations. However, if only a part of the components of the high rank coupling blocks is relevant, we can decrease the growth speed of the reduction bases. To do so, we first need to determine, which components of the coupling are important and should be kept, and which ones can be neglected. One of the ways to make this decision, is to apply the generalized singular value decomposition (GSVD) to the coupling matrices $\mathbf{A}_{12}$ and $\mathbf{A}_{21}$. The GSVD should be applied to the pairs $(\mathbf{A}_{11}^T, \mathbf{A}_{12}^T)$ and $(\mathbf{A}_{22}^T, \mathbf{A}_{21}^T)$. One then has

$$\mathbf{A}_{12}^T = \mathbf{V}_1\mathbf{S}_1\mathbf{X}_1^T \quad \text{and} \quad \mathbf{A}_{21}^T = \mathbf{V}_2\mathbf{S}_2\mathbf{X}_2^T$$

which results in the expressions for the coupling blocks

$$\mathbf{A}_{12} = \mathbf{X}_1\mathbf{S}_1^T\mathbf{V}_1^T \tag{1.43}$$

$$\mathbf{A}_{21} = \mathbf{X}_2\mathbf{S}_2^T\mathbf{V}_2^T. \tag{1.44}$$

Note, that here the matrices $\mathbf{C}_1$ and $\mathbf{C}_2$ are not used to denote external output matrices, but components of the GSVD. Assuming that the coupling blocks are of the form (1.39), since $\mathbf{S}_1$ and $\mathbf{S}_2$ are real-valued non-negative diagonal, we can define the input and output matrices as following products

$$\tilde{\mathbf{B}}_3 = \mathbf{X}_1\mathbf{S}_1^{1/2}, \qquad \tilde{\mathbf{C}}_4 = \mathbf{V}_1\mathbf{S}_1^{1/2} \tag{1.45}$$

$$\tilde{\mathbf{B}}_4 = \mathbf{X}_2\mathbf{S}_2^{1/2}, \qquad \tilde{\mathbf{C}}_3 = \mathbf{V}_2\mathbf{S}_2^{1/2}. \tag{1.46}$$

Since $\mathbf{S}_1$ and $\mathbf{S}_2$ are diagonal matrices with non-negative entries their square roots are diagonal matrices with entries $\sqrt{[\mathbf{S}_1]_{ii}}$ and $\sqrt{[\mathbf{S}_2]_{ii}}$. Constructing the inputs and outputs as in (1.45) and (1.46), all of $\mathbf{B}_i$ and $\mathbf{C}_i$, $i = 3,4$ are scaled by $\sqrt{\mathbf{S}_1}$ or $\sqrt{\mathbf{S}_2}$.

According to the Theorems 1.3 and 1.4, $\mathscr{K}_p(\mathbf{A}_{11}, \mathbf{B}_3\mathbf{C}_4^T) = \mathscr{K}_p(\mathbf{A}_{11}, \tilde{\mathbf{B}}_3\tilde{\mathbf{C}}_4^T)$ and $\mathscr{K}_p(\mathbf{A}_{22}, \mathbf{B}_4\mathbf{C}_3^T) = \mathscr{K}_p(\mathbf{A}_{11}, \tilde{\mathbf{B}}_4\tilde{\mathbf{C}}_3^T)$. Moreover, using a type of the decomposition that orders the components with respect to their importance has an additional benefit. It makes it possible to approximate the inputs and outputs leaving only the most relevant components and, as a result, reduces the dimensions of the blocks. In some cases, this reduction is sufficient to allow for an efficient application of the SBR algorithm.

Let us now compare the procedures of building the standard and GSVD-based Krylov subspaces. Here, we will limit the discussion to the case of creation of a Krylov space based on inputs of the sub-system (1.7), but a similar analysis applies to all the other cases, i.e. input-based Krylov subspace for (1.8) and output-based Krylov subspaces for both sub-systems, (1.7) and (1.8). As defined in Chap. 1.3, matrices $\mathbf{A}_{11} \in \mathbb{R}^{N_1 \times N_1}$, $\mathbf{B}_1 \in \mathbb{R}^{N_1 \times m_1}$ and $\mathbf{B}_3 \in \mathbb{R}^{N_1 \times m_3}$. Assume, that $\mathbf{B}_3$ has full column rank $m_3$ and that application of GSVD to the pair $(\mathbf{A}_{11}^T, \mathbf{A}_{12}^T)$ leads to

$$\tilde{\mathbf{B}}_3 = \mathbf{X}_1\mathbf{S}_1^{1/2} = [\tilde{\mathbf{b}}_1, .., \tilde{\mathbf{b}}_{m_3}] \in \mathbb{R}^{N_1 \times m_3}.$$

where both $\mathbf{X}_1$ and $\mathbf{S}_1$ in (1.45) are of full column rank. Next, let $\hat{\mathbf{B}}_3 = \mathbf{X}_1^{(k)}(\mathbf{S}_1^{(k)})^{1/2}$ approximate $\tilde{\mathbf{B}}_3$ with the use of $k$ dominant components. Then

$$\hat{\mathbf{B}}_3 = [\tilde{\mathbf{b}}_1, .., \tilde{\mathbf{b}}_k] \in \mathbb{R}^{N_1 \times k}. \tag{1.47}$$

For simplicity, we assume that $m_1 + m_3$ is a multiple of $m_1 + k$ (this may not be the case in general), so there exists $\lambda \in \mathbb{N}$ such that

$$m_1 + m_3 = \lambda(m_1 + k). \tag{1.48}$$

The $p$th Krylov subspace created by the SBR algorithm for the sub-system (1.7) for $s_0 \in \mathbb{C}$ is

$$\mathscr{K}_p(\mathbf{P}_1, \mathbf{R}_1) = \text{colspan}\{\mathbf{R}_1, \mathbf{P}_1\mathbf{R}_1, \dots, \mathbf{P}_1^{p-1}\mathbf{R}_1\},$$

where

$$\mathbf{P}_1 = (s_0\mathbf{E}_{11} - \mathbf{A}_{11})^{-1}\mathbf{E}_{11} \quad \text{and} \quad \mathbf{R}_1 = (s_0\mathbf{E}_{11} - \mathbf{A}_{11})^{-1}[\mathbf{B}_1\ \mathbf{B}_3]$$

and consists of $p(m_1 + m_3)$ columns (assuming that no linear dependence occurs).

Likewise,

$$\mathscr{K}_{\lambda p}(\mathbf{P}_1, \hat{\mathbf{R}}_1) = \text{colspan}\{\hat{\mathbf{R}}_1, \mathbf{P}_1\hat{\mathbf{R}}_1, \dots, \mathbf{P}_1^{\lambda p-1}\hat{\mathbf{R}}_1\}, \tag{1.49}$$

where

$$\mathbf{P}_1 = (s_0\mathbf{E}_{11} - \mathbf{A}_{11})^{-1}\mathbf{E}_{11} \quad \text{and} \quad \hat{\mathbf{R}}_1 = (s_0\mathbf{E}_{11} - \mathbf{A}_{11})^{-1}[\mathbf{B}_1\ \hat{\mathbf{B}}_3]$$

consists also of $p(m_1 + m_3)$ columns, but *approximately* matches $\lambda$ as many moments of the original transfer function.

Projecting system (1.7) onto a subspace $\mathscr{K}_{\lambda p}(\mathbf{P}_1, \hat{\mathbf{R}}_1)$ in (1.49) does not preserve the moments of the transfer function of this sub-system. However, if the column span of the matrix $\hat{\mathbf{B}}_3$ gives a good approximation of the column span of the matrix $\mathbf{B}_3$ we can expect that the reduced-order system obtained by projection onto the space (1.49) will give an accurate approximation of the appropriate number of moments of the transfer function of the original system. Moreover, if the matrix $\mathbf{B}_3$ can be approximated by $\hat{\mathbf{B}}_3$ with a significantly smaller number of columns, $\lambda$ times more steps may be used during the Krylov procedure (to approximate a higher number of moments) or one can use more expansion points, keeping the reduced-order model still relatively small.

In the next section, we present the results of some numerical tests that show the advantage of using the low-rank approximation based SBR algorithm for a system with a high order of coupling.

## 1.5 Numerical Examples

In this section, we present two examples of the application of the SBR method combined with low rank approximations for the coupling blocks. The first example is a simple and small example, yet exhibiting interesting behaviour as far as coupling is concerned. The second example is described in much more detail, as this is an industrial benchmark problem and needs some preliminary steps before the methods described in this chapter can be applied.

### *1.5.1 A Simple Example*

In this section, we consider a simple example. The difficulty of this test case is that here the coupling blocks of the system are of rank 10 (the coupled system has 10 internal inputs and 10 internal outputs), while each of the sub-systems contains only 40 degrees of freedom (80 degrees of freedom in total). In this case, the standard SBR algorithm generates too many columns to be competitive. However, the use of low-rank approximations makes the SBR algorithm more competitive. Fig. 1.10 shows the magnitude plots with respect to the frequency of the original and reduced-order frequency response functions. In case of the two-sided BSP method and the two-sided SBR algorithm based reduced-order systems, the original system was reduced to 42 degrees of freedom. The low-rank approximation based two-sided SBR algorithm created the reduction bases for rank 3 approximations of the coupling blocks, i.e. the internal input and output matrices $\mathbf{B}_i, \mathbf{C}_i \in \mathbb{R}^{40\times 10}$, $i = 3,4$ were approximated by $\hat{\mathbf{B}}_i, \hat{\mathbf{C}}_i \in \mathbb{R}^{40\times 3}$, $i = 3,4$. Hence, every Krylov step was adding 4 new columns to the reduction basis (3 corresponding to $\hat{\mathbf{B}}_3$ or $\hat{\mathbf{C}}_4$ and 1 corresponding to $\mathbf{B}_1$ or $\mathbf{C}_1$) in case of the sub-system $S_1$ and 3 new columns (corresponding to $\hat{\mathbf{B}}_4$ or $\hat{\mathbf{C}}_3$) in case of the sub-system $S_2$. To construct the reduced-order system of dimension 42, the low-rank approximation based SBR algorithm performed 6 iterations for each sub-system (for both, input and output related bases). Figure 1.11 shows the magnitude plots of the relative errors of the reduced-order frequency response functions with respect to the original one. Note that the two-sided SBR algorithm

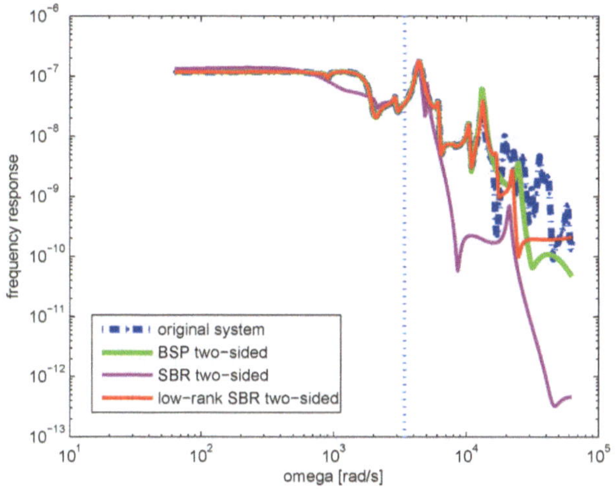


**Fig. 1.10.** Magnitude plots of the frequency response functions of the original and reduced-order systems

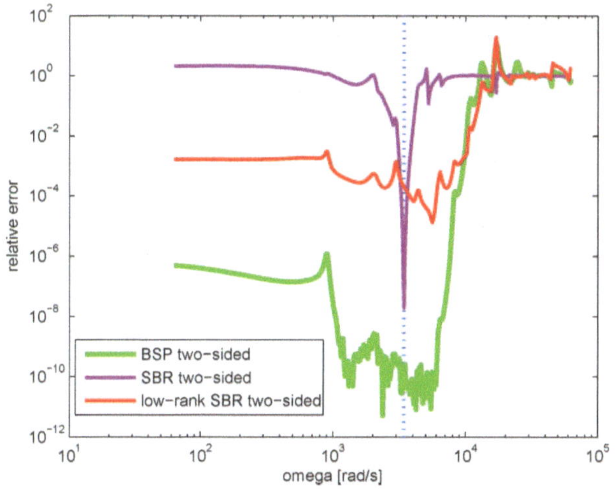


**Fig. 1.11.** Magnitude plots of the relative errors of the reduced-order frequency response functions with respect to the original frequency response function

based on low-rank approximations of the internal inputs and outputs leads to much better results than the SBR algorithm applied to the original coupling blocks. The two-sided low-rank based reduced-order transfer function $\mathbf{H}_{\text{low-rank-SBR}}$ approximates $\mathbf{H}$ less accurate than the standard two-sided BSP transfer function but in the neighborhood of the expansion point $s$ the relative error is still below 2%. Table 1.1 shows that not only the first 6 derivatives are matched but also the $7^{th}$ one is well approximated.

**Table 1.1.** Derivatives of the original and low-rank approximation based reduced-order transfer functions for the expansion point ($s$) for the second example, multiplied by $10^7$

| $i$ | $\partial^i \mathbf{H}(s)$ | $\partial^i \mathbf{H}_{\text{low-rank-SBR}}(s)$ |
|---|---|---|
| 0 | −0.349984611544531 | −0.349975323605725 |
| 1 | 0.000580754070987 | 0.000580770193275 |
| 2 | −0.000001928114532 | −0.000001928787960 |
| 3 | 0.000000012770698 | 0.000000012766510 |
| 4 | −0.000000000067912 | −0.000000000068062 |
| 5 | 0.000000000000859 | 0.000000000000859 |
| 6 | −0.000000000000014 | −0.000000000000014 |
| 7 | 0.000000000000000 | 0.000000000000000 |
| 8 | -0.000000000000000 | −0.000000000000000 |
| 9 | 0.000000000000000 | 0.000000000000000 |

## *1.5.2 Industrial Benchmark Problem*

The benchmark system treated in this chapter is a model of a printhead delivered by Océ Technologies B.V. in the Netherlands. It is a MEMS (micro-electro-mechanical-system) based design, containing a large number of individual channels integrated into a single chip. A schematic overview of a single channel (a side and bottom view) is shown in Fig. 1.12. The dotted line depicts the ink flow; the ink, coming from the reservoir, enters through a restriction (1), from which it flows into the actuation

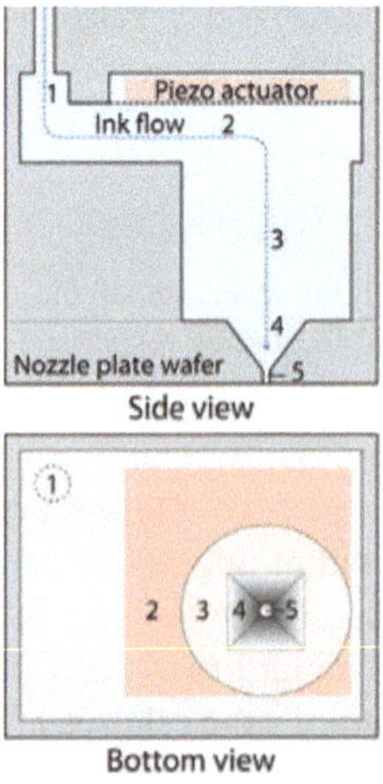


**Fig. 1.12.** A schematic overview of a single channel (courtesy of Herman Wijshoff)

chamber (2). Below the actuation chamber, a 300 $\mu$m long feed-through is placed (3), after which the nozzle plate is reached. The nozzle plate is 75 $\mu$m thick and consists of a pyramid shaped funnel (4) and a nozzle (5) with a radius of 11 $\mu$m.

The main goal is to suppress acoustic pressure waves, which can be generated in a number of ways, such as the non-continuous ink supply by many thousands of ink channels, residual vibrations at the inlet of the ink channels, fast movement of the printhead, resonance of the whole structure, etc.

The models of such devices used for simulations can reach large dimensions, hence application of the model order reduction techniques is often required, to decrease the simulation time. In this chapter, we study the application of the GSVD based approximations for the coupling blocks in the model of the printhead.

### 1.5.3 The Second and First Order System

The related system of equations is a second order system. Let $n_1, n_2 \in \mathbb{N}$ and $n = n_1 + n_2$. The *second order system* of interest is

$$\begin{cases} \mathbf{M}\mathbf{x}'' + \mathbf{K}\mathbf{x} = \mathbf{b} \\ \mathbf{y} = \mathbf{c}\mathbf{x} \end{cases} \tag{1.50}$$

with $(n_1+n_2)\times(n_1+n_2)$, $2\times 2$ block-matrices

$$\mathbf{M} = \begin{bmatrix} \mathbf{M}_{11} & \mathbf{0} \\ \mathbf{M}_{22} & \mathbf{M}_{22} \end{bmatrix}, \qquad \mathbf{K} = \begin{bmatrix} \mathbf{K}_{11} & \mathbf{K}_{12} \\ \mathbf{0} & \mathbf{K}_{22} \end{bmatrix} \tag{1.51}$$

and $\mathbf{M}_{21} = -\rho\mathbf{K}_{12}^{\mathrm{T}}$. The first sub-system corresponds to the displacement of the structure and the second sub-system describes the pressure of the fluid. The related *Laplace transformation*

$$\begin{cases} \tilde{w}^2\mathbf{M}\mathbf{X} + \mathbf{K}\mathbf{X} = \mathbf{B} \\ \mathbf{Y} = \mathbf{c}\mathbf{X} \end{cases}$$

leads to transfer function

$$H(w) = \mathbf{c}(\mathbf{K} + \tilde{w}^2\mathbf{M})^{-1}\mathbf{b}, \qquad \tilde{w} \in \mathbb{C}.$$

Searching for purely *oscillatory modes* implies that the related $\tilde{w}$ is purely imaginary, i.e., that one is interested in positive real values $w$ of:

$$H(w) = \mathbf{c}(\mathbf{K} - w^2\mathbf{M})^{-1}\mathbf{b}, \qquad w \in \mathbb{R}. \tag{1.52}$$

Let $\mathbf{x}_2 = \mathbf{x}_1'$. Then the *first order system* reformulation of (1.50) is

$$\begin{cases} \mathbf{x}_2 = \mathbf{x}_1' \\ \mathbf{M}\mathbf{x}_2' + \mathbf{K}\mathbf{x}_1 = \mathbf{b} \end{cases} \Longrightarrow \begin{cases} \mathbf{x}_1' - \mathbf{x}_2 = \mathbf{0} \\ \mathbf{M}\mathbf{x}_2' + \mathbf{K}\mathbf{x}_1 = \mathbf{b} \end{cases}$$

which implies

$$\begin{cases} \underbrace{\begin{bmatrix} \mathbf{I} & \\ & \mathbf{M} \end{bmatrix}}_{\mathbf{E}} \begin{bmatrix} \mathbf{x}_1 \\ \mathbf{x}_2 \end{bmatrix}' = \underbrace{\begin{bmatrix} & \mathbf{I} \\ -\mathbf{K} & \end{bmatrix}}_{\mathbf{A}} \begin{bmatrix} \mathbf{x}_1 \\ \mathbf{x}_2 \end{bmatrix} + \underbrace{\begin{bmatrix} \mathbf{0} \\ \mathbf{b} \end{bmatrix}}_{\mathbf{B}} \\ \begin{bmatrix} \mathbf{y}_1 \\ \mathbf{y}_2 \end{bmatrix} = \begin{bmatrix} \mathbf{c} & \mathbf{0} \\ \mathbf{0} & \mathbf{0} \end{bmatrix} + \begin{bmatrix} \mathbf{x}_1 \\ \mathbf{x}_2 \end{bmatrix}. \end{cases}$$

Its related transfer function is

$$H(s) = \mathbf{C}(s\mathbf{E} - \mathbf{A})^{-1}\mathbf{B}. \tag{1.53}$$

Solution of $\mathbf{FX} = \mathbf{B}$:

$$\begin{cases} s\mathbf{x}_1 - \mathbf{x}_2 = \mathbf{0} \\ s\mathbf{M}\mathbf{x}_2 + \mathbf{K}\mathbf{x}_1 = \mathbf{b} \end{cases} \Longrightarrow \begin{cases} s\mathbf{x}_1 - \mathbf{x}_2 = \mathbf{0} \\ s^2\mathbf{M}\mathbf{x}_1 + \mathbf{K}\mathbf{x}_1 = \mathbf{b} \end{cases} \Longrightarrow \begin{cases} \mathbf{x}_1 = \left(s^2\mathbf{M} + \mathbf{K}\right)^{-1}\mathbf{b} \\ \mathbf{x}_2 = s\mathbf{x}_1. \end{cases}$$

This implies that

$$\mathbf{y}_1 = \mathbf{c}\left(s^2\mathbf{M} + \mathbf{K}\right)^{-1}\mathbf{b}$$

is identical to $\mathbf{y}$ if and only if $s = iw$, $w \in \mathbb{R}$.

In the sequel we will examine the second order system.

### 1.5.4 Sparsity Patterns and Magnitudes of the Blocks of M, K

There are three available discretizations for the *OCE application*: *coarse*: `1188_1050`, *medium*: `4752_5304` and *fine*: `20748_35775`. The numbers relate to the amount of degrees of freedom as follows: Case `4752_5304` implies $n_1 = 4752$ and $n_2 = 5304$. Extracted from ANSYS, the blocks $\mathbf{M}_{11}, \mathbf{M}_{21}, \mathbf{M}_{22}, \mathbf{K}_{11}, \mathbf{K}_{12}, \mathbf{K}_{22}$ in (1.51) are very differently scaled: For instance, for the medium case their absolute value greatest resp. smallest entries (*magnitude*) are of the order

$$\begin{aligned} \mathbf{M} &= O(\begin{bmatrix} 10^{-10} & \mathbf{0} \\ 10^{-5} & 10^{-18} \end{bmatrix}), \quad \mathbf{K} = O(\begin{bmatrix} 10^{+8} & 10^{-8} \\ \mathbf{0} & 10^{-4} \end{bmatrix}), \\ \mathbf{M} &= O(\begin{bmatrix} 10^{-12} & \mathbf{0} \\ 10^{-6} & 10^{-20} \end{bmatrix}), \quad \mathbf{K} = O(\begin{bmatrix} 10^{-12} & 10^{-9} \\ \mathbf{0} & 10^{-6} \end{bmatrix}). \end{aligned} \tag{1.54}$$

For the calculation of the transfer function furthermore note that $w \in [0, 2\pi * 1500]$. Thus approximately, $w^2 \in [0, 10^8]$. The use of the standard `MATLAB` '\' operations to solve $(\mathbf{K} - w^2\mathbf{M})\mathbf{x} = \mathbf{b}$ leads to error messages and abortions, not to solutions. An alternative, the use of the `MATLAB` package `Factorize`, alleviates this problem, but (too) severe round-off remains. Furthermore, the '\' operation turns out to be very slow for this poorly scaled problem. Investigation shows that that $\mathbf{K}_{11}$ contains entries in $[10^{-12}, 10^{+8}]$. The use of standard double precision floating point *IEEE arithmetic* involved in matrix operations such as matrix multiplication is bound to round-away contributions of the smaller entries.

Further investigation shows that all diagonal blocks but $\mathbf{K}_{11}$ are symmetric. For the results shown in this chapter the slightly non-symmetric ANSYS block $\mathbf{K}_{11}$ has been used as is. The results would be the same if one had instead used its symmetric part $(\mathbf{K}_{11}+\mathbf{K}_{11}^{\mathrm{T}})/2$ (tested). It has also been shown that indeed $\mathbf{M}_{21}=-\rho\mathbf{K}_{12}^{\mathrm{T}}$ for all three examples, where $\rho=1090$.

Observe that the determination of the smallest absolute value positive entry of a sparse `MATLAB` matrix with `MATLAB` is not trivial: The smallest entry of a sparse matrix usually is zero (since the default entry has value zero), MATHWORKS and other sources do not provide an on-the-shelf solution. To obtain the smallest non-zero entry we have written a `MATLAB` function `vfilter` which for a full or sparse matrix $\mathbf{X}$ writes all entries $\mathbf{X}_{ij}$ such that $|\mathbf{X}_{ij}|>\varepsilon\geq 0$ column-wise into a full vector. The use of this function applied to matrix `X` and $\varepsilon=0$ in combination with `min` provides the smallest absolute value entry of `X`.

Naturally, small entries should only be discarded if they are not relevant to the system of interest, i.e., if the the system is properly scaled, which is the topic of discussion of the next subsection.

### *1.5.5 Scaling the Second Order System*

We need to scale the matrices **K** and **M** (**E** and **A**) to obtain a *numerically robust* solution of the system

$$\mathbf{F}(w)\mathbf{x}=\mathbf{b}\iff(\mathbf{K}-w^2\mathbf{M})\mathbf{x}=\mathbf{b}\iff\begin{bmatrix}\mathbf{K}_{11}-w^2\mathbf{M}_{11} & \mathbf{K}_{12}\\ \rho w^2\mathbf{K}_{12}^{\mathrm{T}} & \mathbf{K}_{22}-w^2\mathbf{M}_{22}\end{bmatrix}\mathbf{x}=\mathbf{b},$$

which depends on $w$. For the problem of interest we expect symmetric blocks $\mathbf{M}_{11},\mathbf{M}_{22},\mathbf{K}_{11}$ and $\mathbf{K}_{22}$, and $\mathbf{M}_{21}=-\rho\mathbf{K}_{12}^{\mathrm{T}}$. This implies that this system could be scaled (preconditioned) into a symmetric one (*symmetry scaling*), for which efficient linear solvers exist. This can be done as follows: Observe that for a two by two matrix

$$\mathbf{A}=\begin{bmatrix}a & d\\ c & b\end{bmatrix},\quad\mathbf{D}_1=\begin{bmatrix}1 & \\ & \sqrt{c/d}\end{bmatrix}\implies\mathbf{D}_1^{-1}\mathbf{A}\mathbf{D}_1=\begin{bmatrix}a & \sqrt{cd}\\ \sqrt{cd} & b\end{bmatrix}$$

can be scaled to a symmetric one. Hence, based on $c=\rho w^2$ and $d=1$, define

$$\mathbf{D}_1=\begin{bmatrix}\mathbf{I}_{n_1} & \\ & \sqrt{\rho w^2}\mathbf{I}_{n_2}\end{bmatrix}.$$

Furthermore, to better scale the entries inside and between blocks (create diagonal elements of magnitude 1), define

$$\mathbf{D}_2=\mathrm{diag}(1/\sqrt{[\mathbf{D}_1^{-1}\mathbf{F}\mathbf{D}_1]_{11}},\ldots,1/\sqrt{[\mathbf{D}_1^{-1}\mathbf{F}\mathbf{D}_1]_{nn}}).$$

We now scale with a *diagonal scaling*:

$$\hat{\mathbf{M}} := \underbrace{\mathbf{D}_2\mathbf{D}_1^{-1}}_{\mathbf{Q}}\mathbf{M}\underbrace{\mathbf{D}_1\mathbf{D}_2}_{\mathbf{P}},$$
$$\hat{\mathbf{K}} := \mathbf{QKP}, \hat{\mathbf{b}} := \mathbf{Qb}, \hat{\mathbf{c}} := \mathbf{cP},$$

which, by invariance under inputs and outputs transformations means that

$$\hat{\mathbf{H}}(w) := \mathbf{c}(\hat{\mathbf{K}} - w^2\hat{\mathbf{M}})^{-1}\hat{\mathbf{b}}$$

is identical to $\mathbf{H}$ in (1.52) for all $w$. Obviously $\mathbf{D}_1$ is non-singular except for $w = 0$ and $\mathbf{D}_2$ exists and is non-singular when all diagonal entries of $\mathbf{D}_1^{-1}\mathbf{F}\mathbf{D}_1$ are non-zero.

The factors $\mathbf{P} = \mathbf{P}(w)$ and $\mathbf{Q} = \mathbf{Q}(w)$ depend on $w$. This is fine for the construction of Krylov spaces to match moments. However, to plot the transfer function $\mathbf{H}$ one needs to evaluate $\mathbf{c}(\mathbf{K} - w_k^2\mathbf{M})$ for many $w_k \in [0, 10^8]$. Repeated calculation of $\mathbf{P}(w_k)$ and $\mathbf{Q}(w_k)$ would be (too) costly, so we decided to use the $w$-independent factors $\mathbf{P} := \mathbf{P}(\hat{w})$ $\mathbf{Q} := \mathbf{Q}(\hat{w})$ for all $w$ where $\hat{w}$ is the average of all $w_k$. For the OCE example, to plot the transfer functions, we sample the provided region of interest: $w_k = 5\pi \cdot k$, $k = 0, \ldots, 600$. The value of $\hat{w}$ turns out to be $w_{301}$ which is close to but not too close to a *pole* of $\mathbf{H}$ and such that all diagonal entries of $\mathbf{D}_1^{-1}\mathbf{F}\mathbf{D}_1$ are non-zero.

## *1.5.6 The Structure and the GSVD of* $\mathbf{K}_{12}$

Here we briefly comment on the GSVD of the scaled $\mathbf{K}_{12}^{\mathrm{T}}$. Figure 1.13 and numerical investigation show that $\mathbf{K}_{12} \in \mathbb{R}^{1188\times 1050}$ is a *sparse matrix* which contains a small *non-zero sub-block* of size $295 \times 175$ (window $(3,\ldots,297) \times (561,\ldots,735)$). This is typical for applications where the different physical quantities are defined in bordering sub-domains and are coupled via the mutual boundary – if one numbers the degrees of freedom on the mutual boundary consecutively. Since $\mathbf{K}_{12}^{\mathrm{T}}$ has this structure it is of the required type. This means that also $\mathbf{V}$ has all its non-zero entries in the same sub-block, i.e., it only has possible non-zero entries from row 561 to 735. This information is of importance, because the standard GSVD implementations such as `MATLAB`'s do not use this information and generate $\mathbf{V}$ which contains round-off (non-zero) entries outside the window, as can be seen in Fig. 1.14. For the medium test case the results are worse, as to be expected: For $p = 5$ and $\varepsilon = 0$, $\mathbf{K}_{12}^{(p)}$ (definitions, see below) is a full matrix.

To work around this problem we have written a `MATLAB` function `spfilter` which for a full or sparse matrix $\mathbf{X}$ copies all entries $\mathbf{X}_{ij}$ such that $|\mathbf{X}_{ij}| > M(\mathbf{X}) \cdot \varepsilon$ into a sparse matrix $\mathbf{Y}$, where $M(\mathbf{X}) := \max\{|X|_{ij}\}_{i,j}$. This way, using $\varepsilon = 10^{-11}$, both $\mathbf{K}_{12} = \mathbf{X}\mathbf{S}^{\mathrm{T}}\mathbf{V}^{\mathrm{T}}$ and all of its dominant parts $\mathbf{K}_{12}^{(p)} := \mathbf{X}^{(p)}\mathbf{S}^{(p)}\mathbf{V}^{(p)}$ (for some $p \le n$) have similar sparsity patterns.

In `MATLAB` there are different but equivalent manners for the filtering of entries from a matrix. However, most of them do not terminate or lead to out of memory

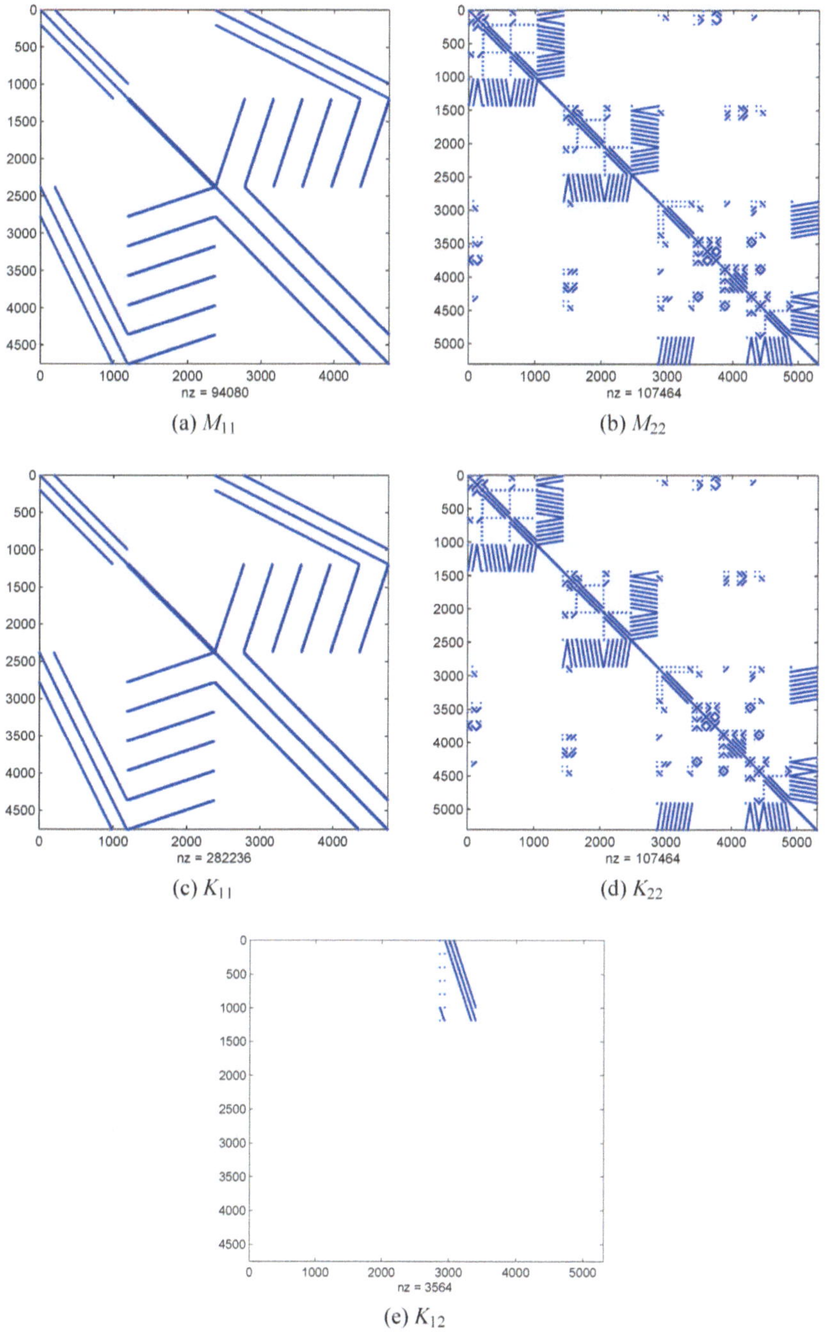


**Fig. 1.13.** Sparsity pattern of matrix blocks

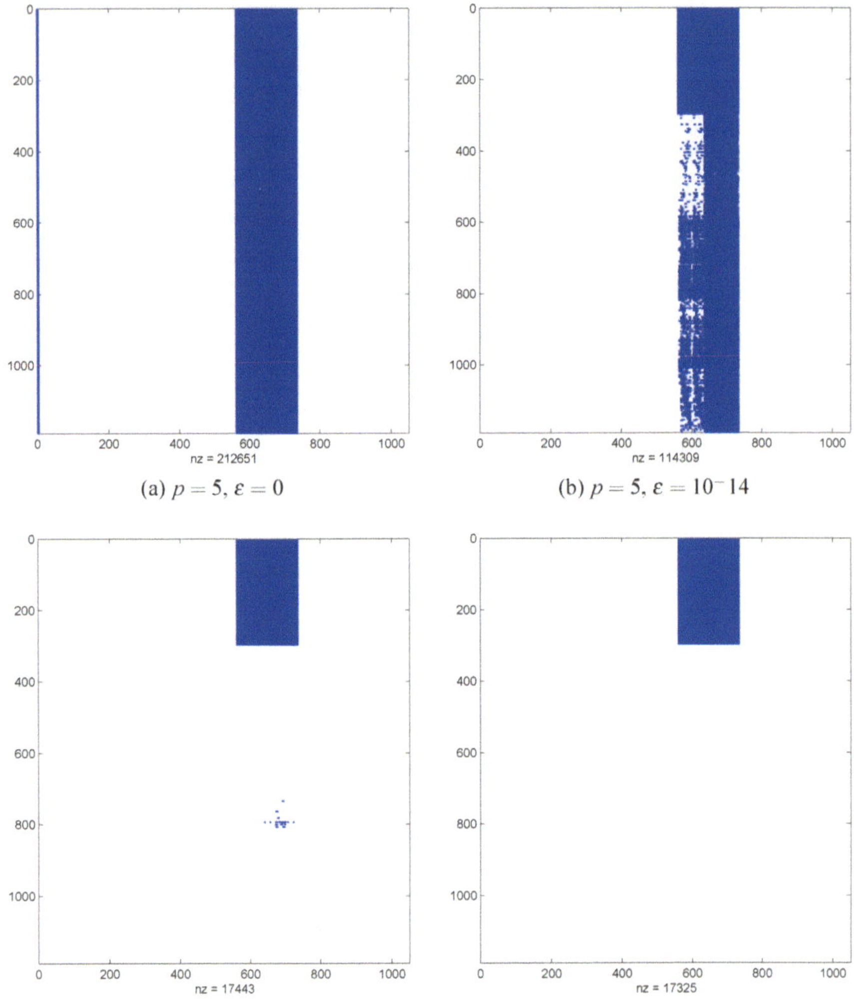


(a) $p = 5$, $\varepsilon = 0$

(b) $p = 5$, $\varepsilon = 10^{-14}$

(c) $p = 5$, $\varepsilon = 10^{-12}$

(d) $p = 5$, $\varepsilon = 10^{-10}$

**Fig. 1.14.** Entries of $\mathbf{K}_{12}^{(p)}$ greater than $\varepsilon \cdot M(\mathbf{K}_{12}^{(p)})$, small case

errors even for the small case. Functions `vfilter` and `spfilter` contain information on manners which somehow do not lead to the desired result.

Explicit multiplication with factors $\mathbf{X}^{(p)}$, $\mathbf{S}^{(p)}$ and $\mathbf{V}^{(p)}$ for the multiplication with $\mathbf{x} \mapsto \mathbf{Kx}$ is likely to be the more efficient then the use of multiplication with $\mathbf{K}_{12}^{(p)}$.

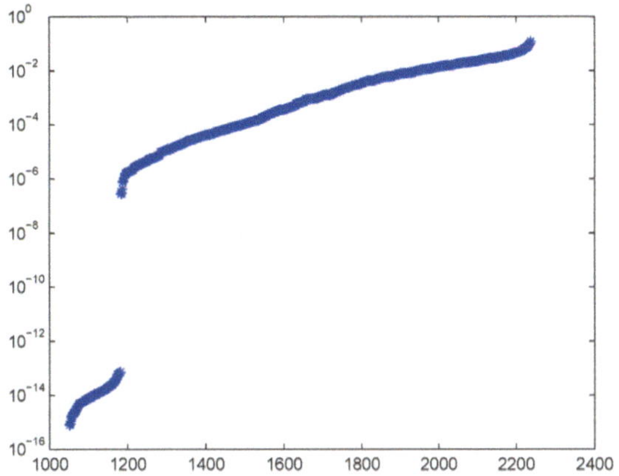


**Fig. 1.15.** Entries of $\mathbf{x}$ in (1.55), sorted

### *1.5.7 A GSVD-Based Approximation of* $\mathbf{K}_{12}$

In this subsection we analyse how the GSVD based approximation of $\mathbf{K}_{12}$ influences the solution of the static problem

$$\begin{bmatrix} \mathbf{K}_{11} & \mathbf{K}_{12} \\ \mathbf{0} & \mathbf{K}_{22} \end{bmatrix} \begin{bmatrix} \mathbf{x}_1 \\ \mathbf{x}_2 \end{bmatrix} = \begin{bmatrix} \mathbf{b}_1 \\ \mathbf{b}_2 \end{bmatrix} \tag{1.55}$$

Based on the definition of $\mathbf{K}_{12}^{(p)}$ the approximation leads to system

$$\begin{bmatrix} \mathbf{K}_{11} & \mathbf{K}_{12}^{(p)} \\ \mathbf{0} & \mathbf{K}_{22} \end{bmatrix} \begin{bmatrix} \mathbf{y}_1^{(p)} \\ \mathbf{y}_2^{(p)} \end{bmatrix} = \begin{bmatrix} \mathbf{b}_1 \\ \mathbf{b}_2 \end{bmatrix}. \tag{1.56}$$

We intend to estimate

$$|| \left( |x_i - y_i^{(p)}| / |x_i| \right)_i ||_\infty \tag{1.57}$$

over the set of indices $i$ for which $x_i$ is non-zero (outside round-off region). To determine this set, we first solved (1.55) and made a log-plot of its sorted entries, shown in Fig. 1.15. Based on this plot we decided to omit all entries smaller than $10^{-7}$ and obtained the results in Table 1.2. The accuracy does not seem to be (very) sensitive to the amount of principal components used, which is due to the fact that the scaled $\mathbf{K}_{12}$ block is still of magnitude $10^5$ smaller than the scaled diagonal blocks $\mathbf{K}_{11}$ and $\mathbf{K}_{22}$. However, Sect. 1.5.8 shows that different amounts of principal components do have a remarkable effect on the related transfer function.

### *1.5.8 The* $\mathbf{K}_{12}{}^{(p)}$ *GSVD-Approximation Based Transfer Function*

The aim is to determine a principal component analysis (*PCA*) based *rank-revealing* factorization $\mathbf{K}_{12} \doteq \mathbf{B}\mathbf{C}^{\mathrm{T}}$ where $\mathbf{B}$ and $\mathbf{C}$ are constructed with the use of the first

**Table 1.2.** Relative errors due to use of the GSVD approximation

| $p$ | $\|\|x(i) - y^{(p)}(i)\|/\|x(i)\|\|_\infty$ |
|---|---|
| 1 | 3.750080647e-008 |
| 2 | 1.427208120e-007 |
| 3 | 1.119493657e-007 |
| 4 | 1.468582269e-007 |
| 5 | 1.500944068e-007 |

$p \leq n$ *principal components*, based on the scaled versions of **K** (and if needed **M**) as constructed above.

To the scaled matrix **K** (which depends on $\hat{w}$) we apply a GSVD to $\mathbf{K}_{11}^{\mathrm{T}}$ and $\mathbf{K}_{12}^{\mathrm{T}}$ such that $\mathbf{K}_{11}^{\mathrm{T}} = \mathbf{UCX}^{\mathrm{T}}$ and $\mathbf{K}_{12}^{\mathrm{T}} = \mathbf{VSX}^{\mathrm{T}}$. Hence,

$$\mathbf{K}_{12} = \mathbf{XS}^{\mathrm{T}}\mathbf{V}^{\mathrm{T}} = \underbrace{\mathbf{X}\sqrt{\mathbf{S}^{\mathrm{T}}}}_{\mathbf{B}}\underbrace{\sqrt{\mathbf{S}^{\mathrm{T}}}\mathbf{V}^{\mathrm{T}}}_{\mathbf{C}^{\mathrm{T}}}.$$

Figure 1.16 shows all of the diagonal values of the matrix $S$ and Fig. 1.17 shows the first 1000 of them. Next, for $p = 1, \ldots, 5$ we approximate $\mathbf{K}_{12}$ by the contribution of its $p$ most dominant modes

$$\mathbf{K}^{(p)} = (\mathbf{X}^{(p)}\sqrt{\mathbf{S}^{(p)}})(\sqrt{\mathbf{S}^{(p)}}\mathbf{V}^{(p)})$$

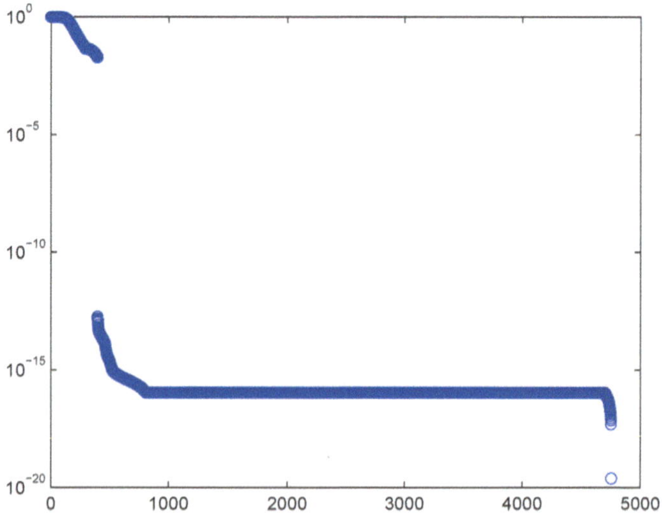


**Fig. 1.16.** Diagonal elements of $S$

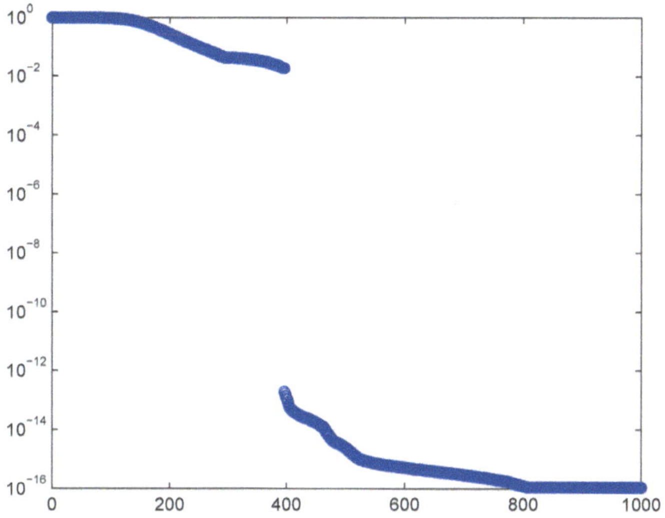


**Fig. 1.17.** First 1000 diagonal elements of $S$

and plot the related transfer functions, together with the transfer function related to $\mathbf{K}_{12}$ (blue) in Fig. 1.18. One can observe that the transfer function related to $\mathbf{K}^{(p)}$ closely approximates $p$ peaks of the original transfer function (the one for $\mathbf{K}_{12}$).

### *1.5.9 The GSVD Approximation of* $\mathbf{M}_{11}^{-1}\mathbf{K}_{12}$

In fact, we need to apply the GSVD to $\mathbf{M}_{11}^{-1}\mathbf{K}_{12}$ rather than $\mathbf{K}_{12}$. Fortunately, there is a straightforward relation between the GSVD of $(\mathbf{K}_{11}, \mathbf{K}_{12})$ and $(\mathbf{M}_{11}^{-1}\mathbf{K}_{11}, \mathbf{M}_{11}^{-1}\mathbf{K}_{12})$. To see this, abbreviate $\mathbf{K} := \mathbf{K}_{12}$ and $\mathbf{M} := \mathbf{M}_{11}$ and observe that

$$
\begin{aligned}
\mathbf{K}^{\mathrm{T}} &= \mathbf{V}\mathbf{S}\mathbf{X}^{\mathrm{T}} \implies \\
\mathbf{K} &= \mathbf{X}\mathbf{S}^{\mathrm{T}}\mathbf{V}^{\mathrm{T}} \implies \\
\mathbf{M}^{-1}\mathbf{K} &= \mathbf{M}^{-1}\mathbf{X}\mathbf{S}^{\mathrm{T}}\mathbf{V}^{\mathrm{T}} \implies \\
\mathbf{M}^{-1}\mathbf{K} &= \underbrace{(\mathbf{M}^{-1}\mathbf{X})\sqrt{\mathbf{S}^{\mathrm{T}}}}_{\mathbf{Y}}\underbrace{\sqrt{\mathbf{S}^{\mathrm{T}}}\mathbf{V}^{\mathrm{T}}}_{\mathbf{Z}}
\end{aligned}
$$

which leads to the principal component based approximation:

$$
\mathbf{M}^{-1}\mathbf{K} \doteq \mathbf{M}^{-1}\mathbf{X}^{(p)}\mathbf{S}^{(p)}\mathbf{V}^{(p)}.
$$

One first rewrites (1.53) to produce the term $s\mathbf{I}$, for instance as follows:

$$
\begin{aligned}
H(w) &= \mathbf{c}(\mathbf{K} - w^2\mathbf{M})^{-1}\mathbf{b} \implies \\
H(w) &= \mathbf{c}(\mathbf{M}^{-1}\mathbf{K} - w^2\mathbf{I})\mathbf{M}^{-1}\mathbf{b} \implies \\
H(w) &= -\mathbf{c}(w^2\mathbf{I} - \mathbf{M}^{-1}\mathbf{K})\mathbf{M}^{-1}\mathbf{b}.
\end{aligned}
\tag{1.58}
$$

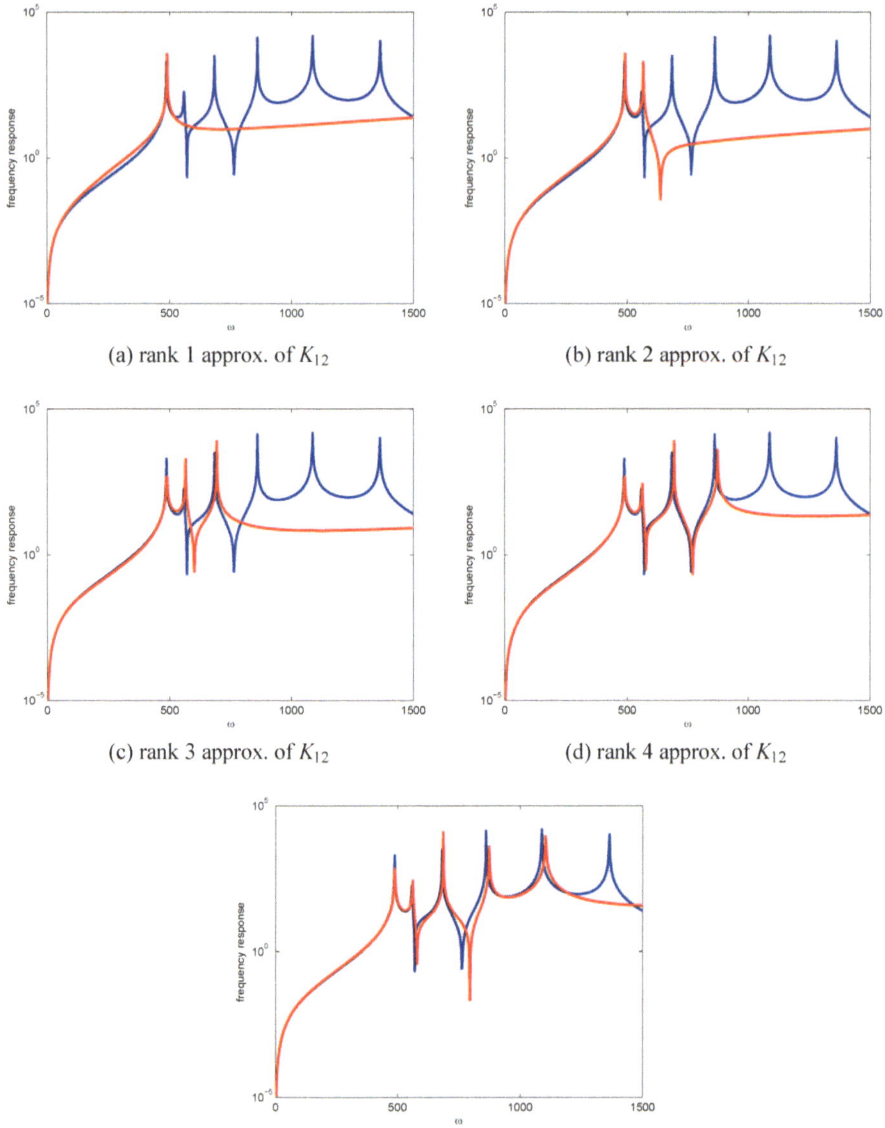


(a) rank 1 approx. of $K_{12}$

(b) rank 2 approx. of $K_{12}$

(c) rank 3 approx. of $K_{12}$

(d) rank 4 approx. of $K_{12}$

(e) rank 5 approx. of $K_{12}$

**Fig. 1.18.** Low-rank approximations of block $K_{12}$

Observe that the inverse of block-matrix $\mathbf{M}$ in (1.51) is

$$\mathbf{M}^{-1} = \begin{bmatrix} \mathbf{M}_{11}^{-1} & \mathbf{0} \\ -\mathbf{M}_{22}^{-1}\mathbf{M}_{21}\mathbf{M}_{11}^{-1} & \mathbf{M}_{22}^{-1} \end{bmatrix}$$

whence

$$\mathbf{M}^{-1}\mathbf{K} = \begin{bmatrix} \mathbf{M}_{11}^{-1}\mathbf{K}_{11} & \mathbf{M}_{11}^{-1}\mathbf{K}_{12} \\ -\mathbf{M}_{22}^{-1}\mathbf{M}_{21}\mathbf{M}_{11}^{-1}\mathbf{K}_{11} & -\mathbf{M}_{22}^{-1}\mathbf{M}_{21}\mathbf{M}_{11}^{-1}\mathbf{K}_{12} + \mathbf{M}_{22}^{-1}\mathbf{K}_{22} \end{bmatrix}$$

Now, SBR applied to the first row of this system leads to the approximation

$$\mathbf{M}^{-1}\mathbf{K} \doteq \begin{bmatrix} \mathbf{M}_{11}^{-1}\mathbf{K}_{11} & \mathbf{M}_{11}^{-1}\mathbf{X}^{(p)}\mathbf{S}^{(p)}\mathbf{V}^{(p)} \\ -\mathbf{M}_{22}^{-1}\mathbf{M}_{21}\mathbf{M}_{11}^{-1}\mathbf{K}_{11} & -\mathbf{M}_{22}^{-1}\mathbf{M}_{21}\mathbf{M}_{11}^{-1}\mathbf{K}_{12} + \mathbf{M}_{22}^{-1}\mathbf{K}_{22} \end{bmatrix}$$

which shows that one can use the GSVD-based approximation

$$H(w) \doteq \mathbf{c}(\mathbf{K}^{(p)} - w^2\mathbf{M})^{-1}\mathbf{b}$$

where

$$\mathbf{K} = \begin{bmatrix} \mathbf{K}_{11} & \mathbf{X}^{(p)}\mathbf{S}^{(p)}\mathbf{V}^{(p)} \\ \mathbf{0} & \mathbf{K}_{22} \end{bmatrix}.$$

## 1.6 Conclusions

We proposed a new model order reduction technique for coupled systems. Our method, called the Separate Bases Reduction (SBR) algorithm, belongs to the family of block-structure preserving (BSP) reduction techniques based on the uncoupled formulation of the coupled problem. However, unlike other reduction approaches dealing with the separate sub-system representation, the SBR algorithm can be applied to a wide category of coupled systems, including strongly coupled systems and interconnected systems with many interconnections. This is due to the fact that for such cases we avoid a too fast growth of the reduction bases and related reduced-order model, as long as the coupling can be well approximated by a relatively small number of GSVD principal components. Examples of such strongly coupled systems are systems with an interface coupling, for instance systems describing interactions between a fluid and a solid wall, or systems which for instance describe an electromagnetic-structural coupling in an electronic device. Another advantage of the proposed technique is that it is computationally cheaper than the more common BSP reduction methods which deal with the coupled formulation of the system.

For the initial version of the SBR algorithm (without low-rank approximations of the couplings), we proved the moment matching property. The GSVD based approximation of the couplings only approximates the moments, but numerical experiments show that taking a sufficient number of dominant components still results in accurately approximated moments. What makes the SBR algorithm universal, is the fact, that it can be applied even if the internal input and output matrices are not known explicitly. We show, that having at our disposal only the coupled system's matrices, external inputs and outputs, and the dimensions of the sub-systems, we are able to create appropriate Krylov subspaces for each sub-system. This property of the reduction method is desirable when dealing with industrial problems for which the separate sub-systems' information may not be available.

The SBR method has been designed keeping in mind the practical use in an industrial environment. It is fairly straightforward to adapt existing software modules and make them suitable for application of SBR. This is certainly not the case for the BSP type methods. Although the reduced-order models obtained by application of the BSP methods frequently show a bit better approximation accuracy, the SBR algorithm is much more beneficial from the point of view of the computational time. This property is especially valuable in case of large industrial applications.

## References


1. Bai, Z.: Krylov subspace techniques for reduced-order modeling of large-scale dynamical systems. Applied Numerical Mathematics **43**(1–2), 9–44 (2002)
2. Bai, Z., Li, B., Su, Y.: A unified Krylov projection framework for structure-preserving model reduction. In: Schilders, W.H.A., van der Vorst, H.A., Rommes, J. (eds) Model Order Reduction: Theory, Research Aspects and Applications. Mathematics in Industry, Vol. 13, pp. 75–93. Springer-Verlag, Berlin Heidelberg (2008)
3. Bisseling, R.H.: Parallel Scientific Computation, A Structured Approach using BSP and MPI. Oxford Scholarship Online (2007)
4. Doris, A., van de Wouw, N., Heemels, W.P.M.H., Nijmeijer, H.: A disturbance attenuation approach for continuous piecewise affine systems: Control design and experiments. J. Dynamic Systems, Measurement and Control 132(4), 044502-1– 044502-7 (2010)
5. Fernández Villena, J., Schilders, W.H.A., Miguel Silveira, L.: Order reduction techniques for coupled multi-domain electromagnetic based models. CASA report (2008)
6. Freund, R.W.: SPRIM: structure-preserving reduced-order interconnect macromodeling. In: Proceedings of the IEEE/ACM International Conference on Computer Aided Design, pp. 80–87 (2004). DOI 10.1109/ICCAD.2004.1382547
7. Géradin, M., Rixen, D.J.: Mechanical Vibrations and Structural Dynamics, 2nd ed. John Wiley and Sons, Chichester (1997)
8. Grimme, E.J.: Krylov projection methods for model reduction. PhD thesis, University of Illinois (1997)
9. He, L., Yu, H., Tan, S.X.D.: Block structure preserving model order reduction. BMAS – IEEE Behavioral Modeling and Simulation Workshop (2005)
10. Henderson, H.V., Searle, S.R.: On deriving the inverse of a sum of matrices. SIAM Review **23**(1), 53–60 (1981)
11. Heres, P.J., Deschrijver, P.J., Schilders, W.H.A., Dhaene, T.: Combining krylov subspace methods and identification-based methods for model order reduction. International Journal of Numerical Modelling: Electronic Networks, Devices and Fields **20**(6), 271–282 (2007)
12. Ionutiu, R.: Model Order Reduction for Multi-terminals Systems with Applications to Circuit Simulation. PhD thesis, Eindhoven University of Technology, Eindhoven, The Netherlands (2011)
13. Lutowska, A.: Model Order Reduction for Coupled Systems using Low-rank Approximations. PhD thesis, Eindhoven University of Technology (2012)

14. Pavlov, A.V., van de Wouw, N., Nijmeijer, H.: Uniform Output Regulation of Nonlinear Systems. Birkhäuser, Bostgon (2005)
15. Rochus, V., Rixen, D.J., Golinval, J.-C.: Electrostatic coupling of mems structures: transient simulations and dynamic pull-in. Nonlinear Analysis **63**(5–7), 1619–1633 (2005)
16. Saad, Y., Schulz, M.H.: GMRES: a generalized minimal residual algorithm for solving nonsymmetric linear systems. SIAM Journal on Scientific and Statistical Computing **7**, 856–869 (1986)
17. Salimbahrami, B., Lohmann, B.: Krylov subspace methods in linear model order reduction: Introduction and invariance properties. Sci. Rep. Inst. of Automation (2002)
18. Schilders, W.H.A., van der Vorst, H.A., Rommes, J.: Model Order Reduction: Theory, Research Aspects and Applications. Mathematics in Industry, Vol. 13. Springer-Verlag Berlin Heidelberg (2008)
19. Skogestad, S., Postlethwaite, I.: Multivariable feedback control, Analysis and Design. John Wiley and Sons, Philadelphia (2005)
20. Sonneveld, P.: CGS, a fast Lanczos-type solver for nonsymmetric linear systems. SIAM Journal on Scientific and Statistical Computing **10**, 36–52 (1989)
21. Ugryumova, M.V.: Model Order Reduction for Multi-terminals Systems with Applications to Circuit Simulation. PhD thesis, Eindhoven University of Technology, Eindhoven, The Netherlands (2000)
22. van der Vorst, H.A., Sleijpen, G.L.G., Fokkema, D.R.: Bicgstab(l) and other hybrid bi-cg methods. Numerical Algorithms **7**, 75–109 (1994)
23. van de Wouw, N., de Kraker, A., van Campen, D.H. Nijmeijer, H.: Non-linear dynamics of a stochastically excited beam system with impact. Int. J. Non-Linear Mech. **38**(5), 767–779 (2003)
24. Vandendorpe, A., Van Dooren, P.: Model reduction of interconnected systems. In: Schilders, W.H.A., van der Vorst, H.A., Rommes, J. (eds) Model Order Reduction: Theory, Research Aspects and Applications. of Mathematics in Industry, Vol. 13, pp. 305–321. Springer-Verlag, Berlin Heidelberg (2008)
25. Wesseling, P.: An Introduction to Multigrid Methods. John Wiley & Sons, Chichester (1992)

2

# Case Study: Parametrized Reduction Using Reduced-Basis and the Loewner Framework

**Antonio C. Ionita and Athanasios C. Antoulas**

**Abstract** In this case study, we compare two methods for model reduction of parametrized systems, namely, Reduced-Basis and Loewner rational interpolation.

While having the same goal of constructing reduced-order models for large-scale parameter-dependent systems, the two methods follow fundamentally different approaches. On the one hand, the well known Reduced-Basis method takes a time-domain approach, using offline snapshots of the full-order system combined with a rigorous error bound. On the other hand, the recently introduced Loewner matrix framework takes a frequency-domain approach that constructs rational interpolants of transfer function measurements, and has the flexibility of allowing different reduced-orders for each of the frequency and parameter variables.

We apply the two methods to a parametrized partial differential equation modeling the transient temperature evolution near the surface of a cylinder immersed in fluid. Then, we compare the resulting reduced-order models with the full-order finite element system by running both time- and frequency-domain simulations.

## 2.1 Introduction

The growing need for highly accurate modeling of physical phenomena often leads to large-scale dynamical systems. For example, accurate simulations involving partial

A.C. Ionita (✉)
Department of Electrical and Computer Engineering, Rice University, 6100 Main St, MS-366, Houston, TX 77005, USA
e-mail: aci1@rice.edu

A.C. Antoulas
Department of Electrical and Computer Engineering, Rice University, 6100 Main St, MS-366, Houston, TX 77005, USA
e-mail: aca@rice.edu
School of Engineering and Science, Jacobs University, Campus Ring 1, 28759 Bremen, Germany
† Supported by NSF through Grant CCF-1017401 and the DFG through Grant AN-693/1-1

A. Quarteroni, G. Rozza (eds.): *Reduced Order Methods for Modeling and Computational Reduction*, MS&A 9. DOI 10.1007/978-3-319-02090-7_2, 

differential equations require taking fine spatial discretizations that, in turn, lead to dynamical systems of large dimensions. Hence, high accuracy comes at a steep price. Simulating such large-scale systems is a prohibitively expensive task that requires long simulation times and large data storage.

Model reduction seeks to overcome these obstacles by constructing models of low dimension that have short simulation times, require low data storage, *but* still accurately capture the behavior of the large-scale system.

In the case of systems that do not depend on parameters, reduced-order models can be obtained using an extensive array of model reduction methods [1]. For instance, we can follow SVD-based approaches such as the proper orthogonal decomposition (POD) [27] for non-linear systems and Balanced Truncation [8] for linear systems. Alternatively, we can follow rational interpolation approaches such as (iterative) Rational Krylov [9, 11]. These methods are well understood and known to give accurate reduced-order models in various practical applications [1,5]. However, in the case of systems that depend on parameters, there is a limited choice of available model reduction methods. The main obstacle is the fact that, in the presence of parameters, approaches like Balanced Truncation or iterative Rational Krylov are difficult to generalize.

Nevertheless, in recent years, a number of efficient methods have emerged to form the so-called Reduced-Basis framework for parametrized model reduction [13, 14, 18, 22–24]. Reduced-Basis methods extend the POD approach to the case of parametrized systems by relying on an offline space that contains snapshots of state trajectories of the full-order system. An error bound is used to iteratively enrich this space and extract a reduced-basis that yields accurate reduced-order models.

More recently, the rational interpolation approach has also been generalized to the case of parametrized systems [3]. Here, we apply this recent approach to construct reduced-order models that interpolate transfer function measurements of the full-order system. The key of the rational interpolation approach is the Loewner matrix, which allows the flexibility of choosing different reduced orders for each of the frequency and parameter variables. The reduced-order models are efficiently computed using a rational barycentric formula together with the null space of a generalized two-variable Loewner matrix.

In this case study, we compare the Reduced-Basis approach and the Loewner matrix approach, i.e., we compare a time-domain, POD-based method with a frequency-domain, rational interpolation method. In Sect. 2.2, we review the two methods, showcasing their common traits and differences. Then, in Sect. 2.3, we present a numerical example involving a parametrized partial differential equation modeling the transient temperature evolution near the surface of a cylinder immersed in fluid. After applying the Reduced-Basis and Loewner frameworks, we compare the resulting reduced-order models in both time- and frequency-domain simulations.

## 2.2 Parametrized Model Reduction

We begin with a short introduction to model reduction of parametrized systems, followed by an overview of the two reduction methods compared in this study.

We define a parametrized linear dynamical system of order $n$ in terms of state-space equations that depend on parameters $\mathbf{p} \in \mathbb{R}^d$:

$$\begin{aligned} \mathbf{E}(\mathbf{p})\,\dot{\mathbf{x}}(t) &= \mathbf{A}(\mathbf{p})\,\mathbf{x}(t) + \mathbf{B}(\mathbf{p})\,\mathbf{u}(t), \\ \mathbf{y}(t) &= \mathbf{C}(\mathbf{p})\,\mathbf{x}(t) + \mathbf{D}(\mathbf{p})\,\mathbf{u}(t), \end{aligned} \tag{2.1}$$

where $\mathbf{x}(t) \in \mathbb{R}^n$ denotes the system's state, $\mathbf{u}(t) \in \mathbb{R}$ the input, $\mathbf{y}(t) \in \mathbb{R}$ the output, and $\mathbf{E}(\mathbf{p}), \mathbf{A}(\mathbf{p}) \in \mathbb{R}^{n\times n}$, $\mathbf{B}(\mathbf{p}), \mathbf{C}^T(\mathbf{p}) \in \mathbb{R}^n$, $\mathbf{D}(\mathbf{p}) \in \mathbb{R}$ are the parameter-dependent system matrices. Notice that the system's state $\mathbf{x}(t)$ and output $\mathbf{y}(t)$ also depend on the parameters $\mathbf{p}$ as the system evolves in time; however, for notational simplicity, we only depict their dependence on time $t$.

Model order reduction methods seek models of order $k$

$$\begin{aligned} \widehat{\mathbf{E}}(\mathbf{p})\,\dot{\widehat{\mathbf{x}}}(t) &= \widehat{\mathbf{A}}(\mathbf{p})\,\widehat{\mathbf{x}}(t) + \widehat{\mathbf{B}}(\mathbf{p})\,\mathbf{u}(t), \\ \widehat{\mathbf{y}}(t) &= \widehat{\mathbf{C}}(\mathbf{p})\,\widehat{\mathbf{x}}(t) + \widehat{\mathbf{D}}(\mathbf{p})\,\mathbf{u}(t), \end{aligned} \tag{2.2}$$

with $\widehat{\mathbf{E}}(\mathbf{p})$, $\widehat{\mathbf{A}}(\mathbf{p}) \in \mathbb{R}^{k\times k}$, $\widehat{\mathbf{B}}(\mathbf{p}), \widehat{\mathbf{C}}^T(\mathbf{p}) \in \mathbb{R}^k$, such that

- the new state $\widehat{\mathbf{x}}(t)$ has *reduced* dimension $k \ll n$;
- the *reduced-order model* (2.2) accurately captures the behavior of the full-order system (2.1), by introducing a small time-domain approximation error $|\mathbf{y}(t) - \widehat{\mathbf{y}}(t)|$, or a small frequency-domain approximation error $|\mathbf{H}(s,\mathbf{p}) - \widehat{\mathbf{H}}(s,\mathbf{p})|$, for

$$\mathbf{H}(s,\mathbf{p}) = \mathbf{C}(\mathbf{p})\left(s\mathbf{E}(\mathbf{p}) - \mathbf{A}(\mathbf{p})\right)^{-1}\mathbf{B}(\mathbf{p}) + \mathbf{D}(\mathbf{p}), \tag{2.3}$$

  denoting a system's transfer function.

We now review the two methods that approach the model reduction problem from different perspectives, but, as we shall ultimately see, both lead to accurate reduced-order models.

### 2.2.1 *Reduced-Basis Approach*

Since their introduction in [18,23], Reduced-Basis methods have become a reliable tool for obtaining accurate parametrized reduced-order models. Here, we summarize the Reduced-Basis approach along the lines of the presentation given in [14].

Reduced-Basis methods construct the reduced state $\widehat{\mathbf{x}}(t)$ by means of a judiciously chosen Petrov-Galerkin projection $\mathbf{V}\mathbf{W}^T$. The reduced-order model (2.2) is obtained by projecting the system matrices:

$$\begin{aligned} &\widehat{\mathbf{E}}(\mathbf{p}) = \mathbf{W}^T\mathbf{E}(\mathbf{p})\,\mathbf{V}, \quad \widehat{\mathbf{A}}(\mathbf{p}) = \mathbf{W}^T\mathbf{A}(\mathbf{p})\,\mathbf{V}, \\ &\widehat{\mathbf{B}}(\mathbf{p}) = \mathbf{W}^T\mathbf{B}(\mathbf{p}), \quad \widehat{\mathbf{C}}(\mathbf{p}) = \mathbf{C}(\mathbf{p})\,\mathbf{V}, \quad \widehat{\mathbf{D}}(\mathbf{p}) = \mathbf{D}(\mathbf{p}), \end{aligned}$$

with initial conditions $\widehat{\mathbf{x}}(0) = \mathbf{W}^T\mathbf{x}(0)$, and the reduced state defined as

$$\widehat{\mathbf{x}}(t) := \mathbf{W}^T\mathbf{x}(t).$$

However, before discussing how to choose the projection $\mathbf{V}\mathbf{W}^T$ in a Reduced-Basis setting, we outline an error analysis [14] that is valid for any general projection with $\mathbf{W}^T\mathbf{V} = \mathbf{I}_k$. Consider the error introduced by the projection framework when approximating the full-order state:

$$\mathbf{e}(t) := \mathbf{x}(t) - \mathbf{V}\widehat{\mathbf{x}}(t).$$

Next, we derive a bound for this error that can be efficiently computed for different values of the parameters $\mathbf{p}$ and time $t$. Towards this end, we define the residual vector

$$\mathbf{R}(t,\mathbf{p}) := \mathbf{A}(\mathbf{p})\mathbf{V}\widehat{\mathbf{x}}(t) + \mathbf{B}(\mathbf{p})\mathbf{u}(t) - \mathbf{E}(\mathbf{p})\mathbf{V}\dot{\widehat{\mathbf{x}}}(t).$$

that depends only on the reduced state and the input, and satisfies by construction the orthogonality condition $\mathbf{W}^T\mathbf{R}(t,\mathbf{p}) = 0$.

Then, it is easily checked that the error satisfies the following evolution equation

$$\mathbf{E}(\mathbf{p})\dot{\mathbf{e}}(t) = \mathbf{A}(\mathbf{p})\mathbf{e}(t) + \mathbf{R}(t,\mathbf{p}).$$

In most practical applications, the matrix $\mathbf{E}(\mathbf{p})$ is invertible for all parameter values inside a domain of interest, and, therefore, we can define $\tilde{\mathbf{A}}(\mathbf{p}) = \mathbf{E}(\mathbf{p})^{-1}\mathbf{A}(\mathbf{p})$ and $\tilde{\mathbf{R}}(t,\mathbf{p}) = \mathbf{E}(\mathbf{p})^{-1}\mathbf{R}(t,\mathbf{p})$ to obtain

$$\dot{\mathbf{e}}(t) = \tilde{\mathbf{A}}(\mathbf{p})\mathbf{e}(t) + \tilde{\mathbf{R}}(t,\mathbf{p}),$$

which has the solution

$$\mathbf{e}(t) = e^{t\tilde{\mathbf{A}}(\mathbf{p})}\mathbf{e}(0) + \int_0^t e^{(t-\tau)\tilde{\mathbf{A}}(\mathbf{p})}\tilde{\mathbf{R}}(\tau,\mathbf{p})d\tau.$$

Then, it immediately follows that the output error can be bounded by

$$\|\mathbf{y}(t) - \widehat{\mathbf{y}}(t)\| \le \|\mathbf{C}e^{t\tilde{\mathbf{A}}(\mathbf{p})}\| \left( \|\mathbf{e}(0)\| + \int_0^t \|\tilde{\mathbf{R}}(\tau,\mathbf{p})\| d\tau \right),$$

and, assuming that we can bound the matrix exponential of the full-order system $\|\mathbf{C}e^{t\tilde{\mathbf{A}}(\mathbf{p})}\| \le C_1(\mathbf{p})$, the error bound becomes

$$\|\mathbf{y}(t) - \widehat{\mathbf{y}}(t)\| \le C_1(\mathbf{p}) \left( \|\mathbf{e}(0)\| + \int_0^t \|\tilde{\mathbf{R}}(\tau,\mathbf{p})\| d\tau \right). \tag{2.4}$$

Since for fixed $\mathbf{p}$, the residual $\tilde{\mathbf{R}}(t,\mathbf{p})$ depends on the reduced state $\widehat{\mathbf{x}}(t)$ and *not* on the original state $\mathbf{x}(t)$, the bound can be efficiently evaluated at different values of time $t$, by using numerical quadrature [16] to compute the integral.

In practical applications, the bound given in (2.4) can be improved by using a norm $\|\cdot\|_{\mathbf{G}}$ tailored for the specific application, namely, $\|\cdot\|_{\mathbf{G}}$ is the vector norm induced by a problem-specific symmetric positive definite matrix $\mathbf{G}$, $\|\mathbf{z}\|_{\mathbf{G}}^2 = \mathbf{z}^T\mathbf{G}\mathbf{z}$. In addition, the bound can be further improved by considering the so-called dual problem. For details concerning the dual problem, such as the additional computational cost involved, we direct the reader to the discussion given in [12, 22].

The error bound shown in (2.4), or its tighter dual, represents a key result for the *offline* stage of Reduced-Basis methods. In this stage, we obtain the most computationally intensive quantities needed in the Reduced-Basis approach, and, usually, it may take arbitrarily long to complete. Nevertheless, the benefits of the offline computational effort become clear in the *online* stage, when we perform fast simulations of the reduced-order model, with simulation times independent of the full order $n$.

In the offline stage, we compute the right-hand term $C_1(\mathbf{p})$ in (2.4), i.e., we bound the matrix exponential of the full-order system (2.1). In a large-scale setting, this task has its own well known challenges, and it often requires great computational effort [20, 21]. Then, we compute the so-called offline space

$$\mathbf{X} = [\ldots\ ,\ \mathbf{x}(t_i, \mathbf{p}_j)\ ,\ \ldots] \in \mathbb{R}^{n \times (NM)} \tag{2.5}$$

which is a collection of full-order state snapshots obtained for a user-selected time grid $t_i$, $i = 1:N$, and parameter grid $\mathbf{p}_j$, $j = 1:M$. Computing the offline space requires solving the large-scale equations (2.1), leading to significant computational effort. In practice, the snapshots are obtained by discretizing the time $t$ and then employing an Euler scheme [16].

The next step in the offline stage is to extract a *reduced-basis* $\mathbf{V}$ from the offline space $\mathbf{X}$, i.e., to compute the projection matrix $\mathbf{V} \in \mathbb{R}^{n \times k}$ such that *column span* $\mathbf{V} \subset$ *column span* $\mathbf{X}$. In short, there are various ways of choosing an appropriate $\mathbf{V}$, such as using a combination of POD, greedy algorithms and adaptive approaches [12, 13, 22, 24]. The main idea behind these iterative approaches is to start with an initial reduced-basis $\mathbf{V} = \mathbf{V}_0$, then evaluate the error bound (2.4) and search for additional basis components $\mathbf{V}_1$ to obtain an enriched reduced-basis $\mathbf{V} = [\mathbf{V}_0, \mathbf{V}_1]$ that in turn gives a new lower error bound. In this case study, the Reduced-Basis model shown in Sect. 2.3 is obtained using the greedy scheme presented in [22].

Once the reduced-basis $\mathbf{V}$ is computed, we can obtain the reduced-order model. It is assumed that the parameter dependence of the full-order system (2.1) is separable into sums of constant matrices weighted by scalar functions of the parameters:

$$\begin{aligned}
\mathbf{E}(\mathbf{p}) &= \sum_{i=1}^{m_{\mathbf{E}}} \varepsilon_i(\mathbf{p})\, \mathbf{E}_i\ , \quad & \mathbf{A}(\mathbf{p}) &= \sum_{i=1}^{m_{\mathbf{A}}} \alpha_i(\mathbf{p})\, \mathbf{A}_i\ , \\
\mathbf{B}(\mathbf{p}) &= \sum_{i=1}^{m_{\mathbf{B}}} \beta_i(\mathbf{p})\, \mathbf{B}_i\ , \quad & \mathbf{C}(\mathbf{p}) &= \sum_{i=1}^{m_{\mathbf{C}}} \gamma_i(\mathbf{p})\, \mathbf{C}_i\ .
\end{aligned}$$

Then, the reduced-order parameter-dependent matrices result from projecting the constant matrices $\widehat{\mathbf{E}}_i = \mathbf{W}^T \mathbf{E}_i \mathbf{V}$, $\widehat{\mathbf{A}}_i = \mathbf{W}^T \mathbf{A}_i \mathbf{V}$, $\widehat{\mathbf{B}}_i = \mathbf{W}^T \mathbf{B}_i$, $\widehat{\mathbf{C}}_i = \mathbf{C}_i \mathbf{V}$, namely

$$\begin{aligned}
\widehat{\mathbf{E}}(\mathbf{p}) &= \sum_{i=1}^{m_{\mathbf{E}}} \varepsilon_i(\mathbf{p})\, \widehat{\mathbf{E}}_i\ , \quad & \widehat{\mathbf{A}}(\mathbf{p}) &= \sum_{i=1}^{m_{\mathbf{A}}} \alpha_i(\mathbf{p})\, \widehat{\mathbf{A}}_i\ , \\
\widehat{\mathbf{B}}(\mathbf{p}) &= \sum_{i=1}^{m_{\mathbf{B}}} \beta_i(\mathbf{p})\, \widehat{\mathbf{B}}_i\ , \quad & \widehat{\mathbf{C}}(\mathbf{p}) &= \sum_{i=1}^{m_{\mathbf{C}}} \gamma_i(\mathbf{p})\, \widehat{\mathbf{C}}_i\ .
\end{aligned} \tag{2.6}$$

The final step in Reduced-Basis methods is the *online* stage, where the pre-computed reduced-order model from the offline stage is used for fast simulations of the output $\widehat{\mathbf{y}}(t)$ for different input signals $\mathbf{u}(t)$ and parameter values $\mathbf{p}$. The reduced matrices (2.6) can be evaluated for different $\mathbf{p}$ in real time, since this operation only requires evaluation of scalar functions $\varepsilon_i(\mathbf{p}), \alpha_i(\mathbf{p}), \beta_i(\mathbf{p})$ and $\gamma_i(\mathbf{p})$. Then, the output $\widehat{\mathbf{y}}(t)$ is computed using an Euler scheme involving only the reduced-order matrices, resulting in an overall computational complexity of the online phase that is independent of the full-order $n$.

## 2.2.2 Loewner Matrix Approach

Next, we construct parametrized reduced-order models using a two-variable rational interpolation approach. The discussion summarizes the recent results in [3], where a Loewner matrix framework was introduced for constructing rational interpolants for frequency-domain measurements of systems with one parameter $p$.

The Loewner approach starts from measurements of the full-order parametrized transfer function (2.3):

$$\phi_{i,j} = \mathbf{H}(s_i, p_j), \tag{2.7}$$

$i = 1 : N$, $j = 1 : M$, and constructs a two-variable rational function $\widehat{\mathbf{H}}(s,p)$ that interpolates these measurements, $\widehat{\mathbf{H}}(s_i, p_j) = \phi_{i,j}$.

In the Loewner framework, the order of the reduced model $\widehat{\mathbf{H}}(s,p)$ is a pair $(k,q)$, where $k$ is the reduced order in the frequency variable $s$, and $q$ is the reduced order in the parameter variable $p$, with $k$ not necessarily equal to $q$. Therefore, we can choose *different* orders for $s$ and $p$, resulting in greater flexibility and a better understanding of the structure of the underlying interpolant $\widehat{\mathbf{H}}(s,p)$.

The first step consists in identifying the reduced order $(k,q)$ directly from the given measurements $\phi_{i,j}$, by computing the ranks of appropriate one-variable Loewner matrices [2, 19]. Hence, consider the pairs $(x_i, f_i)$, $i = 1 : T$, which we partition in any two disjoint sets

$$\begin{aligned} \{x_i\} &= \{\lambda_1, \ldots, \lambda_r\} \cup \{\mu_1, \ldots, \mu_\ell\}, \\ \{f_i\} &= \{\mathbf{w}_1, \ldots, \mathbf{w}_r\} \cup \{\mathbf{v}_1, \ldots, \mathbf{v}_\ell\}, \end{aligned} \tag{2.8}$$

such that $r + \ell = T$. Then, the one-variable Loewner matrix $\mathbb{L}$ associated with $(x_i, f_i)$ and the partitioning in (2.8) is defined as

$$\mathbb{L} = \begin{bmatrix} \frac{\mathbf{v}_1 - \mathbf{w}_1}{\mu_1 - \lambda_1} & \cdots & \frac{\mathbf{v}_1 - \mathbf{w}_r}{\mu_1 - \lambda_r} \\ \vdots & \ddots & \vdots \\ \frac{\mathbf{v}_\ell - \mathbf{w}_1}{\mu_\ell - \lambda_1} & \cdots & \frac{\mathbf{v}_\ell - \mathbf{w}_r}{\mu_\ell - \lambda_r} \end{bmatrix}. \tag{2.9}$$

Using this definition, we introduce the following one-variable Loewner matrices associated with the two-variable measurements given in (2.7):

$$\begin{aligned} \mathbb{L}_{p_j} &= \mathbb{L} \text{ associated with } (s_i, \phi_{i,j}), \quad j = 1 : M, \\ \mathbb{L}_{s_i} &= \mathbb{L} \text{ associated with } (p_j, \phi_{i,j}), \quad i = 1 : N, \end{aligned}$$

where the index $p_j$ $(s_i)$ indicates that $\mathbb{L}_{p_j}$ $(\mathbb{L}_{s_i})$ corresponds to measurements given by constant $p = p_j$ $(s = s_i)$. Then, the ranks of these Loewner matrices give the order $(k,q)$ of the underlying interpolant $\widehat{\mathbf{H}}(s,p)$:

$$\begin{aligned} k &= \max_j \ \operatorname{rank} \mathbb{L}_{p_j}, \qquad j = 1:M, \\ q &= \max_i \ \operatorname{rank} \mathbb{L}_{s_i}, \qquad i = 1:N. \end{aligned} \tag{2.10}$$

Next, we construct the rational interpolant $\widehat{\mathbf{H}}(s,p)$ of order $(k,q)$ by computing the null space of an appropriate two-variable Loewner matrix $\mathbb{L}_{2D}$. Towards this end, we partition the frequency and parameter grids (2.7) into any disjoint sets

$$\begin{aligned} \{s_i\} &= \{\lambda_1,\ldots,\lambda_{n'}\} \cup \{\mu_1,\ldots,\mu_{N-n'}\}, \\ \{p_j\} &= \{\pi_1,\ldots,\pi_{m'}\} \cup \{\nu_1,\ldots,\nu_{M-m'}\}, \end{aligned} \tag{2.11}$$

using the following notation for the corresponding partitioned measurements

$$\left[\phi_{i,j}\right] =: \left[\begin{array}{ccc|ccc} \mathbf{w}_{1,1} & \cdots & \mathbf{w}_{1,m'} & \phi_{1,m'+1} & \cdots & \phi_{1,M} \\ \vdots & \ddots & \vdots & \vdots & \ddots & \vdots \\ \mathbf{w}_{n',1} & \cdots & \mathbf{w}_{n',m'} & \phi_{n',m'+1} & \cdots & \phi_{n',M} \\ \hline \phi_{n'+1,1} & \cdots & \phi_{n'+1,m'} & \mathbf{v}_{1,1} & \cdots & \mathbf{v}_{1,M-m'} \\ \vdots & \ddots & \vdots & \vdots & \ddots & \vdots \\ \phi_{N,1} & \cdots & \phi_{N,m'} & \mathbf{v}_{N-n',1} & \cdots & \mathbf{v}_{N-n',M-m'} \end{array}\right] = \left[\begin{array}{c|c} \Phi_{11} & \Phi_{12} \\ \hline \Phi_{21} & \Phi_{22} \end{array}\right] = \Phi \tag{2.12}$$

namely $\Phi_{11}$ contains $\mathbf{w}_{i,j} := \mathbf{H}(\lambda_i, \pi_j)$ for $i = 1:n'$, $j = 1:m'$, while $\Phi_{22}$ contains $\mathbf{v}_{i,j} := \mathbf{H}(\mu_i, \nu_j)$, for $i = 1:(N-n')$, $j = 1:(M-m')$.

Then, from this partitioning, we define the *two-variable Loewner matrix*

$$\mathbb{L}_{2D}(i,j) = \frac{\mathbf{v}_{e(i),f(i)} - \mathbf{w}_{\widehat{e}(j),\widehat{f}(j)}}{\left(\mu_{e(i)} - \lambda_{\widehat{e}(j)}\right)\left(\nu_{f(i)} - \pi_{\widehat{f}(j)}\right)}, \tag{2.13}$$

of dimension $(N-n')(M-m') \times (n'm')$ and with indices $e, \widehat{e}, f, \widehat{f}$ having the following Kronecker structure

$$\begin{aligned} e &= [1:N-n'] \otimes \mathbf{1}_{n'} &&= [1,\ldots,1,\ 2,\ldots,2,\ldots,\ N-n',\ldots,N-n'], \\ \widehat{e} &= [1:n'] \otimes \mathbf{1}_{N-n'} &&= [1,\ldots,1,\ 2,\ldots,2,\ldots,\ n',\ldots,n'], \\ f &= \mathbf{1}_{m'} \otimes [1:M-m'] &&= [1,\ldots,M-m',1,\ldots,M-m',\ldots,1,\ldots,M-m'], \\ \widehat{f} &= \mathbf{1}_{M-m'} \otimes [1:m'] &&= [1,\ldots,m',\ 1,\ldots,m',\ldots,\ 1,\ldots,m'], \end{aligned}$$

for $\mathbf{1}_{n'} \in \mathbb{R}^{1 \times n'}$ a row vector with all entries equal to 1.

The main feature of the Loewner matrix $\mathbb{L}_{2D}$ is that its rank *encodes the order* $(k,q)$ of the underlying rational interpolant $\widehat{\mathbf{H}}(s,p)$, and, furthermore, $\widehat{\mathbf{H}}(s,p)$ can be easily constructed from its null space.

**Theorem 2.1 (Two-variable rational interpolation [3])** *If $(k,q)$ is given by (2.10) and $n' > k$, $m' > q$, then the two-variable Loewner matrix (2.13) is singular, with*

$$\operatorname{rank} \mathbb{L}_{2D} = n'm' - (n'-k)(m'-q).$$

*In addition, if we set $(n',m') = (k+1,q+1)$ in (2.11), then the rational function $\widehat{\mathbf{H}}(s,p)$ of order $(k,q)$ that interpolates all given measurements $\phi_{i,j}$ has the form*

$$\widehat{\mathbf{H}}(s,p) = \frac{\displaystyle\sum_{i=1}^{k+1}\sum_{j=1}^{q+1} \frac{c_{i,j}\mathbf{w}_{i,j}}{(s-\lambda_i)(p-\pi_j)}}{\displaystyle\sum_{i=1}^{k+1}\sum_{j=1}^{q+1} \frac{c_{i,j}}{(s-\lambda_i)(p-\pi_j)}}, \tag{2.14}$$

*with $\mathbf{c} = [c_{1,1}, c_{1,2}, \ldots, c_{2,1}, c_{2,2}, \ldots, c_{k+1,q+1}]$ in the null space of $\mathbb{L}_{2D}$, i.e., $\mathbb{L}_{2D}\,\mathbf{c} = \mathbf{0}$.*

Notice that $\widehat{\mathbf{H}}(s,p)$ is given in terms of a *rational barycentric* formula that depends on the two-variables $s$ and $p$, and is, in fact, a generalization of the one-variable rational barycentric formula [2,4]. It is easily checked that if we multiply both the numerator and denominator in (2.14) with $\prod_{i=1}^{k+1}\prod_{j=1}^{q+1}(s-\lambda_i)(p-\pi_j)$, then, after simplification, we obtain two polynomials having the highest degree in $s$ equal to $k$ and the highest degree in $p$ equal to $q$. Hence, $\widehat{\mathbf{H}}(s,p)$ is a two-variable rational function of order $(k,q)$.

The barycentric formula allows us to write down the interpolant in terms of the two-variable Lagrange basis $(s-\lambda_i)(p-\pi_j)$, $i = 1:n'$, $j = 1:m'$, which is formed directly from the partitioned frequency and parameter grids in (2.11). The Kronecker structure of the Lagrange basis dictates the Kronecker structure of the denominator in each entry of $\mathbb{L}_{2D}$. As a result, the rank of $\mathbb{L}_{2D}$ is not fixed, but it depends on the order $(k,q)$ of the underlying interpolant and on the dimensions $(n',m')$ of the partitioning. To obtain $\widehat{\mathbf{H}}(s,p)$ of order $(k,q)$, we choose $(n',m') = (k+1,q+1)$.

Furthermore, the barycentric formula in (2.14) cannot be directly evaluated at the grid points $\lambda_i$ and $\pi_j$ as it requires dividing by zero. However, just like in the case of evaluating a one-variable barycentric formula [4], we use the convention that $\widehat{\mathbf{H}}(\lambda_i,\pi_j) = c_{i,j}\mathbf{w}_{i,j}/c_{i,j} = \mathbf{w}_{i,j}$. Therefore, $\widehat{\mathbf{H}}(s,p)$ interpolates the measurements $\mathbf{w}_{i,j}$ contained in $\Phi_{11}$ by construction. Then, we force interpolation of the remaining measurements $\Phi_{12}, \Phi_{21}, \Phi_{22}$ by computing the barycentric coefficients $\mathbf{c}$ such that $\mathbb{L}_{2D}\mathbf{c} = \mathbf{0}$.

In practice, it is possible to obtain models of even lower order $(k,q)$ than the one given by (2.10). Choosing $k <$ max rank $\mathbb{L}_{p_j}$ and $q <$ max rank $\mathbb{L}_{s_i}$, results in a Loewner matrix $\mathbb{L}_{2D}$ that is full rank. However, if $\mathbb{L}_{2D}$ is close to being singular, we can still compute barycentric coefficients such that $\mathbb{L}_{2D}\mathbf{c} \approx \mathbf{0}$. In this case, the coefficients $\mathbf{c}$ give a rational function $\widehat{\mathbf{H}}(s,p)$ (2.14) that interpolates $\Phi_{11}$ by construction, and, approximates the entries in $\Phi_{12}, \Phi_{21}, \Phi_{22}$ with small error, i.e., we get a two-variable rational *approximant*, instead of an *interpolant*.

Finally, notice that equation (2.14) gives $\widehat{\mathbf{H}}(s,p)$ in transfer function form as a ratio of barycentric sums. Therefore, evaluating $\widehat{\mathbf{H}}(s,p)$ for a particular frequency $s$

and parameter $p$, can be efficiently implemented using *only* $O(kq)$ operations. Nevertheless, in practical applications, we also need to have $\widehat{\mathbf{H}}(s,p)$ expressed in terms of state-space matrices, as in (2.2). Next, we present two simple state-space realizations for $\widehat{\mathbf{H}}(s,p)$.

**Lemma 2.1 (State-space realization)** *The rational barycentric form* $\widehat{\mathbf{H}}(s,p)$ *in equation (2.14) has the following state-space realization*

$$\widehat{\mathbf{H}}(s,p) = \widehat{\mathbf{C}}(p)\left(s\widehat{\mathbf{E}} - \widehat{\mathbf{A}}(p)\right)^{-1}\widehat{\mathbf{B}} \tag{2.15}$$

*with the system matrices defined as*

$$\widehat{\mathbf{E}} = \begin{bmatrix} 1 & -1 & & \\ \vdots & & \ddots & \\ 1 & & & -1 \\ 0 & 0 & \cdots & 0 \end{bmatrix}, \qquad \widehat{\mathbf{A}}(p) = \begin{bmatrix} -\lambda_1 & \lambda_2 & & \\ \vdots & & \ddots & \\ -\lambda_1 & & & \lambda_{k+1} \\ \alpha_1(p) & \alpha_2(p) & \cdots & \alpha_{k+1}(p) \end{bmatrix},$$

$$\widehat{\mathbf{C}}(p) = \left[\beta_1(p),\ \ldots,\ \beta_{k+1}(p)\right],\quad \widehat{\mathbf{B}} = \left[0,\ \ldots,\ 0,\ 1\right]^T,$$

$$\beta_i(p) = \sum_{j=1}^{q+1} \frac{\mathbf{c}_{i,j}\mathbf{w}_{i,j}}{p-\pi_j}, \qquad \alpha_i(p) = \sum_{j=1}^{q+1} \frac{\mathbf{c}_{i,j}}{p-\pi_j},$$

*and the convention that* $\beta_i(\pi_j) = \mathbf{c}_{i,j}\mathbf{w}_{i,j}$ *and* $\alpha_i(\pi_j) = \mathbf{c}_{i,j}$.

The proof of this result relies on exploiting the non-zero structure of the matrices together with a cofactor expansion to show that $\det\left(s\widehat{\mathbf{E}} - \widehat{\mathbf{A}}(p)\right)$ equals the denominator in (2.14). For simplicity, the full details are omitted here.

The above state-space realization uses system matrices of dimension $k+1$ and has no $\mathbf{D}(p)$ term. The parameter dependencies are present only in the $\widehat{\mathbf{C}}(p)$ and $\widehat{\mathbf{A}}(p)$ matrices, and take the form of barycentric sums involving the parameter $p$. In contrast to standard state-space realizations that use a companion matrix $\widehat{\mathbf{A}}(p)$ and coefficients $\alpha_i(p)$ that are polynomials in $p$ [1], the realization given in (2.15) is better suited for practical implementations, since the coefficients $\alpha_i(p)$ do not contain powers of $p$.

Furthermore, we can also avoid barycentric sums by using the following result.

**Lemma 2.2 (State-space realization [3])** *The rational barycentric form* $\widehat{\mathbf{H}}(s,p)$ *in equation* (2.14) *has the following state-space realization*

$$\widehat{\mathbf{H}}(s,p) = \widehat{\mathbf{C}}\,\Theta(s,p)^{-1}\,\widehat{\mathbf{B}} \tag{2.16}$$

*with the system matrices defined as*

$$\Theta(s,p) = \begin{bmatrix} \mathbf{J}(s,\lambda,k) & \mathbf{0} & \mathbf{0} \\ \mathbb{A} & \mathbf{J}^*(p,\pi,q) & \mathbf{0} \\ \mathbb{B} & \mathbf{0} & [\mathbf{J}^*(p,\pi,q),\tau] \end{bmatrix}, \widehat{\mathbf{C}} = \begin{bmatrix} \mathbf{0}\ \mathbf{0}\ -\mathbf{e}_{q+1}^* \end{bmatrix}, \widehat{\mathbf{B}} = \begin{bmatrix} \mathbf{0} \\ \tau \\ \mathbf{0} \end{bmatrix},$$

$$\mathbb{A} = \begin{bmatrix} \mathbf{c}_{1,1} & \cdots & \mathbf{c}_{k+1,1} \\ \vdots & \ddots & \vdots \\ \mathbf{c}_{1,q+1} & \cdots & \mathbf{c}_{k+1,q+1} \end{bmatrix}, \quad \mathbb{B} = \begin{bmatrix} \mathbf{c}_{1,1}\mathbf{w}_{1,1} & \cdots & \mathbf{c}_{k+1,1}\mathbf{w}_{k+1,1} \\ \vdots & \ddots & \vdots \\ \mathbf{c}_{1,q+1}\mathbf{w}_{1,q+1} & \cdots & \mathbf{c}_{k+1,q+1}\mathbf{w}_{k+1,q+1} \end{bmatrix},$$

$$\mathbf{J}(s,\lambda,k) = \begin{bmatrix} s-\lambda_1 & \lambda_2 - s & & \\ s-\lambda_1 & & \lambda_3 - s & \\ \vdots & & & \ddots \\ s-\lambda_1 & & & \lambda_{k+1} - s \end{bmatrix}, \quad \tau(i) = \frac{1}{\prod_{j=1,j\neq i}^{q+1}(\pi_i - \pi_j)}.$$

Unlike the realization of Lemma 2.1, the parameter $p$ enters only linearly in the resolvent $\Theta(s,p) = s\widehat{\mathbf{E}} - \widehat{\mathbf{A}}(p)$. However, having such a simple parameter dependence results in a realization of dimension $k+2(q+1)$.

We also remark that the existence of state-space realizations with linear dependence in *p and* minimal dimension $k$ is still an open problem [6, 7, 17, 25]. Such minimal realizations are known only for the special case of $\widehat{\mathbf{H}}(s,p)$ having a separable denominator, i.e., the denominator can be factored as the product of two one-variable polynomials in $s$ and $p$ [10]. Nevertheless, the realizations provided in this section are useful in practical applications, since their dimensions are close to the minimal dimension $k$ in a reduced-order setting.

### 2.2.3 Discussion

Next, we discuss the common traits and differences between the two methods. We begin with the computational effort required for each. Notice that the most computationally intensive part of the Reduced-Basis approach is the offline stage. Its computational cost depends on the number of operations needed for obtaining the snapshots and on the algorithm used for assembling the reduced basis. For the Loewner approach, the computational effort consists in computing the full-order transfer function measurements, the reduced-order $(k,q)$ and the null space of $\mathbb{L}_{2D}$. In practice, computing $(k,q)$ does not require the ranks of all Loewner matrices $\mathbb{L}_{p_j}$ and $\mathbb{L}_{s_i}$; in fact, the ranks of only a few of these matrices usually give a good indication for appropriate values of $(k,q)$. The most computationally intensive part is computing the full-order measurements $\mathbf{H}(s_i,p_j)$, since it involves the full-order matrices and (2.3). In most practical applications, the resolvent $s\mathbf{E}(\mathbf{p}) - \mathbf{A}(\mathbf{p})$ has sparse structure; hence, we can use sparse linear system solvers [26] in (2.3) to efficiently compute the measurements.

The use of explicit transfer function measurements $\mathbf{H}(s_i,p_j)$ has another advantage. Suppose we do not have a model of the full-order system (2.1), but we only have access to its transfer function measurements; for instance, suppose we use a device to take frequency response measurements of a system. Then, we can still obtain a reduced-order model by applying the Loewner approach; i.e., we *identify* a reduced-order model directly from the available measurements.

We also remark that the results given in [3] developed the Loewner approach for the case of systems that depend on a scalar parameter $p$, unlike the Reduce-Basis approach which can accommodate a vector of parameters $\mathbf{p}$. However, since the publication of [3], the authors of this case study have generalized the Loewner approach to a vector of parameters $\mathbf{p}$. A detailed discussion of this case is scheduled for publication [15].

Perhaps the most obvious difference between the two methods is the possibility of choosing different reduced-orders for $s$ and $p$ in the Loewner approach. This is a direct consequence of using the two-variable Lagrange basis, and, in practice, it can prove useful to differentiate between $s$ and $p$, since some systems have an inherently low order dependence on the parameter $p$. This feature is discussed in detail in the example given in Sect. 2.3.

The common trait of the two methods is the fact that they both offer ways of efficiently evaluating the reduced-order models for different values of $p$. The Reduced-Basis approach achieves this in the online stage using equation (2.6), while the Loewner approach uses the rational barycentric formula (2.14).

Finally, after these theoretical remarks, we are ready to see how these methods compare in a practical application. In the next section, we give such an example.

## 2.3 Numerical Experiments

In this section, we compare the Reduced-Basis approach and the Loewner rational interpolation approach through a numerical example treating a parameter-dependent partial differential equation. This parametrized system models the transient evolution of the temperature field near the surface of a cylinder immersed in fluid. For details on deriving the state-space matrices (2.1) using a finite element spatial discretization, we direct the reader to the book [22] and its software package.

The parameter dependence is present only in the $\mathbf{A}$ matrix as

$$\mathbf{A}(p) = \mathbf{A}_1 + p^{-1}\mathbf{A}_2, \tag{2.17}$$

with the parameter $p \in [0.1, 100]$ representing the Péclet number. The dimension of the full-order state-space matrices (2.1) is $n = 878$, and the output matrix $\mathbf{C}$ is highly sparse with dimension $919 \times 878$, as it maps the $n = 878$ system states to the 919 nodes in the spatial discretization.

### *2.3.1 The Reduced-Order Models*

First, we obtain a Reduced-Basis model (2.6) of order $k = 11$. This particular reduced-order model is already available as part of the software package included with [22]. The Reduced-Basis $\mathbf{V}$ is computed using a greedy approach and an offline

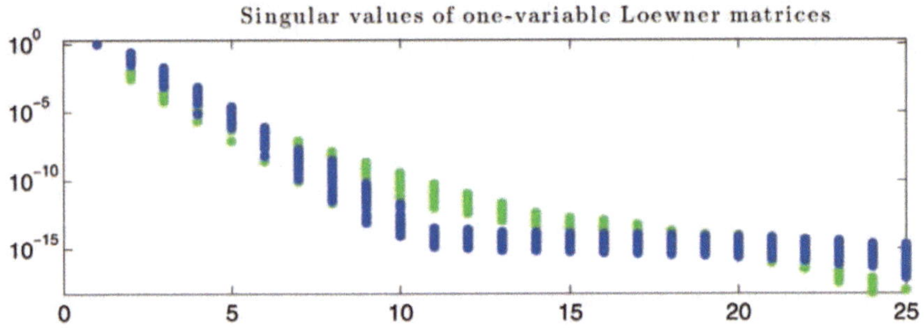


**Fig. 2.1.** Singular values of $\mathbb{L}_{p_j}$ (green dots) and $\mathbb{L}_{s_i}$ (blue dots), $i = 1:N,\ j = 1:M$

space (2.5) generated from a parameter grid $p \in [0.1, 100]$ and a time grid $t \in [0,1]$ having time step $\delta t = 0.01$. We denote the Reduced-Basis model with $\mathbf{\Sigma}_{RB}$.

Next, to obtain the Loewner model (2.15), we consider a frequency grid of $N = 50$ frequencies $s_i$ logarithmically spaced in $[10^{-2}, 10^2]$, and a parameter grid of $M = 50$ parameters $p_j$ logarithmically spaced in $[0.1, 100]$. We then compute the associated transfer function measurements $\phi_{i,j} = \mathbf{H}(s_i, p_j)$.

The crucial step of the Loewner approach is to determine the reduced order $(k,q)$, with $k$ representing the order in the frequency variable $s$, and $q$ the order in the parameter variable $p$. Therefore, in Fig. 2.1, we plot the singular values of the one-variable Loewner matrices $\mathbb{L}_{p_j}$ and $\mathbb{L}_{s_i}$, and, from (2.10), the maximum rank of $\mathbb{L}_{p_j}$ gives $k$ and the maximum rank of $\mathbb{L}_{s_i}$ gives $q$. Then, by Theorem 2.1, the two-variable Loewner matrix $\mathbb{L}_{2D}$ is singular and the barycentric coefficients $\mathbf{c}$ in its null space, $\mathbb{L}_{2D}\mathbf{c} = \mathbf{0}$, give a model $\widehat{\mathbf{H}}(s,p)$ (2.14) that interpolates all given measurements $\phi_{i,j}$.

However, for the purpose of comparing the Loewner model with the Reduced-Basis model, we select $k$ and $q$ lower than the ranks of $\mathbb{L}_{p_j}$ and $\mathbb{L}_{s_i}$, namely, we take $k = 11$, the same value as for the Reduced-Basis model. In addition, we take $q = 7$ to showcase that the order in $p$ can be chosen to be different from the order in $s$.

As a result of this choice of $(k,q) = (11,7)$, the two-variable Loewner matrix $\mathbb{L}_{2D}$ is not singular. However, its smallest singular value is equal to $3 \cdot 10^{-8}$, i.e., $\mathbb{L}_{2D}$ is close to being singular, and we can still compute barycentric coefficients $\mathbf{c}$ such that $\mathbb{L}_{2D}\mathbf{c} \approx \mathbf{0}$. Thus, $\widehat{\mathbf{H}}(s,p)$ in (2.14) approximates the given measurements $\phi_{i,j}$, instead of interpolating them. The final step consists in forming a state-space realization using either (2.15) or (2.16). We denote the Loewner model with $\mathbf{\Sigma}_{\mathbb{L}}$.

### *2.3.2 Comparison of the Reduced-Order Models*

We now compare $\mathbf{\Sigma}_{RB}$, the Reduced-Basis model, and $\mathbf{\Sigma}_{\mathbb{L}}$, the Loewner model. Before presenting their time- and frequency-domain behavior, we briefly discuss their reduced orders.

Notice that the parameter dependence (2.17) of the full-order system has a rational form, present only in the $\mathbf{A}(p)$ matrix; therefore, the resolvent $(s\mathbf{E} - \mathbf{A}(p))^{-1} \in \mathbb{C}^{878 \times 878}$ is also rational in both $s$ and $p$. Hence, the system's transfer function $\mathbf{H}(s,p)$

(2.3) is a two-variable rational function with the highest degree in $s$ equal to 878 and highest degree in $p$ equal to 878, i.e., $\mathbf{H}(s,p)$ has order $(878,878)$.

Since the projection framework (2.6) preserves the structure of the parameter dependence, the Reduced-Basis model $\boldsymbol{\Sigma}_{RB}$ is also rational in both $s$ and $p$, and has order $(11,11)$, given by the dimension of the reduced-order system matrices (2.6). On the other hand, $\boldsymbol{\Sigma}_{\mathbb{L}}$ has the flexibility of differentiating between the orders of $s$ and $p$. Therefore, we have selected a lower order for the parameter $p$, resulting in a Loewner model $\boldsymbol{\Sigma}_{\mathbb{L}}$ of order $(11,7)$.

Next, we compare the frequency-domain behavior of $\boldsymbol{\Sigma}_{RB}$ and $\boldsymbol{\Sigma}_{\mathbb{L}}$. In Fig. 2.2, we plot the frequency response of the two models for 4 different values of the parameter $p \in \{0.1, 1, 10, 100\}$. The models have one input and 919 outputs, with frequency responses $\widehat{\mathbf{H}}_{RB}(\mathrm{j}\omega_i, p)$, $\widehat{\mathbf{H}}_{\mathbb{L}}(\mathrm{j}\omega_i, p) \in \mathbb{C}^{919\times 1}$. To get a single line plot for each parameter value, we show the average of each frequency response.

On one hand, Fig. 2.2 shows that $\boldsymbol{\Sigma}_{RB}$ provides a loose approximation of the full-order system frequency-domain behavior. This was to be expected, since the Reduced-Basis method is tailored for approximation of time-domain snapshots. On the other hand, $\boldsymbol{\Sigma}_{\mathbb{L}}$ accurately matches the full-order system, since the Loewner approach is a bespoke frequency-domain method. Nevertheless, for this particular example of a parametrized partial differential equation, the frequency-domain behavior has secondary importance. Our primary goal is to accurately match the time-domain transient behavior using reduced-order models.

Therefore, we now simulate the transient behavior of the temperature field when the system is excited by the input $\mathbf{u}(t) = 10t$ for $t \in [0,1]$. Figure 2.3 shows the temperature field around the cylinder at final time $t = 1$ when the simulation is run for the parameter value $p = 0.1$. Because of the problem's symmetry, we plot only half of the rectangular domain and half of the cylinder.

As expected, the Reduced-Basis approach gives an accurate approximation of the temperature field, with the relative error $|\mathbf{y}(t) - \widehat{\mathbf{y}}(t)| / |\mathbf{y}(t)|$ bellow $10^{-2}$. In

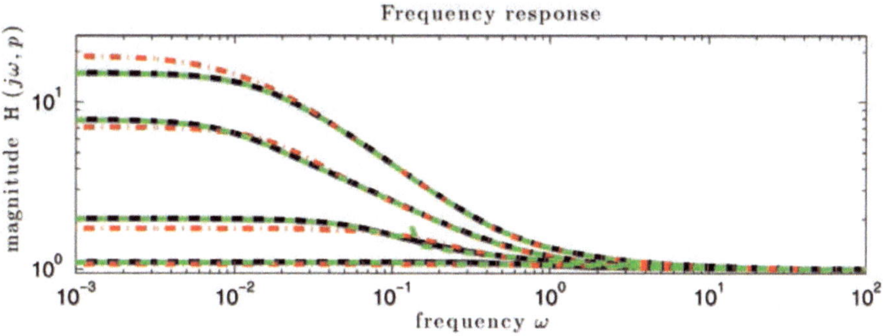


**Fig. 2.2.** Frequency responses of: full-order system $\mathbf{H}(\mathrm{j}\omega,p)$ (black) of order $(878,878)$, reduced-order model $\boldsymbol{\Sigma}_{RB}$ (red) of order $(11,11)$, and $\boldsymbol{\Sigma}_{\mathbb{L}}$ (green) of order $(11,7)$, for $p \in \{0.1,1,10,100\}$

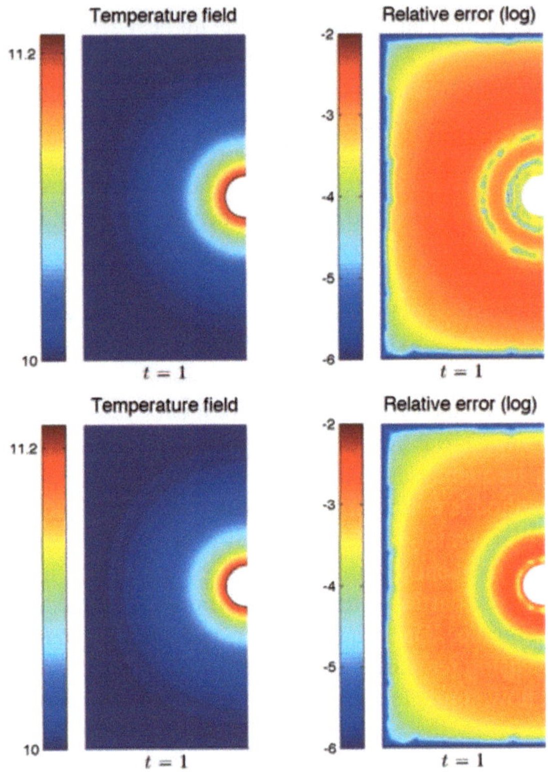


**Fig. 2.3.** Temperature field at time $t = 1$ for input $\mathbf{u}(t) = 10t$ and $p = 0.1$. Upper pane: Reduced-Basis model $\mathbf{\Sigma}_{RB}$. Lower pane: the Loewner model $\mathbf{\Sigma}_{\mathbf{L}}$. Right-hand side: the relative error (in logarithmic scale) between the reduced-order models and the full-order finite element model

addition, the lower half of Fig. 2.3 shows that the Loewner approach produces similar levels of accuracy.

Therefore, through this numerical example, we have seen that, although they approach the problem from different perspectives, both methods produce accurate reduced-order models.

## 2.4 Conclusions

Motivated by the ever increasing need for accurate, low dimension models of parameter-dependent systems, this case study is one of the first efforts to compare different approaches for parametrized model reduction. More precisely, we compared the well known Reduced-Basis approach with the recently introduced Loewner matrix approach for rational interpolation.

We saw that the main difference between the two is the fact that Reduced-Basis uses time-domain snapshots, while the Loewner approach uses frequency-domain transfer function measurements. Furthermore, the key feature of Reduced-Basis is an error bound; while for the Loewner approach, it is the possibility of choosing different reduced orders for the frequency and parameter variables.

Although different in their approach, both methods proved successful at computing accurate reduced-order models in a numerical example involving a parametrized partial differential equation.

## References

1. Antoulas, A.C.: Approximation of large-scale dynamical systems. Advances in Design and Control, DC-06. SIAM, Philadelphia (2005, second printing: Summer 2008)
2. Antoulas, A.C., Anderson, B.D.O.: On the scalar rational interpolation problem. IMA J. of Mathematical Control and Information **3**, 61–88 (1986)
3. Antoulas, A.C., Ionita, A.C., Lefteriu, S.: On two-variable rational interpolation. Linear Algebra and its Applications **436**(8), 2889–2915 (2012). DOI 10.1016/j.laa.2011.07.017
4. Berrut, J.P., Trefethen, L.N.: Barycentric Lagrange Interpolation. SIAM Review **46**(3), 501–517 (2004). DOI 10.1137/S0036144502417715
5. Chahlaoui, Y., Van Dooren, P.: A collection of benchmark examples for model reduction of linear time invariant dynamical systems. SLICOT Working Note 2002-2 (2002)
6. Eising, R.: Realization and stabilization of 2-d systems. Automatic Control, IEEE Transactions on **23**(5), 793–799 (1978). DOI 10.1109/TAC.1978.1101861
7. Fornasini, E., Marchesini, G.: Doubly-indexed dynamical systems: State-space models and structural properties. Theory of Computing Systems **12**, 59–72 (1978). DOI 10.1007/BF01776566
8. Glover, K.: All optimal Hankel-norm approximations of linear multivariable systems and their $L^\infty$-error bounds. International Journal of Control **39**(6), 1115–1193 (1984). DOI 10.1080/00207178408933239
9. Grimme, E.J.: Krylov projection methods for model reduction. Ph.D. Thesis, ECE Dept., U. of Illinois, Urbana, Champaign, IL, USA (1997)
10. Gu, G., Aravena, J.L., Zhou, K.: On minimal realization of 2-D systems. IEEE Transactions on Circuits and Systems **38**(10), 1228–1233 (1991). DOI 10.1109/31.97545
11. Gugercin, S., Antoulas, A.C., Beattie, C.: $\mathscr{H}_2$ Model Reduction for Large-Scale Linear Dynamical Systems. SIAM Journal on Matrix Analysis and Applications **30**(2), 609–638 (2008). DOI 10.1137/060666123
12. Haasdonk, B.: Reduzierte-Basis-Methoden. Preprint 2011/004 – Institute of Applied Analysis and Numerical Simulation, University of Stuttgart, Stuttgart (2011)
13. Haasdonk, B., Ohlberger, M.: Reduced basis method for finite volume approximations of parametrized linear evolution equations. ESAIM: Mathematical Modelling and Numerical Analysis **42**(02), 277–302 (2008). DOI 10.1051/m2an:2008001
14. Haasdonk, B., Ohlberger, M.: Efficient reduced models and a posteriori error estimation for parametrized dynamical systems by offline/online decomposition. Mathematical and Computer Modelling of Dynamical Systems **17**(2), 145–161 (2011). DOI 10.1080/13873954.2010.514703

15. Ionita A.C., Antoulas A.C.: Data-driven parametrized model reduction in the Loewner framework. Submitted to SIAM Journal on Scientific Computing (2013)
16. Kahaner, D., Moler, K., Nash, S.: Numerical Methods and Software. Prentice Hall, NJ, USA (1989)
17. Kung, S.Y., Levy, B.C., Morf, M., Kailath, T.: New results in 2-D systems theory, part II: 2-D state-space models - Realization and the notions of controllability, observability, and minimality. Proceedings of the IEEE **65**(6), 945–961 (1977). DOI 10.1109/PROC.1977.10592
18. Machiels, L., Maday, Y., Oliveira, I.B., Patera, A.T., Rovas, D.V.: Output bounds for reduced-basis approximations of symmetric positive definite eigenvalue problems. Comptes Rendus de l'Académie des Sciences, Series I, Mathematics **331**(2), 153–158 (2000). DOI 10.1016/S0764-4442(00)00270-6
19. Mayo, A.J., Antoulas, A.C.: A framework for the solution of the generalized realization problem. Linear Algebra and its Applications **425**(2–3), 634–662 (2007). DOI 10.1016/j.laa.2007.03.008
20. Moler, C., Van Loan, C.: Nineteen Dubious Ways to Compute the Exponential of a Matrix. SIAM Review **20**(4), 801–836 (1978). DOI 10.1137/1020098
21. Moler, C., Van Loan, C.: Nineteen Dubious Ways to Compute the Exponential of a Matrix, Twenty-Five Years Later. SIAM Review **45**(1), 3–49 (2003). DOI 10.1137/S00361445024180
22. Patera, A.T., Rozza, G.: Reduced Basis Approximation and A Posteriori Error Estimation for Parametrized Partial Differential Equations. MIT Pappalardo Graduate Monographs in Mechanical Engineering. Cambridge, MA (2007). Available from http://augustine.mit.edu/methodology/methodology_book.htm
23. Prudhomme, C., Rovas, D.V., Veroy, K., Machiels, L., Maday, Y., Patera, A.T., Turinici, G.: Reliable Real-Time Solution of Parametrized Partial Differential Equations: Reduced-Basis Output Bound Methods. Journal of Fluids Engineering **124**(1), 70 (2002). DOI 10.1115/1.1448332
24. Quarteroni, A., Rozza, G., Manzoni, A.: Certified reduced basis approximation for parametrized partial differential equations and applications. Journal of Mathematics in Industry **1**(1), 3 (2011). DOI 10.1186/2190-5983-1-3
25. Roesser, R.: A discrete state-space model for linear image processing. Automatic Control, IEEE Transactions on **20**(1), 1–10 (1975). DOI 10.1109/TAC.1975.1100844
26. Saad, Y.: Iterative Methods for Sparse Linear Systems, 2nd Ed. SIAM (2003)
27. Volkwein, S.: Proper orthogonal decomposition and singular value decomposition. Technical Report SFB-153, Institut für Mathematik, Universität Graz. Appeared also as part of the author's Habilitationsschrift, Institut für Mathematik, Universität Graz, Graz (2001) (1999)

3

# Comparison of Some Reduced Representation Approximations

**Mario Bebendorf, Yvon Maday and Benjamin Stamm**

**Abstract** In the field of numerical approximation, specialists considering highly complex problems have recently proposed various ways to simplify their underlying problems. In this field, depending on the problem they were tackling and the community that are at work, different approaches have been developed with some success and have even gained some maturity, the applications can now be applied to information analysis or for numerical simulation of PDE's. At this point, a crossed analysis and effort for understanding the similarities and the differences between these approaches that found their starting points in different backgrounds is of interest. It is the purpose of this paper to contribute to this effort by comparing some constructive reduced representations of complex functions. We present here in full details the Adaptive Cross Approximation (ACA) and the Empirical Interpolation Method (EIM) together with other approaches that enter in the same category.

## 3.1 Introduction

This paper deals with the economical representation of *dedicated* sets of data, that are currently – and more and more importantly – available stemming out of various

M. Bebendorf
Institute for Numerical Simulation, University of Bonn, Wegelerstraße 6, 53115 Bonn, Germany
e-mail: bebendorf@ins.uni-bonn.de

Y. Maday
UPMC Univ. Paris 06, UMR 7598 LJLL, Paris, F-75005 France;
Institut Universitaire de France and Division of Applied Mathematics, Brown University, Providence, RI, USA
e-mail: maday@ann.jussieu.fr

B. Stamm (✉)
UPMC Univ. Paris 06, UMR 7598 LJLL, Paris, F-75005 France;
CNRS, UMR 7598 LJLL, Paris, F-75005 France
e-mail: stamm@ann.jussieu.fr

A. Quarteroni, G. Rozza (eds.): *Reduced Order Methods for Modeling and Computational Reduction*, MS&A 9. DOI 10.1007/978-3-319-02090-7_3, 

experiences or given by formal expressions. The amount of information that can be derived out of a given massive set of data is far much smaller than the size of the data itself, therefore, parallel to the increasing size of data acquisition and storage available on computer architectures, an effort for post processing and economically represent, analyze and derive pertinent information out of the data has been done during the last century. The main idea starts from the translation of the fact that the data are *dedicated* to some phenomenon and thus, there exists a certain amount of *coherence* in these data which can be separated into two classes: deterministic or statistical. Among them have been proposed: regularity, sparsity, small $n$-width etc. that can be either assumed, verified or proven.

The data themselves can be known in different ways, either (i) completely explicitly, like for instance (i-1) from an analytic representation or at least access to the values at every point, (i-2) or only given on a large set of points, (i-3) or also given through various global measures like moments, or (ii) given implicitly through a model like a partial differential equation (PDE). The range of applications is huge, examples can be found in statistics, image and information process, learning process, experiments in mechanics, meteorology, earth sciences, medicine, biology, etc. and the challenge is in computationally processing such a large amount of high-dimensional data so as to obtain low-dimensional descriptions and capture much of the phenomena of interest.

We consider the following problem formulation: Let us assume that we are given a (presumably large) set $\mathscr{F}$ of functions $\varphi \in \mathscr{F}$ defined over $\Omega_{\mathbf{x}} \subset \mathbb{R}^{d_x}$ (with $d_x \geq 1$). Our aim is to find some functions $h_1, h_2, \ldots, h_Q : \Omega_{\mathbf{x}} \to \mathbb{R}$ such that every $\varphi \in \mathscr{F}$ can be well approximated as follows

$$\varphi(x) \approx \sum_{q=1}^{Q} \hat{\varphi}_q h_q(x),$$

where $Q \ll \dim(\mathrm{span}\{\mathscr{F}\})$. As said above, the ability for $\mathscr{F}$ to posses this property is an assumption. It is precisely stated under the notion of small Kolmogorov $n$-width, defined as follows:

Let $\mathscr{F}$ be a subset of some Banach space $\mathscr{X}$ and $\mathbb{V}_Q$ be a generic $Q$-dimensional subspace of $\mathscr{X}$. The angle between $\mathscr{F}$ and $\mathbb{V}_Q$ is

$$E(\mathscr{F};\mathbb{V}_Q) := \sup_{\varphi \in \mathscr{F}} \inf_{v_Q \in \mathbb{V}_Q} \|\varphi - v_Q\|_{\mathscr{X}}.$$

The Kolmogorov $n$-width of $\mathscr{F}$ in $\mathscr{X}$ is given by

$$d_Q(\mathscr{F},\mathscr{X}) := \inf\{E(\mathscr{F};\mathbb{V}_Q) \,|\, \mathbb{V}_Q \text{ a } Q\text{-dimensional subspace of } \mathscr{X}\}.$$

The $n$-width of $\mathscr{F}$ thus measures to what extent the set $\mathscr{F}$ can be approximated by a $n$-dimensional subspace of $\mathscr{X}$.

This assumption of small Kolmogorov $n$-width can be taken for granted, but there are also reasons on the elements of $\mathscr{F}$ that can lead to such a smallness such as *regularity* of the functions $\varphi \in \mathscr{F}$. As an example, we can quote, in the periodic settings, the well-known Fourier series. Small truncated Fourier series are good approximations of the full expansion if the decay rate of the Fourier coefficients is fast enough, i.e. if the functions $\varphi$ have enough continuous derivatives. In this case, the basis is actually multipurpose since it is not dedicated to the particular set $\mathscr{F}$. Fourier series are indeed adapted to any set of regular enough functions, the more regular they are, the better the approximation is. Another property for $\mathscr{F}$ to have a small Kolmogorov $n$-width is that it satisfies the principle of *transform sparsity*, i.e., we assume that the functions $\varphi \in \mathscr{F}$ are expressed in a *sparse* way when written in some orthonormal basis set $\{\psi_i\}$, e.g. an orthonormal wavelet basis, a Fourier basis, or a local Fourier basis, depending on the application: this means that the coefficients $\hat{\varphi}_i = \langle \varphi, \psi_i \rangle$ satisfy, for some $p$, $0 < p < 2$, and some $R$:

$$\|\varphi\|_{\ell^p} = \Big(\sum_i |\hat{\varphi}_i|^p\Big)^{1/p} \leq R.$$

A key implication of this assumption is that if we denote by $\varphi_N$ the sum of the $N$ largest contributions then

$$\exists C(R,p),\ \forall \varphi \in \mathscr{F}, \quad \|\varphi - \varphi_N\|_{\ell^2} \leq C(R,p)(N+1)^{1/2-1/p},$$

i.e. there exists a contracted representation of such a $\varphi$. Note that the representation is adaptive and tuned to each $\varphi$ (it is what is called a nonlinear approximation). However, under these assumptions, and if $\mathscr{F}$ is finite dimensional (with a dimension that is much larger than $N$), the theory of compressed sensing (see [29]), at the price of having a slight logarithmic degradation of the convergence rate, allows to propose a non-adaptive recovery of $\ell^p$ functions, with $p \leq 1$, that is almost optimal. We refer to [29] and the references therein for more details on this question. Anyway, these are cases where the set of basis functions $\{h_i\}$ does not constitute a multipurpose approximation set, all the contrary: it is tuned to that choice of $\mathscr{F}$ and will not have any good property for another one.

The difficulty is of course to find the basis set $\{h_i\}$. Note additionally that, from the definition of the small Kolmogorov $n$-width, except in a Hilbertian framework, the optimal elements need not even be in span$\{\mathscr{F}\}$.

Let us proceed and propose a way to better identify the various elements in $\mathscr{F}$: we consider that they are parametrized with $y \in \Omega_{\mathrm{y}} \subset \mathbb{R}^{d_y}$ (with $d_y \geq 1$), so that $\mathscr{F}$ consists of the parametrized functions $f : \Omega_{\mathrm{x}} \times \Omega_{\mathrm{y}} \to \mathbb{R}$. In what follows, we denote the function $f$ as a function of $x$ for some fixed parameter value $y$ as $f_y := f(\cdot, y)$. However, the role of $x$ and $y$ could be interchanged and both $x$ and $y$ will be considered equally as variables of the same level or as variable and parameter in all what follows.

In this paper, we present a survey of algorithms that search for an affine decomposition of the form

$$f(x,y) \approx \sum_{q=1}^{Q} g_q(y)\, h_q(x). \tag{3.1}$$

We focus on the case where the decomposition is chosen in an optimal way (in terms of sparse representation) and additionally we focus on methods with minimal computational complexity. It is assumed that we have a priori some or all the knowledge on functions $f$ in $\mathscr{F}$, i.e. they are not implicitly defined by a PDE. In that "implicit" case there exists a family of reduced modeling approaches such as the reduced basis method; see e.g. [62].

Note that the domains $\Omega_x$ and $\Omega_y$ can be with finite cardinality $M$ and $N$, in which case the functions can be written as matrices, then, the above algorithms can often be stated as a low-rank approximation: Given a matrix $\mathrm{M} \in \mathbb{R}^{M \times N}$, find a decomposition of the matrix M:

$$\mathrm{M} \approx \mathrm{U}\,\mathrm{V}^T$$

where U is of size $M \times Q$ and V of size $N \times Q$.

In this completely discrete setting, the Singular Value Decomposition (SVD), or the related Proper Orthogonal Decomposition (POD), yields an optimal (in terms of approximability with respect to the $\|\cdot\|_{\ell^2}$-norm) solution, but is rather expensive to compute. After presenting the POD in a general setting in Sect. 3.2, we present two alternatives, the Adaptive Cross Approximation (ACA) in Sect. 3.3 and the Empirical Interpolation Method (EIM), in Sect. 3.4, which originate from completely different backgrounds. We give a comparative overview of features and existing results of those approaches which are computationally much cheaper and yield in practice similar approximation results. The relation between ACA and the EIM is studied in Sect. 3.5. Section 3.6 is devoted to a projection method based on incomplete data known as Gappy POD or Missing Point Estimation, which in some cases can be interpreted as an interpolation scheme.

## 3.2 Proper Orthogonal Decomposition

Let us start by assuming that we have an unlimited knowledge of the data set and that we have unlimited computer resources – coming back at the end of this section to more realistic matter of facts. The first approach is known under the generic concept of Proper Orthogonal Decomposition (POD) which is a mathematical technique that stands at the intersection of various horizons that have actually been developed independently and concomitantly in various disciplines and is thus known under various names, including:

- Proper Orthogonal Decomposition (POD): a term used in turbulence;
- Singular Value Decomposition (SVD): a term used in algebra;
- Principal Component Analysis (PCA): a term used in statistics for discrete random processes;
- the discrete Karhunen-Loeve transform (KLT): a term used in statistics for continuous random processes;
- the Hotelling transform: a term used in image processing;
- Principal Orthogonal Direction (POD): a term used in geophysics;

- Empirical Orthogonal Functions (EOFs): a term used in meteorology and geophysics.

All these somewhat equivalent approaches aim at obtaining low-dimensional approximate descriptions of high-dimensional processes, therefore eliminating information which has little impact on the overall understanding.

### 3.2.1 Historical Overview

As stated above, the POD is present under various forms in many contributions.

The original SVD was established for real-square matrices in the 1870's by Beltrami and Jordan, for complex square matrices in 1902 by Autonne, and for general rectangular matrices in 1936 by Eckart and Young; see also the generalization to unitarily invariant norms by Mirsky [58]. The SVD can be viewed as the extension of the eigenvalue decomposition for the case of non-symmetric matrices and non-square matrices.

The PCA is a statistical technique. The earliest descriptions of the technique were given by Pearson [63] and Hotelling [44]. The purpose of the PCA is to identify the dependence structure behind a multivariate stochastic observation in order to obtain a compact description of it.

Lumley [51] traced the idea of the POD back to independent investigations by Kosambi [47], Loève [50], Karhunen [46], Pougachev [64] and Obukhov [59].

These methods aim at providing a set of orthonormal basis functions that allow to express approximately and optimally any function in the data set. The equivalence between all these approaches has been also investigated by many authors, among them [48, 56, 71].

### 3.2.2 Algorithm

Let us now present the POD algorithm in a semi-discrete framework, that is, we consider a finite family of functions $\{f_y\}_{y\in\Omega_{\mathrm{y}}^{\mathtt{train}}}$ where $f_y : \Omega_{\mathrm{x}} \to \mathbb{R}$ for each $y \in \Omega_{\mathrm{y}} = \Omega_{\mathrm{y}}^{\mathtt{train}}$ where $\Omega_{\mathrm{y}}^{\mathtt{train}}$ is finite with cardinality $N$. In this context, the goal is to define an approximation $P_Q[f_y]$ to $f_y$ defined by

$$P_Q[f_y](x) = \sum_{q=1}^{Q} g_q(y)\, h_q(x) \tag{3.2}$$

with $Q \ll N$. The POD actually incorporates a scalar product, for functions depending on $x \in \Omega_{\mathrm{x}}$, and the above projection is then an orthogonal projection on the $Q$-dimensional vectorial space $\mathrm{span}\{h_q, q = 1, \ldots, Q\}$.

The question is now to select properly the functions $h_q$. With a scalar product, orthonormality is useful, since we would like that these modes are selected in order that they carry as much of the information that exists in the $\{f_y\}_{y\in\Omega_{\mathrm{y}}^{\mathtt{train}}}$, i.e. the first function $h_1$ should be selected such that it provides the best one-term approximation similarly, then $h_q$ should be selected so that, with $h_1, h_2, \ldots, h_{q-1}$ it gives the best $q$-

**Scheme 3.1.** Proper orthogonal decomposition (POD)

a. Let $\Omega_{\mathrm{y}}^{\mathtt{train}} = \{\hat{y}_1, \dots, \hat{y}_N\}$ be a $N$-dimensional dicrete representation of $\Omega_{\mathrm{y}}$.
b. Construct the correlation matrix

$$\mathrm{C}_{i,j} = \frac{1}{N}\left(f_{\hat{y}_j}, f_{\hat{y}_i}\right)_{\Omega_{\mathrm{x}}}, \quad 1 \le i,j \le N,$$

where $(\cdot,\cdot)_{\Omega_{\mathrm{x}}}$ denotes a scalar product of functions depending on $\Omega_{\mathrm{x}}$.
c. Then, solve for the $Q$ largest eigenvalue-eigenvector pairs $(\lambda_q, \mathrm{v}_q)$ such that

$$\mathrm{C}\,\mathrm{v}_q = \lambda_q \mathrm{v}_q, \quad 1 \le q \le Q. \tag{3.3}$$

d. The orthogonal POD basis functions $\{h_1, \dots, h_Q\}$ such that $\mathbb{V}_Q = \mathrm{span}\{h_1, \dots, h_Q\}$ are then given by the linear combinations

$$h_q(x) = \sum_{n=1}^{N} (\mathrm{v}_q)_n f(x, \hat{y}_n), \quad 1 \le q \le Q, \quad x \in \Omega_{\mathrm{x}},$$

and where $(\mathrm{v}_q)_n$ denotes the $n$-th coefficient of the eigenvector $\mathrm{v}_q$.

**Approximation.** The approximation $P_Q[f_y]$ to $f_y : \Omega_{\mathrm{x}} \to \mathbb{R}$, for any $y \in \Omega_{\mathrm{y}}$, is then given by

$$P_Q[f_y](x) = \sum_{q=1}^{Q} g_q(y)\, h_q(x), \quad x \in \Omega_{\mathrm{x}},$$

with $g_q(y) = \frac{(f_y, h_q)_{\Omega_{\mathrm{x}}}}{(h_q, h_q)_{\Omega_{\mathrm{x}}}}$.

term approximation. The best $q$-term above is understood in the sense that the mean square error over all $y \in \Omega_{\mathrm{y}}^{\mathtt{train}}$ is the smallest. Such specially ordered orthonormal functions are called the proper orthogonal modes for the function $f(x,y)$. With these functions, the expression (3.2) is called the POD of $f$ and the algorithm is given in Table 3.1.

**Proposition 3.1** *The approximation error*

$$d_2^{\mathrm{POD}}(Q) = \sqrt{\frac{1}{N} \sum_{y \in \Omega_{\mathrm{y}}^{\mathtt{train}}} \|f_y - P_Q[f_y]\|_{\Omega_{\mathrm{x}}}^2}$$

*minimizes the mean square error* $\sqrt{\frac{1}{N} \sum_{y \in \Omega_{\mathrm{y}}^{\mathtt{train}}} \|f_y - \mathscr{P}_Q[f_y]\|_{\Omega_{\mathrm{x}}}^2}$ *over all projection operators* $\mathscr{P}_Q$ *onto a space of dimension* $Q$. *It is given by*

$$d_2^{\mathrm{POD}}(Q) = \sqrt{\sum_{q=Q+1}^{N} \lambda_q}, \tag{3.4}$$

*where* $\{\lambda_{Q+1},\ldots,\lambda_N\}$ *denotes the set of the* $N-Q$ *smallest eigenvalues of the eigenvalue problem (3.3).*

*Remark 3.1 (Relation to SVD)* If the scalar product $(\cdot,\cdot)_{\Omega_x}$ is approximated in the sense of $\ell^2$ on a discrete set of points $\Omega_x^{\mathtt{train}} = \{\hat{x}_1,\ldots,\hat{x}_M\} \subset \Omega_x$, i.e.

$$(v,w)_{\Omega_x^{\mathtt{train}}} = \frac{|\Omega_x|}{M}\sum_{i=1}^{M} v(\hat{x}_i)\,w(\hat{x}_i),$$

then we see that $\mathrm{C} = \mathrm{A}^T\,\mathrm{A}$ where A is the matrix defined by $A_{i,j} = \sqrt{\frac{|\Omega_x|}{NM}} f_{\hat{y}_j(\hat{x}_i)}$. And thus, the square roots of the eigenvalues (3.3) are singular values of A.

*Remark 3.2 (Infinite dimensional version)* In the case where the POD is processed by leaving the parameter $y$ continuous in $\Omega_y$, the correlation matrix becomes an operator $C: L^2(\Omega_y) \to L^2(\Omega_y)$ with kernel $C(y_1,y_2) = (f_{y_1},f_{y_2})_{\Omega_x}$ that acts on functions of $y \in \Omega_y$ as follows

$$(C\phi)(y) = (C(y,\cdot),\phi)_{\Omega_y}, \quad \phi \in L^2(\Omega_y).$$

Assuming that $f \in L^2(\Omega_x \times \Omega_y)$, by the results obtained in [67] (that generalize Mercer's theorem to more general domains) there exists a sequence of positive real eigenvalues (that can be ranked in decreasing order) and associated orthonormal eigenvectors, which can be used to construct best $L^2$-approximations (3.1).

The infinite dimensional version is important to understand the generality of the approach, e.g. how the various POD algorithms are linked together. In essence, this boils down to spectral theory of self-adjoint operators, either finite (in the matrix case) or infinite (for integral operator defined with symmetric kernels). Such operators have positive real eigenvalues and the corresponding eigenvectors can be ranked in decreasing order of eigenvalues. The approximation is based on considering the only eigenmodes that corresponds to the largest eigenvalues, they are those that carry the maximum information.

In practice though, both in the $x$ and the $y$ variables, sample sets $\Omega_x^{\mathtt{train}}$ and $\Omega_y^{\mathtt{train}}$ are devised. Depending on the size of $N$, the solution of the eigenvalue problem (3.3) can be prohibitively expensive. Most of the time though, there is not much hint on the way these training points should be chosen and they are generally quite large sets with $N \gg Q$.

We finally remind that the original goal is to approximate any function $f(x,y)$ for all $x \in \Omega_x$ and $y \in \Omega_y$. In this regard, the error bound (3.4) only provides an upper error estimate for functions $f_y$ with $y \in \Omega_y^{\mathtt{train}}$ and no certified error bound for functions $f_y$ with $y \in \Omega_y \setminus \Omega_y^{\mathtt{train}}$ can be provided.

## 3.3 Adaptive Cross Approximation

In order to cope with the difficulty of implementation of the POD algorithms, let us present here the *Adaptive Cross Approximation*. The approximation leading to (3.1) is

$$f(x,y) \approx \mathfrak{I}_Q[f_y](x) := \begin{bmatrix} f(x,y_1) \\ \vdots \\ f(x,y_Q) \end{bmatrix}^T \mathrm{M}_Q^{-1} \begin{bmatrix} f(x_1,y) \\ \vdots \\ f(x_Q,y) \end{bmatrix} \tag{3.5}$$

with points $x_q$, $y_q$, $q = 1,\ldots,Q$, chosen such that the matrix

$$\mathrm{M}_Q := \begin{bmatrix} f(x_1,y_1) & \ldots & f(x_1,y_Q) \\ \vdots & & \vdots \\ f(x_Q,y_1) & \ldots & f(x_Q,y_Q) \end{bmatrix} \in \mathbb{R}^{Q\times Q}$$

is invertible. Notice that while $P_Q$ used in the construction of the POD is an orthogonal projector, $\mathfrak{I}_Q : C^0(\Omega_{\mathrm{x}}) \to \mathbb{V}_Q$ is an interpolation operator from the space of continuous functions $C^0(\Omega_{\mathrm{x}})$ onto the system $\mathbb{V}_Q := \mathrm{span}\{f_{y_1},\ldots,f_{y_Q}\}$, i.e.

$$\mathfrak{I}_Q[f_y](x_q) = f(x_q,y) \quad \text{for all } y \text{ and } q = 1,\ldots,Q.$$

Due to the symmetry of $x$ and $y$ in (3.5), we also have $\mathfrak{I}_Q[f_{y_q}](x) = f(x,y_q)$ for all $x$ and $q = 1,\ldots,Q$.

### 3.3.1 Historical Overview

Approximations of type (3.5) were first considered by Micchelli and Pinkus in [57]. There, it was proved for so-called totally positive functions $f$, i.e. continuous functions $f : [0,1]\times[0,1] \to \mathbb{R}$ with non-negative determinants

$$\left| \begin{bmatrix} f(\xi_1,\upsilon_1) & \ldots & f(\xi_1,\upsilon_q) \\ \vdots & & \vdots \\ f(\xi_q,\upsilon_1) & \ldots & f(\xi_q,\upsilon_q) \end{bmatrix} \right|$$

for all $0 \le \xi_1 < \ldots < \xi_q \le 1$, $0 \le \upsilon_1 < \ldots < \upsilon_q \le 1$, and $q = 1,\ldots,Q$, that such approximations are optimal with respect to the $L^1$-norm, i.e.

$$\min_{u_q,v_q} \int_0^1 \int_0^1 \left| f(x,y) - \sum_{q=1}^{Q} u_q(x)v_q(y) \right| \mathrm{d}y\,\mathrm{d}x = \int_0^1 \int_0^1 \left| f(x,y) - \mathfrak{I}_Q[f_y](x) \right| \mathrm{d}y\,\mathrm{d}x,$$

where $\mathfrak{I}_Q$ is defined at implicitly known nodes $x_1,\ldots,x_Q$ and $y_1,\ldots,y_Q$; see [57] for an additional technical assumption.

Instead of $L^1$-estimates, it is usually required to obtain $L^\infty$-estimates. The obvious estimate

$$\|f_y - \mathfrak{I}_Q[f_y]\|_{L^\infty(\Omega_x)} \le (1+\sigma_1[f]) \inf_{v\in\mathbb{V}_Q} \|f_y - v\|_{L^\infty(\Omega_x)}$$

contains the expression

$$\sigma_1[f] := \sup_{x\in\Omega_x} \left\| \mathrm{M}_Q^{-T} \begin{bmatrix} f(x,y_1) \\ \vdots \\ f(x,y_Q) \end{bmatrix} \right\|_{\ell^1}.$$

Since there is usually no estimate on the previous infimum (note that $\mathbb{V}_Q$ also depends on $\mathscr{F} = \{f_y\}_{y\in\Omega_y}$), one tries to relate $f_y - \mathfrak{I}_Q[f_y]$ with the interpolation error in another system $\mathbb{W}_Q = \operatorname{span}\{w_1,\dots,w_Q\}$ of functions (e.g. polynomials, spherical harmonics, etc.); cf. [6, 12]. Assume that the determinant of the Vandermonde matrix $\mathrm{W}_Q := [w_i(x_j)]_{i,j=1,\dots,Q}$ does not vanish and let $L : \Omega_x \to \mathbb{R}^Q$ be the vector consisting of Lagrange functions $L_i \in \mathbb{W}_Q$, i.e. $L_i(x_j) = \delta_{ij}$, $i,j = 1,\dots,Q$. Then, the interpolation operator $\mathfrak{I}'_Q$ defined over $C^0(\Omega_x)$ with values in $\mathbb{W}_Q$ can be represented as

$$\mathfrak{I}'_Q[\varphi](x) = \begin{bmatrix} \varphi(x_1) \\ \vdots \\ \varphi(x_Q) \end{bmatrix}^T L(x), \quad \varphi \in C^0(\Omega_x),$$

and we obtain

$$\begin{aligned} & f_y(x) - \mathfrak{I}_Q[f_y](x) \\ &= f_y(x) - \begin{bmatrix} f(x_1,y) \\ \vdots \\ f(x_Q,y) \end{bmatrix}^T L(x) - \left( \begin{bmatrix} f(x,y_1) \\ \vdots \\ f(x,y_Q) \end{bmatrix} - \mathrm{M}_Q^T L(x) \right)^T \mathrm{M}_Q^{-1} \begin{bmatrix} f(x_1,y) \\ \vdots \\ f(x_Q,y) \end{bmatrix} \\ &= f_y(x) - \mathfrak{I}'_Q[f_y](x) - \begin{bmatrix} f_{y_1}(x) - \mathfrak{I}'_Q[f_{y_1}](x) \\ \vdots \\ f_{y_Q}(x) - \mathfrak{I}'_Q[f_{y_Q}](x) \end{bmatrix}^T \mathrm{M}_Q^{-1} \begin{bmatrix} f(x_1,y) \\ \vdots \\ f(x_Q,y) \end{bmatrix}. \end{aligned}$$

Hence, for any $y \in \Omega_y$

$$\|f_y - \mathfrak{I}_Q[f_y]\|_{L^\infty(\Omega_x)} \le (1+\sigma_2[f]) \max_{z\in\{y,y_1,\dots,y_Q\}} \|f_z - \mathfrak{I}'_Q[f_z]\|_{L^\infty(\Omega_x)}, \tag{3.6}$$

where

$$\sigma_2[f] := \sup_{y\in\Omega_y} \left\| \mathrm{M}_Q^{-1} \begin{bmatrix} f(x_1,y) \\ \vdots \\ f(x_Q,y) \end{bmatrix} \right\|_{\ell^1}.$$

### 3.3.2 Construction of Interpolation Nodes

The assumption that the determinant of the Vandermonde matrix $\mathrm{W}_Q$ does not vanish, can be guaranteed by the choice of $x_1,\dots,x_Q$. To this end, let $Q$ linearly independent functions $w_1,\dots,w_Q$ be given as above. As in [8], we construct linearly independent functions $\ell_1,\dots,\ell_Q$ satisfying $\ell_q(x_p)=0$, $p<q$, and $\operatorname{span}\{\ell_1,\dots,\ell_Q\}=\mathbb{W}_Q$, $q\le Q$, in the following way. Let $\ell_1=w_1$ and $x_1\in\Omega_{\mathrm{x}}$ be a maximum of $|\ell_1|$. Assume that $\ell_{Q-1}$ has already been constructed. For the construction of $\ell_Q$ define $\ell_{Q,0}:=w_Q$ and

$$\ell_{Q,q}:=\ell_{Q,q-1}-\ell_{Q,q-1}(x_q)\frac{\ell_q}{\ell_q(x_q)},\quad q=1,\dots,Q-1.$$

Then $\ell_{Q,Q-1}(x_q)=0$, $q<Q$, and $\operatorname{span}\{\ell_{Q,0},\dots,\ell_{Q,Q-1}\}=\operatorname{span}\{\ell_1,\dots,\ell_{Q-1},w_Q\}$. Hence, we set $\ell_Q:=\ell_{Q,Q-1}$ and choose

$$x_Q:=\arg\sup_{x\in\Omega_{\mathrm{x}}}|\ell_Q(x)|. \tag{3.7}$$

The previous construction guarantees unisolvency at the nodes $x_q$, $q=1,\dots,Q$.

**Lemma 3.1** *It holds that* $\det\mathrm{W}_Q\neq 0$.

*Proof* Since $\operatorname{span}\{\ell_1,\dots,\ell_Q\}=\operatorname{span}\{w_1,\dots,w_Q\}$ it follows that there is a non-singular matrix $\mathrm{T}\in\mathbb{R}^{Q\times Q}$ such that

$$\begin{bmatrix}\ell_1\\ \vdots\\ \ell_Q\end{bmatrix}=\mathrm{T}\begin{bmatrix}w_1\\ \vdots\\ w_Q\end{bmatrix}.$$

Hence, $\mathrm{R}_Q=\mathrm{T}\,\mathrm{W}_Q$ where $\mathrm{R}_Q:=[\ell_i(x_j)]_{i,j=1}^{Q}$ is upper triangular. The assertion follows from

$$\det\mathrm{R}_Q=\ell_1(x_1)\cdot\ldots\cdot\ell_Q(x_Q)\neq 0.$$

As an example, we choose $\mathbb{W}_Q=\Pi_{Q-1}$ the space of polynomials of degree at most $Q-1$. Then, it follows from (3.6) that ACA converges if, e.g., $f$ is analytic with respect to $x$, and the speed of convergence is determined by the decay of $f$'s derivatives or the elliptical radius of the ellipse in which $f$ has a holomorphic extension. Furthermore, it can be seen that

$$\ell_Q(x)=\prod_{q=1}^{Q-1}(x-x_q).$$

Hence, the choice (3.7) of $x_Q$ is a generalization of a construction that is due to Leja [49]. Leja recursively defines a sequence of nodes $\{x_1,\dots,x_Q\}$ for polynomial interpolation in a compact set $K\subset\mathbb{C}$ as follows. Let $x_1\in K$ be arbitrary. Once $x_1,\dots,x_{Q-1}$ have been found, choose $x_Q\in K$ so that

$$\prod_{q=1}^{Q-1}|x_Q-x_q|=\max_{x\in K}\prod_{q=1}^{Q-1}|x-x_q|.$$

In [68] it is proved that Lebesgue constants associated with Leja points are subexponential for fairly general compact sets in $\mathbb{C}$; see also [65]. Hence, analyticity is required in general for the convergence of the interpolation process.

The expression $\sigma_2[f]$ on the right-hand side of (3.6) can be controlled by the choice of the points $y_1,\ldots,y_Q \in \Omega_y$. Due to Laplace's theorem

$$\left( \mathrm{M}_Q^{-1} \begin{bmatrix} f(x_1,y) \\ \vdots \\ f(x_Q,y) \end{bmatrix} \right)_q = \frac{\det \mathrm{M}_q(y)}{\det \mathrm{M}_Q}, \quad q=1,\ldots,Q,$$

where $\mathrm{M}_q(y)$ arises from replacing the $q$-th column of $\mathrm{M}_Q$ by the vector $[f(x_1,y),\ldots,f(x_Q,y)]^T$, we obtain that $\sigma_2[f] \le Q$ if $y_1,\ldots,y_Q$ are chosen such that

$$|\det \mathrm{M}_Q| \ge |\det \mathrm{M}_q(y)|, \quad q=1,\ldots,Q,\ y \in \Omega_y. \tag{3.8}$$

In connection with the so-called *maximum volume condition* (3.8), we also refer to the error estimates in [66] which are based on the technique of *exact annihilators* (see [2, 3]) in order to provide similar results as (3.6).

### 3.3.3 Incremental Construction

The maximum volume condition (3.8) is difficult to satisfy by an a-priori choice of $y_1,\ldots,y_Q$. Therefore, the following incremental construction of approximations (3.5), which is called *Adaptive Cross Approximation* (ACA) [6], has turned out to be practically more relevant. Let $r_0(x,y) := f(x,y)$ and define the sequence of remainders as

$$r_q(x,y) := r_{q-1}(x,y) - \frac{r_{q-1}(x,y_q)\, r_{q-1}(x_q,y)}{r_{q-1}(x_q,y_q)}, \quad q=1,\ldots,Q, \tag{3.9}$$

where $x_q$ and $y_q$ are chosen such that $r_{q-1}(x_q,y_q) \neq 0$. Then, the algorithm is summarized in Table 3.2.

Since $r_{q-1}(x_q,y_q)$ coincides with the $q$-th diagonal entry of the upper triangular factor of the LU decomposition of $\mathrm{M}_Q$, we obtain that $\det \mathrm{M}_Q \neq 0$. In [12], it is shown that

$$f(x,y) = \mathfrak{I}_Q[f_y](x) + r_Q(x,y) \tag{3.10}$$

and

$$\mathfrak{I}_Q[f_y](x) = \sum_{q=1}^{Q} r_{q-1}(x,y_q) \frac{r_{q-1}(x_q,y)}{r_{q-1}(x_q,y_q)}.$$

This method is used in [21] (see also [23]) under the name *Geddes-Newton series expansion* for the numerical integration of bivariate functions, where instead of the maximum volume condition (3.8) $(x_q,y_q)$ is found from maximizing $|r_{q-1}|$. This choice of $(x_q,y_q)$ is usually referred to as *global pivoting*. Another pivoting strategy

**Scheme 3.2.** Bivariate Adaptive Cross Approximation (ACA2)

Set $q := 1$.
While $\texttt{err} < \texttt{tol}$

a. Define the remainder $r_{q-1} = f - \sum_{i=1}^{q-1} c_i$ and choose $(x_q, y_q) \in \Omega_\mathrm{x} \times \Omega_\mathrm{y}$ such that

$$r_{q-1}(x_q, y_q) \neq 0.$$

b. Define the next tensor product by

$$c_q(x,y) = \frac{r_{q-1}(x, y_q)\, r_{q-1}(x_q, y)}{r_{q-1}(x_q, y_q)}.$$

c. Define the error level by

$$\texttt{err} = \|r_{q-1}\|_{L^\infty(\Omega_\mathrm{x} \times \Omega_\mathrm{y})}$$

and set $q := q+1$.

is the so-called *partial pivoting*, i.e., $y_q$ is chosen in the $q$-th step such that

$$|r_{q-1}(x_q, y_q)| \geq |r_{q-1}(x_q, y)| \text{ for all } y \in \Omega_\mathrm{y}$$

for $x_q \in \Omega_\mathrm{x}$ chosen by (3.7). For the latter condition (and in particular for the stronger global pivoting) the conservative bound $\sigma_2[f] \leq 2^Q - 1$ can be guaranteed; see [6]. The actual growth of $\sigma_2[f]$ with respect to $Q$ is, however, typically significantly weaker.

### *3.3.4 Application to Matrices*

Approximations of the form (3.5) are particularly useful when they are applied to large-scale matrices $\mathrm{A} \in \mathbb{R}^{M \times N}$. In this case, (3.5) becomes

$$\mathrm{A} \approx \tilde{\mathrm{A}} := \mathrm{A}_{:,\sigma}\, \mathrm{A}_{\tau,\sigma}^{-1}\, \mathrm{A}_{\tau,:}, \tag{3.11}$$

where $\tau := \{i_1, \ldots, i_Q\}$ and $\sigma := \{j_1, \ldots, j_Q\}$ are sets of row and column indices, respectively, such that $\mathrm{A}_{\tau,\sigma} \in \mathbb{R}^{Q \times Q}$ is invertible. Here and in the following, we use the notation $\mathrm{A}_{\tau,:}$ for the rows $\tau$ and $\mathrm{A}_{:,\sigma}$ for the columns $\sigma$ of A. Notice that the approximation $\tilde{\mathrm{A}}$ has rank at most $Q$ and is constructed from few of the original matrix entries. Such kind of approximations were investigated by Eisenstat and Gu [37] and Tyrtyshnikov et al. [35] in the context of the maximum volume condition. Again, the approximation can be constructed incrementally by the sequence of remainders $\mathrm{R}^{(0)} := \mathrm{A}$ and

$$\mathrm{R}^{(q)} := \mathrm{R}^{(q-1)} - \frac{\mathrm{R}^{(q-1)}_{:,j_q}\, \mathrm{R}^{(q-1)}_{i_q,:}}{\mathrm{R}^{(q-1)}_{i_q,j_q}}, \quad q = 1, \ldots, Q,$$

where the index pair $(i_q, j_q)$ is chosen such that $R^{(q-1)}_{i_q j_q} \neq 0$. The previous condition guarantees that $A_{\tau,\sigma}$ is invertible, and we obtain

$$\tilde{A} = \sum_{q=1}^{Q} \frac{R^{(q-1)}_{:,j_q} R^{(q-1)}_{i_q,:}}{R^{(q-1)}_{i_q,j_q}}.$$

If A arises from evaluating a smooth function at given points, then $R^{(q)}$ can be estimated using (3.6).

In order to avoid the computation of each entry of the remainders $R^{(q)}$, it is important to notice that only the entries in the $i_q$-th row and the $j_q$-th column of $R^{(q-1)}$ are required for the construction of $\tilde{A}$. Therefore, the following algorithm computes the column vectors $u_q := R^{(q-1)}_{:,j_q}$ and row vectors $v_q := R^{(q-1)}_{i_q,:}$ resulting in

$$\tilde{A} = \sum_{q=1}^{Q} \frac{u_q v_q^T}{(v_q)_{j_q}}. \tag{3.12}$$

The iteration stops after $Q$ steps if the error satisfies

$$\|A - \tilde{A}\|_{\ell^2} = \|R^{(Q)}\|_{\ell^2} < \varepsilon \tag{3.13}$$

with given accuracy $\varepsilon > 0$. The previous condition cannot be evaluated with linear complexity. Since the next rank-1 term $(v_{Q+1})^{-1}_{j_{Q+1}} u_{Q+1} v^T_{Q+1}$ approximates $R^{(Q)}$, we replace (3.13) with the error indicator

$$\frac{\|u_{Q+1} v^T_{Q+1}\|_{\ell^2}}{|(v_{Q+1})_{j_{Q+1}}|} = \frac{\|u_{Q+1}\|_{\ell^2} \|v_{Q+1}\|_{\ell^2}}{|(v_{Q+1})_{j_{Q+1}}|} < \varepsilon.$$

The algorithm is presented in Table 3.3.

*Remark 3.3* Notice that almost no condition has been imposed on the row index $i_q$. The following three methods are commonly used to choose $i_q$. In addition to choosing $i_q$ randomly, $i_q$ can be found as

$$i_q := \underset{i=1,\dots,M}{\arg\max} |(u_{q-1})_i|,$$

which leads to a cyclic pivoting strategy. If A stems from the evaluation of a function at given nodes, then the construction of Sect. 3.3.2 should be used in order to guarantee the well-posedness of the interpolation operator $\mathfrak{I}'_Q$ and exploit the error estimate (3.6).

In some cases (see [15]), it is required to put more effort in the choice of $i_q$ to guarantee a well-suited approximation space $\mathrm{span}\{A_{i_1,:}, \dots, A_{i_Q,:}\}$; cf. [7].

Instead of the $M \cdot N$ entries of A, we only have to compute $Q(M+N)$ entries of A for the approximation by $\tilde{A}$. The construction of (3.12) requires $\mathcal{O}(Q^2(M+N))$ arithmetic operations, and $\tilde{A}$ can be stored with $Q(M+N)$ units of storage.

**Scheme 3.3.** Adaptive Cross Matrix Approximation

Set $q := 1$.
While $\texttt{err} < \texttt{tol}$

a. Choose $i_q$ such that

$$v_q := A^T_{i_q,:} - \sum_{\ell=1}^{q-1} \frac{(u_\ell)_{i_q}}{(v_\ell)_{j_\ell}} v_\ell$$

is nonzero and $j_q$ such that $|(v_q)_{j_q}| = \max_{j=1,\dots,N} |(v_q)_j|$.

b. Compute the vector

$$u_q := A_{:,j_q} - \sum_{\ell=1}^{q-1} \frac{(v_\ell)_{j_q}}{(v_\ell)_{j_\ell}} u_\ell.$$

c. Compute the error indicator

$$\texttt{err} = |(v_q)_{j_q}|^{-1} \|u_q\|_{\ell^2} \|v_q\|_{\ell^2}$$

and set $q := q+1$.

Possible redundancies among the vectors $u_q$, $v_q$, $q = 1,\dots,Q$, can be removed via orthogonalization.

The origin of this matrix version of ACA is the construction of so-called hierarchical matrices [7,39,40] for the efficient treatment of integral formulations of elliptic boundary value problems. Hierarchical matrices allow to treat discretizations of such non-local operators with logarithmic-linear complexity. To this end, subblocks $A_{t,s}$ from a suitable partition of large-scale matrices A are approximated by low-rank matrices.

A form that is slightly different from (3.11) and which looks more complicated at first glance is

$$A_{t,s} \approx \hat{A}_{t,s} := A_{:,\sigma_t} A^{-1}_{\tau_t,\sigma_t} A_{\tau_t,\sigma_s} A^{-1}_{\tau_s,\sigma_s} A_{\tau_s,:}$$

with suitable index sets $\tau_t$, $\sigma_t$, $\tau_s$, and $\sigma_s$ depending on the respective index $t$ or $s$ only. Notice that in contrast to $\tilde{A}$, $\hat{A}$ does not interpolate A on the "cross" but rather at single points specified by the indices $\tau_t, \sigma_s$, i.e. $\hat{A}_{\tau_t,\sigma_s} = A_{\tau_t,\sigma_s}$. The advantage of this approach is the fact that the large parts $A_{:\sigma_t} A^{-1}_{\tau_t,\sigma_t}$ and $A^{-1}_{\tau_s,\sigma_s} A_{\tau_s,:}$ depend only on either one of the two index sets $t$ or $s$, while only the small matrix $A_{\tau_t,\sigma_s}$ depends on both. This allows to further reduce the complexity of hierarchical matrix approximations by constructing so-called nested bases approximations [13], which are mandatory to efficiently treat high-frequency Helmholtz problems; see [11].

### *3.3.5 Relation with Gaussian Elimination*

Without loss of generality, we may assume for the moment that $i_q = j_q = q$, $q = 1,\dots,Q$. Otherwise, interchange the rows and columns of the original matrix $R^{(0)}$.

Then

$$\mathrm{R}^{(q)} = \left(\mathrm{I} - \frac{\mathrm{R}^{(q-1)}\mathrm{e}_q\mathrm{e}_q^T}{\mathrm{e}_q^T\mathrm{R}^{(q-1)}\mathrm{e}_q}\right)\mathrm{R}^{(q-1)} = \mathrm{L}^{(q)}\mathrm{R}^{(q-1)},$$

where $\mathrm{L}^{(q)} \in \mathbb{R}^{M\times N}$ is the matrix

$$\mathrm{L}^{(q)} = \begin{bmatrix} 1 & & & & & \\ & \ddots & & & & \\ & & 1 & & & \\ & & & 0 & & \\ & & & -\frac{\mathrm{e}_{q+1}^T\mathrm{R}^{(q-1)}\mathrm{e}_q}{\mathrm{e}_q^T\mathrm{R}^{(q-1)}\mathrm{e}_q} & 1 & \\ & & & \vdots & & \ddots \\ & & & -\frac{\mathrm{e}_M^T\mathrm{R}^{(q-1)}\mathrm{e}_q}{\mathrm{e}_q^T\mathrm{R}^{(q-1)}\mathrm{e}_q} & & & 1 \end{bmatrix},$$

which differs from a Gaussian matrix only in the position $(q,q)$; cf. [6]. This relation was exploited in [41] for the convergence analysis of ACA in the case of positive definite matrices A.

Furthermore, it is an interesting observation that ACA reduces the rank of the remainder in each step, i.e. $\operatorname{rank}\mathrm{R}^{(q)} = \operatorname{rank}\mathrm{R}^{(q-1)} - 1$. This was first discovered by Wedderburn in [69, p. 69]; see also [6, 26]. Hence, ACA may be regarded as a rank revealing LU factorization [22,45]. As we know, it is possible that the elements grow in the LU decomposition algorithm; cf. [34]. Thus the exponential bound $2^Q$ on $\sigma_2[f]$ is not a result of overestimation.

### *3.3.6 Generalizations of ACA*

The Adaptive Cross Approximation can easily be generalized to a linear functional setting. Instead the evaluation of the remainders at the chosen points $x_q$, $y_q$, $q = 1,\dots,Q$, one considers the recursive construction

$$r_q(x,y) := r_{q-1}(x,y) - \frac{\langle r_{q-1}(x,\cdot),\psi_q\rangle\,\langle\varphi_q, r_{q-1}(\cdot,y)\rangle}{\langle\varphi_q, r_{q-1},\psi_q\rangle}, \quad q = 1,\dots,Q.$$

Here, $\varphi_q$ and $\psi_q$ denote given linear functionals acting on $x$ and $y$, respectively. It is easy to show (see [10]) that

$$\langle\varphi_i, r_q(\cdot,y)\rangle = 0 = \langle r_q(x,\cdot),\psi_i\rangle \quad \text{for all } i \le q, x \in \Omega_{\mathrm{x}} \text{ and } y \in \Omega_{\mathrm{y}}. \tag{3.14}$$

Hence, $r_q$ vanishes for an increasing number of functionals and

$$\mathfrak{I}''_Q[f_y](x) := \sum_{q=1}^{Q}\langle r_{q-1}(x,\cdot),\psi_q\rangle\frac{\langle\varphi_q, r_{q-1}(\cdot,y)\rangle}{\langle\varphi_q, r_{q-1},\psi_q\rangle}$$

gradually interpolates $f_y$ (in the sense of functionals). The Adaptive Cross Approximation (3.9) is obtained from choosing the Dirac functionals $\varphi_q := \delta_{x_q}$ and $\psi_q := \delta_{y_q}$.

The benefits of the separation of variables resulting form (3.5) are even more important for multivariate functions $f$. We present two ways to generalize (3.9) to functions depending on $d$ variables. An obvious idea is to group the set of variables into two parts each containing $d/2$ variables; see [10] for a method that uses the covariance of $f$ to construct this separation. Each of the two parts can be treated as a single new variable. Then, the application of (3.9) results in a sequence of less-dimensional functions which inherit the smoothness of $f$. Hence, (3.9) can be applied again until only univariate functions are left. Due the nestedness of the construction, the constructed approximation cannot be regarded as an interpolation. Error estimates for this approximation were derived in [8] for $d = 3,4$. The application to tensors of order $d > 2$ was presented in [4, 60, 61].

A more sophisticated way to generalize ACA to multivariate functions is presented in [9]. For the case $d = 3$, the sequence of remainders is constructed as

$$r_q(x,y,z) := r_{q-1}(x,y,z) - \frac{r_{q-1}(x,y,z_q)\, r_{q-1}(x,y_q,z)\, r_{q-1}(x_q,y,z)\, r_{q-1}(x_q,y_q,z_q)}{r_{q-1}(x,y_q,z_q)\, r_{q-1}(x_q,y,z_q)\, r_{q-1}(x_q,y_q,z)}$$

instead of (3.9). Notice that this kind of approximation requires that $x_q, y_q, z_q$ can be found such that the denominator $r_{q-1}(x,y_q,z_q)\, r_{q-1}(x_q,y,z_q)\, r_{q-1}(x_q,y_q,z) \neq 0$. On the other hand, the advantage of this generalization is that it is equi-directional in contrast to the aforementioned idea, i.e., none of the variables is preferred to the others. Hence, similar to (3.14) we obtain for all $x,y,z$

$$r_q(x,y,z_i) = r_q(x,y_i,z) = r_q(x_i,y,z) = 0, \quad i \leq q.$$

## 3.4 Empirical Interpolation Method

### *3.4.1 Historical Overview*

The Empirical Interpolation Method (EIM) [5] originates from reduced order modeling and its application to the resolution of parameter dependent partial differential equations. We are thus in the context where the set of solutions $u(\cdot,y)$ to the PDE generates a manifold, parametrized by $y$ (the parameter is generally called $\mu$ in these applications) that possesses a small Kolmogorov $n$-width. In the construction stage of the reduced basis method, the reduced basis is constructed from a greedy approach where each new basis function, that is a solution to the PDE associated to an optimally chosen parameter, is incorporated recursively. The selection criteria of the parameter is based on maximal (a posteriori) error estimates over the parameter space. This construction stage can be expensive: indeed it requires an initial accurate classical discretization method of finite element, spectral or finite volume type and every solution associated to a parameter that is optimally selected, needs to be approximated during this stage by the classical method. Once the preliminary stage is performed off-line, all the approximations of solutions corresponding to a new parameter are performed as a linear combination of the (few) basis functions constructed during the first phase. This second on-line stage is very cheap. This is due to two facts. The first one is related to the fact that the greedy approach is proven to be quite optimal [14, 16, 28], for exponential or polynomial decay of the Kolmogorov $n$-width, the greedy method provides a basis set that has the same feature.

The second fact is related to the approximation process. A Galerkin approximation in this reduced space indeed provides very good approximations, and if $Q$ modes are used, a linear PDE can be simulated by inverting $Q \times Q$ matrices only, i.e. much smaller complexity than the classical approaches.

In order that the same remains true for nonlinear PDE's, a strategy, similar to the pseudo-spectral approximation for high-order Fourier or polynomial approximations has been sought. This involves the use of an interpolation operator. In order to be coherent, an approximation $u_Q(\cdot,y) = \sum_{i=1}^{Q} \alpha_i(y)\, u(\cdot,y_i)$ being given (where the $y_i$ are the parameters that define the reduced basis snapshots) we want to approximate $\mathscr{G}(u_Q(\cdot,y))$ ($\mathscr{G}$ being a nonlinear functional) as a linear combination

$$\mathscr{G}(u_Q(\cdot,y)) \approx \sum_{i=1}^{Q} \beta_i(y)\mathscr{G}(u(\cdot,y_i)).$$

The derivation of the set $\{\beta_i\}_i$ from $\{\alpha_i\}_i$ needs to be very fast, it is defined by interpolation through the Empirical Interpolation Method defined in the following section. This has been extensively used for different types of equations in [36] and has led to the definition of general interpolation techniques and rapid derivation of the associated points.

The approach having a broader scope than only the use in reduced basis approximation, a dedicated analysis of the approximation properties for sets with small-Kolmogorov $n$-width has been presented in [54]. This approach for nonlinear problems has actually also been used for problems where the dependency in the parameter is involved (the so called "non-affine problems") and has boosted the domain of application of reduced order approximations.

### *3.4.2 Motivation*

As said above and in the introduction, we are in a situation where the set $\mathscr{F} = \{f(\cdot,y)\}_{y\in\Omega_\mathrm{y}}$ denotes a family of parametrized functions with small Kolmogorov $n$-width. We therefore do not identify $\Omega_\mathrm{x}$ with $\Omega_\mathrm{y}$. In addition, for a given parameter $y$, $f(\cdot,y)$ is supposed to be accessible at all values in $\Omega_\mathrm{x}$.

The EIM is designed to find approximations to members of $\mathscr{F}$ through an interpolation operator $I_q$ that interpolates the function $f_y = f(\cdot,y)$ at some particular points in $\Omega_\mathrm{x}$. That is, given an interpolatory system defined by a set of basis functions $\{h_1,\dots,h_q\}$ (linear combination of particular "snapshots" $f_{y_1},\dots,f_{y_q}$) and interpolation points $\{x_1,\dots,x_q\}$, the interpolant $I_q[f_y]$ of $f_y$ with $y \in \Omega_\mathrm{y}$ written as

$$I_q[f_y](x) = \sum_{j=1}^{q} g_j(y)h_j(x), \quad x \in \Omega_\mathrm{x}, \tag{3.15}$$

**Scheme 3.4.** Empirical Interpolation Method

Set $q = 1$. Do while $\texttt{err} < \texttt{tol}$:

a. Pick the sample point

$$y_q = \arg\sup_{y\in\Omega_y} \|f_y - I_{q-1}[f_y]\|_{L^p(\Omega_x)}, \tag{3.18}$$

and the corresponding interpolation point

$$x_q = \arg\sup_{x\in\Omega_x} |f_{y_q}(x) - I_{q-1}[f_{y_q}](x)|. \tag{3.19}$$

b. Define the next basis function as

$$h_q = \frac{f_{y_q} - I_{q-1}[f_{y_q}]}{f_{y_q}(x_q) - I_{q-1}[f_{y_q}](x_q)}. \tag{3.20}$$

c. Define the error level by

$$\texttt{err} = \|\texttt{err}_p\|_{L^\infty(\Omega_y)} \quad \text{with} \quad \texttt{err}_p(y) = \|f_y - I_{q-1}[f_y]\|_{L^p(\Omega_x)},$$

and set $q := q + 1$.

is defined by

$$I_q[f_y](x_i) = f_y(x_i), \quad i = 1, \ldots, q. \tag{3.16}$$

Thus, (3.16) is equivalent to the following linear system

$$\sum_{j=1}^{q} g_j(y) h_j(x_i) = f_y(x_i), \quad i = 1, \ldots, q. \tag{3.17}$$

One of the problems is to ensure that the system above is uniquely solvable, i.e. that the matrix $(h_j(x_i))_{i,j}$ is invertible, which will be considered in the design of the interpolation scheme.

### *3.4.3 Algorithm*

The construction of the basis functions and interpolation points is based on a greedy algorithm. Note that the EIM is defined with respect to a given norm on $\Omega_x$ and we consider here $L^p(\Omega_x)$-norms for $1 \le p \le \infty$. The algorithm is given in Table 3.4.

*Remark 3.4* Note that whenever $\dim(\text{span}\{\mathscr{F}\}) = q^\star$, the algorithm finishes for $q = q^\star$.

As long as $q \le q^\star$, note that the basis functions $\{h_1, \ldots, h_q\}$ and the snapshots $\{f_{y_1}, \ldots, f_{y_q}\}$ span the same space, i.e.,

$$\mathbb{V}_q = \text{span}\{h_1, \ldots, h_q\} = \text{span}\{f_{y_1}, \ldots, f_{y_q}\}.$$

The former are preferred to the latter due to the following properties

$$h_i(x_i) = 1, \quad \forall i = 1,\dots,q \qquad \text{and} \quad h_j(x_i) = 0, \quad 1 \le i < j \le q. \tag{3.21}$$

*Remark 3.5* It is easy to show that the interpolation operator $I_q$ is the identity if restricted to the space $\mathbb{V}_q$, i.e.,

$$I_q[f_{y_i}](x) = f_{y_i}(x), \quad i = 1,\dots,q, \quad x \in \Omega_{\mathrm{x}}.$$

*Remark 3.6* The construction of the interpolating functions and the associated interpolation points follows a greedy approach: we add the function in $\mathscr{F}$ that is the worse approximated by the current interpolation operator and the interpolation point is where the error is the largest. The construction is thus recursive which, in turn, means that it is of low computational cost.

*Remark 3.7* As explained in [5], the algorithm can be reduced to the selection of the interpolation points only, in the case where the family of interpolating functions $\{f_{y_1},\dots,f_{y_q},\dots\}$ is preexisting. This can be the case for instance if a POD strategy has been used previously or when one considers a set that has a canonical basis and ordering (like the set of polynomials).

Note that solving the interpolation system (3.17) can be written as a linear system $\mathrm{B}\,\mathrm{g}_y = \mathrm{f}_y$ with $q$ unknowns and equations where

$$\mathrm{B}_{i,j} = h_j(x_i), \quad (\mathrm{f}_y)_i = f_y(x_i), \quad i,j = 1,\dots,q,$$

such that the interpolant is defined by

$$I_q[f_y](x) = \sum_{j=1}^{q} (\mathrm{g}_y)_j h_j(x), \quad x \in \Omega_{\mathrm{x}}.$$

This construction of the basis functions and interpolation points satisfies the following theoretical properties (see [5]):

- the basis functions $\{h_1,\dots,h_q\}$ consist of linearly independent functions;
- the interpolation matrix $\mathrm{B}_{i,j}$ is lower triangular with unity diagonal by (3.21) and hence invertible, the remaining entries belong to $[-1,1]$;
- the empirical interpolation procedure is well-posed in $L^p(\Omega_{\mathrm{x}})$, as long as $q \le q^\star$.

If the $L^\infty(\Omega_{\mathrm{x}})$-norm ($p = \infty$) is considered, the error analysis of the interpolation procedure classically involves the Lebesgue constant $\Lambda_q = \sup_{x\in\Omega_{\mathrm{x}}} \sum_{i=1}^{q} |L_i(x)|$ where $L_i \in \mathbb{V}_q$ are the Lagrange functions satisfying $L_i(x_j) = \delta_{ij}$. The following bound holds [5]

$$\|f_y - I_q[f_y]\|_{L^\infty(\Omega_{\mathrm{x}})} \le (1+\Lambda_q) \inf_{v_q\in\mathbb{V}_q} \|f_y - v_q]\|_{L^\infty(\Omega_{\mathrm{x}})}.$$

An (in practise very pessimistic) upper bound (cf. [54]) of the Lebesque constant is given by

$$\Lambda_q \le 2^q - 1,$$

which in turn results in the following estimate. Assume that $\mathscr{F} \subset \mathscr{X} \subset L^\infty(\Omega_\mathrm{x})$ and that there exists a sequence of finite dimensional spaces

$$\mathbb{Z}_1 \subset \mathbb{Z}_2 \subset \ldots, \quad \dim(\mathbb{Z}_q) = q, \quad \text{and} \quad \mathbb{Z}_q \subset \mathscr{F},$$

such that there exists $c > 0$ and $\alpha > \log(4)$ with

$$\inf_{v_q \in \mathbb{Z}_q} \|f_y - v_q\|_{\mathscr{X}} \leq ce^{-\alpha q}, \quad y \in \Omega_\mathrm{y},$$

then

$$\|f_y - I_q[f_y]\|_{L^\infty(\Omega_\mathrm{x})} \leq ce^{-(\alpha - \log(4))q}.$$

*Remark 3.8* The worst-case situation where the Lebesgue constant scales indeed like $\Lambda_q \leq 2^q - 1$ is rather artificial and in all implementations we have done so far involving functions belonging to some reasonable set with small Kolmogorov $n$-width, the growth of the Lebesgue constant is much more reasonable and in most of the times a linear growth is observed. Note that, the points that are generated by the EIM using polynomial basis functions (in increasing order of degree) on $[-1,1]$ are exactly the Leja points as indicated in the frame of the EIM by A. Chkifa[1] and the discussion in Sect. 3.3.2 in the case of ACA. On the other hand, if one considers the Leja points on a unit circle and then project them onto the interval $[-1,1]$ a linear growth is shown in [25].

### 3.4.4 Practical Implementation

In the practical implementation of the EIM one encounters the following problem. Finding the supremum respectively the arg sup in (3.18) and (3.19) is not feasible if any kind of approximation is effected. The least difficult way, but not the only one, is to consider representative point-sets $\Omega_\mathrm{x}^{\mathtt{train}} = \{\hat{x}_1, \hat{x}_2, \ldots, \hat{x}_M\}$ of $\Omega_\mathrm{x}$ and $\Omega_\mathrm{y}^{\mathtt{train}} = \{\hat{y}_1, \hat{y}_2, \ldots, \hat{y}_N\}$ of $\Omega_\mathrm{y}$. Then, the EIM is written as in Table 3.5.

This possible implementation of the EIM is sometimes referred to as the Discrete Empirical Interpolation Method (DEIM) [24].

*Remark 3.9* Different strategies have been reported in [38,55] to successively enrich the training set $\Omega_\mathrm{y}^{\mathtt{train}}$. The main idea is to start with a small number of training points and enrich the set during the iterations of the algorithm and obtain a very fine discretization only towards the end of the algorithm. One can also think of enriching the training set $\Omega_\mathrm{x}^{\mathtt{train}}$ simultaneously.

*Remark 3.10* Using representative pointsets $\Omega_\mathrm{x}^{\mathtt{train}}$ and $\Omega_\mathrm{y}^{\mathtt{train}}$ is only one way to discretize the problem. Alternatively, one can think of using optimization methods to find the maximum over $\Omega_\mathrm{x}$ and $\Omega_\mathrm{y}$. Such a strategy has been reported in [18, 19] in the context of the reduced basis method, which, as well as the EIM, is based on a greedy algorithm.

[1] personal communication.

**Scheme 3.5.** Empirical Interpolation Method (possible implementation of EIM)

Set $q = 1$. Do while $\texttt{err} < \texttt{tol}$:

a. Pick the sample point

$$y_q = \underset{y \in \Omega_y^{\texttt{train}}}{\arg\max} \|f_y - I_{q-1}[f_y]\|_{L^p(\Omega_x)}, \tag{3.22}$$

and the corresponding interpolation point

$$x_q = \underset{x \in \Omega_x^{\texttt{train}}}{\arg\max} |f_{y_q}(x) - I_{q-1}[f_{y_q}](x)|.$$

b. Define the next basis function as

$$h_q = \frac{f_{y_q} - I_{q-1}[f_{y_q}]}{f_{y_q}(x_q) - I_{q-1}[f_{y_q}](x_q)}.$$

c. Define the error level by

$$\texttt{err} = \|\texttt{err}_p\|_{L^\infty(\Omega_y)} \quad \text{with} \quad \texttt{err}_p(y) = \|f_y - I_{q-1}[f_y]\|_{L^p(\Omega_x)}$$

and set $q := q + 1$.

### 3.4.5 Practical Implementation Using the Matrix Representation of the Function

One can define an implementation of the EIM in a completely discrete setting using the representative matrix of $f$ defined by $\mathrm{M}_{i,j} = f(x_i, y_j)$ for $1 \le i \le M$ and $1 \le j \le N$. For the sake of short notation we recall the notation $\mathrm{M}_{:,j}$ used for the $j$-th column of M.

Assume that we are given a set of basis vectors $\{\mathrm{h}_1, \ldots, \mathrm{h}_q\}$ and interpolation indices $i_1, \ldots, i_q$, the discrete interpolation operator $\mathrm{I}_q : \mathbb{R}^N \to \mathbb{R}^N$ of column vectors is given in the span of the basis vectors $\{\mathrm{h}_j\}_{j=1}^q$, i.e. by $\mathrm{I}_q[\mathrm{r}] = \sum_{j=1}^q g_j(\mathrm{r})\mathrm{h}_j$ for some scalars $g_j(\mathrm{r})$, such that

$$(\mathrm{I}_q[\mathrm{r}])_{i_k} = \sum_{j=1}^{q} g_j(\mathrm{r})\,(\mathrm{h}_j)_{i_k} = \mathrm{r}_{i_k}, \quad \mathrm{r} \in \mathbb{R}^N, \quad k = 1, \ldots, q.$$

Using this notation, we then present the matrix version of the EIM in Table 3.6.

This procedure allows to define an approximation of any coefficient of the matrix M. In some cases however, one would like to obtain an approximation of $f(x,y)$ for *any* $(x,y) \in \Omega_x \times \Omega_y$. After running the implementation, one can still construct the continuous interpolant $I_Q[f](x,y)$ for any $(x,y) \in \Omega_x \times \Omega_y$. Indeed, the interpolation points $x_1, \ldots, x_Q$ are provided by $x_q = \hat{x}_{i_q}$. The construction of the (continuous) basis functions $h_q$ is based on mimicking part b of the discrete algorithm but in a continuous context. Therefore, during the discrete version one saves the following

**Scheme 3.6.** Empirical Interpolation Method (implementation based on representative matrix M of $f$)

Set $q = 1$. Do while `err` $<$ `tol`

a. Pick the sample index

$$j_q = \underset{j=1,\dots,M}{\arg\max} \|\mathrm{M}_{:,j} - \mathrm{I}_{q-1}[\mathrm{M}_{:,j}]\|_{\ell^p},$$

and the corresponding interpolation index

$$i_q = \underset{i=1,\dots,N}{\arg\max} |\mathrm{M}_{i,j_q} - (\mathrm{I}_{q-1}[\mathrm{M}_{:,j_q}])_i|.$$

b. Define the next approximation column by

$$\mathrm{h}_q = \frac{\mathrm{M}_{:,j_q} - \mathrm{I}_{q-1}[\mathrm{M}_{:,j_q}]}{\mathrm{M}_{i_q,j_q} - (\mathrm{I}_{q-1}[\mathrm{M}_{:,j_q}])_{i_q}}.$$

c. Define the error level by

$$\mathtt{err} = \max_{j=1,\dots,M} \|\mathrm{M}_{:,j} - \mathrm{I}_{q-1}[\mathrm{M}_{:,j}]\|_{\ell^p}$$

and set $q := q+1$.

data

$$s_{q,j} = g_j(\mathrm{M}_{:,j_q}), \qquad \text{from} \quad \mathrm{I}_{q-1}[\mathrm{M}_{:,j_q}] = \sum_{j=1}^{q-1} g_j(\mathrm{M}_{:,j_q})\mathrm{h}_j,$$

$$s_{q,q} = \mathrm{M}_{i_q,j_q} - (\mathrm{I}_{q-1}[\mathrm{M}_{:,j_q}])_{i_q}.$$

Then, the continuous basis functions can be recovered by the following recursive formula

$$h_q = \frac{f_{y_q} - \sum_{j=1}^{q-1} s_{q,j} h_j}{s_{q,q}}$$

using the notation $y_q = \hat{y}_{i_q}$.

## 3.4.6 Generalizations of the EIM

In the following, we present some generalizations of the core concept behind the EIM.

### 3.4.6.1 Generalized Empirical Interpolation Method (gEIM)

We have seen that the EIM-interpolation operator $I_q[f_y]$, $y \in \Omega_{\mathrm{y}}$, interpolates the function $f_y$ at some empirically constructed points $x_1, \dots, x_q$. The EIM can be gen-

**Scheme 3.7.** Generalized Empirical Interpolation Method (gEIM)

Set $q = 1$. Do while `err` $<$ `tol`:

a. Pick the sample point

$$y_q = \arg\sup_{y\in\Omega_y} \|f_y - J_{q-1}[f_y]\|_{L^p(\Omega_x)},$$

and the corresponding interpolation moment

$$\sigma_q = \arg\sup_{\sigma\in\Sigma} |\sigma(f_{y_q} - J_{q-1}[f_{y_q}])|.$$

b. Define the next basis function as

$$h_q = \frac{f_{y_q} - J_{q-1}[f_{y_q}]}{\sigma_q(f_{y_q} - J_{q-1}[f_{y_q}])}.$$

c. Define the error level by

$$\texttt{err} = \|\texttt{err}_p\|_{L^\infty(\Omega_y)} \quad \text{with} \quad \texttt{err}_p(y) = \|f_y - I_{q-1}[f_y]\|_{L^p(\Omega_x)}$$

and set $q := q+1$.

eralized in the following sense as proposed in [52]. Let $\Sigma$ be a dictionary of linear continuous forms (say for the $L^2(\Omega_x)$-norm) acting on functions $f_y$, $y \in \Omega_y$. Then, the gEIM consists in providing a set of basis functions $h_1, \ldots, h_q$, such that $\mathbb{V}_q = \text{span}\{h_1, \ldots, h_q\} = \text{span}\{f_{y_1}, \ldots, f_{y_q}\}$ for some empirically chosen $\{y_1, \ldots, y_q\} \subset \Omega_y$, and a set of linear forms, or moments, $\{\sigma_1, \ldots, \sigma_q\} \subset \Sigma$. The generalized interpolant then takes the form

$$J_q[f_y] = \sum_{j=1}^{q} g_j(y) h_j(x), \quad x \in \Omega_x, \quad y \in \Omega_y,$$

and is defined in the following way

$$\sigma_i(J_q[f_y]) = \sigma_i(f_y), \quad i = 1, \ldots, q,$$

which will define the coefficients $g_j(y)$ for each $y \in \Omega_y$. We note that if the linear forms are Dirac functionals $\delta_x$ with $x \in \Omega_x$, then the gEIM reduces to the plain EIM. The algorithm is given in Table 3.7. This constructive algorithm satisfies the following theoretical properties (see [52]):

- the set $\{h_1, \ldots, h_q\}$ consists of linearly independent functions;
- the generalized interpolation matrix $(\mathrm{B})_{ij} = \sigma_i(h_j)$ is lower triangular with unity diagonal (hence invertible) with other entries $s \in [-1, 1]$;
- the generalized empirical interpolation procedure is well-posed in $L^2(\Omega_x)$.

In order to quantify the error of the interpolation procedure, like in the standard interpolation procedure, we introduce the Lebesgue constant in the $L^2$-norm:

$$\Lambda_q = \sup_{y \in \Omega_y} \frac{\|J_q[f_y]\|_{L^2(\Omega_x)}}{\|f_y\|_{L^2(\Omega_x)}},$$

i.e. the $L^2$-operator norm of $J_q$. Thus, the interpolation error satisfies:

$$\|f_y - J_q[f_y]\|_{L^2(\Omega)} \leq (1 + \Lambda_q) \inf_{v_q \in \mathbb{V}_q} \|f_y - v_q\|_{L^2(\Omega)}.$$

Again, a (very pessimistic) upper-bound for $\Lambda_q$ is:

$$\Lambda_q \leq 2^{q-1} \max_{i=1,\ldots,q} \|h_i\|_{L^2(\Omega)},$$

indeed, the Lebesgue constant is, in many cases, uniformly bounded in this generalized case. The following result proves that the greedy construction is quite optimal [53].

1. Assume that the Kolmogorov $n$-width of $\mathscr{F}$ in $L^2(\Omega_x)$ is upper bounded by $C_0 n^{-\alpha}$ for any $n \geq 1$, then the interpolation error of the gEIM greedy selection process satisfies for any $f \in \mathscr{F}$ the inequality $\|f(\cdot,y) - J_Q[f(\cdot,y)]\|_{L^2(\Omega_x)} \leq C_0(1+\Lambda_Q)^3 Q^{-\alpha}$.
2. Assume that the Kolmogorov $n$-width of $\mathscr{F}$ in $L^2(\Omega_x)$ is upper bounded by $C_0 e^{-c_1 n^\alpha}$ for any $n \geq 1$, then the interpolation error of the gEIM greedy selection process satisfies for any $f \in \mathscr{F}$ the inequality $\|f(\cdot,y) - J_Q[f(\cdot,y)]\|_{L^2(\Omega_x)} \leq C_0(1+\Lambda_Q)^3 e^{-c_2 Q^\alpha}$ for a positive constant $c_2$ slightly smaller than $c_1$.

#### 3.4.6.2 *hp*-EIM

If the Kolmogov $n$-width is only decaying slowly with respect to $n$ and the resulting number of basis functions and associated integration points is larger than desired, a remedy consists of partitioning the space $\Omega_y$ into different elements $\Omega_y^1, \ldots, \Omega_y^P$ on which a separate interpolation operator $I_{q_p} : \{f_y\}_{y \in \Omega_x^p} \to \mathbb{V}_{q_p}$ with $p = 1, \ldots, P$ is constructed. That is, for each element $\Omega_y^p$ a standard EIM as described above is performed. The choice of creating the partition is subject to some freedom and different approaches have been presented in [30, 32].

A somewhat different approach is presented in [55], although in the framework of a projection method, where the idea of a strict partition of the space $\Omega_y$ is abandoned. Instead, given a set of sample points $y_1, \ldots, y_K$ for which the basis functions $f(\cdot, y_1), \ldots, f(\cdot, y_K)$ are known (or have been computed) a local approximation space for any $y \in \Omega_y$ is constructed by considering the $N$ basis functions whose parameter values are closest to $y$. In addition, the distance function, measuring the distance between two points in $\Omega_y$, can be empirically built in order to represent local anisotropies in the parameter space $\Omega_y$. Further, the distance function can also be used to define the training set $\Omega_{\mathrm{y}}^{\mathtt{train}}$ which can be uniformly sampled with respect to the problem dependent distance function.

#### *3.4.6.3 Curse of High-Dimensionality*

Several approaches have been presented in cases where $\Omega_{\mathrm{y}}$ is high-dimensional ($\dim(\Omega_{\mathrm{y}}) \approx 10$). In such cases, finding the maximizer in (3.22) becomes a challenge. Since the discrete set $\Omega_{\mathrm{y}}^{\mathtt{train}}$ should be representative of $\Omega_{\mathrm{y}}$, we require that $\Omega_{\mathrm{y}}^{\mathtt{train}}$ consists of a very large number of training points. Finding the maximum over this huge set is therefore prohibitive expensive as a result of the curse of dimensionality.

In [42], the authors propose a computational approach that randomly samples the space $\Omega_{\mathrm{y}}$ with a feasible number of training points, that is however changing over the iterations. Therefore, counting all tested training points over all iterations is still a very large number, at each iteration though finding the maximum is a feasible task.

In [43], the authors use, in the framework of the reduced basis method, an ANOVA expansion based on sparse grid quadrature in order to identify the sensitivity of each dimension in $\Omega_{\mathrm{y}}$. Then, once unimportant dimensions in $\Omega_{\mathrm{y}}$ are identified, the values of the unimportant dimensions are fixed to some reference value and the variation of $y$ in $\Omega_{\mathrm{y}}$ is then restricted to the important dimensions. Finally, a greedy-based algorithm is used to construct a low-order approximation.

## 3.5 Comparison of ACA versus EIM

In the previous sections, we have given independent presentations of the basics of the ACA and the EIM type methods. As was explained, the backgrounds and the applications are different. In addition, we have also presented the results of the convergence analysis of these approximations yielding another fundamental difference between the two approaches. The frame for the convergence of the ACA is a comparison to any other interpolating system, such as the polynomial approximation and the existence of derivatives for the family of functions $f_y, y \in \Omega_{\mathrm{y}}$ is then the reason for convergence. The convergence of the EIM is compared with respect to the $n$-width expressed by the Kolmogorov small dimension.

Nevertheless, despite there differences in origins, it is clear that some link exist between these two constructive approximation methods. We show now the relation between the ACA and the EIM in a particular case.

**Theorem 3.1** *The Bivariate Adaptive Cross Approximation with global pivoting is equivalent to the Empirical Interpolation Method using the $L^\infty(\Omega_{\mathrm{x}})$-norm.*

*Proof* We proceed by induction. Our affirmation $\mathtt{A}_q$ at the $q$-th step is:

$(\mathtt{A}_q)_1$: the interpolation points $\{x_1,\ldots,x_q\}$ and $\{y_1,\ldots,y_q\}$ of the EIM and ACA are identical;

$(\mathtt{A}_q)_2$: $g_q(y) = r_{q-1}(x_q,y), \quad y \in \Omega_{\mathrm{y}}$;

$(\mathtt{A}_q)_3$: $I_q[f_y](x) = \mathfrak{I}_q[f_y](x), \quad (x,y) \in \Omega_{\mathrm{x}} \times \Omega_{\mathrm{y}}$.

*Induction base ($q = 1$):* First, we note that $r_0 = f$ and thus

$$(x_1,y_1) = \underset{(x,y)\in\Omega_x\times\Omega_y}{\arg\sup} |r_0(x,y)| = \underset{(x,y)\in\Omega_x\times\Omega_y}{\arg\sup} |f(x,y)|.$$

Then, from (3.20) we conclude that $h_1(x) = \frac{f(x,y_1)}{f(x_1,y_1)}$ and by (3.17) we obtain that $g_1(y) = \frac{f(x_1,y)}{h_1(x_1)} = f(x_1,y) = r_0(x_1,y)$ since $h_1(x_1) = 1$. Further, using additionally (3.15), we get

$$I_1[f_{y_1}](x) = g_1(y)h_1(x) = r_0(x_1,y)\frac{f(x,y_1)}{f(x_1,y_1)} = \frac{r_0(x_1,y)r_0(x,y_1)}{r_0(x_1,y_1)} = \mathfrak{I}_1[f_y](x),$$

for all $(x,y) \in \Omega_x \times \Omega_y$ and $\mathtt{A}_1$ holds in consequence.

*Induction step ($q > 1$):* Let us assume $\mathtt{A}_{q-1}$ to be true and we first note that

$$r_{q-1}(x,y) = f(x,y) - I_{q-1}[f_y](x) \tag{3.23}$$

by (3.10) and $(\mathtt{A}_{q-1})_3$. Therefore, the selection criteria for the points $(x_q,y_q)$ are identical for the EIM wit $p = \infty$ and the ACA with global pivoting. In consequence, the chosen sample points $(x_q,y_q)$ are identical. Further, combining (3.20) and (3.23) yields

$$h_q(x) = \frac{f_{y_q}(x) - I_{q-1}[f_{y_q}](x)}{f_{y_q}(x_q) - I_{q-1}[f_{y_q}](x_q)} = \frac{r_{q-1}(x,y_q)}{r_{q-1}(x_q,y_q)}. \tag{3.24}$$

By (3.17) for $i = q$, using that $h_q(x_q) = 1$ and (3.23), we obtain $(\mathtt{A}_q)_2$:

$$g_q(y) = f(x_q,y) - \sum_{j=1}^{q-1} g_j(y)h_j(x_q) = f(x_q,y) - I_{q-1}[f_y](x_q) = r_{q-1}(x_q,y). \tag{3.25}$$

Finally, combining (3.24) and (3.25) in addition to $(\mathtt{A}_{q-1})_3$, we conclude that

$$\begin{aligned} I_q[f_y](x) &= I_{q-1}[f_y](x) + g_q(y)h_q(x) = \mathfrak{I}_{q-1}[f_y](x) + r_{q-1}(x_q,y)\frac{r_{q-1}(x,y_q)}{r_{q-1}(x_q,y_q)} \\ &= \mathfrak{I}_q[f_y](x) \end{aligned}$$

and the proof is complete.

## 3.6 Gappy POD

In the following, we present a completion to the POD method called *Gappy POD* [17, 31, 70] or *Missing Point Estimation* [1]. We refer to it as the Gappy POD in the following. It is a projection based method (thus not an interpolation based method although in some particular cases it can be interpreted as an interpolation scheme). However, the projection matrix is approximated by a low-rank approximation that in turn is based on partial or incomplete ("gappy") data of the functions under consideration. In a first turn, we present the method as introduced in [17, 70] and we generalize it in a second turn.

### 3.6.1 The Gappy POD Algorithm

We start from the conceptual idea that a set of basis functions $\{h_1,\dots,h_Q\}$, that can – but does not need to – be obtained through a POD procedure, is given. We first introduce the idea of Gappy POD in the context of Remark 3.1 where functions are represented by a vector containing its pointwise values on a given grid $\Omega_{\mathrm{x}}^{\mathtt{train}} = \{\hat{x}_1,\dots,\hat{x}_M\}$. We remind that the projection $P_Q[f_y]$ of $f_y$ with $y \in \Omega_{\mathrm{y}}$ onto the space spanned by $\{h_1,\dots,h_Q\}$ is defined by

$$(P_Q[f_y],h_q)_{\Omega_{\mathrm{x}}^{\mathtt{train}}} = (f_y,h_q)_{\Omega_{\mathrm{x}}^{\mathtt{train}}}, \quad q=1,\dots,Q.$$

Next, assume that we only dispose of some incomplete data of $f_y$. That is, we are given say $L$ ($< M$) distinct points $\{x_1,\dots,x_L\}$ among $\Omega_{\mathrm{x}}^{\mathtt{train}}$ where $f_y(x_i)$ is available. Then, we define the gappy scalar product by

$$(v,w)_{L,\Omega_{\mathrm{x}}^{\mathtt{train}}} = \frac{|\Omega_{\mathrm{x}}|}{L}\sum_{i=1}^{L} v(x_i)\,w(x_i),$$

which only takes into account available data of $f_y$. We can compute the gappy projection defined by

$$(P_{Q,L}[f_y],h_q)_{L,\Omega_{\mathrm{x}}^{\mathtt{train}}} = (f_y,h_q)_{L,\Omega_{\mathrm{x}}^{\mathtt{train}}}, \quad q=1,\dots,Q.$$

Observe that the basis functions $\{h_1,\dots,h_Q\}$ are no longer orthonormal for the gappy scalar product and that the stability of the method mainly depends on the properties of the mass matrix $\mathrm{M}_{h,L}$ defined by

$$(\mathrm{M}_{h,L})_{i,j} = (h_j,h_i)_{L,\Omega_{\mathrm{x}}^{\mathtt{train}}}.$$

To summarize, in the above presentation we assumed that the data of $f_y$ at some given points was available and then defined a "best approximation" with respect to the available but incomplete data. For instance, the data can be assimilated by physical experiments and the Gappy POD allows to reconstruct the solution in the whole domain $\Omega_{\mathrm{x}}^{\mathtt{train}}$ assuming that it can be accurately represented by the basis functions $\{h_1,\dots,h_Q\}$.

We now change the viewpoint and ask the question: If we can place $L$ sensors at the locations $\{x_i\}_{i=1}^L \subset \Omega_{\mathrm{x}}$ at which we have access to the data $f_y(x_i)$ (through measurements), where would we place the points $\{x_i\}_{i=1}^L$?

One might consider different criteria to chose the sensors. In [70] the placement of $L$ sensors is stated as a minimization problem

$$\min \kappa(\mathrm{M}_{h,L}) \quad \text{where} \quad \mathrm{M}_{h,L} \text{ is based on } L \text{ points } \{x_1,\dots,x_L\}$$

**Scheme 3.8.** Sensor placement algorithm with Gappy POD and minimal condition number

For $1 \leq l \leq L$:

$$x_l = \arg\min_{x \in \Omega_x} \kappa(\mathrm{M}_{h,l}(x))$$

where

$$(\mathrm{M}_{h,l}(x))_{i,j} = \frac{|\Omega_x|}{l} \left[ \sum_{k=1}^{l-1} h_i(x_k) h_j(x_k) + h_i(x) h_j(x) \right], \quad 1 \leq i,j \leq \min(Q,l).$$

**Scheme 3.9.** Sensor placement algorithm with Gappy POD and minimal error

For $1 \leq l \leq L$:

$$x_l = \arg\max_{x \in \Omega_x} \| P_{Q,l-1}[f_y] - f_y \|_{L^p(\Omega_x)}$$

where $P_{Q,l-1}[f_y]$ is the gappy projection of $f_y$ onto the span of $\{h_1, \ldots, h_{\min(Q,l-1)}\}$ based on the pointwise information at $\{x_1, \ldots, x_{l-1}\}$.

and $\kappa(\mathrm{M}_{h,L})$ denotes the condition number of $\mathrm{M}_{h,L}$. We report in Table 3.8 a slight modification of the algorithm presented in [1, 70] to construct a sequence of sensor placements $\{x_1, \ldots, x_L\}$ (with $L \geq Q$) based on an incremental greedy algorithm.

This natural algorithm actually seems to have some difficulties at the beginning, for small values of $l$. It is thus recommended to start with the algorithm presented in Table 3.9.

This criterion is actually the one that is used in the Gappy POD method presented in [20] in the frame of the GNAT approach that allows a stabilized implementation of the gappy method for a challenging CFD problem. Further, we have the following link between the gappy projection and the EIM as noticed in [33].

**Lemma 3.2** *Let $\{h_1, \ldots, h_Q\}$ and $\{x_1, \ldots, x_Q\}$ be given basis functions and interpolation nodes. If the interpolation is well-defined (i.e. the interpolation matrix being invertible), then the interpolatory system based on the basis functions $\{h_1, \ldots, h_Q\}$ and the interpolation nodes $\{x_1, \ldots, x_Q\}$ is equivalent to the gappy projection system based on the basis functions $\{h_1, \ldots, h_Q\}$ with available data at the points $\{x_1, \ldots, x_Q\}$, that is, for any $y \in \Omega_y$ the unique interpolant $I_Q[f_y] \in span\{h_1, \ldots, h_Q\}$ such that*

$$I_Q[f_y](x_q) = f_y(x_q), \quad q = 1, \ldots, Q, \tag{3.26}$$

*is equivalent to the unique gappy projection $P_{Q,L}[f_y]$ defined by*

$$(P_{Q,L}[f_y], h_q)_{Q,\Omega_x^{\mathtt{train}}} = (f_y, h_q)_{Q,\Omega_x^{\mathtt{train}}}, \quad q = 1, \ldots, Q. \tag{3.27}$$

*Proof* Multiply (3.26) by $\frac{|\Omega_x|}{Q} h_i(x_q)$ and take the sum over all $q = 1,\ldots,Q$ to obtain

$$\frac{|\Omega_x|}{Q} \sum_{q=1}^{Q} I_Q[f_y](x_q) h_i(x_q) = \frac{|\Omega_x|}{Q} \sum_{q=1}^{Q} f_y(x_q) h_i(x_q), \quad i = 1,\ldots,Q,$$

which is equivalent to $(I_Q[f_y], h_i)_{L,\Omega_x^{\mathtt{train}}} = (f_y, h_i)_{L,\Omega_x^{\mathtt{train}}}$ for all $i = 1,\ldots,Q$. On the other hand, if $P_{Q,L}[f_y]$ is the solution of (3.27), then there holds that

$$\sum_{q=1}^{Q} P_{Q,L}[f_y](x_q) h_i(x_q) = \sum_{q=1}^{Q} f_y(x_q) h_i(x_q), \quad i = 1,\ldots,Q. \tag{3.28}$$

Since the interpolating system is well-posed, the interpolation matrix $\mathrm{B}_{i,j} = h_j(x_i)$ is invertible and thus there exists a vector $\mathrm{u}_j$ such that $\mathrm{B}\,\mathrm{u}_j = \mathrm{e}_j$ for some $j = 1,\ldots,Q$ where $\mathrm{e}_j$ is the canonical basis vector. Then, multiply (3.28) by $(\mathrm{u}_j)_i$ and sum over all $i$:

$$\sum_{i,q=1}^{Q} P_{Q,L}[f_y](x_q)\,(\mathrm{u}_j)_i\,\mathrm{B}_{qi} = \sum_{i,q=1}^{Q} f_y(x_q)\,(\mathrm{u}_j)_i\,\mathrm{B}_{q,i}, \quad j = 1,\ldots,Q,$$

to get

$$P_{Q,L}[f_y](x_j) = f_y(x_j), \quad j = 1,\ldots,Q.$$

Thus, the gappy projection satisfies the interpolation scheme.

One feature of the sensor placement algorithm based on the Gappy POD framework is that the basis functions $\{h_1,\ldots,h_q\}$ are given and the sensors are chosen accordingly. As a consequence of the interpretation of the gappy projection as an interpolation scheme if the number of basis functions and sensors coincide, one might combine the Gappy POD approach with the EIM in the following way in order to construct basis functions and redundant sensor locations simultaneously:

1. use the EIM to construct simultaneously $Q$ basis functions $\{h_q\}_{q=1}^{Q}$ and interpolation points $\{x_q\}_{q=1}^{Q}$ until a sufficiently small error is achieved;
2. use the gappy projection framework as outlined above to add interpolation points (sensors) to enhance the stability of the scheme.

### *3.6.2 Generalization of Gappy POD*

In the previous algorithm the functions were represented by their nodal values at some points $\hat{x}_1,\ldots,\hat{x}_M$. That is, we can introduce for each point $\hat{x}_i$ a functional $\hat{\sigma}_i = \delta_{\hat{x}_i}$ ($\delta_x$ denoting the Dirac functional associated to the point $x$) such that the interpolant of any continuous function $f$ onto the space $\mathbb{V}_M$ of piecewise linear and globally continuous functions can be written as

$$\sum_{m=1}^{M} \hat{\sigma}_m(f)\,\hat{\varphi}_m,$$

where $\{\hat{\varphi}_m\}_{m=1}^M$ denotes the Lagrange basis of $\mathbb{V}_M$ with respect to the points $\hat{x}_1, \ldots, \hat{x}_M$.

We present a generalization where we allow a more general discrete space $\mathbb{V}_M$. Therefore, let $\mathbb{V}_M$ be a $M$-dimensional discrete space spanned by a set of basis functions $\{\hat{\varphi}_i\}_{i=1}^M$ such as for example the finite element hat-functions, Fourier-basis or polynomial basis functions. In the context of the theory of finite elements, cf. [27], we are given $M$ functionals $\{\hat{\sigma}_m\}_{m=1}^M$, associated with the basis set $\{\hat{\varphi}_i\}_{i=1}^M$, which determine the degrees of freedom of a function. That is, for $f$ regular enough such that all degrees of freedom $\hat{\sigma}_m(f)$ are well-defined, the following interpolation scheme

$$f \to \sum_{m=1}^{M} \hat{\sigma}_m(f)\hat{\varphi}_m$$

defines a function in $\mathbb{V}_M$ that interpolates the degrees of freedom.

We start with noting that the scalar product between two functions $f, g$ in $\mathbb{V}_M$ is given by

$$(f,g)_{\Omega_\mathrm{x}} = \sum_{n,m=1}^{M} \hat{\sigma}_n(f)\, \hat{\sigma}_m(g)\, (\hat{\varphi}_n, \hat{\varphi}_m)_{\Omega_\mathrm{x}}.$$

In this framework, the meaning of "gappy" data is generalized. We speak of gappy data if only partial data of degrees of freedom, i.e. the $\hat{\sigma}_m(f)$ is available. Thus, in this generalized context, the degrees of freedom are not necessarily nodal values, i.e. the functionals being Dirac functionals, and depend on the choice of the basis functions.

Assume that we are given $Q$ basis functions $h_1, \ldots, h_Q$ that describe a subspace in $\mathbb{V}_M$ and $L \geq Q$ degrees of freedom $\sigma_l = \hat{\sigma}_{i_l}$, for $l = 1, \ldots, L$ (chosen among all $M$ degrees of freedom $\hat{\sigma}_1, \ldots, \hat{\sigma}_L$). Denoting by $\varphi_l = \hat{\varphi}_{i_l}$ the corresponding $L$ basis functions, we then define a gappy scalar product

$$(f,g)_{L,\Omega_\mathrm{x}} = \frac{M}{L} \sum_{l,k=1}^{L} \sigma_l(f)\, \sigma_k(g)\, (\varphi_l, \varphi_k)_{\Omega_\mathrm{x}}.$$

Given any $f_y, y \in \Omega_\mathrm{y}$, the gappy projection $P_{Q,L}[f_y] \in \operatorname{span}\{h_1, \ldots, h_Q\}$ is defined by

$$(P_{Q,L}[f_y], h_q)_{L,\Omega_\mathrm{x}} = (f_y, h_q)_{L,\Omega_\mathrm{x}}, \quad q = 1, \ldots, Q.$$

Then, the sensor placement algorithm introduced in the previous section can easily be generalized to this setting.

*Remark 3.11* If the mass matrix $\hat{\mathrm{M}}_{i,j} = (\hat{\varphi}_j, \hat{\varphi}_i)_{\Omega_\mathrm{x}}$ associated with the basis set $\{\hat{\varphi}_i\}_{i=1}^M$ satisfies the following orthogonality property $\hat{\mathrm{M}}_{i,j} = (\hat{\varphi}_j, \hat{\varphi}_i)_{\Omega_\mathrm{x}} = \frac{|\Omega_\mathrm{x}|}{M}\delta_{ij}$ either by construction of the basis functions or using mass lumping (in the case of finite elements) and if the basis functions $\{\hat{\varphi}_i\}_{i=1}^M$ are nodal basis functions associated with the set of points $\Omega_\mathrm{x}^{\mathtt{train}} = \{\hat{x}_1, \ldots, \hat{x}_M\}$, then the original Gappy POD method is established.

*Remark 3.12* If the mass matrix $(\hat{\mathrm{M}})_{i,j} = (\hat{\varphi}_j, \hat{\varphi}_i)_{\Omega_\mathrm{x}}$ associated with the selected functions $\{\varphi_i\}_{i=1}^L$ is orthonormal, then the gappy projection $P_{Q,L}[f_y]$ is solution to the following quadratic minimization problem

$$\min_{f\in\mathbb{V}_Q} \sum_{l=1}^{L} |\sigma_l(f_y) - \sigma_l(f)|^2.$$

Since $L > Q$ in a general setting, this means that the gappy projection fits the selected degrees of freedom optimally in a least-squares sense. In the general case, $P_{Q,L}[f_y]$ is solution to the following minimization problem

$$\min_{f\in\mathbb{V}_Q} \sum_{l,k=1}^{L} (\sigma_l(f_y) - \sigma_l(f))(\varphi_l, \varphi_m)_{\Omega_\mathbf{x}}(\sigma_k(f_y) - \sigma_k(f)).$$

**Acknowledgements** This work was supported by the research grant ApProCEM-FP7-PEOPLE-PIEF-GA-2010-276487.

## References


1. Astrid, P., Weiland, S., Willcox, K., Backx T.: Missing Point Estimation in Models Described by Proper Orthogonal Decomposition. IEEE Transactions on Automatic Control, **53**(10), 2237–2251 (2008)
2. Babaev, M.-B.A.: Best approximation by bilinear forms. Mat. Zametki **46**(2), 21–33, 158 (1989)
3. Babaev, M.-B.A.: Exact annihilators and their applications in approximation theory. Trans. Acad. Sci. Azerb. Ser. Phys.-Tech. Math. Sci. **20**(1, Math. Mech.), 17–24, 233 (2000)
4. Ballani, J., Grasedyck, L., Kluge, M.: Black Box Approximation of Tensors in Hierarchical Tucker Format. Linear Algebra and its Applications **438** 639–657 (2013)
5. Barrault, M., Maday, Y., Nguyen, N.C., Patera, A.T.: An 'empirical interpolation' method: application to efficient reduced-basis discretization of partial differential equations. Comptes Rendus de l'Académie des Sciences. Série I. Mathématique **339**(9), 667–672 (2004)
6. Bebendorf, M.: Approximation of boundary element matrices. Numer. Math. **86**(4), 565–589 (2000)
7. Bebendorf, M.: Hierarchical Matrices: A Means to Efficiently Solve Elliptic Boundary Value Problems. Lecture Notes in Computational Science and Engineering (LNCSE) 63. Springer-Verlag, Berlin Heidelberg (2008)
8. Bebendorf, M.: Adaptive cross approximation of multivariate functions. Constr. Appr., **34**(2), 149–179 (2011)
9. Bebendorf, M., Kühnemund, A., Rjasanow, S.: A symmetric generalization of adaptive cross approximation for higher-order tensors. Technical Report 503, SFB611, University of Bonn, Bonn (2011)
10. Bebendorf, M., Kuske, C.: Separation of variables for function generated high-order tensors. Technical Report 1303, INS, University of Bonn, Bonn (2013)
11. M. Bebendorf, C. Kuske, and R. Venn. Wideband nested cross approximation for Helmholtz problems. Technical report, SFB 611 Preprint (2012)

12. Bebendorf, M., Rjasanow, S.: Adaptive low-rank approximation of collocation matrices. Computing **70**(1), 1–24 (2003)
13. Bebendorf, M., Venn, R.: Constructing nested bases approximations from the entries of non-local operators. Numer. Math. **121**(4), 609–635 (2012)
14. Binev, R., Cohen, A., Dahmen, W., DeVore, R., Petrova, G., Wojtaszczyk, P.: Convergence rates for greedy algorithms in reduced basis methods. SIAM Journal on Mathematical Analysis **43**(3), 1457–1472 (2011)
15. Börm, S., Grasedyck, L.: Hybrid cross approximation of integral operators. Numer. Math. **101**(2), 221–249 (2005)
16. Buffa, A., Maday, Y., Patera, A.T., Prudhomme, C., Turinici, G.: A priori convergence of the greedy algorithm for the parametrized reduced basis method. ESAIM: Mathematical Modelling and Numerical Analysis **46**(03), 595–603 (2012)
17. Bui-Thanh, T., Damodaran, M., Willcox, K.E.: Aerodynamic data reconstruction and inverse design using proper orthogonal decomposition. AIAA journal **42**(8), 1505–1516 (2004)
18. Bui-Thanh, T., Willcox, K., Ghattas, O.: Model reduction for large-scale systems with high-dimensional parametric input space. SIAM J. Sci. Comput. **30**(6), 3270–3288 (2008)
19. Bui-Thanh, T., Willcox, K., Ghattas, O., van Bloemen Waanders, B.: Goal-oriented, model-constrained optimization for reduction of large-scale systems. J. Comput. Phys. **224**(2), 880–896 (2007)
20. Carlberg, K., Farhat, C., Cortial, J., Amsallem, D.: The gnat method for nonlinear model reduction: effective implementation and application to computational fluid dynamics and turbulent flows. Journal of Computational Physics (2013)
21. Carvajal, O.A., Chapman, F.W., Geddes, K.O.: Hybrid symbolic-numeric integration in multiple dimensions via tensor-product series. ISSAC'05, pp. 84–91 (electronic). ACM, New York (2005)
22. Chan, T.F.: On the existence and computation of $LU$-factorizations with small pivots. Math. Comp. **42**(166), 535–547 (1984)
23. Chapman, F.W.: Generalized orthogonal series for natural tensor product interpolation. PhD thesis, University of Waterloo, Waterloo (2003)
24. Chaturantabut, S., Sorensen, D. C.: Discrete empirical interpolation for nonlinear model reduction. In Decision and Control, 2009, held jointly with the 2009 28th Chinese Control Conference. CDC/CCC 2009. Proceedings of the 48th IEEE Conference on, pp. 4316–4321. IEEE (2009)
25. Chkifa, A.: On the Lebesgue constant of Leja sequences for the complex unit disk and of their real projection. Journal of Approximation Theory (2012)
26. Chu, M.T., Funderlic, R.E., Golub, G. H.: A rank-one reduction formula and its applications to matrix factorizations. SIAM Review **37**(4), 512–530 (1995)
27. Ciarlet, P.G.: The finite element method for elliptic problems, vol. 4. North Holland, Amsterdam–New York–Oxford (1978)
28. DeVore, R., Petrova, G., Wojtaszczyk, P.: Greedy algorithms for reduced bases in banach spaces. Constructive Approximation, 1–12 (2012)
29. Donoho, D.L.: Compressed sensing. Information Theory, IEEE Transactions on, 52(4):1289–1306 (2006)
30. Eftang, J.L., Stamm, B.: Parameter multi-domain 'hp' empirical interpolation. Int. J. Numer. Meth. Eng. **90**(4), 412–428 (2012)
31. Everson, R., Sirovich, L.: Karhunen–loeve procedure for gappy data. JOSA A **12**(8), 1657–1664 (1995)

32. Fares, M., Hesthaven, J.S., Maday, Y., Stamm, B.: The reduced basis method for the electric field integral equation. Journal of Computational Physics **230**(14), 5532–5555 (2011)
33. Galbally, D., Fidkowski, K., Willcox, K., Ghattas, O.: Non-linear model reduction for uncertainty quantification in large-scale inverse problems. International journal for numerical methods in engineering **81**(12), 1581–1608 (2010)
34. Golub, G.H., Van Loan, C.F.: Matrix computations, 3rd Ed. Johns Hopkins University Press, Baltimore, MD (1996)
35. Goreinov, S.A., Tyrtyshnikov, E.E., Zamarashkin, N.L.: A theory of pseudoskeleton approximations. Linear Algebra Appl. **261**, 1–21 (1997)
36. Grepl, M.A., Maday, Y., Nguyen, N.C., Patera, A.T.: Efficient reduced-basis treatment of nonaffine and nonlinear partial differential equations. ESAIM: Mathematical Modelling and Numerical Analysis **41**(03), 575–605 (2007)
37. Gu, M., Eisenstat, S.C.: Efficient algorithms for computing a strong rank-revealing QR factorization. SIAM J. Sci. Comput. **17**(4), 848–869 (1996)
38. Haasdonk, B., Dihlmann, M., Ohlberger, M.: A training set and multiple bases generation approach for parameterized model reduction based on adaptive grids in parameter space. Mathematical and Computer Modelling of Dynamical Systems. Methods, Tools and Applications in Engineering and Related Sciences **17**(4), 423–442 (2011)
39. Hackbusch, W.: A sparse matrix arithmetic based on $\mathscr{H}$-matrices. Part I: Introduction to $\mathscr{H}$-matrices. Computing **62**(2), 89–108 (1999)
40. Hackbusch, W., Khoromskij, B.N.: A sparse $\mathscr{H}$-matrix arithmetic. Part II: Application to multi-dimensional problems. Computing **64**(1), 21–47 (2000)
41. Harbrecht, H., Peters, M., Schneider, R.: On the low-rank approximation by the pivoted cholesky decomposition. Technical report, 2011. to appear in APNUM
42. Hesthaven, J.S., Stamm, B., Zhang, S.: Efficient greedy algorithms for high-dimensional parameter spaces with applications to empirical interpolation and reduced basis methods. Technical report, Providence, RI, USA (2011)
43. Hesthaven, J.S., Zhang, S.: On the use of ANOVA expansions in reduced basis methods for high-dimensional parametric partial differential equations. Technical Report 2011-31, Scientific Computing Group, Brown University, Providence, RI, USA (Dec. 2011)
44. Hotelling, H.: Analysis of a complex of statistical variables into principal components. The Journal of educational psychology, 498–520 (1933)
45. Hwang, T.M., Lin, W.W., Yang, E.K.: Rank revealing *LU* factorizations. Linear Algebra Appl. **175**, 115–141 (1992)
46. Karhunen, K.: Zur spektraltheorie stochastischer prozesse. Suomalainen tiedeakatemia (1946)
47. Kosambi, D.: Statistics in function space. J. Indian Math. Soc **7**(1), 76–88 (1943)
48. Kunisch, K., Volkwein, S.: Control of the burgers equation by a reduced-order approach using proper orthogonal decomposition. Journal of Optimization Theory and Applications **102**(2), 345–371 (1999)
49. Leja, F.: Sur certaines suites liées aux ensembles plans et leur application à la représentation conforme. Ann. Polon. Math. **4**, 8–13 (1957)
50. Loève, M.: Fonctions aléatoires de second ordre. CR Acad. Sci. Paris **220**, 380 (1945)
51. Lumley, J.L.: Stochastic tools in turbulence. Courier Dover Publications, USA (2007)
52. Maday, Y., Mula, O.: A generalized empirical interpolation method: application of reduced basis techniques to data assimilation. Analysis and Numerics of Partial Differential Equations **XIII**, 221–236 (2013)

53. Maday, Y., Mula, O., Turinici, G.: A priori convergence of the generalized empirical interpolation method. http://hal.archives-ouvertes.fr/docs/00/79/81/14/PDF/bare_conf.pdf
54. Maday, Y., Nguyen, N.C., Patera, A.T., Pau, G.S.H.: A general multipurpose interpolation procedure: the magic points. Communications on Pure and Applied Analysis **8**(1), 383–404 (2009)
55. Maday, Y., Stamm, B.: Locally adaptive greedy approximations for anisotropic parameter reduced basis spaces. arXiv.org (Apr. 2012)
56. Mees, A., Rapp, P., Jennings, L.: Singular-value decomposition and embedding dimension. Physical Review A **36**(1), 340 (1987)
57. Micchelli, C.A., Pinkus, A.: Some problems in the approximation of functions of two variables and $n$-widths of integral operators. J. Approx. Theory **24**(1), 51–77 (1978)
58. Mirsky, L.: Symmetric gauge functions and unitarily invariant norms. Quart. J. Math. Oxford Ser. (2), 11:50–59 (1960)
59. Obukhov, A.M.: Statistical description of continuous fields. Trudy Geophys. Inst. Akad. Nauk. SSSR **24**(151), 3–42 (1953)
60. Oseledets, I.V., Savostyanov, D.V., Tyrtyshnikov, E.E.: Linear algebra for tensor problems. Computing **85**(3), 169–188 (2009)
61. Oseledets, I.V., Tyrtyshnikov, E.E.: TT-Cross Approximation for Multidimensional Arrays. Linear Algebra Appl. **432**(5), 70–88 (2010)
62. Patera, A.T., Rozza, G.: Reduced Basis Approximation and A Posteriori Error Estimation for Parametrized Partial Differential Equations. MIT Pappalardo Graduate Monographs in Mechanical Engineering. Cambridge, MA (2007). Available from http://augustine.mit.edu/methodology/methodology_book.htm
63. Pearson, K.: On lines and planes of closest fit to systems of points in space. The London, Edinburgh, and Dublin Philosophical Magazine and Journal of Science **2**(11), 559–572 (1901)
64. Pougachev, V.S.: General theory of the correlation of random functions. Izv. Akad. Nauk. SSSR, Ser Mat **17**, 401 (1953)
65. Reichel, L.: Newton interpolation at Leja points. BIT **30**(2), 332–346 (1990)
66. Schneider, J.: Error estimates for two-dimensional Cross Approximation. J. Approx. Theory **162**(9), 1685–1700 (2010)
67. Šimša, J.: The best $L^2$-approximation by finite sums of functions with separable variables. Aequationes Math. **43**(2–3), 248–263 (1992)
68. Taylor, R.: Lagrange interpolation on Leja points. PhD thesis, University of South Florida (2008)
69. Wedderburn, J.H.M.: Lectures on matrices. Dover Publications Inc., New York (1964)
70. Willcox, K.: Unsteady flow sensing and estimation via the gappy proper orthogonal decomposition. Computers & Fluids **35**(2), 208–226 (2006)
71. Wu, C., Liang, Y., Lin, W., Lee, H., Lim, S.: A note on equivalence of proper orthogonal decomposition methods. Journal of Sound Vibration **265**, 1103–1110 (2003)

4

# Application of the Discrete Empirical Interpolation Method to Reduced Order Modeling of Nonlinear and Parametric Systems*

**Harbir Antil, Matthias Heinkenschloss and Danny C. Sorensen**

**Abstract** Projection based methods lead to reduced order models (ROMs) with dramatically reduced numbers of equations and unknowns. However, for nonlinear or parametrically varying problems the cost of evaluating these ROMs still depends on the size of the full order model and therefore is still expensive. The Discrete Empirical Interpolation Method (DEIM) further approximates the nonlinearity in the projection based ROM. The resulting DEIM ROM nonlinearity depends only on a few components of the original nonlinearity. If each component of the original nonlinearity depends only on a few components of the argument, the resulting DEIM ROM can be evaluated efficiently at a cost that is independent of the size of the original problem. For systems obtained from finite difference approximations, the $i$th component of the original nonlinearity often depends only on the $i$th component of the argument. This is different for systems obtained using finite element methods, where the dependence is determined by the mesh and by the polynomial degree of the finite element subspaces. This paper describes two approaches of applying DEIM in the finite element context, one applied to the assembled and the other to the unassembled form of the nonlinearity. We carefully examine how the DEIM is applied in each case, and the substantial efficiency gains obtained by the DEIM. In addition, we demonstrate how to apply DEIM to obtain ROMs for a class of parameterized system that arises, e.g., in shape optimization. The evaluations of the DEIM ROMs are substantially faster than those of the standard projection based ROMs. Additional

H. Antil (✉)
Department of Mathematical Sciences, George Mason University Fairfax, VA 22030
e-mail: hantil@gmu.edu

M. Heinkenschloss · D.C. Sorensen
Department of Computational and Applied Mathematics, Rice University, Houston, TX 77005-1892
e-mail: heinken@rice.edu, sorensen@rice.edu

* This research was supported in part by AFOSR grant FA9550-12-1-0155 and NSF grants CCF-1017401, DMS-0915238, DMS-1115345.

A. Quarteroni, G. Rozza (eds.): *Reduced Order Methods for Modeling and Computational Reduction*, MS&A 9. DOI 10.1007/978-3-319-02090-7_4, 

gains are obtained with the DEIM ROMs when one has to compute derivatives of the model with respect to the parameter.

## 4.1 Introduction

Projection based reduced order models systematically extract the features of very large-scale systems to approximate these systems by substantially smaller ones. However, if the original system is parameter dependent or is semilinear, then, although the new small system involves substantially fewer equations and unknowns than the original one, the computational cost of its numerical solution can be essentially the same as that of the original large-scale system. The discrete empirical interpolation method (DEIM) of [7] further approximates projection based reduced order models to obtain small systems that capture the solution of the original large-scale system and that can also be solved at a computational cost that depends only on the size of the small system, provided each component of the original semilinear function depends only on a few components of its argument. So far, the DEIM has been primarily applied to finite difference discretizations of semilinear PDEs where the $i$th component of the nonlinearity depends only on the $i$th component of the argument. This is different in finite element discretizations, where the dependence of the nonlinear function is determined by the mesh as well as by the polynomial degree used to construct the finite element spaces. Therefore results from DEIM applied to finite difference approximations of PDEs do not necessarily carry over to DEIM applied to finite element approximations of PDEs. One purpose of this paper is demonstrate two approaches to apply DEIM to finite element discretizations of semilinear PDEs and numerically study their computational cost. The two approaches apply DEIM at different stages of the finite element assembly process. The size of the nonlinear function as well as its dependence on the argument are different at each stage of the assembly process, which impacts the computational efficiency of the resulting DEIM reduced models. The second purpose of this paper is to demonstrate how to apply DEIM to a class of parameter dependent systems that arise, e.g., in shape optimization.

Discretizations of parameterized semilinear elliptic partial differential equations (PDEs) lead to large scale nonlinear algebraic systems of the form

$$\mathbf{A}(\theta)\mathbf{y} + \mathbf{F}(\mathbf{y};\theta) = \mathbf{b}(\theta), \tag{4.1}$$

where the parameters $\theta \in \Theta \subset \mathbb{R}^p$ and for each parameter $\theta$ the matrix $\mathbf{A}(\theta) \in \mathbb{R}^{N\times N}$ and the vectors $\mathbf{F}(\mathbf{y};\theta)$ and $\mathbf{b}(\theta) \in \mathbb{R}^N$. Projection based model reduction techniques [1,19,24,29] generate matrices $\mathbf{V}_\ell$ and $\mathbf{V}_r \in \mathbb{R}^{N\times n}$ with $n \ll N$ and replace (4.1) with the reduced system

$$\mathbf{V}_\ell^T\mathbf{A}(\theta)(\overline{\mathbf{y}} + \mathbf{V}_r\widehat{\mathbf{y}}) + \mathbf{V}_\ell^T\mathbf{F}(\overline{\mathbf{y}} + \mathbf{V}_r\widehat{\mathbf{y}};\theta) = \mathbf{V}_\ell^T\mathbf{b}(\theta). \tag{4.2}$$

While the reduced order system (4.2) is much smaller than the original one, the cost of computation of $\theta \mapsto \mathbf{V}_\ell^T\mathbf{A}(\theta)\mathbf{V}_r$, $\theta \mapsto \mathbf{V}_\ell^T\mathbf{b}(\theta)$, and $(\widehat{\mathbf{y}},\theta) \mapsto \mathbf{V}_\ell^T\mathbf{F}(\overline{\mathbf{y}} + \mathbf{V}_r\widehat{\mathbf{y}};\theta)$

still depends on $N$. Therefore, additional approximations are needed to obtain reduced order models that capture the original system as well as evaluate with a computational complexity that depends only on the reduced order system size $n$ but is independent of the full order model size $N \gg n$.

The empirical interpolation method (EIM) of [2] and the DEIM of [7] generate reduced order models from (4.2) that approximate the full order model within desired error bounds and that can be numerically solved at a cost that essentially depends only on the reduced order system size. While the EIM is applied to the variational formulation that leads to the nonlinear algebraic system (4.1), its derivative, the DEIM is applied directly to discrete systems. Applications of EIM and DEIM to nonlinear finite element computations are also discussed, e.g., in [9, 15, 17, 22]. We will focus on discrete systems (4.1) and therefore consider the DEIM. Especially, we carefully expose the dependency of the computational complexity of the DEIM on the polynomial degree of the finite element method.

One purpose of this paper is the study of DEIM to nonlinear systems $\mathbf{A}\mathbf{y}+\mathbf{F}(\mathbf{y}) = \mathbf{b}$ obtained from finite element discretizations. The DEIM reduced order model is of the form $\mathbf{V}_\ell^T\mathbf{A}(\overline{\mathbf{y}}+\mathbf{V}_r\widehat{\mathbf{y}})+\widehat{\mathbf{F}}(\widehat{\mathbf{y}}) = \mathbf{V}_\ell^T\mathbf{b}$, where $\widehat{\mathbf{F}}$ depends only on $m$ components of the original nonlinearity $\mathbf{F}$. As we have mentioned before, the efficiency with which the DEIM reduced order model can be applied depends on how many components of the argument are needed to evaluate $m$ components of the original nonlinearity $\mathbf{F}$. For systems obtained from finite element discretizations the dependence of $\mathbf{F}$ on its argument is determined by the mesh, as well as by the polynomial degree used to construct the finite element spaces. One can apply DEIM at different stages of the finite element assembly process. This effects the structure of the nonlinearity. We demonstrate how to apply DEIM to finite element discretizations of nonlinear PDEs in the assembled and in the unassembled form, and we numerically study the computational cost of the resulting reduced order models. Either version of the DEIM is preferable over the naive application of projection based model reduction as in (4.2). For large systems, the application of the DEIM to the so-called unassembled form of the nonlinearity leads to additional gains in the on-line cost of the reduced order models.

A second focus of this paper is the application of DEIM to generate reduced order models for parametrically dependent PDEs $\mathbf{A}(\theta)\mathbf{y} = \mathbf{b}(\theta)$, where $\mathbf{A}(\theta) = \sum_{i=1}^M \mathbf{g}_i(\theta)\mathbf{A}_i$ and $\mathbf{b}(\theta) = \sum_{i=1}^M \mathbf{l}_i(\theta)\mathbf{b}_i$. For large $M$ the complexity of evaluating the reduced order matrix $\mathbf{V}_\ell^T\mathbf{A}(\theta)\mathbf{V}_r = \sum_{i=1}^M \mathbf{g}_i(\theta)\mathbf{V}_\ell^T\mathbf{A}_i\mathbf{V}_r$ is still high. The DEIM can be used to obtain an approximation that allows more pre-computation of matrices and that can be evaluated more efficiently in the on-line phase. Additional benefits arise when derivatives of the matrix with respect to the parameter $\theta$ have to be computed, and we illustrate these gains in the context of shape optimization.

The next section describes two model problems, a semilinear elliptic advection reaction diffusion equation and the Stokes equations on a parameterized domain, and their finite element discretizations. These problems will be used to demonstrate the application of the DEIM, and to numerically evaluate the computational costs required to solve the full and the reduced order models. Section 4.3 reviews approaches to construct the reduced order subspaces spanned by the columns of the matrices $\mathbf{V}_\ell$

and $\mathbf{V}_r \in \mathbb{R}^{N\times n}$, and it reviews the DEIM. The main contributions of this paper are presented in Sects. 4.4 and 4.5.

In Sect. 4.4 we discuss the application of the DEIM to finite element discretizations of semilinear PDEs. We illustrate how $i$th component of the nonlinearity depends on the components of its arguments for piecewise linear and piecewise quadratic elements, and we demonstrate how this dependence impacts the efficiency of the DEIM. In addition, we discuss the application of DEIM to the fully assembled system, as well as the unassembled form of the nonlinearity. The latter was originally suggested by [9, 33]. The nonlinear vectors are larger, but each component depends on fewer components of the argument. We describe both version of the DEIM and computationally compare them on the semilinear elliptic advection reaction diffusion model equation of Sect. 4.2.1. As we have mentioned before, either version of the DEIM is preferable over the naive application of projection based model reduction as in (4.2). For large systems, the application of the DEIM to the unassembled form of the nonlinearity is more expensive in the off-line cost, but leads to additional gains in the on-line cost of the reduced order models.

The application of the DEIM to obtain efficient reduced order models for systems with parameterized matrices $\mathbf{A}(\theta) = \sum_{i=1}^{M} \mathbf{g}_i(\theta)\mathbf{A}_i$ and vectors $\mathbf{b}(\theta) = \sum_{i=1}^{M} \mathbf{l}_i(\theta)\mathbf{b}_i$ is demonstrated in Sect. 4.5. We numerically illustrate the efficiency gains achieved by the DEIM reduced order model using the Stokes equation on parameterized domains introduced in Sect. 4.2.2. The DEIM not only leads to reduced order models that can be evaluated efficiently, but in addition it also leads to reduced order models where derivatives with respect to the parameter $\theta$ can be computed efficiently. Both efficiency gains are crucial, e.g., for shape optimization.

## 4.2 Model Problems

### *4.2.1 Semilinear Advection-Diffusion-Reaction PDE*

Our first model problem is a semilinear advection diffusion reaction equation. Let $\Omega \subset \mathbb{R}^d$, $d \in \{2,3\}$ be an open, bounded Lipschitz domain with boundary $\partial\Omega = \Gamma_D \cup \Gamma_N$, where $\Gamma_D$ and $\Gamma_N$ corresponds to Dirichlet and Neumann parts. Given a diffusion coefficient $\nu > 0$, an advection vector $\beta \in \mathbb{R}^d$, a nonlinear function $f: \mathbb{R}\times\mathbb{R}^p \to \mathbb{R}$, and Dirichlet data $h$, the semilinear advection diffusion reaction equation is given by

$$-\nabla\cdot(\nu\nabla y) + \beta\cdot\nabla y + f(y,\theta) = 0, \qquad \text{in } \Omega, \tag{4.3a}$$

$$y = h, \qquad \text{on } \Gamma_D, \tag{4.3b}$$

$$\nabla y\cdot n = 0, \qquad \text{on } \Gamma_N. \tag{4.3c}$$

We consider the specific nonlinearity

$$f(y,\theta) = Ay(C-y)e^{-E/(D-y)} \tag{4.4}$$

used e.g., in [11]. Here $C$, $D$ are known constants and $\theta = (\ln(A), E)$ are system parameters that can vary within the parameter domain $\Theta = [5.00, 7.25] \times [0.05, 0.15] \subset \mathbb{R}^2$.

The weak form of (4.3) is given as follows. Find $y \in H^1(\Omega)$ with $y = h$ on $\Gamma_D$ such that

$$\int_{\Omega} \nu \nabla y \cdot \nabla v dx + \int_{\Omega} \beta \cdot \nabla y v dx + \int_{\Omega} f(y, \theta) v dx = 0 \tag{4.5}$$

for all $v \in H^1(\Omega)$ with $v = 0$ on $\Gamma_D$. Existence results for linear and nonlinear advection diffusion equations can be found, e.g., in [26, 32] and [23], [27, Sec. 6.3]

We discretize the equations using an SUPG (streamline upwind/Petrov-Galerkin) stabilized FEM [5, 10, 25]. The Dirichlet boundary conditions are implemented via interpolation. Let $\{\Omega_e\}_{e=1}^{n_e}$ be a conforming triangulation of the domain $\Omega$. Furthermore, let $\{\phi_j\}_{j=1}^{N}$ be the piecewise polynomial nodal basis functions. To simplify the presentation, we assume that nodes with indices $1, \ldots, N_F$ are in $\overline{\Omega} \setminus \Gamma_D$ and that the nodes with indices $N_F + 1, \ldots, N_F + N_D$ are in $\Gamma_D$. We define

$$\begin{aligned} a_h(y_h, \phi) = & \int_{\Omega} \nu \nabla y_h(x) \cdot \nabla \phi(x) + \beta \cdot \nabla y_h(x) \phi(x) dx \\ & + \sum_{e=1}^{n_e} \int_{\Omega_e} \tau_e \, \beta \cdot \nabla \phi(x) \, (-\nabla \cdot (\nu \nabla y_h(x)) + \beta \cdot \nabla y_h(x)) dx, \end{aligned} \tag{4.6a}$$

$$F_h(y_h, \phi; \theta) = \int_{\Omega} f(y_h(x), \theta) \phi dx + \sum_{e=1}^{n_e} \int_{\Omega_e} \tau_e \, \beta \cdot \nabla \phi(x) \, f(y_h(x), \theta) dx. \tag{4.6b}$$

If we let $h_e$ denote the length of largest side of each element $\Omega_e$ and $P_e = h_e \|\beta\| / (2\nu)$ the mesh Péclet number, then the SUPG stabilization parameter is defined as

$$\tau_e = \frac{h_e}{2\|\beta\|} \left(1 - \frac{1}{P_e}\right).$$

The solution $y$ of (4.5) is approximated by

$$y_h(x) = \sum_{j=1}^{N_F + N_D} y_j \phi_j(x) \tag{4.7}$$

where $y_h$ satisfies

$$a_h(y_h, \phi_i) + F_h(y_h, \phi_i; \theta) = 0, \qquad i = 1, \ldots, N_F, \tag{4.8a}$$

$$y_h(x_{N_F+i}) = h(x_{N_F+i}), \qquad i = 1, \ldots, N_D. \tag{4.8b}$$

To state the nonlinear algebraic system corresponding to (4.8), we define

$$\mathbf{y}_F = (y_1, \ldots, y_{N_F})^T, \quad \mathbf{y}_D = (y_{N_F+1}, \ldots, y_{N_F+N_D})^T,$$

$$\mathbf{h} = (h(x_{N_F+1}), \ldots, h(x_{N_F+N_D}))^T,$$

and partition the matrices and vectors into submatrices and subvectors corresponding to the free variables $\mathbf{y}_F$ and those determined by the Dirichlet boundary conditions,

(4.8) leads to a system of algebraic equations of the type

$$\begin{pmatrix} \mathbf{A}_{FF} & \mathbf{A}_{FD} \\ \mathbf{0} & \mathbf{I} \end{pmatrix} \begin{pmatrix} \mathbf{y}_F \\ \mathbf{y}_D \end{pmatrix} + \begin{pmatrix} \mathbf{F}_F(\mathbf{y}_F, \mathbf{y}_D; \theta) \\ \mathbf{0} \end{pmatrix} = \begin{pmatrix} \mathbf{0} \\ \mathbf{h} \end{pmatrix}. \tag{4.9}$$

Of course, this is equivalent to $\mathbf{A}_{FF}\mathbf{y}_F + \mathbf{F}_F(\mathbf{y}_F, \mathbf{h}; \theta) + \mathbf{A}_{FD}\mathbf{h} = \mathbf{0}$. If we set $\mathbf{b} = -\mathbf{A}_{FD}\mathbf{h}$, $N = N_F$, if we drop the subscript $F$, and if we drop the constant $\mathbf{h}$ from the arguments of the nonlinearity $\mathbf{F}$, then we arrive at the $N \times N$ system

$$\mathbf{A}\mathbf{y} + \mathbf{F}(\mathbf{y}; \theta) = \mathbf{b}, \tag{4.10}$$

which is a special case of (4.2). In this model problem, the matrix $\mathbf{A}$ and the vector $\mathbf{b}$ do not depend on the parameter $\theta$. For later reference, we note that the matrix $\mathbf{A}$, the function $\mathbf{F}$, and the vector $\mathbf{b}$ are given by

$$\mathbf{A}_{ij} = a_h(\phi_j, \phi_i) \qquad i,j = 1, \dots N, \tag{4.11a}$$

$$\mathbf{F}_i(\mathbf{y}; \theta) = F_h\Big(\sum_{j=1}^{N} y_j \phi_j + \sum_{j=N+1}^{N+N_D} h(x_j)\phi_j, \phi_i; \theta\Big), \qquad i = 1, \dots N, \tag{4.11b}$$

$$\mathbf{b}_i = b_h(\phi_i) := a_h\Big(\sum_{j=N+1}^{N+N_D} h(x_j)\phi_j, \phi_i\Big), \qquad i = 1, \dots N. \tag{4.11c}$$

### 4.2.2 The Stokes Equations on Parameterized Domains

As our second model problem we consider the Stokes equations posed on a family of parameterized domains $\Omega(\theta) \subset \mathbb{R}^2$, where $\theta \in \Theta \subset \mathbb{R}^p$. Since our numerical examples are 2D problems we describe the approach for parameterized domains in $\mathbb{R}^2$. However, everything can be easily generalized to the Stokes equations on parameterized domains in $\mathbb{R}^3$. The boundary $\partial\Omega = \Gamma_D \cup \Gamma_{\text{out}}$ is decomposed into an outflow boundary $\Gamma_{\text{out}}$ and $\Gamma_D = \partial\Omega \setminus \Gamma_{\text{out}}$. We assume that the parameterized domains $\Omega(\theta)$ can be mapped onto a reference domain $\widetilde{\Omega} \subset \mathbb{R}^2$. That is we assume that for each $\theta \in \Theta$ there exists a diffeomorphism $\Phi(\cdot; \theta)$ with

$$\Omega(\theta) = \Phi(\widetilde{\Omega}; \theta). \tag{4.12}$$

The Stokes equations for the velocity $u$ and the pressure $p$ are

$$-\nu\Delta u(x) + \nabla p(x) = f(x), \qquad \text{in } \Omega(\theta) \tag{4.13a}$$

$$\nabla \cdot u(x) = 0, \qquad \text{in } \Omega(\theta) \tag{4.13b}$$

$$u(x) = h(x), \qquad \text{on } \Gamma_D(\theta) \tag{4.13c}$$

$$(\nu\nabla u(x) - p(x)) \cdot n(x) = 0, \qquad \text{on } \Gamma_{\text{out}}(\theta), \tag{4.13d}$$

where $f \in (L^2(\Omega(\theta)))^2$. The weak form of (4.13) is given as follows: Find $u \in (H^1(\Omega(\theta)))^2$ with $u = h$ on $\Gamma_D(\theta)$ and $p \in L^2(\Omega(\theta))$ such that

$$\int_{\Omega(\theta)} \nu \nabla u(x) : \nabla v(x) - \int_{\Omega(\theta)} \nabla \cdot v(x) p(x) = \int_{\Omega(\theta)} f(x) v(x), \tag{4.14a}$$

$$- \int_{\Omega(\theta)} \nabla \cdot u(x) q(x) = 0, \tag{4.14b}$$

for all $v \in \{\phi \in (H^1(\Omega(\theta)))^2 \,:\, \phi = 0 \text{ on } \Gamma_D(\theta)\}$ and $q \in L^2(\Omega(\theta))$. Existence results for the Stokes equations can be found, e.g., in [12, 13].

We approximate (4.14) using Taylor-Hood P2-P1 finite elements [10]. We triangulate the reference domain $\widetilde{\Omega}$ and use (4.12). Let $N_v$ be the number of velocity nodes in $\widetilde{\Omega} \cup \widetilde{\Gamma}_{\text{out}}$ and let $N_p$ be the number of pressure nodes in $\overline{\widetilde{\Omega}}$. If the piecewise quadratic basis functions for the velocities on the reference domain are $\widetilde{\phi}_j$, $j = 1, \ldots, N_v$, and the piecewise linear basis functions for the pressure on the reference domain are $\widetilde{\psi}_j$, $j = 1, \ldots, N_p$, then the basis functions for velocities and pressure on the domain $\Omega(\theta)$ are

$$\begin{aligned} \phi_j(\cdot;\theta) &= \widetilde{\phi}_j \circ \Phi^{-1}(\cdot;\theta), \quad j = 1, \ldots, N_v, \\ \psi_j(\cdot;\theta) &= \widetilde{\psi}_j \circ \Phi^{-1}(\cdot;\theta), \quad j = 1, \ldots, N_p. \end{aligned} \tag{4.15}$$

The Taylor-Hood P2-P1 finite element discretization of (4.14) leads to

$$\mathbf{S}(\theta)\mathbf{y} = \mathbf{b}(\theta), \tag{4.16}$$

where

$$\mathbf{S}(\theta) = \begin{pmatrix} \mathbf{A}(\theta) & 0 & \mathbf{B}^{(1)}(\theta)^T \\ 0 & \mathbf{A}(\theta) & \mathbf{B}^{(2)}(\theta)^T \\ \mathbf{B}^{(1)}(\theta) & \mathbf{B}^{(2)}(\theta) & 0 \end{pmatrix}, \ \mathbf{y} = \begin{pmatrix} \mathbf{u}_1 \\ \mathbf{u}_2 \\ \mathbf{p} \end{pmatrix}, \ \mathbf{b}(\theta) = \begin{pmatrix} \mathbf{b}^{(1)}(\theta) \\ \mathbf{b}^{(2)}(\theta) \\ \mathbf{b}^{(3)}(\theta) \end{pmatrix}, \tag{4.17}$$

with

$$\mathbf{A}(\theta)_{ij} = \int_{\Omega(\theta)} \nu \nabla \phi_i^T \nabla \phi_j \, dx, \qquad 1 \le i, j \le N_v,$$

and

$$\mathbf{B}^{(1)}(\theta)_{ij} = -\int_{\Omega(\theta)} \frac{\partial \phi_j}{\partial x_1} \psi_i \, dx, \quad \mathbf{B}^{(2)}(\theta)_{ij} = -\int_{\Omega(\theta)} \frac{\partial \phi_j}{\partial x_2} \psi_i \, dx,$$

$1 \le j \le N_v$, $1 \le i \le N_p$.

We use the integral transformation as well as the structure (4.12) of the basis functions to compute

$$\mathbf{A}(\theta)_{ij} = \int_{\widetilde{\Omega}} \nu \, \widetilde{\nabla} \widetilde{\phi}_i(\widetilde{x})^T (D\Phi(\widetilde{x};\theta))^{-1} (D\Phi(\widetilde{x};\theta))^{-T} \widetilde{\nabla} \widetilde{\phi}_j(\widetilde{x}) \, |\det(D\Phi(\widetilde{x};\theta))| \, d\widetilde{x}$$

for $1 \le i, j \le N_v$, and

$$\begin{pmatrix} \mathbf{B}^{(1)}(\theta)_{ij} \\ \mathbf{B}^{(2)}(\theta)_{ij} \end{pmatrix} = -\int_{\widetilde{\Omega}} (D\Phi(\widetilde{x};\theta))^{-T} \widetilde{\nabla} \widetilde{\phi}_j(\widetilde{x}) \, \widetilde{\psi}_i(\widetilde{x}) \, |\det(D\Phi(\widetilde{x};\theta))| \, d\widetilde{x},$$

for $1 \le j \le N_v$, $1 \le i \le N_p$.

Finally, we approximate the integrals by a quadrature rule with nodes $\widetilde{x}_i$ and weights $\omega_i$, $i = 1,\dots,M$. To keep the presentation simple, we assume that the same quadrature rule is used for all integrals. If we define functions $\mathbf{g}_k : \Theta \to \mathbb{R}^M$, $k = 1,\dots,7$, component-wise as follows

$$\begin{pmatrix} (\mathbf{g}_1(\theta))_\ell & (\mathbf{g}_2(\theta))_\ell \\ (\mathbf{g}_2(\theta))_\ell & (\mathbf{g}_3(\theta))_\ell \end{pmatrix} = \nu\, \omega_\ell\, (D\Phi(\widetilde{x}_\ell;\theta))^{-1}(D\Phi(\widetilde{x}_\ell;\theta))^{-T}\, |\det(D\Phi(\widetilde{x}_\ell;\theta))|,$$

$$\begin{pmatrix} (\mathbf{g}_4(\theta))_\ell & (\mathbf{g}_5(\theta))_\ell \\ (\mathbf{g}_6(\theta))_\ell & (\mathbf{g}_7(\theta))_\ell \end{pmatrix} = \omega_\ell\, (D\Phi(\widetilde{x}_\ell;\theta))^{-T}\, |\det(D\Phi(\widetilde{x}_\ell;\theta))|,$$

then, if the integrals are replaced by quadrature, the matrices in the Stokes system can be written as

$$\mathbf{A}(\theta)_{ij} = \sum_{\ell=1}^{M} \widetilde{\nabla}\widetilde{\phi}_i(\widetilde{x}_\ell)^T \begin{pmatrix} (\mathbf{g}_1(\theta))_\ell & (\mathbf{g}_2(\theta))_\ell \\ (\mathbf{g}_2(\theta))_\ell & (\mathbf{g}_3(\theta))_\ell \end{pmatrix} \widetilde{\nabla}\widetilde{\phi}_j(\widetilde{x}_\ell), \quad 1 \le i,j \le N_v$$

$$\begin{pmatrix} \mathbf{B}^{(1)}(\theta)_{ij} \\ \mathbf{B}^{(2)}(\theta)_{ij} \end{pmatrix} = \sum_{\ell=1}^{M} \widetilde{\psi}_i(\widetilde{x}_\ell) \begin{pmatrix} (\mathbf{g}_4(\theta))_\ell & (\mathbf{g}_5(\theta))_\ell \\ (\mathbf{g}_6(\theta))_\ell & (\mathbf{g}_7(\theta))_\ell \end{pmatrix} \widetilde{\nabla}\widetilde{\phi}_j(\widetilde{x}_\ell), \; 1 \le j \le N_v,\; 1 \le i \le N_p.$$

If we insert this representation into (4.17), then

$$\mathbf{S}(\theta) = \sum_{\ell=1}^{M}\sum_{k=1}^{7} (\mathbf{g}_k)_\ell(\theta)\mathbf{S}_{\ell k}. \tag{4.18}$$

Similarly, if we replace the integrals in the right hand side vectors

$$\mathbf{b}^{(k)}(\theta)_i = \int_{\Omega(\theta)} f_k(x)\phi_i(x)\, dx = \int_{\widetilde{\Omega}} f_k(\Phi(\widetilde{x};\theta))\, \widetilde{\phi}_i(\widetilde{x})\, |\det(D\Phi(\widetilde{x};\theta))|\, d\widetilde{x}, \;\; k = 1,2,$$

by quadrature rules, then

$$\mathbf{b}^{(k)}(\theta)_i = \sum_{\ell=1}^{M} \widetilde{\phi}_i(\widetilde{x}_\ell)\, (\mathbf{g}_{7+k}(\theta))_\ell, \quad k = 1,2,$$

where $(\mathbf{g}_{7+k}(\theta))_\ell = \omega_\ell\, f_k(\Phi(\widetilde{x}_\ell;\theta))\, |\det(D\Phi(\widetilde{x}_\ell;\theta))|$, $k = 1,2$.

## 4.3 Projection Based Reduced Order Models

### *4.3.1 Generating the Reduced Order Model Subspaces*

The computation of the matrices $\mathbf{V}_\ell, \mathbf{V}_r \in \mathbb{R}^{N\times n}$ is crucial for the accuracy of the resulting reduced order model and involves some sort of sampling of the solutions to the full order model. Commonly used methods to generate these matrices include the greedy algorithm (see, e.g., [4, 6, 24, 29]), proper orthogonal decomposition (POD) (see, e.g., [19]), and, for time dependent linear problems, balanced POD (see, e.g., [1, 21, 28]). Since emphasis of this paper is the efficient evaluation of the reduced order model (4.2) using DEIM, it does not matter how $\mathbf{V}_\ell, \mathbf{V}_r \in \mathbb{R}^{N\times n}$ have been generated. We assume these matrices have been generated by a suitable method. In our numerical examples, we generate $\mathbf{V} = \mathbf{V}_\ell = \mathbf{V}_r \in \mathbb{R}^{N\times n}$ using a simple sampling strategy and proper orthogonal decomposition. This often results in good reduced order models, although more sophisticated sampling strategies might have provided equally good reduced order models using fewer samples.

Since we will refer to the proper orthogonal decomposition (POD) later, we provide a few details on this method. First by POD we mean the construction of a $k$ dimensional subspace that best approximates given samples $\mathbf{s}_1,\ldots,\mathbf{s}_K$. Thus, selection of these samples is not part of POD. We assume $\mathbf{s}_1,\ldots,\mathbf{s}_K \in \mathbb{R}^N$, but in general these samples could be vectors in a Hilbert space. See, e.g., [19]. Given the samples $\mathbf{s}_1,\ldots,\mathbf{s}_K$ the POD successively computes vectors $\mathbf{v}_1,\ldots,\mathbf{v}_k$ as the solution of

$$\text{minimize} \sum_{j=1}^{K} \|\mathbf{s}_j - \sum_{i=1}^{\ell} \mathbf{v}_i\,\mathbf{v}_i^T\,\mathbf{s}_j\|_2^2 \tag{4.19a}$$

$$\text{subject to } \mathbf{v}_i^T\mathbf{v}_j = \delta_{ij}, \quad i,j = 1,\ldots,k, \tag{4.19b}$$

where $\delta_{ij}$ is the Kronecker delta, or in matrix notation

$$\text{minimize } \|\mathbf{S} - \mathbf{V}_k\,\mathbf{V}_k^T\mathbf{S}\|_F^2 \tag{4.20a}$$

$$\text{subject to } \mathbf{V}_k^T\mathbf{V}_k = \mathbf{I}_k, \tag{4.20b}$$

where $\mathbf{I}_k \in \mathbb{R}^{k\times k}$ is the identity. Is is well known that the solution can be computed via the singular value decomposition (SVD) of $\mathbf{S}$, $\mathbf{S} = \mathbf{V\Sigma W}^T$. In fact, since $\mathbf{W}$ is orthogonal, $\|\mathbf{S} - \mathbf{V}_k\,\mathbf{V}_k^T\mathbf{S}\|_F^2 = \|\mathbf{V\Sigma} - \mathbf{V}_k\,\mathbf{V}_k^T\mathbf{V\Sigma}\|_F^2$. If $\mathbf{V}_k \in \mathbb{R}^{N\times k}$ is submatrix consisting of the first $k$ columns of $\mathbf{V} \in \mathbb{R}^{N\times N}$, and if $\Sigma_k \in \mathbb{R}^{N\times K}$ is obtained by replacing the singular values $\sigma_{k+1}, \sigma_{k+2},\ldots$ in $\Sigma \in \mathbb{R}^{N\times K}$ by zero, then

$$\begin{aligned}\|\mathbf{S} - \mathbf{V}_k\,\mathbf{V}_k^T\mathbf{S}\|_F^2 &= \|\mathbf{V\Sigma} - \mathbf{V}_k\,\mathbf{V}_k^T\mathbf{V\Sigma}\|_F^2 = \|\mathbf{V\Sigma} - \mathbf{V\Sigma}_k\|_F^2 = \|\Sigma - \Sigma_k\|_F^2 \\ &= \sum_{j=k+1}^{\min\{K,N\}} \sigma_j^2. \end{aligned} \tag{4.21}$$

---

**Algorithm 4.1 (POD)**

---

**Input**: Samples $\mathbf{S} = (\mathbf{s}_1,\ldots,\mathbf{s}_K) \in \mathbb{R}^{N\times K}$ and tolerance $\tau > 0$.
**Output**: $\mathbf{V}_k = (\mathbf{v}_1,\ldots,\mathbf{v}_k) \in \mathbb{R}^{N\times k}$.

1. Compute the singular value decomposition $\mathbf{S} = \mathbf{V\Sigma W}^T$.
2. Find smallest index $k$ such that the singular values satisfy $\sigma_{k+1} < \tau\sigma_1$.
3. Return the first $k$ columns $\mathbf{V}_k = (\mathbf{v}_1,\ldots,\mathbf{v}_k) \in \mathbb{R}^{N\times k}$ of $\mathbf{V}$.

---

Given the bound (4.21), the index $k$ is often chosen to be the smallest index such that $\sum_{j=k+1}^{\min\{K,N\}} \sigma_j^2 < \tau$. This requires computation of all singular values, which can be expensive. Therefore, we use the smallest index $k$ such that $\sigma_{k+1} < \tau\sigma_1$. This alternative provides a bound on the relative error in the two-norm: $\|\mathbf{S} - \mathbf{V}_k\,\mathbf{V}_k^T\mathbf{S}\|_2 \leq$

$\tau\|\mathbf{S}\|_2$. In our examples, the matrix of samples $\mathbf{S} \in \mathbb{R}^{N\times K}$ satisfies $K \ll N$ and we compute the so-called economy-sized SVD. In the large scale setting, we can use an iterative method (e.g. ARPACK) to compute just the largest $k$ singular values without computing all of them.

We note that often the snapshots do not have to be approximated in the Euclidean norm sense as in (4.19), but instead using a weighted dot product $\mathbf{v}_i^T\mathbf{M}\mathbf{s}_j$ and corresponding norm $\|\mathbf{s}\|_{\mathbf{M}}^2 = \mathbf{s}^T\mathbf{M}\mathbf{s}$, respectively, where $\mathbf{M} \in \mathbb{R}^{N\times N}$ is a symmetric positive definite matrix. This is for example the case when the snapshots $s_j(x) = \sum_{i=1}^N \mathbf{s}_{ij}\phi_i(x)$ belong to the Hilbert space $H_0^1(\Omega)$. In this case, $\mathbf{M}$ is the stiffness matrix. See, e.g., [19]. This can be accomplished by modifying the SVD.

### *4.3.2 The DEIM*

In this section we review the DEIM to approximate a function $\mathbf{G} : \mathbb{R}^k \to \mathbb{R}^N$. We require a subspace with basis $\{\mathbf{u}_1,\dots,\mathbf{u}_m\}$ such that $\mathbf{G}(\mathbf{z})$ is approximately contained in span$\{\mathbf{u}_1,\dots,\mathbf{u}_m\}$ for the arguments $\mathbf{z}$ of interest. Typically, one samples $\mathbf{G}$ and then applies the POD to the samples to obtain an orthonormal basis $\{\mathbf{u}_1,\dots,\mathbf{u}_m\}$. To obtain a computationally efficient DEIM approximation of $\mathbf{G}$ one needs that $m \ll N$.

The DEIM [7] can be viewed as variant of the empirical interpolation method of [2] (see also [14]) applied to large scale finite dimensional systems.

The DEIM computes indices $p_1,\dots,p_m$ in $\{1,\dots,N\}$ and an approximation $\widehat{\mathbf{G}} : \mathbb{R}^k \to \mathbb{R}^N$ of the function $\mathbf{G}$ which satisfies

$$\widehat{\mathbf{G}}_{p_i}(\mathbf{z}) = \mathbf{G}_{p_i}(\mathbf{z}) \quad \text{for } i = 1,\dots,m \tag{4.22}$$

moreover, for each $\mathbf{z}$ the computation of $\widehat{\mathbf{G}}(\mathbf{z})$ only requires the $m$ components $\mathbf{G}_{p_1}(\mathbf{z}),\dots,\mathbf{G}_{p_m}(\mathbf{z})$ of the original function $\mathbf{G}$. More specifically, if $\mathbf{e}_i$ is the $i$th unit vector in $\mathbb{R}^N$, $\mathbf{P} = [\mathbf{e}_{p_1},\dots,\mathbf{e}_{p_m}] \in \mathbb{R}^{N\times m}$ is the submatrix of the identity obtained by extracting the columns $p_1,\dots,p_m$, and $\mathbf{U} = [\mathbf{u}_1,\dots,\mathbf{u}_m]$, then the DEIM approximation of $\mathbf{G}$ is

$$\widehat{\mathbf{G}} = \mathbf{U}(\mathbf{P}^T\mathbf{U})^{-1}\mathbf{P}^T\mathbf{G} : \mathbb{R}^k \to \mathbb{R}^N.$$

Clearly, $\mathbf{P}^T\widehat{\mathbf{G}} = \mathbf{P}^T\mathbf{G}$, which verifies the interpolation property (4.22), and $\mathbf{P}^T\mathbf{G} = (\mathbf{G}_{p_1},\dots,\mathbf{G}_{p_m})^T$, which means that only the components $p_1,\dots,p_m$ of $\mathbf{G}$ are needed to compute the approximation $\widehat{\mathbf{G}}$. This is the source of the complexity reduction provided by the DEIM.

Before we review how DEIM computes the indices $p_1,\dots,p_m$ and the DEIM error bounds, we discuss when the DEIM approximation is useful. For example, in model reduction we have to evaluate the nonlinearity $(\widehat{\mathbf{y}};\theta) \mapsto \mathbf{V}_\ell^T\mathbf{F}(\overline{\mathbf{y}}+\mathbf{V}_r\widehat{\mathbf{y}};\theta)$, where $\mathbf{F} : \mathbb{R}^N \times \mathbb{R}^p \to \mathbb{R}^N$ and $\mathbf{V}_\ell$ and $\mathbf{V}_r \in \mathbb{R}^{N\times n}$ with $n \ll N$. As we have mentioned, this requires the computation of $\overline{\mathbf{y}}+\mathbf{V}_r\widehat{\mathbf{y}}$, the evaluation of the nonlinearity $\mathbf{F}(\overline{\mathbf{y}}+\mathbf{V}_r\widehat{\mathbf{y}};\theta)$ and the projection $\mathbf{V}_\ell^T\mathbf{F}(\overline{\mathbf{y}}+\mathbf{V}_r\widehat{\mathbf{y}};\theta)$. All of these operations depend on the size $N$ of the full system and, therefore, the evaluation of the reduced order model is almost expensive as that of the full order model. The complexity of the reduced order model can be made independent of the full order problem size $N$ using the

DEIM approximation. If we compute a DEIM approximation

$$\widehat{\mathbf{F}} = \mathbf{U}(\mathbf{P}^T\mathbf{U})^{-1}\mathbf{P}^T\mathbf{F},$$

then we can approximate the nonlinearity $\mathbf{V}_\ell^T\mathbf{F}(\overline{\mathbf{y}}+\mathbf{V}_r\widehat{\mathbf{y}};\theta)$ by

$$\mathbf{V}_\ell^T\widehat{\mathbf{F}}(\overline{\mathbf{y}}+\mathbf{V}_r\widehat{\mathbf{y}};\theta) = \left(\mathbf{V}_\ell^T\mathbf{U}(\mathbf{P}^T\mathbf{U})^{-1}\right)\mathbf{P}^T\mathbf{F}(\overline{\mathbf{y}}+\mathbf{V}_r\widehat{\mathbf{y}};\theta). \tag{4.23}$$

Typically in problems arising from spatial discretization of a PDE, the $i$th component of $\mathbf{F}$ depends only on a few components of $\mathbf{y}$. Hence, the evaluation of the $m$ components $\mathbf{P}^T\mathbf{F}(\overline{\mathbf{y}}+\mathbf{V}_r\widehat{\mathbf{y}};\theta)$ of the nonlinearity requires only a few, say $O(m)$ components of $\overline{\mathbf{y}}+\mathbf{V}_r\widehat{\mathbf{y}}$. Hence we do not need to compute $\overline{\mathbf{y}}+\mathbf{V}_r\widehat{\mathbf{y}}$ at a cost of $2Nn+N$ flops (we count multiplication and addition as a flop), but only some components of this vector at a cost of $O(mn)$. Furthermore, the matrix $\mathbf{V}_\ell^T\mathbf{U}(\mathbf{P}^T\mathbf{U})^{-1} \in \mathbb{R}^{n\times m}$ can be precomputed so that afterwards the evaluation of $(\widehat{\mathbf{y}};\theta) \mapsto \mathbf{V}_\ell^T\widehat{\mathbf{F}}(\overline{\mathbf{y}}+\mathbf{V}_r\widehat{\mathbf{y}};\theta)$ defined in (4.23) requires only $O(mn)$ operations. In finite difference approximations, the $i$th component of the nonlinearity $\mathbf{F}$ typically depends only on the $i$th component of the argument $\mathbf{y}$. Finite difference approximations are used, e.g., in the examples in [7,8]. If finite element methods are used, the $i$th component of the nonlinearity $\mathbf{F}$ depends on more than the $i$th component of the argument. The dependency of the $i$th component of $\mathbf{F}$ on the components of the argument depends on the polynomial order used in the finite element method, on the mesh, and also in what stage of the finite element assembly process the DEIM is applied. We will explore this in Sec. 4.4.

---

**Algorithm 4.2 (DEIM)**

---

**Input**: Linearly independent vectors $\mathbf{u}_1,\ldots,\mathbf{u}_m$.
**Output**: Indices $p_1,\ldots,p_m$.

1. $[\rho, p_1] = \max\{|\mathbf{u}_1|\}$
2. Set $\mathbf{U} = [\mathbf{u}_1]$, $\mathbf{P} = [\mathbf{e}_{p_1}]$, $\mathbf{p} = [p_1]$
3. For $i = 2,\ldots,m$ do
   a. Solve $(\mathbf{P}^T\mathbf{U})\mathbf{c} = \mathbf{P}^T\mathbf{u}_i$ for $\mathbf{c}$
   b. $\mathbf{r}_i = \mathbf{u}_i - \mathbf{U}\mathbf{c}$
   c. $[\rho, p_i] = \max\{|\mathbf{r}_i|\}$
   d. Update $\mathbf{U} = [\mathbf{U} \quad \mathbf{u}_i]$, $\mathbf{P} = [\mathbf{P} \quad \mathbf{e}_{p_i}]$, $\mathbf{p} = [\mathbf{p}^T \quad p_i]^T$

---

We next state an error estimate from [7] for the DEIM approximation

$$\widehat{\mathbf{G}} = \mathbf{U}(\mathbf{P}^T\mathbf{U})^{-1}\mathbf{P}^T\mathbf{G}$$

to $\mathbf{G}$. If $\mathbf{U} \in \mathbb{R}^{N\times m}$ has ortho-normal columns, then

$$\|\mathbf{G}-\widehat{\mathbf{G}}\|_2 \le \|(\mathbf{P}^T\mathbf{U})^{-1}\|_2\|(\mathbf{I}-\mathbf{U}\mathbf{U}^T)\mathbf{G}\|_2. \tag{4.24}$$

This result indicates that very little accuracy is lost when the orthogonal projection of POD is replaced by the DEIM interpolatory projection so long as $\|(\mathbf{P}^T\mathbf{U})^{-1}\|_2$ is of modest size. In practice, we simply compute this quantity and use it as an a-posteriori estimate. The greedy DEIM index selection actually limits the growth of $\|(\mathbf{P}^T\mathbf{U})^{-1}\|_2$ and typically it has remained on the order of 100 or less in all of the examples we have considered. Finally, we must emphasize that the DEIM does not improve the accuracy of the POD reduced model. The sole benefit of the DEIM is to greatly reduce the complexity of evaluating the reduced model.

## 4.4 Evaluation of Nonlinear Functions Arising in Finite Element Methods Using DEIM

We study the application of the DEIM for the evaluation of nonlinear terms in finite element models. As noted in Sect. 4.3.2, the main issue here is the computational complexity of the DEIM reduced model. It depends on how may components of the argument influence a component of the nonlinearity, and it is determined by the finite elements used. We present two ways of applying the DEIM. One approach applies DEIM to the assembled form of the nonlinear term, the other approach, originally suggested by Dedden et al. [9, 33], to the unassembled form.

We use the semilinear advection diffusion reaction equation from Sect. 4.2.1 and continuous finite element approximations. However, the approaches can easily be extended to other equations and discontinuous Galerkin methods.

### *4.4.1 The Reduced Order Model*

We consider the finite element discretization of the semilinear advection diffusion reaction equation discussed in Sect. 4.2.1. To simplify our notation, we assume that the boundary data $h(x) = 0$ in (4.3). The finite element discretization of (4.3) leads to the $N \times N$ system of nonlinear equations

$$\mathbf{A}\mathbf{y} + \mathbf{F}(\mathbf{y};\theta) = \mathbf{b}, \tag{4.25}$$

where $\mathbf{A} \in \mathbb{R}^{N\times N}$ and $\mathbf{F} : \mathbb{R}^N \times \mathbb{R}^p \to \mathbb{R}^N$ are given by (4.11). Note that since $h(x) = 0$, the vector $\mathbf{b} = \mathbf{0} \in \mathbb{R}^N$.

Assume we have generated $\mathbf{V}_\ell$ and $\mathbf{V}_r \in \mathbb{R}^{N\times n}$ with $n \ll N$. Then the reduced order model of (4.25) is

$$\mathbf{V}_\ell^T\mathbf{A}(\overline{\mathbf{y}} + \mathbf{V}_r\widehat{\mathbf{y}}) + \mathbf{V}_\ell^T\mathbf{F}(\overline{\mathbf{y}} + \mathbf{V}_r\widehat{\mathbf{y}};\theta) = \mathbf{V}_\ell^T\mathbf{b}. \tag{4.26}$$

As we have mentioned before, $\mathbf{V}_\ell^T\mathbf{A}\mathbf{V}_r$, $\mathbf{V}_\ell^T\mathbf{A}\overline{\mathbf{y}}$ and $\mathbf{V}_\ell^T\mathbf{b}$ can be precomputed, but since the nonlinearity depends on $\widehat{\mathbf{y}}$ and $\theta$ the term $\mathbf{V}_\ell^T\mathbf{F}(\overline{\mathbf{y}} + \mathbf{V}_r\widehat{\mathbf{y}};\theta)$ needs to be evaluated whenever $\widehat{\mathbf{y}}$ or $\theta$ changes, and the cost of evaluating this nonlinearity still depends on the size $N$ of the full order model.

To reduce the complexity of the nonlinear term, we apply the DEIM. The DEIM reduced order model is given by

$$\mathbf{V}_\ell^T\mathbf{A}(\overline{\mathbf{y}} + \mathbf{V}_r\widehat{\mathbf{y}}) + \left(\mathbf{V}_\ell^T\mathbf{U}(\mathbf{P}^T\mathbf{U})^{-1}\right)\mathbf{P}^T\mathbf{F}(\overline{\mathbf{y}} + \mathbf{V}_r\widehat{\mathbf{y}};\theta) = \mathbf{V}_\ell^T\mathbf{b}. \tag{4.27}$$

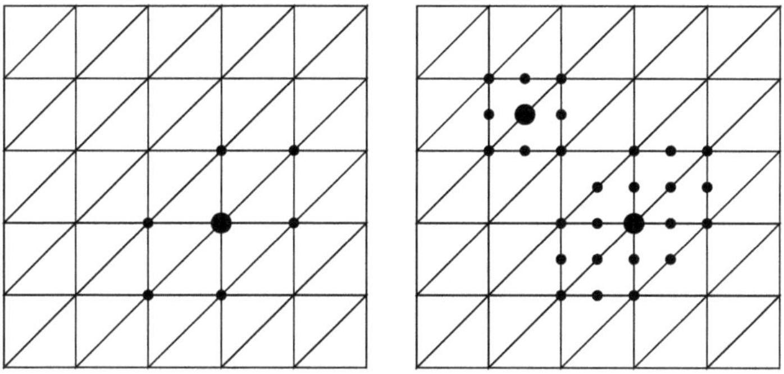

**Fig. 4.1.** Left plot: Piecewise linear finite elements on triangles. If the DEIM index $p_i$ corresponds to the vertex indicated by the large dot, then the $p_i$th component of the nonlinear function depends on the seven adjacent vertices indicated by dots. Right plot: Piecewise quadratic finite elements on triangles. If the DEIM index $p_i$ corresponds to the vertex indicated by the large dot, then the $p_i$th component of the nonlinear function depends on nineteen adjacent nodes indicated by dots. If the DEIM index $p_i$ corresponds to the midpoint indicated by the large dot, then the $p_i$th component of the nonlinear function depends on nine adjacent nodes indicated by dots

The $n \times m$ matrix $\mathbf{V}_\ell^T \mathbf{U}(\mathbf{P}^T \mathbf{U})^{-1}$ can be precomputed once. We still need to study the complexity of the evaluation of the nonlinearity

$$\mathbf{P}^T \mathbf{F}(\overline{\mathbf{y}} + \mathbf{V}_r \widehat{\mathbf{y}}; \theta) = \Big( \mathbf{F}_{p_1}(\overline{\mathbf{y}} + \mathbf{V}_r \widehat{\mathbf{y}}; \theta), \ldots, \mathbf{F}_{p_m}(\overline{\mathbf{y}} + \mathbf{V}_r \widehat{\mathbf{y}}; \theta) \Big)^T \in \mathbb{R}^m$$

in the DEIM reduced model (4.27). The $i$th component $\mathbf{F}_i$ of the nonlinearity depends on all components $\mathbf{y}_j$ for which the intersection of the support of basis functions $\phi_i$ and $\phi_j$ does not have measure zero. See (4.11b).

This is illustrated in Fig. 4.1 for piecewise linear (left plot) and piecewise quadratic (right plot) basis functions $\phi_i$ on triangles. In the case of piecewise linear basis functions, there are $N = 36$ degrees of freedom, which correspond to the vertices. If the DEIM index $p_i$ corresponds to the vertex indicated by the large dot, then the $p_i$th component of $\mathbf{F}$ depends on seven components of $\mathbf{y}$, which corresponds to the vertices indicated by dots. If piecewise quadratic basis functions are used, then there are $N = 121$ degrees of freedom, which correspond to the vertices and edge midpoints. If the DEIM index $p_i$ corresponds to the vertex indicated by the large dot, then the $p_i$th component of $\mathbf{F}$ depends on nineteen components of $\mathbf{y}$, which corresponds to the vertices and edge midpoints indicated by dots in the bottom right part of the right plot in Fig. 4.1. On the other hand, if the DEIM index $p_i$ corresponds to a midpoint, then this midpoint is shared by only two triangles, and the $p_i$th component of $\mathbf{F}$ depends on nine components of $\mathbf{y}$, which corresponds to the vertices and edge midpoints indicated by dots in the top left part of the right plot in Fig. 4.1.

An alternative DEIM reduced order model is obtained when we consider the unassembled nonlinearity. As we have mentioned earlier, this was first suggested and explored by [9,33]. Since $\int_\Omega = \sum_{e=1}^{n_e} \int_{\Omega_e}$, we can write (4.11b) as

$$\mathbf{F}_i(\mathbf{y};\theta) = \sum_{e=1}^{n_e} \int_{\Omega_e} f\Big(\sum_{j=1}^{N} y_j\phi_j;\theta\Big)\phi_i + \tau_e\,(\beta\cdot\nabla\phi_i)\, f\Big(\sum_{j=1}^{N} y_j\phi_j;\theta\Big)\,dx.$$

When the intersection of the supports of the basis functions $\phi_i$ and $\phi_j$ and of the element $\Omega_e$ has measure zero, the integral $\int_{\Omega_e} f(\sum_{j=1}^N y_j\phi_j;\theta)\phi_i + \tau_e\,(\beta\cdot\nabla\phi_i) f(\sum_{j=1}^N y_j\phi_j;\theta)\,dx$ is zero. Therefore for nodal basis functions, this integral can only be nonzero when the indices $i$ and $j$ correspond to nodes in $\overline{\Omega_e}$. For each of the $n_e$ elements $\Omega_e$ we can compute $n_p$ integrals

$$\mathbf{F}_i^e(\mathbf{y};\theta) = \int_{\Omega_e} f\Big(\sum_{j=1}^{N} y_j\phi_j;\theta\Big)\phi_i + \tau_e\,(\beta\cdot\nabla\phi_i)\, f\Big(\sum_{j=1}^{N} y_j\phi_j;\theta\Big)\,dx, \tag{4.28a}$$

where $n_p$ is the number of degrees of freedom per element and the indices $i$ corresponds to nodes in the element $\overline{\Omega_e}$. This gives a function $\mathbf{F}^e(\mathbf{y};\theta) : \mathbb{R}^{n_e n_p} \times \mathbb{R}^p \to \mathbb{R}^{n_e n_p}$. Then we can assemble the element information into the global vector of unknowns $\mathbf{F}$. This can be expressed as

$$\mathbf{F}(\mathbf{y};\theta) = \mathbf{Q}\,\mathbf{F}^e(\mathbf{y};\theta) \tag{4.28b}$$

where $\mathbf{Q} \in \mathbb{R}^{N\times(n_e n_p)}$. The size of the unassembled nonlinearity $\mathbf{F}^e$ is larger than that of the assembled one $\mathbf{F}$. If the $i$th component of the unassembled nonlinearity belongs to element $\Omega_e$, then $\mathbf{F}_i^e$ only depends on the unknowns $y_j$ with indices $j$ corresponding to nodes in the element $\Omega_e$, see Fig. 4.2. Consequently, a component

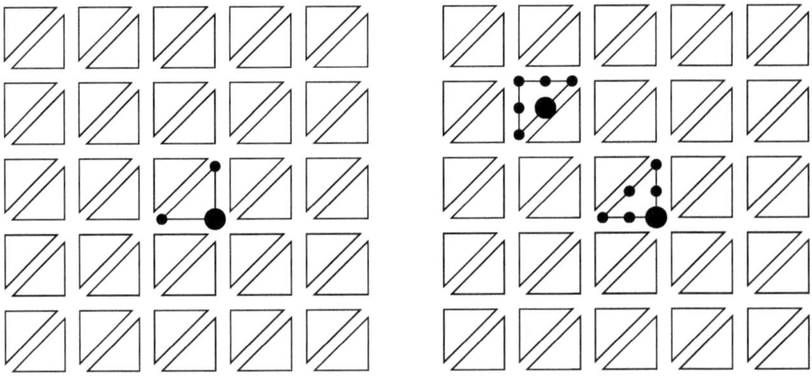

**Fig. 4.2.** If DEIM is applied to unassembled piecewise linear elements, then the $p_i$-th component of the unassembled nonlinearity only depends on values at nodes in the element that contains the node $p_i$. Left plot: For piecewise linear elements on triangles, the $p_i$-th component of the unassembled nonlinearity only depends on the values at the three vertices, indicated by dots, of one triangle. Right plot: For piecewise quadratic elements on triangles, the $p_i$-th component of the unassembled nonlinearity only depends on the values at the vertices and edge midpoints, indicated by dots, of one triangle

of unassembled nonlinearity depends on fewer components than a component of assembled nonlinearity does.

The reduced order model (4.26) can now be written as

$$\mathbf{V}_\ell^T \mathbf{A}(\bar{\mathbf{y}} + \mathbf{V}_r \widehat{\mathbf{y}}) + \mathbf{V}_\ell^T \mathbf{Q} \mathbf{F}^e(\bar{\mathbf{y}} + \mathbf{V}_r \widehat{\mathbf{y}}; \theta) = \mathbf{V}_\ell^T \mathbf{b}. \tag{4.29}$$

We can apply DEIM to the unassembled nonlinearity. Let the columns of $\mathbf{U}^e = [\mathbf{u}_1^e, \ldots, \mathbf{u}_{m^e}^e]$ be a basis of a subspace that approximately contains $\mathbf{F}^e(\mathbf{y}; \theta)$ for the arguments $\mathbf{y}$ and $\theta$ of interest. The DEIM approximation of the unassembled nonlinearity is given by

$$\widehat{\mathbf{F}}^e(\mathbf{y}; \theta) = \mathbf{U}^e \big((\mathbf{P}^e)^T (\mathbf{U}^e)\big)^{-1} (\mathbf{P}^e)^T \mathbf{F}^e(\mathbf{y}; \theta).$$

Here $\mathbf{P}^e$ is the sub matrix of the identity generated using the indices $p_1^e, \ldots, p_{m^e}^e$ generated by the DEIM applied to $\mathbf{u}_1^e, \ldots, \mathbf{u}_{m^e}^e$.

If we insert this into (4.29) we arrive at the DEIM reduced order model

$$\mathbf{V}_\ell^T \mathbf{A} \mathbf{V}_r(\bar{\mathbf{y}} + \mathbf{V}_r \widehat{\mathbf{y}}) + \left(\mathbf{V}_\ell^T \mathbf{Q} \mathbf{U}^e \big((\mathbf{P}^e)^T (\mathbf{U}^e)\big)^{-1}\right) (\mathbf{P}^e)^T \mathbf{F}^e(\bar{\mathbf{y}} + \mathbf{V}_r \widehat{\mathbf{y}}; \theta) = \mathbf{V}_\ell^T \mathbf{b}. \tag{4.30}$$

The $n \times m^e$ matrix $\mathbf{V}_\ell^T \mathbf{Q} \mathbf{U}^e \big((\mathbf{P}^e)^T (\mathbf{U}^e)\big)^{-1}$ can be precomputed.

The advantage of the DEIM reduced order model (4.30) over (4.27) is that each component of the unassembled nonlinearity in (4.30) depends on fewer components of the argument than the nonlinearity in (4.27) does. Hence, if the dimension of the subspace $\mathscr{R}(\mathbf{U})$ containing the image of $\mathbf{F}$ is roughly equal to dimension of the subspace $\mathscr{R}(\mathbf{U}^e)$ containing the image of $\mathbf{F}^e$, i.e., if $m \approx m^e$, then the evaluation of (4.30) is computationally less expensive than that of (4.27). This is illustrated in Fig. 4.2. If a DEM point $p_i$ corresponds to a node in a triangle, the the $p_i$th nonlinearity depends on all components of the argument that correspond to nodes in the triangle. The left plot in Fig. 4.2 illustrates this for one point when piecewise linear elements are used, whereas the right plot in Fig. 4.2 illustrates this when piecewise quadratic elements are used. Note, that if the unassembled form of the nonlinearity is used, the connectivity is the same no matter whether the DEIM point corresponds to an vertex or an edge midpoint.

The disadvantage of the DEIM reduced order model (4.30) compared to (4.27) is that the size of the unassembled nonlinearity $\mathbf{F}^e(\mathbf{y}; \theta)$ is significantly larger than the size $N$ of the nonlinearity $\mathbf{F}(\mathbf{y}; \theta)$. The size $n_e n_p$ of the unassembled nonlinearity $\mathbf{F}^e(\mathbf{y}; \theta)$ now depends on the number $n_e$ of elements and the number $n_p$ of degrees of freedom $n_p$ per element. For example, if we use piecewise linear basis functions on the mesh in the left plot in Fig. 4.1, there are $N = 36$ vertices, where as $n_e n_p = 150$. If we use piecewise quadratic basis functions on the mesh in the right plot in Fig. 4.1, then there are $N = 121$ degrees of freedom, whereas $n_e n_p = 300$. Since the vectors $\mathbf{u}_1, \ldots, \mathbf{u}_m$ and $\mathbf{u}_1^e, \ldots, \mathbf{u}_{m^e}^e$ are typically computed from a POD of samples of the nonlinearities $\mathbf{F}$ and $\mathbf{F}^e$, respectively, the computation of the vectors $\mathbf{u}_1^e, \ldots, \mathbf{u}_{m^e}^e$ is more expensive than the computation of $\mathbf{u}_1, \ldots, \mathbf{u}_m$. However, this computation is done in the off-line phase.

## 4.4.2 Numerical Examples

We apply DEIM reduced order models to approximate the semilinear advection diffusion reaction equation (4.3) with nonlinearity (4.4). The full order model is obtained using the SUPG stabilized finite elements reviewed in Sect. 4.2.1. The diffusivity is $\nu = 5 \cdot 10^{-6}$, and the parameters $C = 0.2$ and $D = 0.4$ in (4.4) are fixed and $\theta = (\ln(A), E)$ vary within $\Theta \equiv [5.00, 7.25] \times [0.05, 0.15] \subset \mathbb{R}^2$.

To construct the reduced basis matrices $\mathbf{V} = \mathbf{V}_\ell = \mathbf{V}_r$, we sample the finite element solution of (4.4) at 25 parameters. We denote these solutions by $\mathbf{y}(\theta_1), \ldots, \mathbf{y}(\theta_{25})$. We compute the mean $\bar{\mathbf{y}} = \frac{1}{25}\sum_{i=1}^{25} \mathbf{y}(\theta_i)$, and generate the reduced basis matrices $\mathbf{V} = \mathbf{V}_\ell = \mathbf{V}_r$ by applying the POD, Algorithm 4.1 to the samples $\mathbf{y}(\theta_1) - \bar{\mathbf{y}}, \ldots, \mathbf{y}(\theta_{25}) - \bar{\mathbf{y}}$ with tolerance $\tau = 10^{-4}$. To construct $\mathbf{U} = [\mathbf{u}_1, \ldots, \mathbf{u}_m]$ and $\mathbf{U}^e = [\mathbf{u}_1^e, \ldots, \mathbf{u}_{m^e}^e]$ for the DEIM approximation, we sample the nonlinearities $\mathbf{F}(\mathbf{y}(\theta); \theta)$ and $\mathbf{F}^e(\mathbf{y}(\theta); \theta)$, respectively, at the same parameters used to construct $\mathbf{V}$, and then we apply the POD with tolerance $\tau = 10^{-4}$ to obtain $\mathbf{U}$ and $\mathbf{U}^e$, respectively.

All computations in this subsections were done using Matlab on MacBook Air with 8GB of memory and 1.8 GHz Intel Core i5 processor. The nonlinear full order or reduced order models are solved using Newton's method. The linear systems in Newton's method are solved using the Matlab backslash command.

#### 4.4.2.1 2D Example

We consider the domain $\Omega \subset \mathbb{R}^2$ shown in Fig. 4.3, taken from [3]. The Dirichlet boundary segments are $\Gamma_D = \{(0, x_2) : x_2 \in (0,2) \cup (2.75, 4.25) \cup (5,7)\}$ and the Dirichlet data $h$ is specified in Fig, 4.3.

To study the computational cost of applying DEIM reduced order models we use three meshes, referred to as Mesh 1 to Mesh 3, of different sizes, and we use piecewise linear and quadratic elements. We compute an approximate solution of (4.3) at the parameter $(\ln(A), E) = (6.4, 0.11)$ not contained in the parameter sample. Figure 4.4 shows the triangulation corresponding to Mesh 2, of medium size, as well as the full order model solution of (4.3). (The reduced order model solutions are indistinguishable from the full order model solution.)

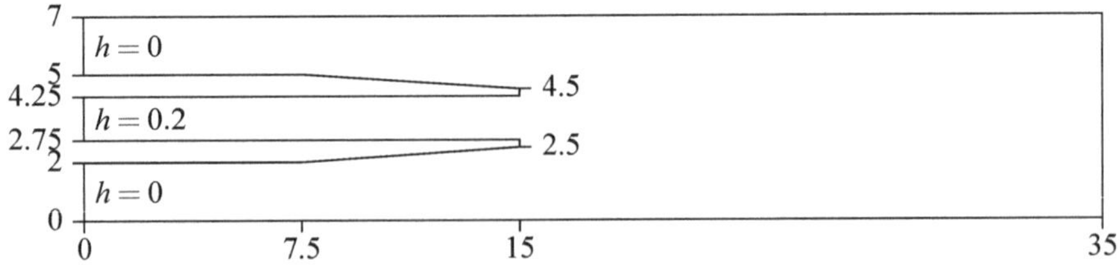


**Fig. 4.3.** 2D Example: The domain $\Omega$ with Dirichlet boundary segments $\Gamma_D = \{(0, x_2) : x_2 \in (0,2) \cup (2.75, 4.25) \cup (5,7)\}$ and Dirichlet data $h$

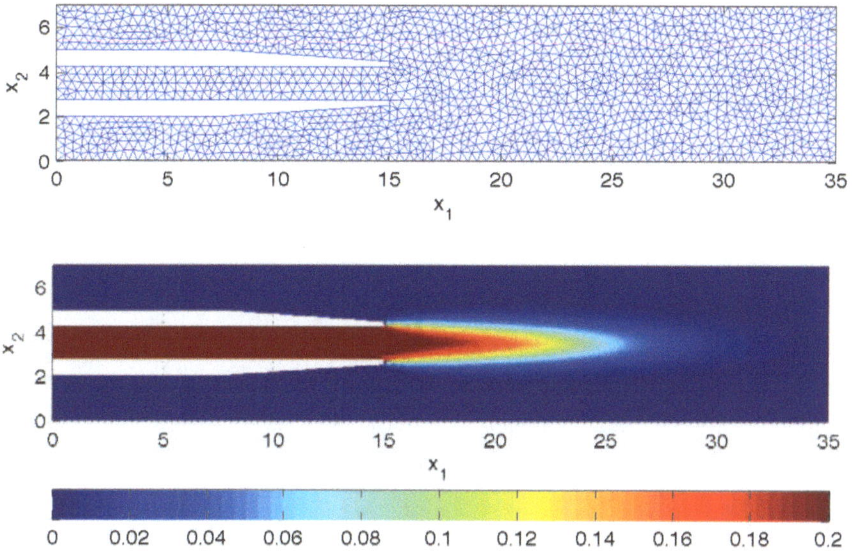


**Fig. 4.4.** 2D Example: A triangulation of the domain $\Omega$ (top plot) and solution of the advection diffusion reaction equation (4.3) with parameter $(\ln(A), E) = (6.4, 0.11)$ (bottom plot)

As we have described in the previous section, the complexity of evaluating DEIM reduced order models depends on the connectivity of the nodes in the finite element mesh. We illustrate this in Fig. 4.5 using Mesh 2. For four different configurations, we plot the triangles that are involved in the evaluation of the DEIM nonlinear term. More precisely, the degrees of freedom corresponding to all nodes in the red solid triangles are needed to evaluate the DEIM nonlinear term. The top two plots correspond to piecewise linear finite elements using the assembled (top plot) and unassembled (second from top plot) form of the nonlinearity. The top two plots in Fig. 4.5 correspond to the schematic plots on the left in Figs. 4.1 and 4.2, respectively.

The bottom two plots in Fig. 4.5 correspond to quadratic finite elements. If we look at the third plot from the top, which colors the triangles involved in the evaluation of the DEIM nonlinear functions (assembled form), then at most two triangles are connected. This means that all DEIM points in this case correspond to edge midpoints (see the right plot in Fig. 4.2). We observed the same for the computations on Mesh 1 and Mesh 3. The bottom plot in Fig. 4.5 corresponds to the unassembled form of the DEIM using quadratic finite elements. In this plot a few adjacent triangles are colored red, which simply means that the DEIM selected points that happen to correspond to nodes in adjacent triangles.

Table 4.1 summarizes the problem size for the different models for the three meshes and piecewise linear and quadratic finite elements. In Tables 4.1 to 4.3, DEIM refers to the DEIM reduced order model (4.27) obtained using the assembled form of the nonlinearity, whereas DEIM-u refers to the DEIM reduced order model (4.30) obtained using the unassembled form of the nonlinearity.

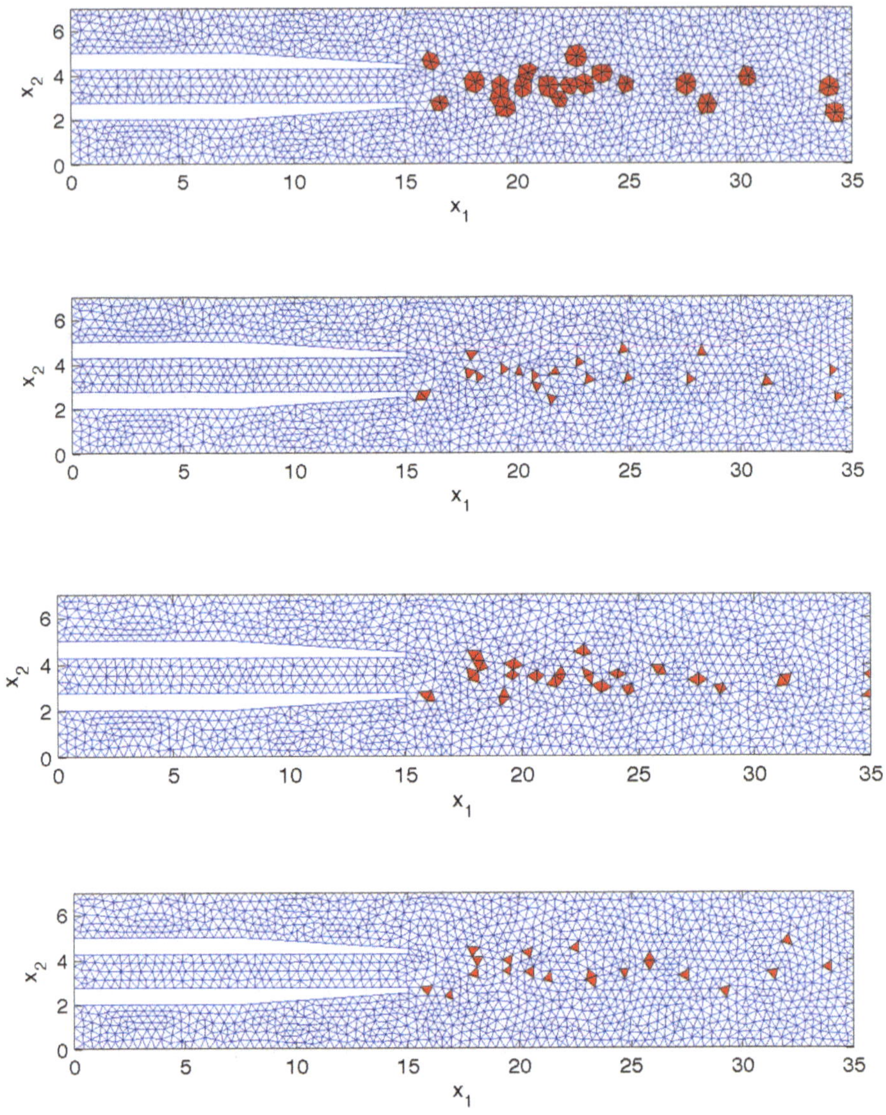


**Fig. 4.5.** 2D Example: The triangles that contain DEIM points are shown in solid red. The different plots correspond to different polynomial degree used in the FEM and application of the DEIM to the assembled or unassembled form of the nonlinearity

The computing times to evaluate the full and the various reduced order models are shown in Table 4.2. The nonlinear systems are solved using Newton's method and the computing times listed are for the Newton solve (and not for one Newton iteration). The number of Newton iterations required are shown in parenthesis. The

**Table 4.1.** 2D Example: The size $N$ of the full order finite element system, the number of POD basis vectors $n$, the number of DEIM points $m$ the number of nodes adjacent to DEIM points, the number of DEIM points $m^e$ when the unassembled (DEIM-u) nonlinearity is used, and the number of nodes adjacent to DEIM points for piecewise linear and quadratic finite elements on three grids. The mesh in Fig. 4.5 correspond to grid number 2

| *Polynomial degree* | $p = 1$ | | | $p = 2$ | | |
|---|---|---|---|---|---|---|
| *Mesh number* | 1 | 2 | 3 | 1 | 2 | 3 |
| number of triangles | 1,437 | 3,213 | 12,976 | 1,437 | 3,213 | 12,976 |
| number of nodes $N$ | 825 | 1,768 | 6,813 | 3,089 | 6,751 | 26,604 |
| number of POD basis vectors $n$ | 17 | 17 | 17 | 17 | 17 | 17 |
| number of DEIM points $m$ | 20 | 20 | 21 | 21 | 21 | 21 |
| number of nodes adjacent to DEIM pts. | 107 | 139 | 166 | 165 | 174 | 186 |
| number of DEIM-u points $m^e$ | 20 | 20 | 21 | 20 | 21 | 21 |
| number of nodes adjacent to DEIM-u pts. | 48 | 56 | 63 | 111 | 117 | 126 |

**Table 4.2.** 2D Example: The computing times (in sec) and the number of Newton iterations (in parenthesis) needed to solve the full order model, the POD reduced order model, the POD-DEIM reduced order model, and the POD-DEIM-u (unassembled) reduced order model for different grid levels and linear and quadratic finite elements

| *Polynomial degree* | $p = 1$ | | | $p = 2$ | | |
|---|---|---|---|---|---|---|
| *Mesh number* | 1 | 2 | 3 | 1 | 2 | 3 |
| Full | 0.55 (4) | 0.41 (4) | 1.49 (4) | 0.85 (4) | 1.36 (4) | 5.75 (4) |
| POD | 0.17 (4) | 0.29 (4) | 1.24 (4) | 0.51 (4) | 1.03 (4) | 3.77 (4) |
| POD-DEIM | 0.04 (4) | 0.04 (4) | 0.02 (4) | 0.08 (4) | 0.07 (4) | 0.12 (4) |
| POD-DEIM-u | 0.11 (8) | 0.05 (5) | 0.04 (5) | 0.13 (5) | 0.07 (4) | 0.08 (4) |

computing times do not include the time needed to compute the matrices $\mathbf{V}$, $\mathbf{U}$, or $\mathbf{U}^e$ via POD.

In this application, the solution of the POD-DEIM-u reduced order model required more Newton iterations in several cases, offsetting the gain in computational complexity of the POD-DEIM-u reduced order model nonlinearity. Another issue that makes computing time comparisons difficult using Matlab is that the computing time is often not determined by how many floating point operations are executed, but instead by how well the code is vectorized. We have made a great effort to vectorize the code for all models as much as possible, this is more effective for the full order and the POD reduced order models because by design the POD-DEIM and POD-DEIM-u reduced order models work with shorter vectors. Therefore the Matlab timings for the smaller problems likely do not accurately reflect what would be observed with, say, C code. However, from the Table 4.2 we can infer that POD reduced order models are only slightly more computationally efficient than the full order model. Applying DEIM for the assembled or unassembled form of the nonlin-

**Table 4.3.** 2D Example: Errors between the full order model solution and the POD reduced order model solution, the POD-DEIM reduced order model solution, and the POD-DEIM-u (unassembled) reduced order model solution

| *Polynomial degree* | $p=1$ | | | $p=2$ | | |
|---|---|---|---|---|---|---|
| *Mesh number* | 1 | 2 | 3 | 1 | 2 | 3 |
| POD | 7.8e-5 | 1.5e-4 | 3.5e-4 | 1.3e-4 | 2.5e-4 | 6.9e-4 |
| POD-DEIM | 7.8e-5 | 9.4e-5 | 4.8e-4 | 2.5e-4 | 2.6e-4 | 7.9e-4 |
| POD-DEIM-u | 1.2e-4 | 1.5e-4 | 2.2e-4 | 1.4e-4 | 1.8e-4 | 6.3e-4 |

earity results in significant computational savings compared to both the full and the POD reduced order models when applied to larger problems. For larger problems the POD-DEIM-u reduced order model nonlinearities can be evaluated more efficiently than the POD-DEIM reduced order model nonlinearities. Different reduced order models may require different numbers of Newton iterations. In this example, the number of Newton iterations needed to solve the POD-DEIM-u reduced order model was at least as large as the number of Newton iterations needed to solve the POD-DEIM reduced order model. If the Newton iterations needed to solve the POD-DEIM-u reduced order model is larger, then the gains in efficiency of evaluating the nonlinearity is offset by the larger number of Newton iterations.

The errors between the full order model solution and the reduced order model solutions shown in Table 4.3 are of the order of the tolerance $\tau = 10^{-4}$ used to construct the bases with the POD.

#### 4.4.2.2 3D Example

The domain is the cube $\Omega = (0,18) \times (0,9) \times (0,9)$ (in [mm]). The left face $\partial\Omega_D = \{0\} \times [0,9] \times [0,9]$ is the Dirichlet boundary, all other faces corresponds to Neumann boundaries $\partial\Omega_N$. On the part $\{0\} \times [3,6] \times \times [3,6]$ of the Dirichlet boundary we impose the Dirichlet conditions $y = 0.2$ and on the remainder of $\partial\Omega_D$ impose $y = 0$. This is the problem setup used in [11].

For the numerical solution, we use SUPG stabilized piecewise linear FEM on tetrahedra. To discretize the domain, $\Omega$ is divided into cubes of size $h \times h \times h$ and then each cube is divided into six tetrahedra. We use three meshes, Mesh 1 to Mesh 3, with $h = 1.125$, $h = 0.5625$, and $h = 0.375$, respectively. Mesh 2 is shown in the left plot in Fig. 4.6. The full order model solution of (4.3) parameter $(\ln(A), E) = (6.4, 0.11)$ is shown in the right plot in Fig. 4.6. (The reduced order model solutions are indistinguishable from the full order model solution.) For reasons explained below, we only apply piecewise quadratic finite elements on Meshes 1 and 2, but not on Mesh 3.

Figure 4.7 shows the tetrahedra in Mesh 2 that contain a node corresponding to a DEIM point. The plots in the left column correspond to the DEIM applied to the assembled form of the nonlinearity. In case of quadratic elements, the nodes are ei-

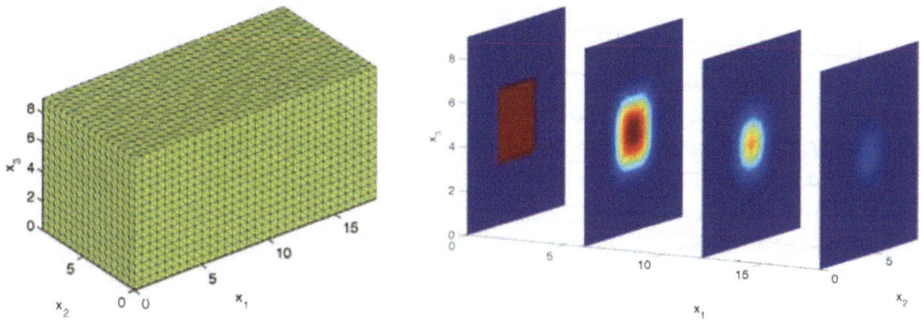

**Fig. 4.6.** 3D Example: Partitioning of the domain $\Omega$ into tetrahedra (left plot) and solution of the advection diffusion reaction equation (4.3) with parameter $(\ln(A), E) = (6.4, 0.11)$ (right plot)

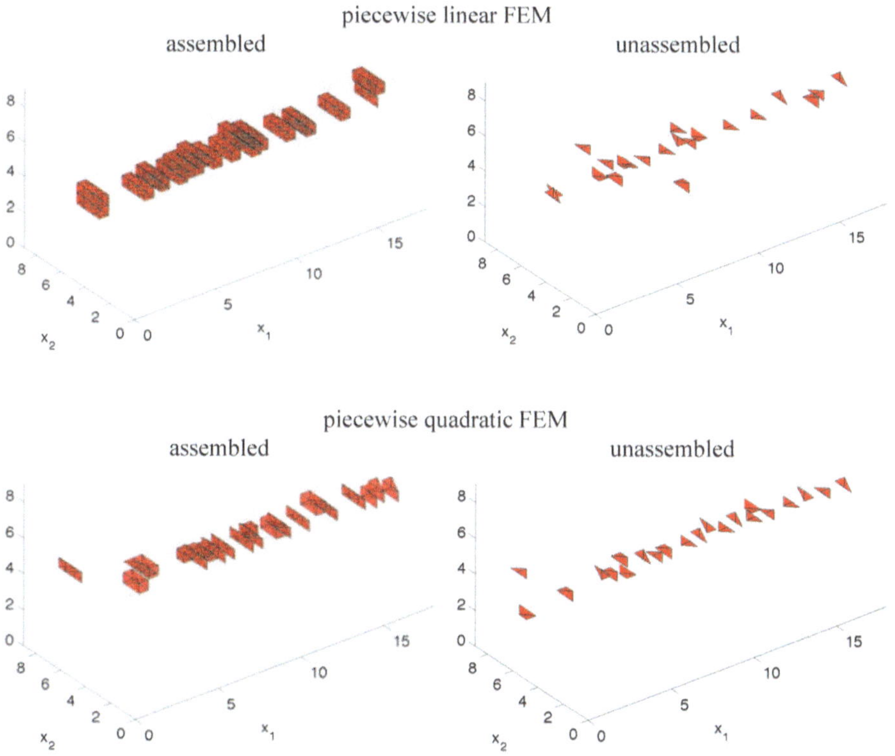


**Fig. 4.7.** 3D Example: The tetrahedra that contain DEIM points are shown. The different plots correspond to different polynomial degree used in the FEM and application of the DEIM to the assembled or unassembled form of the nonlinearity.

ther vertices or are edge midpoints. If we use Mesh 1, only one of the 21 DEIM points corresponds to a vertex. If we use Mesh 2, then none of the 21 DEIM points corresponds to a vertex. Since vertices are shared by more tetrahedra than edge midpoints, this means that DEIM points corresponding to edge midpoints lead to DEIM reduced order nonlinearities that can be evaluated more efficiently.

Table 4.4 summarizes the problem size for the different models for the three meshes and piecewise linear and quadratic finite elements. As before, in Tables 4.4 to 4.6, DEIM refers to the DEIM reduced order model (4.27) obtained using the assembled form of the nonlinearity, whereas DEIM-u refers to the DEIM reduced order model (4.30) obtained using the unassembled form of the nonlinearity.

The computing times to evaluate the full and the various reduced order models are shown in Table 4.5. Again, the computing times listed are for the entire Newton solve (and not for one Newton iteration). The number of Newton iterations required

**Table 4.4.** 3D Example: The size $N$ of the full order finite element system, the number of POD basis vectors $n$, the number of DEIM points $m$ the number of nodes adjacent to DEIM points, the number of DEIM points $m^e$ when the unassembled nonlinearity is used, and the number of nodes adjacent to DEIM points for piecewise linear and quadratic finite elements on three grids. The mesh in Fig. 4.7 corresponds to grid number 2

| *Polynomial degree* | $p=1$ | | | $p=2$ | |
|---|---|---|---|---|---|
| *Mesh number* | 1 | 2 | 3 | 1 | 2 |
| number of tetrahedra | 6,144 | 49,152 | 165,888 | 6,144 | 49,152 |
| number of nodes $N$ | 1,296 | 9,248 | 30,000 | 9,248 | 69,696 |
| number of POD basis vectors $n$ | 19 | 18 | 19 | 18 | 19 |
| number of DEIM points $m$ | 21 | 21 | 22 | 21 | 22 |
| number of nodes adjacent to DEIM pts. | 183 | 271 | 320 | 445 | 559 |
| number of DEIM points $m^e$ | 21 | 21 | 22 | 21 | 22 |
| number of nodes adjacent to DEIM pts. | 67 | 80 | 88 | 193 | 220 |

**Table 4.5.** 3D Example: The computing times (in sec) and the number of Newton iterations (in parenthesis) needed to solve the full order model, the POD reduced order model, the POD-DEIM reduced order model, and the POD-DEIM-u (unassembled) reduced order model for different grid levels and linear and quadratic finite elements

| *Polynomial degree* | $p=1$ | | | $p=2$ | |
|---|---|---|---|---|---|
| *Mesh number* | 1 | 2 | 3 | 1 | 2 |
| Full | 1.78 (4) | 10.60 (3) | 43.30 (3) | 7.80 (3) | 185.00 (3) |
| POD | 1.28 (4) | 8.04 (3) | 23.80 (3) | 4.12 (3) | 38.80 (3) |
| POD-DEIM | 0.15 (4) | 0.10 (3) | 0.21 (4) | 0.21 (3) | 0.40 (3) |
| POD-DEIM-u | 0.16 (9) | 0.07 (4) | 0.10 (4) | 0.01 (4) | 0.18 (4) |

**Table 4.6.** 3D Example: Errors between the full order model solution and the POD reduced order model solution, the POD-DEIM reduced order model solution, and the POD-DEIM-u (unassembled) reduced order model solution

| *Polynomial degree* | | $p=1$ | | $p=2$ | |
|---|---|---|---|---|---|
| *Mesh number* | 1 | 2 | 3 | 1 | 2 |
| POD | 4.7e-5 | 1.7e-4 | 5.2e-4 | 1.1e-4 | 7.3e-4 |
| POD-DEIM | 3.4e-4 | 4.0e-4 | 4.5e-3 | 4.8e-4 | 2.4e-3 |
| POD-DEIM-u | 4.4e-4 | 1.7e-3 | 5.7e-3 | 1.4e-3 | 4.5e-3 |

are shown in (in parenthesis). The computing times do not include the time needed to compute the matrices $\mathbf{V}$, $\mathbf{U}$, or $\mathbf{U}^e$ via POD.

Table 4.4 shows that in the 3D case the DEIM applied to the unassembled form leads to nonlinear terms in the reduced order models which depend on significantly fewer components of the arguments than the nonlinear terms resulting from the DEIM applied to the assembled form. Table 4.5 shows that the POD-DEIM-u reduced order models are computationally more efficient than the POD-DEIM reduced order models, even if their solution required one more Newton iteration. The POD reduced order model leads to greater computational savings over the full order model in the 3D case compared to the 2D case (see Table 4.2). This is due to the computing time needed to solve the sparse linear systems in Newton's method. As before, significant reductions in computing times can only be achieved after DEIM is applied (either to the assembled or the unassembled form of the nonlinearity).

For 3D problems, the cost of solving the large sparse linear systems arising in Newton's method using the sparse LU decomposition is significant, especially for finer meshes and for piecewise quadratic elements. For the larger problems, it is likely beneficial to replace the direct solvers by iterative solvers. For this reason we have not included results for quadratic elements on the fine Mesh 3. The solution of the full order model using the sparse LU decomposition would have made the full order model solution artificially costly. Switching to iterative solvers for some discretizations would have raised the question what the 'best' iterative solver is. Therefore, we have limited our computational tests, to cases where the use of direct solvers still seems to be justifiable.

As in the 2D case, the errors between the full order model solution and the reduced order model solutions shown in Table 4.6 are of the order of the tolerance $\tau = 10^{-4}$ used to construct the bases with POD.

## 4.5 Evaluation of Parameterized Matrices and Vectors in Reduced Order Models Using DEIM

In this section we describe the use of the DEIM for the generation of efficient reduced order models that involve parameterized matrices. We first describe the approach applied to a generic matrix $\mathbf{A}(\theta)$ and afterwards we apply it to the solution of Stokes equation in parameterized domains.

### 4.5.1 The Reduced Order Matrix

We consider a parametrically dependent matrix $\mathbf{A}(\theta)$ that has the representation

$$\mathbf{A}(\theta) = \sum_{i=1}^{M} \mathbf{g}_i(\theta)\mathbf{A}_i \tag{4.31}$$

with functions $\mathbf{g} = (\mathbf{g}_1,\dots,\mathbf{g}_M)^T : \Theta \to \mathbb{R}^M$ and matrices $\mathbf{A}_i \in \mathbb{R}^{N\times N}$, $i = 1,\dots,M$. As we have seen in Sect. 4.2.2 this is, e.g., the case when $\mathbf{A}(\theta)$ is the stiffness matrix of a parametrically varying linear PDE. In this subsection $\mathbf{A}(\theta)$ is a generic matrix. In the next subsection we will apply the reduction technique to the parametrically dependent Stokes system (4.16).

If we have computed the matrices $\mathbf{V}_\ell, \mathbf{V}_r \in \mathbb{R}^{N\times n}$, then the system matrix for the reduced order model is given by

$$\mathbf{V}_\ell^T \mathbf{A}(\theta)\mathbf{V}_r = \sum_{i=1}^{M} \mathbf{g}_i(\theta)\, \mathbf{V}_\ell^T \mathbf{A}_i \mathbf{V}_r. \tag{4.32}$$

If $M$ is small, we can precompute the matrices $\mathbf{V}_\ell^T \mathbf{A}_i \mathbf{V}_r$ and for each $\theta$ we can use (4.32) to compute $\mathbf{V}_\ell^T \mathbf{A}(\theta)\mathbf{V}_r$ in $n^2M$ operations. However, if $M$ is large, which is the case, e.g., in the example in Sect. 4.2.2 an additional approximation is needed to allow for a fast computation of an approximation of $\mathbf{V}_\ell^T \mathbf{A}(\theta)\mathbf{V}_r$. We can apply the DEIM.

The DEIM computes a matrix $\mathbf{U} \in \mathbb{R}^{M\times m}$ of rank $m$ and a function

$$\widetilde{\mathbf{g}} = (\widetilde{\mathbf{g}}_1,\dots,\widetilde{\mathbf{g}}_m)^T = (\mathbf{P}^T\mathbf{U})^{-1}\mathbf{P}^T\mathbf{g} : \Theta \to \mathbb{R}^m \tag{4.33a}$$

such that

$$\mathbf{g}(\theta) \approx \widehat{\mathbf{g}}(\theta) = \mathbf{U}\widetilde{\mathbf{g}}(\theta). \tag{4.33b}$$

We assume $m \ll M$.

If we insert (4.33b) into (4.32), we obtain

$$\begin{aligned}
\mathbf{V}_\ell^T \mathbf{A}(\theta)\mathbf{V}_r &= \sum_{i=1}^{M} \mathbf{g}_i(\theta)\, \mathbf{V}_\ell^T \mathbf{A}_i \mathbf{V}_r \\
&\approx \sum_{i=1}^{M} \widehat{\mathbf{g}}_i(\theta)\, \mathbf{V}_\ell^T \mathbf{A}_i \mathbf{V}_r = \sum_{i=1}^{M}\sum_{j=1}^{m} \mathbf{U}_{ij}\widetilde{\mathbf{g}}_j(\theta)\, \mathbf{V}_\ell^T \mathbf{A}_i \mathbf{V}_r \\
&= \sum_{j=1}^{m} \left( \sum_{i=1}^{M} \mathbf{U}_{ij}\, \mathbf{V}_\ell^T \mathbf{A}_i \mathbf{V}_r \right) \widetilde{\mathbf{g}}_j(\theta).
\end{aligned} \tag{4.34}$$

The matrices $\sum_{i=1}^{M} \mathbf{U}_{ij}\ \mathbf{V}_\ell^T \mathbf{A}_i \mathbf{V}_r \in \mathbb{R}^{n\times n}$, $j = 1,\dots,m$, can be precomputed. Afterwards, for each $\theta$ the reduced order matrix

$$\widehat{\mathbf{A}}(\theta) = \sum_{j=1}^{m} \left( \sum_{i=1}^{M} \mathbf{U}_{ij}\ \mathbf{V}_\ell^T \mathbf{A}_i \mathbf{V}_r \right) \widetilde{\mathbf{g}}_j(\theta) \tag{4.35}$$

can be computed at a cost of $n^2 m \ll n^2 M$ operations. Under the assumption that $\mathbf{b}(\theta)$ has a decomposition similar to (4.31), we can easily extend the ideas from reduction to $\mathbf{M}(\theta)$ in (4.35) to $\mathbf{b}(\theta)$. We have omitted those details to avoid repetition.

We note that the approximation presented previously can be generalized if $\mathbf{A}(\theta)$ is of the form

$$\mathbf{A}(\theta) = \sum_{i=1}^{M} \sum_{j=1}^{K} (\mathbf{g}_j)_i(\theta)\mathbf{A}_{ij} \tag{4.36}$$

by applying the previous techniques to each of the functions $\mathbf{g}_j = (\mathbf{g}_{1j},\dots,\mathbf{g}_{Mj})^T$, $j = 1,\dots,K$.

In many applications we also need to compute the derivative of the matrix $\mathbf{A}(\theta)$ with respect to $\theta$. If the function $\mathbf{g}$ is differentiable, then the derivative of $\mathbf{A}(\theta)$ is given by

$$D_\theta \mathbf{A}(\theta) = \sum_{i=1}^{M} D_\theta \mathbf{g}_i(\theta)\mathbf{A}_i \tag{4.37}$$

and requires the evaluation of the derivative of all $M$ functions $\mathbf{g}_1,\dots,\mathbf{g}_M$. The same is true for the derivative of (4.32). The derivative of the DEIM reduced matrix,

$$D_\theta \widehat{\mathbf{A}}(\theta) = \sum_{j=1}^{m} \left( \sum_{i=1}^{M} \mathbf{U}_{ij}\ \mathbf{V}_\ell^T \mathbf{A}_i \mathbf{V}_r \right) D_\theta \widetilde{\mathbf{g}}_j(\theta) \tag{4.38}$$

only requires the evaluation of the $m \ll M$ functions $\widetilde{\mathbf{g}}_1,\dots,\widetilde{\mathbf{g}}_m$. From (4.33) we have that

$$D_\theta \widetilde{\mathbf{g}} = \begin{pmatrix} D_\theta \widetilde{\mathbf{g}}_{p_1} \\ \vdots \\ D_\theta \widetilde{\mathbf{g}}_{p_m} \end{pmatrix} = (\mathbf{P}^T\mathbf{U})^{-1}\mathbf{P}^T D_\theta \mathbf{g} = (\mathbf{P}^T\mathbf{U})^{-1} \begin{pmatrix} D_\theta \mathbf{g}_{p_1} \\ \vdots \\ D_\theta \mathbf{g}_{p_m} \end{pmatrix},$$

since $\mathbf{P}^T$ just extracts the $m$ rows from $D_\theta \mathbf{g}$ that corresponding to the DEIM indices $p_1,\dots,p_m$. Thus evaluating the derivative $D_\theta \widehat{\mathbf{A}}(\theta)$ of the DEIM reduced matrix, requires the derivative of only $m \ll M$ functions $\mathbf{g}_{p_1},\dots,\mathbf{g}_{p_m}$.

### *4.5.2 Numerical Example*

We illustrate the DEIM approximation of parametrized matrices and vectors on the example of evaluating the objective function and its derivative in shape optimization of Stokes equation.

Suppose we want to minimize the functional

$$J(\theta) = \int_{\Omega(\theta)} l(u(\theta), p(\theta))dx, \tag{4.39}$$

where the velocities $u(\theta)$ and pressure $p(\theta)$ are the solution of the Stokes equation (4.13). The function $l$ will be specified later.

We assume that the domains $\Omega(\theta)$ are obtained by mapping a reference domain as shown in (4.12). Furthermore, we discretize the Stokes equation using P2-P1 finite elements as described in Sect. 4.2.2. The discretized Stokes system is given by

$$\mathbf{S}(\theta)\begin{pmatrix}\mathbf{u}\\ \mathbf{p}\end{pmatrix} = \mathbf{b}(\theta), \tag{4.40}$$

where $\mathbf{S}(\theta)$ has the form (4.18). In our examples, the forcing function $f$ in (4.13) is zero. Therefore, the right hand side $\mathbf{b}$ is determined by the $\mathbf{S}(\theta)$ and the in homogenous Dirichlet data on the velocity in (4.13). The Stokes matrix $\mathbf{S}(\theta)$ in (4.18) has the same structure as the generic matrix $\mathbf{S}$ in (4.36). Since the forcing function $f$ in (4.13) is zero, no additional parameterization of the right hand side $\mathbf{b}(\theta)$ is needed.

Applying the domain mapping, the P2-P1 finite element discretization, and the quadrature formula from Sect. 4.2.2 to the objective (4.39) gives the discrete objective functional

$$J_h(\theta) = \sum_{\ell=1}^{M} \omega_\ell \, l(u_h(\widetilde{x}_\ell), p_h(\widetilde{x}_\ell)) \, |\det(D\Phi(\widetilde{x}_\ell;\theta))|, \tag{4.41}$$

where $u_h$ is the piecewise quadratic FEM approximation of the velocity and $p_h$ is the piecewise linear FEM approximation of the pressure. The objective (4.41) depends on $\theta$ via the function $\mathbf{g}_8 : \Theta \to \mathbb{R}^M$ defined by

$$(\mathbf{g}_8(\theta))_\ell = \omega_\ell |\det(D\Phi(\widetilde{x}_\ell;\theta))|.$$

Since the discretized velocity and pressure $u_h$ and $p_h$ are determined by their coefficients $\mathbf{u}$ and $\mathbf{p}$, we can write the discrete objective functional (4.41) as

$$J_h(\theta) = \sum_{\ell=1}^{M} \mathbf{l}_\ell(\mathbf{u},\mathbf{p}) \, (\mathbf{g}_8(\theta))_\ell. \tag{4.42}$$

Note that its parameter dependence has the same structure as that of the generic matrix and therefore DEIM can be applied to reduce the computational cost. As in Sect. 4.2.2 we set

$$\mathbf{y} = \begin{pmatrix}\mathbf{u}\\ \mathbf{p}\end{pmatrix}.$$

To summarize, we want to minimize

$$J_h(\theta) = \mathbf{l}(\mathbf{y}(\theta))^T \mathbf{g}_8(\theta) = \sum_{\ell=1}^{M} \mathbf{l}_\ell(\mathbf{y}(\theta)) \, (\mathbf{g}_8(\theta))_\ell,$$

where $\mathbf{y}(\theta)$ solves (4.40). In many minimization problems there will be additional constraints on the parameter or on the velocities or pressures. Since we focus on the evaluation of reduced order models, we focus on the evaluation of $J_h(\theta)$ and on its gradient.

The evaluation of the objective function requires the following steps.

$\mathscr{J}_1$: Assemble $\mathbf{S}(\theta)$ and $\mathbf{b}(\theta)$ and solve the state equation $\mathbf{S}(\theta)\mathbf{y} = \mathbf{b}(\theta)$ for $\mathbf{y}(\theta)$.
$\mathscr{J}_2$: Compute $J_h(\theta) = \mathbf{l}(\mathbf{y}(\theta))^T \mathbf{g}_8(\theta)$.

We briefly summarize the computation of the gradient of $J_h(\theta)$ via the adjoint approach, see, e.g., [18, Sec. 1.6]. We define the Lagrangian functional

$$L(\mathbf{y}, \boldsymbol{\lambda}, \theta) = \mathbf{l}(\mathbf{y})^T \mathbf{g}_8(\theta) + \boldsymbol{\lambda}^T (\mathbf{S}(\theta)\mathbf{y} - \mathbf{b}(\theta)).$$

We assume that the objective has already been computed, i.e., that $\mathbf{S}(\theta)$ and $\mathbf{b}(\theta)$ have been assembled and that $\mathbf{y}(\theta)$ has been computed. Then the computation of the gradient requires the following steps.

$\mathscr{G}_1$: Solve the adjoint equation $\mathbf{S}(\theta)^T \boldsymbol{\lambda} = -D_{\mathbf{y}}\mathbf{l}(\mathbf{y}(\theta))^T \mathbf{g}_8(\theta)$ for $\boldsymbol{\lambda}(\theta)$.
$\mathscr{G}_2$: Compute $\nabla J_h(\theta) = \mathbf{l}(\mathbf{y}(\theta))^T D_\theta \mathbf{g}_8(\theta) + \boldsymbol{\lambda}(\theta)^T (D_\theta \mathbf{S}(\theta)\mathbf{y}(\theta) - D_\theta \mathbf{b}(\theta))$.

To construct the reduced basis for the state equations (4.40), we sample the solution of the discrete Stokes (4.40) at $r$ samples in the parameter domain $\Theta$. We then apply the POD Algorithm 4.1 with tolerance $\tau$ individually to the snapshots for the $x_1$- and $x_2$-components of the velocity and the snapshots for the pressures. If $N_v$ are the degrees of freedom for the $x_1$- and $x_2$-components of the velocity and $N_p$ are the degrees of freedom for the pressure, the POD generates matrices $\mathbf{V}_{v_1} \in \mathbb{R}^{N_v \times n_{v_1}}$, $\mathbf{V}_{v_2} \in \mathbb{R}^{N_v \times n_{v_2}}$, and $\mathbf{V}_p \in \mathbb{R}^{N_p \times n_p}$. The reduced order Stokes matrix and right hand side are

$$\begin{aligned}
\mathbf{V}^T \mathbf{S}(\theta)\mathbf{V} &= \begin{pmatrix} \mathbf{V}_{v_1} & 0 & 0 \\ 0 & \mathbf{V}_{v_2} & 0 \\ 0 & 0 & \mathbf{V}_p \end{pmatrix}^T \begin{pmatrix} \mathbf{A}(\theta) & 0 & \mathbf{B}^{(1)}(\theta)^T \\ 0 & \mathbf{A}(\theta) & \mathbf{B}^{(2)}(\theta)^T \\ \mathbf{B}^{(1)}(\theta) & \mathbf{B}^{(2)}(\theta) & 0 \end{pmatrix} \begin{pmatrix} \mathbf{V}_{v_1} & 0 & 0 \\ 0 & \mathbf{V}_{v_2} & 0 \\ 0 & 0 & \mathbf{V}_p \end{pmatrix} \\
&= \begin{pmatrix} \mathbf{V}_{v_1}^T \mathbf{A}(\theta)\mathbf{V}_{v_1} & 0 & \mathbf{V}_{v_1}^T \mathbf{B}^{(1)}(\theta)^T \mathbf{V}_p \\ 0 & \mathbf{V}_{v_2}^T \mathbf{A}(\theta)\mathbf{V}_{v_2} & \mathbf{V}_{v_2}^T \mathbf{B}^{(2)}(\theta)^T \mathbf{V}_p \\ \mathbf{V}_p^T \mathbf{B}^{(1)}(\theta)\mathbf{V}_{v_1} & \mathbf{V}_p^T \mathbf{B}^{(2)}(\theta)\mathbf{V}_{v_2} & 0 \end{pmatrix}
\end{aligned} \tag{4.43}$$

and

$$\mathbf{V}^T \mathbf{b}(\theta) = \begin{pmatrix} \mathbf{V}_{v_1} & 0 & 0 \\ 0 & \mathbf{V}_{v_2} & 0 \\ 0 & 0 & \mathbf{V}_p \end{pmatrix}^T \mathbf{b}(\theta).$$

The preliminary version of the reduced order Stokes system is

$$\mathbf{V}^T \mathbf{S}(\theta)\mathbf{V}\widehat{\mathbf{y}} = \mathbf{V}^T \mathbf{b}(\theta). \tag{4.44}$$

Since the $\mathbf{b}_3$ component of the right hand side (4.16) of the discrete Stokes equation is nonzero, the velocity snapshots are not divergence free (in the discrete sense). Therefore, as already noted in [30], there is no guarantee that the reduced Stokes matrix (4.43) satisfies an inf-sup condition. In [30] a procedure is proposed that enriches the velocity subspaces to guarantee the inf-sup condition. Instead we monitor the inf-sup constant corresponding to (4.43) by computing the singular values of the small matrix $\widehat{\mathbf{B}}(\theta) = (\mathbf{V}_p^T \mathbf{B}^{(1)}(\theta)\mathbf{V}_{v_1}\ ,\ \mathbf{V}_p^T \mathbf{B}^{(2)}(\theta)\mathbf{V}_{v_2})^T$ and found that no enrichment of the velocity space was needed in our example.

The matrix $\mathbf{S}(\theta)$ has the structure (4.18). Since in our examples the forcing function $f$ in (4.13) is zero, no additional parameterization of the right hand side $\mathbf{b}(\theta)$ is needed. We apply the DEIM as described in the previous Sect. 4.5.1 to obtain a DEIM reduced order matrix $\widehat{\mathbf{S}}(\theta)$ and right hand side vector $\widehat{\mathbf{b}}(\theta)$. Specifically, the reduced bases, the matrices $\mathbf{U}$ for the nonlinear terms $\mathbf{g}_1, \ldots, \mathbf{g}_8$ are constructed by sampling these nonlinearities at the same parameters used to construct the reduced basis $\mathbf{V}_{v_1}, \mathbf{V}_{v_2}$ and $\mathbf{V}_p$ and then applying POD with tolerance $\tau$ to get matrices $\mathbf{U}$ for the DEIM. We apply DEIM to each of the eight functions $\mathbf{g}_1, \ldots, \mathbf{g}_8$ separately, i.e., for each of the eight functions we generate a matrix $\mathbf{U}$. Applying the DEIM approximation of Sect. 4.5.1 we obtain the DEIM reduced order system

$$\widehat{\mathbf{S}}(\theta)\widehat{\mathbf{y}} = \widehat{\mathbf{b}}. \tag{4.45}$$

Furthermore, applying DEIM to the function $\mathbf{g}_8$ in the objective, i.e., approximating

$$\mathbf{g}_8(\theta) \approx \widehat{\mathbf{g}}_8(\theta) = \mathbf{U}_8\widetilde{\mathbf{g}}_8(\theta).$$

where

$$\widetilde{\mathbf{g}}_8 = ((\widetilde{\mathbf{g}}_8)_1, \ldots, (\widetilde{\mathbf{g}}_8)_m)^T = (\mathbf{P}_8^T\mathbf{U}_8)^{-1}\mathbf{P}_8^T\mathbf{g}_8 : \Theta \to \mathbb{R}^{m_8}$$

leads to the reduced order objective

$$\widehat{J}_h(\theta) = \mathbf{l}(\mathbf{V}\widehat{\mathbf{y}}(\theta))^T\mathbf{U}_8\widetilde{\mathbf{g}}_8(\theta), \tag{4.46}$$

where $\widehat{\mathbf{y}}(\theta)$ is the solution of (4.45). In our applications, $\mathbf{l}$ is affine linear or quadratic in $\mathbf{y}$, so that fast computation of $\widehat{\mathbf{y}} \mapsto \mathbf{l}(\mathbf{V}\widehat{\mathbf{y}})^T\mathbf{U}_8\widetilde{\mathbf{g}}_8(\theta)$ is possible.

All computations were done using Matlab on a MacBook Pro with 8GB of memory and a 2.53 GHz Intel Core 2 Duo processor. The nonlinear full order or reduced order models are solved using Newton's method. The linear systems are solved using the Matlab backslash command. The $\theta$ derivatives are computed using INTLAB Version 5.5 [31].

#### *4.5.2.1 Evaluation of Drag Generated by Parameterized Airfoil*

The drag on the boundary portion $\Gamma_{\text{drag}}(\theta) \subset \partial\Omega(\theta)$ is defined by

$$C_D = -\frac{2}{U_\infty^2 L}\int_{\Gamma_{\text{drag}}(\theta)} ((\nu\nabla u(x) - p(x)I)n(x)) \cdot \widehat{u}_\infty \, ds, \tag{4.47}$$

where $u_\infty = U_\infty\widehat{u}_\infty$ is the velocity of the incoming flow, $\widehat{u}_\infty$ is the unit vector directed as the incoming flow, $U_\infty$ is constant, and $L$ is the characteristic length of the body. See, e.g., [16,20]. As usual, we use the Stokes equations (4.13) to find an equivalent formula for the drag that avoids integration over the boundary. We use a function $v_\infty \in (H^1(\Omega(\theta)))^2$ with $v_\infty = \widehat{u}_\infty$ on $\Gamma_{\text{drag}}(\theta)$ and $v_\infty = 0$ on $\partial\Omega(\theta) \setminus \Gamma_{\text{drag}}(\theta)$ as a test function in (4.13) to obtain

$$
\begin{aligned}
0 &= -\int_{\partial\Omega(\theta)} ((\nu\nabla u - pI)n)\cdot v_\infty + \int_{\Omega(\theta)} (\nu\nabla u - pI) : \nabla v_\infty - \int_{\Omega(\theta)} f\cdot v_\infty, \\
&= -\int_{\Gamma_{\text{drag}}(\theta)} ((\nu\nabla u - pI)n)\cdot \widehat{u}_\infty + \int_{\Omega(\theta)} (\nu\nabla u - pI) : \nabla v_\infty - \int_{\Omega(\theta)} f\cdot v_\infty.
\end{aligned}
$$

Hence,

$$
C_D = -\frac{2}{U_\infty^2 L}\left(\int_{\Omega(\theta)} (\nu\nabla u(x) - p(x)I) : \nabla v_\infty(x)\,dx - \int_{\Omega(\theta)} f(x)\cdot v_\infty(x)\,dx\right). \tag{4.48}
$$

We use $C_D$ as our objective functional for this example, i.e., in this example $J$ in (4.39) is given by (4.48).

The domain $\Omega(\theta)$ (for $\theta = 0.5$) is sketched in Fig. 4.8 and has the boundary $\partial\Omega(\theta) = \Gamma_{\text{in}} \cup \Gamma_D \cup \Gamma_{\text{drag}}(\theta) \cup \Gamma_{\text{out}}$, where $\Gamma_{\text{in}} = \{-6\} \times (-3,5)$, $\Gamma_D = ((-6,6) \times \{-3\}) \cup ((-6,6) \times \{5\})$, $\Gamma_{\text{out}} = \{6\} \times (-3,5)$ and $\Gamma_{\text{drag}}(\theta)$ is the boundary of airfoil. We specify an inflow velocity $h = 1$, on $\Gamma_{\text{in}}$ and a constant viscosity $\nu = 0.1$. The forcing function $f$ in (4.13) is taken to be zero. We assume that the airfoil is of unit length, and the boundary has the following parameterization,

$$
\Gamma_{drag} = \{(x_1, x_2) \mid 0 \le x \le 1, x_2 = 1 + \eta(\theta)\}
$$

where for $\theta \in [0,2]$

$$
\eta(\theta) = \pm\frac{\theta}{0.2}\left(0.2969\sqrt{x_1} - 0.1260x_1 - 0.3520x_1^2 + 0.2832x_1^3 - 0.1021x_1^4\right).
$$

The diffeomorphism $\Phi$ that is used to map the reference domain $\widetilde{\Omega}$, is given by

$$
\Phi((\widetilde{x}_1,\widetilde{x}_2);\theta) = \begin{pmatrix} \widetilde{x}_1 \\ (1+\eta(\theta))\widetilde{x}_2 \end{pmatrix} = (x_1, x_2)^T.
$$

for $\widetilde{x}_1 \in [0,1]$ and $\Phi((\widetilde{x}_1,\widetilde{x}_2);\theta) = (\widetilde{x}_1,\widetilde{x}_2)$ else. The reference domain is $\widetilde{\Omega} = \Phi((-6,6)\times(-3,5);0.5)(=\Omega(0.5))$ and is shown in Fig. 4.8.

The problem is discretized using P2-P1 Taylor Hood elements as described in Sect. 4.2.2. We compute 25 snapshots each for both the solution to the state equations and the nonlinear terms. The reduced basis are generated using the POD with a tolerance $\tau = 10^{-6}$.

We evaluate $C_D$ (see (4.47)) and its derivative with respect to $\theta$ at an arbitrary point $\theta = \sqrt{2} \in \Theta$, which is not in the snapshot set. Table 4.7 summarizes the size of the full and the reduced order systems for three finite element grids using the full order model, the POD reduced order model and the POD-DEIM reduced order model. The mesh in Fig. 4.8 is the coarse Mesh 1. The DEIM points (quadrature points) chosen are contained in the triangles marked in red.

The computing times to evaluate the objective function (steps $\mathscr{J}_1 + \mathscr{J}_2$) and its gradient (steps $\mathscr{G}_1 + \mathscr{G}_2$ ) for the full order model and the reduced order models are shown in Table 4.8. For the reduced order models the times do not include off-line cost. Most of the computational cost in computing objective functional occurs in Step

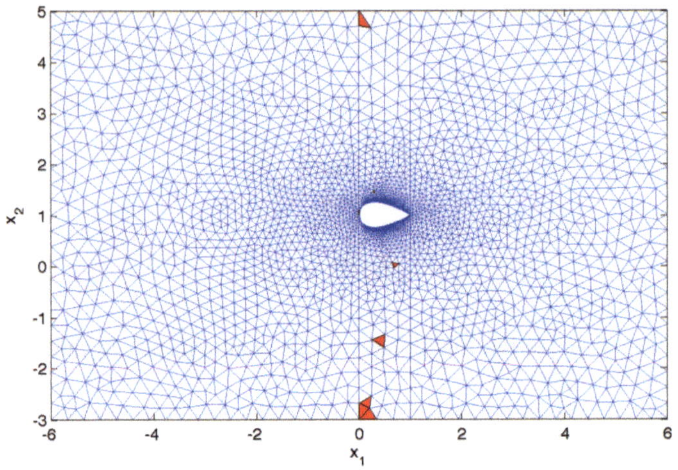


**Fig. 4.8.** The reference domain $\widetilde{\Omega}$ for the NACA airfoil 4-digit family. The DEIM points (quadrature points) are contained in the triangles marked in solid red

$\mathscr{J}_1$. Computing the gradient requires solving the adjoint equations (Step $\mathscr{G}_1$) and the sensitivities of system matrices and the objective functional (Step $\mathscr{G}_2$) with respect to the shape parameter $\theta$. Since this example only involves a scalar parameter, for the full order model Step $\mathscr{G}_1$ is the most expensive step in the evaluation of the gradient of objective functional. The cost of sensitivities increases with the number of parameters, see Sect. 4.5.2.2.

The errors between the full order model (objective functional $C_D$) and the reduced order model solutions shown in Table 4.9 are of the order of the tolerance $\tau = 10^{-6}$ used to construct the bases with the POD. The error in gradient computation is slightly higher, due to the fact that adjoint solutions have not been taken into account to generate the reduced bases. This accuracy can be easily improved by enriching the snapshot set.

**Table 4.7.** The size $N = N_v + N_v + N_p$ of the full order finite element system, the number of POD basis vectors $n = n_{v_1} + n_{v_2} + n_p$, and the number of DEIM points $m = \sum_{\ell=1}^{8} m_\ell$

| *Mesh number* | 1 | 2 | 3 |
|---|---|---|---|
| number of triangles | 6,094 | 8,838 | 24,990 |
| number of nodes $N$ | 27,039 | 39,343 | 111,887 |
| number of POD basis vectors $n$ | 55 | 58 | 63 |
| number of DEIM points $m(=\sum_{\ell=1}^{8} m_\ell)$ | 26 | 26 | 26 |

#### *4.5.2.2 Channel with Parameterized Top and Bottom Wall*

In our second example we consider a channel in which the bottom and top boundaries are parameterized using Bézier curves. The reference domain is $\widetilde{\Omega} = (-1,1)^2$. The bottom and top wall of the channel are parameterized by Bézier curves with $p_T$ control points for the top boundary and $p_B$ control points for the bottom boundary. Thus the physical domain $\Omega(\theta)$ is parameterized by $\theta \in \Theta \subset \mathbb{R}^p$, $p = p_T + p_B$.

The boundary $\partial\Omega(\theta)$ is decomposed into the inflow and outflow boundaries $\Gamma_{\text{in}}(\theta) = \{-1\} \times (-1,1)$, and $\Gamma_{\text{out}}(\theta) = \{1\} \times (-1,1)$, both of which are independent of the parameterization and the top and bottom boundaries $\Gamma_t(\theta)$ and $\Gamma_b(\theta)$. The viscosity is $\nu = 1.0$. On the inflow boundary $\Gamma_{\text{in}}$ we specify an parabolic inflow velocity $h = 8(1+x_2)(1-x_2)$. The velocity $h = 0$ on $\Gamma_t(\theta)$ and $\Gamma_b(\theta)$. The forcing function $f$ in (4.13) is taken to be zero.

We use $p_T = p_B = 2$ Bézier control points to specify the top and the bottom boundary of the variable domain $\Omega(\theta)$. The parameters are in $\Theta = (0.5,3.0) \times (0.5,3.0) \times (-3.0,-0.5) \times (-3.0,-0.5)$

For this example the objective functional is

$$J(\theta) = \int_{\Omega(\theta)} |u - u^d|^2 dx$$

where $u$ are the velocities computed as the solution of the Stokes equations (4.13) on $\Omega(\theta)$. The functions $u^d$ are the desired velocities computed by solving the stokes equation on $\Omega(\theta^d)$ with fixed parameter $\theta^d = (1.0, 0.5, -0.5 - 1.0)^T$.

The problem is discretized using P2-P1 Taylor Hood elements as described in Sect. 4.2.2. To construct the reduced basis, we compute $5^4$ snapshots in the parameter domain $\Theta$ i.e., we take 5 sample points in each direction. Then we apply Algorithm 4.1 with tolerance $\tau = 10^{-4}$ to construct the reduced basis, as before.

We evaluate $J$ and its derivative with respect to $\theta$ at an arbitrary point $\theta = (\sqrt{2}, \sqrt{2}, -\sqrt{2}, -\sqrt{2})^T \in \Theta$, which is not in the snapshot set. Table 4.10 summarizes the size of the full and the reduced order systems for three finite element grids using the full order model, the POD reduced order model and the POD-DEIM reduced order model.

**Table 4.8.** The computing times (in sec) to evaluate the objective functional (Steps $\mathcal{J}_1 + \mathcal{J}_2$), and the gradient of objective functional (Steps $\mathcal{G}_1 + \mathcal{G}_2$) corresponding to the full order model, the POD reduced order model, and the POD-DEIM reduced order model for different meshes

| *Mesh number* | 1 | | 2 | | 3 | |
|---|---|---|---|---|---|---|
| | $\mathcal{J}_1 + \mathcal{J}_2$ | $\mathcal{G}_1 + \mathcal{G}_2$ | $\mathcal{J}_1 + \mathcal{J}_2$ | $\mathcal{G}_1 + \mathcal{G}_2$ | $\mathcal{J}_1 + \mathcal{J}_2$ | $\mathcal{G}_1 + \mathcal{G}_2$ |
| Full | 2.04 | 2.00 | 3.16 | 2.94 | 12.50 | 12.80 |
| POD | 0.83 | 0.89 | 1.23 | 1.07 | 3.49 | 2.95 |
| POD-DEIM | 0.02 | 0.02 | 0.02 | 0.01 | 0.05 | 0.03 |

**Table 4.9.** The errors between the full order and the POD reduced order model (objective functional and its gradient) and errors between the full order and the POD-DEIM reduced order model (objective functional and its gradient) for different meshes

| *Mesh number* | 1 | | 2 | | 3 | |
|---|---|---|---|---|---|---|
| | *objective* | *gradient* | *objective* | *gradient* | *objective* | *gradient* |
| POD | 4.67e-6 | 5.08e-5 | 5.83e-6 | 2.44e-4 | 1.12e-5 | 4.50e-4 |
| POD-DEIM | 5.31e-6 | 1.11e-4 | 6.05e-6 | 2.66e-4 | 1.21e-5 | 3.67e-4 |

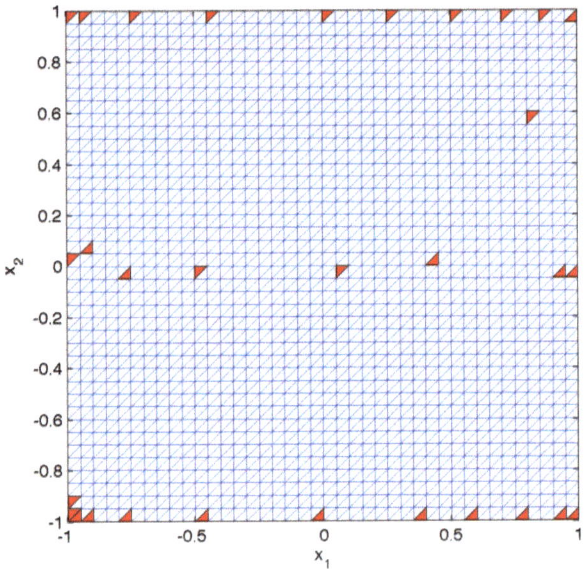


**Fig. 4.9.** The reference domain $\widetilde{\Omega}$ for the channel example. The top $\Gamma_T$ and bottom $\Gamma_B$ boundaries are parameterized by $p_T = 2$ and $p_B = 2$ Bézier control points respectively. The DEIM points (quadrature points) lies in the interior of the triangles marked in solid red

The mesh in Fig. 4.9 is the coarse Mesh 1. The DEIM points (quadrature points) chosen are contained in the triangles marked in red.

The computing times to evaluate the full and the various reduced order objective (Steps $\mathscr{J}_1 + \mathscr{J}_2$) and its gradient (Steps $\mathscr{G}_1 + \mathscr{G}_2$ ) are shown in Table 4.8. For the reduced order models the times do not include off-line cost. As in the previous example, most of the computing cost for the computation of the objective function occurs in step $\mathscr{J}_1$, the assembly and solution of the state equation. Computing the gradient requires solving the adjoint equations (Step $\mathscr{G}_1$) and the sensitivities of system matrices and the objective functional (Step $\mathscr{G}_2$) with respect to the shape parameter $\theta$. We observe that in this example and for the full order model, Step $\mathscr{G}_2$ is the most expensive step in the evaluation of the gradient of the objective functional. This is due to the fact that we have four parameters.

**Table 4.10.** The size $N = N_v + N_v + N_p$ of the full order finite element system, the number of POD basis vectors $n = n_{v_1} + n_{v_2} + n_p$, and the number of DEIM points $m = \sum_{\ell=1}^{8} m_\ell$. The mesh in Fig. 4.8 corresponds to grid number 1

| *Mesh number* | 1 | 2 | 3 |
|---|---|---|---|
| number of triangles | 800 | 3,200 | 7,200 |
| number of nodes $N$ | 3,561 | 14,321 | 32,281 |
| number of POD basis vectors $n$ | 261 | 264 | 265 |
| number of DEIM points $m(=\sum_{\ell=1}^{8} m_\ell)$ | 53 | 53 | 53 |

**Table 4.11.** The computing times (in sec) to evaluate the objective functional (Steps $\mathscr{I}_1 + \mathscr{I}_2$), and the gradient of objective functional (Steps $\mathscr{G}_1 + \mathscr{G}_2$) corresponding to the full order model, the POD reduced order model, and the POD-DEIM reduced order model for different meshes

| *Mesh number* | 1 | | 2 | | 3 | |
|---|---|---|---|---|---|---|
| | $\mathscr{I}_1 + \mathscr{I}_2$ | $\mathscr{G}_1 + \mathscr{G}_2$ | $\mathscr{I}_1 + \mathscr{I}_2$ | $\mathscr{G}_1 + \mathscr{G}_2$ | $\mathscr{I}_1 + \mathscr{I}_2$ | $\mathscr{G}_1 + \mathscr{G}_2$ |
| Full | 0.33 | 0.73 | 1.50 | 3.86 | 5.00 | 13.30 |
| POD | 0.36 | 1.06 | 1.26 | 4.95 | 4.41 | 15.90 |
| POD-DEIM | 0.05 | 0.13 | 0.04 | 0.14 | 0.09 | 0.14 |

**Table 4.12.** The errors between the full order and the POD reduced order model (objective functional and its gradient) and errors between the full order and the POD-DEIM reduced order model (objective functional and its gradient) for different meshes

| *Mesh number* | 1 | | 2 | | 3 | |
|---|---|---|---|---|---|---|
| | objective | gradient | objective | gradient | objective | gradient |
| POD | 2.01e-3 | 2.53e-3 | 1.91e-3 | 1.61e-3 | 1.83e-3 | 1.62e-3 |
| POD-DEIM | 2.03e-3 | 2.57e-3 | 1.93e-3 | 1.71e-3 | 1.82e-3 | 1.63e-3 |

The errors between the full order model (objective functional $C_D$ and its gradient) and the reduced order model solutions shown in Table 4.12 are of the order of the tolerance $\tau = 10^{-4}$ used to construct the bases with the POD.

## 4.6 Conclusions

We have demonstrated the application of the DEIM to compute reduced order models for finite element discretizations of seminar elliptic PDEs and for parameterized linear systems that arise, e.g., in shape optimization, and we have studied the computational efficiency of the resulting reduced order models.

The efficiency with which DEIM reduced order models of discretized semilinar elliptic PDEs can be evaluated is determined by how many components of the argument each component of the nonlinearity depends on. For finite element discretizations this dependence is determined by the mesh, the polynomial degree used in the finite element approximation, but also by whether the nonlinearity is defined in its assembled or unassembled form. For nodal based finite element methods, each component of the unassembled form of the nonlinearity depends only on the components associated with the degrees of freedom corresponding to one element. This is different for the assembled form of the nonlinearity. Here a component of the nonlinearity can depend on the degrees of freedom in several adjacent elements. More precisely, if the component of the nonlinearity corresponds to a node on the boundary of an element, then this component of the nonlinearity depends on all degrees of freedom in the elements that share this node. Because of the dependence of the components of the nonlinearity on the components of its argument, the unassembled form is attractive for DEIM. Since DEIM applied to the different forms of the nonlinearity generates different reduced order models, which require different numbers of Newton iterations to solve, the dependency of the nonlinearity on its argument alone cannot be used to decide which form of the DEIM is favorable. Our numerical examples have shown that either version of the DEIM is preferable over the naive application of projection based model reduction. For large systems, the application of the DEIM to the unassembled form of the nonlinearity led to additional gains in the on-line cost of the reduced order models. The off-line cost of DEIM applied to the unassembled form of the nonlinearity is always higher (and can be significantly higher) since the unassembled form results in a nonlinear vector valued function that has significantly more components than the nonlinear vector valued function arising in the assembled form.

A second focus of this paper was to demonstrate the application of the DEIM to compute reduced order models for an important class of parameterized linear systems. The DEIM not only leads to reduced order models that can be evaluated efficiently, but in addition the derivatives of the reduced order models with respect to the parameter can be computed efficiently. Both efficiency gains are crucial, e.g., for shape optimization. We have demonstrated this numerically using the Stokes equations on parameterized domains.

## References

1. Antoulas, A.C.: Approximation of large-scale dynamical systems, Advances in Design and Control, vol. 6. Society for Industrial and Applied Mathematics, Philadelphia, PA (2005)
2. Barrault, M., Maday, Y., Nguyen, N.D., Patera, A.T.: An 'empirical interpolation' method: application to efficient reduced-basis discretization of partial differential equations. C. R. Math. Acad. Sci. Paris **339**(9), 667–672 (2004)
3. Becker, R., Braack, M., Vexler, B.: Numerical parameter estimation for chemical models in multidimensional reactive flows. Combust. Theory Modelling **8**, 6 (2004)

4. Binev, P., Cohen, A., Dahmen, W., DeVore, R., Petrova, G., Wojtaszczyk, P.: Convergence rates for greedy algorithms in reduced basis methods. SIAM J. Math. Anal. **43**(3), 1457–1472 (2011)
5. Brooks, A.N., Hughes, T.J.R.: Streamline upwind/Petrov–Galerkin formulations for convection dominated flows with particular emphasis on the incompressible Navier–Stokes equations. Comp. Meth. Appl. Mech. Engng. **32**, 199–259 (1982)
6. Buffa, A., Maday, Y., Patera, A.T., Prud'homme, C., Turinici, G.: A priori convergence of the greedy algorithm for the parametrized reduced basis method. ESAIM Math. Model. Numer. Anal. **46**(3), 595–603 (2012)
7. Chaturantabut, S., Sorensen, D.C.: Nonlinear model reduction via discrete empirical interpolation. SIAM Journal on Scientific Computing **32**(5), 2737–2764 (2010)
8. Chaturantabut, S., Sorensen, D.C.: Application of POD and DEIM on dimension reduction of non-linear miscible viscous fingering in porous media. Math. Comput. Model. Dyn. Syst. **17**(4), 337–353 (2011)
9. Dedden, R.J.: Model order reduction using the discrete empirical interpolation method. Master's thesis, Technical University Delft, Netherlands, 2012. Available from http://repository.tudelft.nl (accessed Dec. 31, 2012)
10. Elman, H.C., Silvester, D. J., Wathen, A. J.: Finite Elements and Fast Iterative Solvers with Applications in Incompressible Fluid Dynamics. Oxford University Press, Oxford (2005)
11. Galbally, D., Fidkowski, K., Willcox, K., Ghattas, O.: Nonlinear model reduction for uncertainty quantification in large-scale inverse problems. Internat. J. Numer. Methods Engrg. **81**(12), 1581–1603 (2010)
12. Galdi, G.P., Simader, C.G., Sohr, H.: On the Stokes problem in Lipschitz domains. Ann. Mat. Pura Appl. (4) **167**, 147–163 (1994)
13. Girault, V., Raviart, P.-A.: Finite element methods for Navier-Stokes equations. Theory and algorithms, volume 5 of Springer Series in Computational Mathematics. Springer-Verlag, Berlin Heidelberg (1986)
14. Grepl, M.A., Maday, Y., Nguyen, N.C., Patera, A.T.: Efficient reduced-basis treatment of nonaffine and nonlinear partial differential equations. M2AN Math. Model. Numer. Anal. **41**(3), 575–605 (2007)
15. Grepl, M.A., Patera, A.T.: A posteriori error bounds for reduced-basis approximations of parametrized parabolic partial differential equations. M2AN Math. Model. Numer. Anal. **39**(1), 157–181 (2005)
16. He, J.-W., Glowinski, R., Metcalfe, R., Nordlander, A., Periaux, J.: Active control and drag optimization for flow past a circular cylinder. I. oscillatory cylinder rotation. Journal of Computational Physics **163**, 83–117 (2000)
17. Hinze, M., Kunkel, M.: Discrete empirical interpolation in pod model order reduction of drift-diffusion equations in electrical networks. In: Michielsen, B., Poirier, J.-R. (eds.) Scientific Computing in Electrical Engineering SCEE 2010, Mathematics in Industry, pp. 423–431. Springer-Verlag, Berlin Heidelberg (2012)
18. Hinze, M., Pinnau, R., Ulbrich, M., Ulbrich, S.: Optimization with Partial Differential Equations. of Mathematical Modelling, Theory and Applications, vol. 23. Springer-Verlag, Berlin Heidelberg New York (2009)
19. Hinze, M., Volkwein, S.: Proper orthogonal decomposition surrogate models for nonlinear dynamical systems: Error estimates and suboptimal control. In: Benner, P., Mehrmann, V., Sorensen, D. C. (eds.) Dimension Reduction of Large-Scale Systems, Lecture Notes in Computational Science and Engineering, vol. 45, pp. 261–306. Springer-Verlag, Berlin Heidelberg (2005)

20. John, V.: Reference values for drag and lift of a two-dimensional time-dependent flow around a cylinder. International Journal for Numerical Methods in Fluids **44**(7), 777–788 (2004)
21. Lall, S., Marsden, J.E., Glavaški, S.: A subspace approach to balanced truncation for model reduction of nonlinear control systems. Internat. J. Robust Nonlinear Control **12**(6), 519–535 (2002)
22. Lass, O., Volkwein, S.: POD Galerkin schemes for nonlinear elliptic-parabolic systems. Konstanzer Schriften in Mathematik No. 301, FB Mathematik & Statistik, Universität Konstanz, D-78457 Konstanz, Germany, 2012. To appear in SIAM J. Scientific Computing
23. Lions, P.-L.: On the existence of positive solutions of semilinear elliptic equations. SIAM Rev. **24**(4), 441–467 (1982)
24. Patera, A.T., Rozza, G.: Reduced Basis Approximation and A Posteriori Error Estimation for Parametrized Partial Differential Equations. MIT Pappalardo Graduate Monographs in Mechanical Engineering. Cambridge, MA (2007). Available from http://augustine.mit.edu/methodology/methodology_book.htm
25. Quarteroni, A., Valli, A.: Numerical Approximation of Partial Differential Equations. Springer-Verlag, Berlin Heidelberg New York (1994)
26. Roos, H.G., Stynes, M., Tobiska, L.: Robust Numerical Methods for Singularly Perturbed Differential Equations. Computational Mathematics, Vol. 24, 2nd ed. Springer-Verlag, Berlin Heidelberg (2008)
27. Roubíček, T.: Nonlinear partial differential equations with applications, volume 153 of International Series of Numerical Mathematics. Birkhäuser, Basel (2005)
28. Rowley, C.W.: Model reduction for fluids, using balanced proper orthogonal decomposition. Int. J. on Bifurcation and Chaos **15**(3), 997–1013 (2005)
29. Rozza, G., Huynh, D.B.P., Patera, A.T.: Reduced basis approximation and a posteriori error estimation for affinely parametrized elliptic coercive partial differential equations: application to transport and continuum mechanics. Arch. Comput. Methods Eng. **15**(3), 229–275 (2008)
30. Rozza, G., Veroy, K.: On the stability of the reduced basis method for Stokes equations in parametrized domains. Comput. Methods Appl. Mech. Engrg. **196**(7), 1244–1260 (2007)
31. Rump, S.M.: INTLAB – INTerval LABoratory. In: Csendes, T. (ed.) Developments in Reliable Computing, pp. 77–104. Kluwer Academic Publishers, Dordrecht (1999). http://www.ti3.tu-harburg.de/rump/
32. Stynes, M.: Steady-state convection-diffusion problems. In: Iserles, A. (ed.) Acta Numerica 2005, pp. 445–508. Cambridge University Press, Cambridge, London, New York (2005)
33. Tiso, P., Dedden, R.J., Rixen, D.J.: DEIM for nonlinear structural dynamics model order reduction, 2012. Talk presented at the 10th World Congress on Computational Mechanics, Sao Paulo, Brazil (8–13 July 2012)

# 5

# Greedy Sampling Using Nonlinear Optimization

**Karsten Urban, Stefan Volkwein and Oliver Zeeb**

**Abstract** We consider the reduced basis generation in the offline stage. As an alternative for standard Greedy-training methods based upon a-posteriori error estimates on a training subset of the parameter set, we consider a nonlinear optimization combined with a Greedy method. We define an optimization problem for selecting a new parameter value on a given reduced space. This new parameter is then used –in a Greedy fashion– to determine the corresponding snapshot and to update the reduced basis. We show the well-posedness of this nonlinear optimization problem and derive first- and second-order optimality conditions. Numerical comparisons with the standard Greedy-training method are shown.

## 5.1 Introduction

Reduced Basis Methods (RBM) are nowadays a well-known tool to solve parametric partial differential equations (PPDEs) in cases where the PPDE has to be solved for various values of the parameters (the so-called *multi-query* context, e.g. in optimization) or when the solution for different parameter values has to be computed

K. Urban
Universität Ulm, Institute for Numerical Mathematics, Helmholtzstraße 20, D-89069 Ulm, Germany
e-mail: Karsten.Urban@uni-ulm.de

S. Volkwein (✉)
University of Konstanz, Department of Mathematics and Statistics, Universitätsstraße 10, D-78457 Konstanz, Germany
e-mail: Stefan.Volkwein@uni-konstanz.de

O. Zeeb
Universität Ulm, Institute for Numerical Mathematics, Helmholtzstraße 22, D-89069 Ulm, Germany
e-mail: Oliver.Zeeb@uni-ulm.de

A. Quarteroni, G. Rozza (eds.): *Reduced Order Methods for Modeling and Computational Reduction*, MS&A 9. DOI 10.1007/978-3-319-02090-7_5, 

extremely efficiently (the *realtime* context), see e.g. [15]. A key ingredient is an *offline-online-decomposition.* In the offline stage, detailed and thus expensive simulations (sometimes called *truth*) are computed for a moderate number of the parameters, $\mu_1,\dots,\mu_N$. The arising solutions $u(\mu_i)$, $i=1,\dots,N$, of the PPDE (sometimes called *snapshots*) are stored and are used to form a low-dimensional linear space spanned by the reduced basis. In the online stage, an approximation $u_N(\mu)$ for a new parameter $\mu \neq \mu_i$ is determined as the Galerkin projection onto the reduced space $V_N = \text{span}\{u(\mu_i) : i=1,\dots,N\}$. A whole variety of results for all sorts of problems has been published in the last years so that an even only halfway complete review including a reference list is far beyond the scope of this paper.

The topic of this paper is the generation of the reduced basis in the offline stage, namely the selection of $\mu_1,\dots,\mu_N$ above. It is nowadays basically standard to use a Greedy method, see e.g. [12]. The starting point is an a-posteriori error estimator $\Delta_N(\mu)$ for the quantity of interest on a current reduced space $V_N$. Such an estimator can often be constructed in such a way that the evaluation for a given parameter $\mu$ is highly efficient (in particular independent of the size of the truth system). A training set $\Xi_{\text{train}}$ is defined and the error estimator $\Delta_N(\mu)$ is maximized over $\Xi_{\text{train}}$. The arising maximizer $\mu_{N+1}$ is used to compute the next snapshot $u(\mu_{N+1})$ in order to form the reduced space $V_{N+1}$ of the next higher dimension. We refer to this approach as *Greedy-training.*

Even though this approach obviously has the advantage of being efficiently realizable, it may also suffer from the following fact: The training set $\Xi_{\text{train}}$ needs to be defined. This may be a delicate task since $\Xi_{\text{train}}$ should be small for efficiency reasons and at the same time sufficiently large in order to represent the whole parameter range as well as possible. The performance of the RBM crucially depends on the choice of $\Xi_{\text{train}}$.

This is the starting point of the present paper. Instead of maximizing the error estimator $\Delta_N(\mu)$ over $\Xi_{\text{train}}$, we develop a nonlinear optimization problem w.r.t. $\mu$ on $V_N$ based upon the residual of the primal (and possibly the dual) problem. We show the well-posedness of this optimization problem and derive first-order optimality conditions. The optimization problem is solved numerically by a gradient-type method. This method suffers from the fact that we can only determine local but not global solutions. To overcome this problem we combine the optimization strategy with different choices for the initial value for the optimization.

Let us refer to the work [2,3], where reduced bases are computed for high-dimensional input spaces. In our paper we prove existence of optimal solutions and derive optimality conditions, which can be also applied in the case, where one has to deal with distributed parameter functions; compare [8]. We also mention the recent work [4–6,11], where adaptive strategies are suggested for the Greedy-training to overcome the problem with high-dimensional parameter spaces. In the context of the method of proper orthogonal decomposition (POD) nonlinear optimization is utilized in [10] to determine optimal snapshot locations in order to control the number of snapshots and minimize the error in the POD reduced-order model.

The remainder of the paper is organized as follows. In Sect. 5.2, we review the basic ingredients of the RBM and develop the nonlinear optimization problem (which, in fact, is a minimization problem). We also prove the existence of a solution (Theorem 5.1). Section 5.3 is devoted to the derivation of first order optimality conditions (Theorem 5.2) while second-order conditions are discussed in Sect. 5.4. Finally, in Sect. 5.5 we report on numerical experiments in which we compare the optimization method with the known Greedy-training approach.

## 5.2 Problem Formulation

In this section we introduce our minimization problem and discuss the existence of optimal solutions.

### 5.2.1 The Exact Variational Problem

Let $\mathscr{D} \subset \mathbb{R}^P$ be a given nonempty, closed, bounded and convex parameter domain and $V$ a separable Hilbert space. For given $\ell \in V'$ ($V'$ denotes the space of all bounded and linear functionals defined on $V$ with norm $\|\cdot\|_{V'}$ and scalar product $(\cdot\,,\cdot)_{V'}$), the goal is to find the scalar output

$$s(\mu) := \langle \ell, u(\mu) \rangle_{V',V}, \quad \mu \in \mathscr{D}, \tag{5.1a}$$

where $u(\mu) \in V$ satisfies the variational problem ($f \in V'$ given)

$$a(u(\mu), \varphi; \mu) = \langle f, \varphi \rangle_{V',V} \quad \text{for all } \varphi \in V. \tag{5.1b}$$

In (5.1a), we denote by $\langle \cdot\,,\cdot \rangle_{V',V}$ the dual pairing of the spaces $V'$ and $V$. Furthermore, in (5.1b) the parameter-dependent, bilinear form $a(\cdot\,,\cdot\,;\mu) : V \times V \to \mathbb{R}$ is assumed to have the affine form

$$a(\varphi, \psi; \mu) = \sum_{q=1}^{Q} \vartheta^q(\mu)\, a^q(\varphi, \psi) \quad \text{for } \varphi, \psi \in V \text{ and } \mu \in \mathscr{D}$$

with (twice) continuously differentiable coefficient functions $\vartheta^q : \mathscr{D} \to \mathbb{R}$ and with parameter-independent bounded bilinear forms $a^q : V \times V \to \mathbb{R}$, $1 \le q \le Q$. Moreover, that the parameter-dependent bilinear form $a$ is uniformly bounded and coercive, i.e., there exist constants $\alpha_0 > 0$ and $\gamma > 0$ such that

$$\alpha(\mu) := \inf_{\varphi \in V \setminus \{0\}} \frac{a(\varphi, \varphi; \mu)}{\|\varphi\|_V^2} \ge \alpha_0 > 0 \qquad \text{for all } \mu \in \mathscr{D}, \tag{5.2a}$$

$$\big|a(\varphi, \phi; \mu)\big| \le \gamma \|\varphi\|_V \|\phi\|_V \qquad \text{for all } \varphi, \phi \in V \text{ and } \mu \in \mathscr{D}. \tag{5.2b}$$

Since the bilinear forms $a^q$ are bounded we assume that

$$\big|a^q(\varphi, \phi)\big| \le \gamma \|\varphi\|_V \|\phi\|_V \quad \text{for all } \varphi, \phi \in V \text{ and for } 1 \le q \le Q. \tag{5.3}$$

Notice that (5.2a) implies

$$a(\varphi,\varphi;\mu) \geq \alpha_0 \|\varphi\|_V^2 \quad \text{for all } \varphi \in V \text{ and for all } \mu \in \mathscr{D}. \tag{5.4}$$

Let us mention that we suppose that both $f$ and $\ell$ do not depend on $\mu$ in the affine form only for simplifying the presentation. From (5.2a) it follows by standard arguments that (5.1b) has a unique solution $u(\mu) \in V$ for any $\mu \in \mathscr{D}$.

Due to (5.1a) we require the following dual problem: for given $\mu \in \mathscr{D}$ find $p(\mu) \in V$ solving

$$a(\varphi,z(\mu);\mu) = -\langle \ell,\varphi\rangle_{V',V} \quad \text{for all } \varphi \in V. \tag{5.5}$$

Since the bilinear form $a(\cdot,\cdot;\mu)$ is bounded and uniformly coercive, the dual problem (5.5) possesses a unique solution $z(\mu) \in V$ for any $\mu \in \mathscr{D}$.

### *5.2.2 The Truth Approximation*

Next we introduce a so-called truth approximation for (5.1). For that purpose let $V^{\mathscr{N}} = \text{span}\{\varphi_1,\ldots,\varphi_{\mathscr{N}}\} \subset V$ be a finite dimensional subspace with linearly independent functions $\varphi_i$. The subspace $V^{\mathscr{N}}$ is endowed with the topology of $V$. We think of $\mathscr{N} \gg 1$ being 'large'. Then, for any $\mu \in \mathscr{D}$ we consider the 'truth' output

$$s^{\mathscr{N}}(\mu) := \langle \ell, u^{\mathscr{N}}(\mu)\rangle_{V',V}, \tag{5.6a}$$

where $u^{\mathscr{N}}(\mu) \in V^{\mathscr{N}}$ satisfies the variational equation

$$a(u^{\mathscr{N}}(\mu),\varphi_i;\mu) = \langle f,\varphi_i\rangle_{V',V} \quad \text{for } 1 \leq i \leq \mathscr{N}. \tag{5.6b}$$

We define the discrete coercivity constant

$$\alpha^{\mathscr{N}}(\mu) := \inf_{\varphi^{\mathscr{N}} \in V^{\mathscr{N}}\setminus\{0\}} \frac{a(\varphi^{\mathscr{N}},\varphi^{\mathscr{N}};\mu)}{\|\varphi^{\mathscr{N}}\|_V^2}, \quad \mu \in \mathscr{D}.$$

Unsing $V^{\mathscr{N}} \subset V$ and (5.2a) we find

$$\alpha^{\mathscr{N}}(\mu) \geq \inf_{\varphi \in V\setminus\{0\}} \frac{a(\varphi,\varphi;\mu)}{\|\varphi\|_V^2} \geq \alpha_0 \quad \text{for all } \mu \in \mathscr{D}.$$

Thus, (5.6b) has a unique solution $u^{\mathscr{N}}(\mu) \in V^{\mathscr{N}}$ for every $\mu \in \mathscr{D}$.

### *5.2.3 The Reduced-Order Modelling*

Let us introduce a reduced-order scheme for (5.6). For chosen linearly independent elements $\{\psi_i\}_{i=1}^{N^{\text{pr}}}$ in $V^{\mathscr{N}}$ we define $V_{N^{\text{pr}}} := \text{span}\{\psi_1,\ldots,\psi_{N^{\text{pr}}}\}$. Analogously, for linearly independent $\{\phi_i\}_{i=1}^{N^{\text{du}}}$ in $V^{\mathscr{N}}$ we set $\tilde{V}_{N^{\text{du}}} := \text{span}\{\phi_1,\ldots,\phi_{N^{\text{du}}}\}$. We have that $\max(N^{\text{pr}},N^{\text{du}}) \leq \mathscr{N}$. In the context of reduced-order modeling, $\max(N^{\text{pr}},N^{\text{du}})$ is much smaller than $\mathscr{N}$.

For any $\mu \in \mathscr{D}$ we consider the scalar output

$$\langle \ell, u_N(\mu) \rangle_{V',V}, \tag{5.7a}$$

where $u_N(\mu) \in V_{N^{\mathrm{pr}}}$ satisfies the variational equation

$$a(u_N(\mu), \psi_i; \mu) = \langle f, \psi_i \rangle_{V',V} \quad \text{for } 1 \le i \le N^{\mathrm{pr}}. \tag{5.7b}$$

For notational convenience, we just write $u_N$ instead of $u_{N^{\mathrm{pr}}}$ (also for other quantities) since there should be no misunderstanding. We collect some more or less known facts for later reference.

**Lemma 5.1** *Suppose that the bilinear form $a(\cdot\,,\cdot\,;\mu)$ satisfies* (5.2)*. Further, $f \in V'$ holds. Then, there exists a unique solution $u_N(\mu) \in V_{N^{pr}}$ to* (5.7b) *for every $\mu \in \mathscr{D}$ with*

$$\|u_N(\mu)\|_V \le \frac{\|f\|_{V'}}{\alpha_0} \quad \textit{for all } \mu \in \mathscr{D}. \tag{5.8}$$

*Proof* By assumption, the bilinear form $a(\cdot\,,\cdot\,;\mu)$ is bounded for every $\mu \in \mathscr{D}$. Since $V_{N^{\mathrm{pr}}} \subset V$, the form $a(\cdot\,,\cdot\,;\mu)$ is also uniformly coercive on $V_{N^{\mathrm{pr}}}$. Thus, it follows from the Lax-Milgram theorem that (5.7b) possesses a unique solution $u_N \in V_{N^{\mathrm{pr}}}$ for every $\mu \in \mathscr{D}$. Utilizing (5.4) and (5.7b) and the uniform coercivity, we obtain

$$\|u_N(\mu)\|_V^2 \le \frac{a(u_N(\mu), u_N(\mu); \mu)}{\alpha_0} = \frac{\langle f, u_N(\mu) \rangle_{V',V}}{\alpha_0} \le \frac{\|f\|_{V'}}{\alpha_0} \|u_N(\mu)\|_V,$$

which gives (5.8).

*Remark 5.1* 1) Due to Lemma 5.1 we can define the primal (non-linear) solution operator $\mathscr{S}_N^{\mathrm{pr}} : \mathscr{D} \to V_{N^{\mathrm{pr}}}$, where $u_N(\mu) = \mathscr{S}_N^{\mathrm{pr}}(\mu)$ denotes the unique solution to (5.7b).

2) Let us consider a specific case. Suppose that the bilinear form is given by $a(\cdot\,,\cdot\,;\mu) = \vartheta^1(\mu) a^1(\cdot\,,\cdot)$ (i.e., $Q = 1$) and $\vartheta^1(\mu) \ne 0$ holds for all $\mu \in \mathscr{D}$. Let $u_N^1 = u_N(\mu_1)$ be a solution to (5.7b) for given $\mu_1 \in \mathscr{D}$. Then, the function $u_N^2 = \vartheta^1(\mu_1) u_N^1 / \vartheta^1(\mu_2) \in V^N$ solves (5.7b) for $\mu_2 \in \mathscr{D}$. In fact, we have

$$\begin{aligned} a(u_N^2, \psi_i; \mu_2) &= \vartheta^1(\mu_2) a^1(u_N^2, \psi_i) = \vartheta^1(\mu_1) a^1(u_N^1, \psi_i) = a(u_N^1, \varphi_i; \mu_1) \\ &= \langle f, \psi_i \rangle_{V',V} \quad \text{for } 1 \le i \le N. \end{aligned}$$

Consequently, solutions to different parameter values are linearly dependent. □

For given $\mu \in \mathscr{D}$ the associated dual variable $z_N(\mu)$ solves the dual problem [1], namely

$$a(\phi_i, z_N(\mu); \mu) = -\langle \ell, \phi_i \rangle_{V',V}, \quad 1 \le i \le N^{\mathrm{du}}. \tag{5.9}$$

*Remark 5.2* 1) If the bilinear form satisfies (5.2) and $\ell \in V'$ holds, it follows by similar arguments as in the proof of Lemma 5.1 that (5.9) admits a unique solution $z_N(\mu) \in \tilde{V}_{N^{\mathrm{du}}}$ satisfying

$$\|z_N(\mu)\|_V \le \frac{\|\ell\|_{V'}}{\alpha_0} \quad \text{for all } \mu \in \mathscr{D}. \tag{5.10}$$

2) We define the (non-linear) solution operator $\mathscr{S}_N^{\mathrm{du}} : \mathscr{D} \to \tilde{V}_{N^{\mathrm{du}}}$, where $z_N = \mathscr{S}_N^{\mathrm{du}}(\mu)$ is the unique solution to (5.9). □

Next we define the residuals $r_N^{\mathrm{pr}}(\cdot;\mu), r_N^{\mathrm{du}}(\cdot;\mu) \in (V^{\mathscr{N}})'$ by

$$r_N^{\mathrm{pr}}(\varphi^{\mathscr{N}};\mu) := \langle f, \varphi^{\mathscr{N}} \rangle_{V',V} - a(u_N(\mu), \varphi^{\mathscr{N}};\mu) \qquad \text{for } \varphi \in V^{\mathscr{N}} \text{ and } \mu \in \mathscr{D},$$
$$r_N^{\mathrm{du}}(\varphi^{\mathscr{N}};\mu) := \langle \ell, \varphi^{\mathscr{N}} \rangle_{V',V} + a(\varphi^{\mathscr{N}}, z_N(\mu);\mu) \qquad \text{for } \varphi \in V^{\mathscr{N}} \text{ and } \mu \in \mathscr{D}.$$

It has turned out that the primal-dual output defined as

$$s_N(\mu) := \langle \ell, u_N(\mu) \rangle_{V',V} - r_N^{\mathrm{pr}}(z_N(\mu);\mu),$$

gives rise to favorable output error estimates which take the form (see [15], for instance)

$$\left| s^{\mathscr{N}}(\mu) - s_N(\mu) \right| \le \Delta_N^s(\mu) = \frac{\|r_N^{\mathrm{pr}}(\cdot;\mu)\|_{(V^{\mathscr{N}})'}}{\alpha_0^{1/2}} \frac{\|r_N^{\mathrm{du}}(\cdot;\mu)\|_{(V^{\mathscr{N}})'}}{\alpha_0^{1/2}}. \tag{5.11}$$

*Remark 5.3* 1) From

$$u_N(\mu) = \sum_{j=1}^{N^{\mathrm{pr}}} \mathrm{u}_{N,j}(\mu)\, \psi_j \quad \text{and} \quad z_N(\mu) = \sum_{j=1}^{N^{\mathrm{du}}} \mathrm{z}_{N,j}(\mu)\, \phi_j$$

we infer that

$$\begin{aligned} r_N^{\mathrm{pr}}(\varphi_i;\mu) &= \langle f, \varphi_i \rangle_{V',V} - \sum_{j=1}^{N^{\mathrm{pr}}} \mathrm{u}_{N,j}(\mu)\, a(\psi_j, \varphi_i;\mu) \\ &= \langle f, \varphi_i \rangle_{V',V} - \sum_{j=1}^{N^{\mathrm{pr}}} \mathrm{u}_{N,j}(\mu) \sum_{q=1}^{Q} \vartheta^q(\mu)\, a^q(\psi_j, \varphi_i), \\ r_N^{\mathrm{du}}(\varphi_i;\mu) &= \langle \ell, \varphi_i \rangle_{V',V} + a(\varphi_i, z_N(\mu);\mu) \\ &= \langle \ell, \varphi_i \rangle_{V',V} + \sum_{j=1}^{N^{\mathrm{du}}} \mathrm{z}_{N,j}(\mu) \sum_{q=1}^{Q} \vartheta^q(\mu)\, a^q(\varphi_i, \phi_j) \end{aligned}$$

for $1 \le i \le \mathscr{N}$. These representations of the residuals are utilized to realize an efficient offline-online decomposition for the reduced-order approach, see e.g. [12, 15].

2) Suppose that the bilinear form is given by $a(\cdot,\cdot;\mu) = \vartheta^1(\mu) a^1(\cdot,\cdot)$ (i.e., $Q = 1$) and $\vartheta^1(\mu) \neq 0$ holds for all $\mu \in \mathscr{D}$. Then, solutions to different parameter values are linearly dependent; see Remark 5.1-2). Let $\mu_1, \mu_2 \in \mathscr{D}$ be chosen arbitrarily. By $u_N^i$, $i = 1,2$, we denote the solutions to (5.7b) for parameter $\mu = \mu_i$. From $u_N^2 = \vartheta^1(\mu_1) u_N^1 / \vartheta^1(\mu_2)$ we infer that

$$V' \ni a(u_N^2, \cdot;\mu_2) - f = \frac{\vartheta^1(\mu_1)}{\vartheta^1(\mu_2)} a(u_N^1, \cdot;\mu_2) - f = a(u_N^1, \cdot;\mu_1) - f.$$

Hence, the norm $\|a(u_N(\mu),\cdot;\mu)-f\|_{(V^{\mathcal{N}})'}$ is constant for all $\mu \in \mathscr{D}$, where $u_N(\mu)$ denotes the solution to (5.7b) for the parameter $\mu$. Analogously, we can prove that the norm $\|a(\cdot,z_N(\mu);\mu)+\ell\|_{(V^{\mathcal{N}})'}$ is constant for all $\mu \in \mathscr{D}$, where $z_N(\mu)$ denotes the solution to (5.9) for the parameter $\mu$. □

### 5.2.4 The Minimization Problem

Let $N := (N^{\mathrm{pr}}, N^{\mathrm{du}})$, $Y_N := V_{N^{\mathrm{pr}}} \times \tilde{V}_{N^{\mathrm{du}}}$, $X_N = Y_N \times \mathbb{R}^P$ and $X_N^{ad} = Y_N \times \mathscr{D}$. We endow $X_N$ with the natural product topology. In the Greedy algorithm a new reduced-basis solution $u_N(\bar{\mu})$ associated with a certain parameter value $\bar{\mu}$ is added to the already computed set of ansatz functions provided an a-posteriori error measure $\Delta_N^s(\bar{\mu})$ in (5.11) is maximal. The idea here is to avoid the Greedy method and to determine $\bar{\mu}$ as the solution of a minimization problem. Thus, we introduce the cost functional $J : X_N \to \mathbb{R}$ for $x_N = (u_N, z_N, \mu) \in X_N$ by

$$J(x_N) = -\frac{1}{2}\left(\|f - a(u_N,\cdot;\mu)\|^2_{(V^{\mathcal{N}})'} + \|\ell + a(\cdot,z_N;\mu)\|^2_{(V^{\mathcal{N}})'}\right).$$

To ensure that the objective $J$ is twice continuously differentiable we do not utilize directly the estimator $\Delta_N^s(\mu)$ from (5.11), but a quadratic upper bound. This choice also ensures that both the primal and the dual residual are minimized during the optimization process, whereas in (5.11) it would be sufficient if only one of the factors becomes small. If $J(x_N(\mu)) \geq -\varepsilon\alpha_0$ holds true for $x_N(\mu) := (u_N(\mu), z_N(\mu), \mu)$, we infer by using Young's inequality that

$$\left|s^{\mathcal{N}}(\mu) - s_N(\mu)\right| \leq \frac{\|r_N^{\mathrm{pr}}(\cdot;\mu)\|^2_{(V^{\mathcal{N}})'} + \|r_N^{\mathrm{du}}(\cdot;\mu)\|^2_{(V^{\mathcal{N}})'}}{2\alpha_0} = -\frac{J(x_N(\mu))}{\alpha_0} \leq \varepsilon.$$

Now we consider the following optimization problem:

$$\min_{x_N \in X_N^{ad}} J(x_N) \quad \text{subject to (s.t.)} \quad x_N = (y_N, \mu),\ y_N = \mathscr{S}_N(\mu), \tag{P}$$

where we have set $\mathscr{S}_N = (\mathscr{S}_N^{\mathrm{pr}}, \mathscr{S}_N^{\mathrm{du}}) : \mathscr{D} \to Y_N$, i.e., $y_N = \mathscr{S}_N(\mu)$ means that $y_N = (u_N(\mu), z_N(\mu))$. Introducing the reduced cost functional

$$\hat{J}(\mu) := J(\mathscr{S}_N(\mu), \mu) \quad \text{for } \mu \in \mathscr{D},$$

we can express **(P)** equivalently in the reduced form

$$\min_{\mu \in \mathscr{D}} \hat{J}(\mu). \tag{$\hat{\mathrm{P}}$}$$

If **($\hat{\mathbf{P}}$)** has a local solution $\bar{\mu} \in \mathscr{D}$, then $\bar{x}_N := (\bar{y}_N, \bar{\mu})$ is a local solution to **(P)**, where we set $\bar{y}_N = (\bar{u}_N, \bar{p}_N) := \mathscr{S}_N(\bar{\mu})$. We now give a general existence result.

**Theorem 5.1** *Suppose that the bilinear form $a(\cdot,\cdot;\mu)$ satisfies* (5.2). *Further, $f$ and $\ell$ belong to $V'$. Then, there exists at least one optimal solution $\bar{x}_N = (\bar{y}_N, \bar{\mu})$, $\bar{y}_N = (\bar{u}_N, \bar{z}_N) \in Y_N$, to* **(P)**.

*Proof* Since $\mathscr{D}$ is assumed to be nonempty and $\mathscr{S}_N : \mathscr{D} \to Y_N$ is well-defined, the set of admissible solutions

$$\mathfrak{F}(\mathbf{P}) = \left\{ x_N = (y_N, \mu) \in X_N^{ad} \,\middle|\, y_N = \mathscr{S}_N(\mu) \right\}$$

is nonempty. Let $\{x_N^{(n)}\}_{n\in\mathbb{N}} \subset \mathfrak{F}(\mathbf{P})$, $x_N^{(n)} = (y_N^{(n)}, \mu^{(n)})$ and $y_N^{(n)} = (u_N^{(n)}, z_N^{(n)})$, be a minimizing sequence for $J$:

$$\inf_{x_N \in \mathfrak{F}(\mathbf{P})} J(x_N) = \lim_{n\to\infty} J(x_N^{(n)}).$$

Since $\mathscr{D}$ is bounded and the a-priori bounds (5.8), (5.10) hold, $\inf_{x_N\in\mathfrak{F}(\mathbf{P})} J(x_N)$ is bounded from below. Moreover, from $\mu^{(n)} \in \mathscr{D} \subset \mathbb{R}^P$ for every $n$ we infer that there exists a subsequence $\{\mu^{(n_k)}\}_{k\in\mathbb{N}}$ in $\mathscr{D}$ and an element $\bar{\mu} \in \mathscr{D}$ so that

$$\lim_{k\to\infty} \mu^{(n_k)} = \bar{\mu} \quad \text{in } \mathbb{R}^P.$$

It follows from the a-priori estimates (5.8) and (5.10) that the sequence $\{(u_N^{(n)}, z_N^{(n)})\}_{n\in\mathbb{N}}$ is bounded in $Y_N$. Consequently, there exist a subsequence $\{y_N^{(n_k)}\}_{k\in\mathbb{N}}$ and a pair $\bar{y}_N = (\bar{u}_N, \bar{p}_N) \in Y_N$ such that

$$u_N^{(n_k)} \rightharpoonup \bar{u}_N \text{ for } k \to \infty \text{ in } V_{N^{\mathrm{pr}}} \quad \text{and} \quad z_N^{(n_k)} \rightharpoonup \bar{z}_N \text{ for } k \to \infty \text{ in } \tilde{V}_{N^{\mathrm{du}}}. \tag{5.12}$$

Next we prove that $\bar{y}_N = \mathscr{S}_N(\bar{\mu})$ holds. For $1 \le i \le N^{\mathrm{pr}}$ we have

$$\begin{aligned}
\langle f, \psi_i \rangle_{V',V} - a(\bar{u}_N, \psi_i; \bar{\mu}) &= a(u_N^{(n_k)}, \psi_i; \mu^{(n_k)}) - a(\bar{u}_N, \psi_i; \bar{\mu}) = \\
&= a(u_N^{(n_k)}, \psi_i; \mu^{(n_k)}) - a(u_N^{(n_k)}, \psi_i; \bar{\mu}) + a(u_N^{(n_k)} - \bar{u}_N, \psi_i; \bar{\mu}) \\
&= \sum_{q=1}^{Q} \Big( \big(\vartheta^q(\mu^{(n_k)}) - \vartheta^q(\bar{\mu})\big)\, a^q(u_N^{(n_k)}, \psi_i) \Big) + a(u_N^{(n_k)} - \bar{u}_N, \psi_i; \bar{\mu}).
\end{aligned}$$

Let us define the functionals $F_i \in V' \subset V_N'$ by $\langle F_i, \varphi \rangle_{V',V} := a(\varphi, \psi_i; \bar{\mu})$ for $\varphi \in V$ and $1 \le i \le N^{\mathrm{pr}}$. From (5.12) we infer that

$$a(u_N^{(n_k)} - \bar{u}_N, \psi_i; \bar{\mu}) = F_i(u_N^{(n_k)} - \bar{u}_N) \to 0 \quad \text{for } k \to \infty \text{ and } 1 \le i \le N^{\mathrm{pr}}.$$

Moreover, $\|u_N^{(n_k)}\|_V$ is uniformly bounded and the $\vartheta^q$'s are continuous. Thus,

$$\sum_{q=1}^{Q} \Big( \big(\vartheta^q(\mu^{(n_k)}) - \vartheta^q(\bar{\mu})\big)\, a^q(u_N^{(n_k)}, \psi_i) \Big) \to 0 \quad \text{for } k \to \infty \text{ and } 1 \le i \le Q.$$

Consequently, $\bar{u}_N = \mathscr{S}_N^{\mathrm{pr}}(\bar{\mu})$ holds. Analogously, we find that $\bar{z}_N = \mathscr{S}^{\mathrm{du}}(\bar{\mu})$ holds true. Thus, $\bar{x}_N = (\bar{y}_N, \bar{\mu}) \in \mathfrak{F}(\mathbf{P})$ is satisfied. Next, we show that $\bar{x}_N$ is a minimizer for $J$. Note that with the above arguments

$$\begin{aligned}&\|a(u_N^{(n_k)},\cdot;\bar{\mu})-a(u_N^{(n_k)},\cdot;\mu^{(n_k)})\|_{(V^{\mathcal{N}})'}\\&\qquad\le\sum_{q=1}^{Q}\left|\vartheta^q(\bar{\mu})-\vartheta^q(\mu^{(n_k)})\right|\|a^q(u_N^{(n_k)},\cdot)\|_{(V^{\mathcal{N}})'}\overset{k\to\infty}{\longrightarrow}0.\end{aligned}$$

This and (5.12) imply

$$\begin{aligned}&\lim_{k\to\infty}\|f-a(u_N^{(n_k)},\cdot;\mu^{(n_k)})\|_{(V^{\mathcal{N}})'}=\\&=\lim_{k\to\infty}\|f-a(u_N^{(n_k)},\cdot;\bar{\mu})\|_{(V^{\mathcal{N}})'}+\lim_{k\to\infty}\|a(u_N^{(n_k)},\cdot;\bar{\mu})-a(u_N^{(n_k)},\cdot;\mu^{(n_k)})\|_{(V^{\mathcal{N}})'}\\&=\|f-a(\bar{u}_N,\cdot;\bar{\mu})\|_{(V^{\mathcal{N}})'}.\end{aligned}$$

Analogously, $\lim_{k\to\infty}\|\ell+a(\cdot,z_N^{(n_k)};\mu^{(n_k)})\|_{(V^{\mathcal{N}})'}=\|\ell+a(\cdot,\bar{z}_N;\bar{\mu})\|_{(V^{\mathcal{N}})'}$ and therefore

$$\inf_{x_N\in\mathcal{F}(\mathbf{P})}J(x_N)=\lim_{k\to\infty}J(x_N^{(n_k)})=J(\bar{x}_N),$$

i.e., $\bar{x}_N$ is a solution to **(P)**. □

Before we continue, let us collect some notation that will be needed in the sequel. Let $\bar{x}_N=(\bar{y}_N,\bar{\mu})$, $\bar{y}_N=(\bar{u}_N,\bar{z}_N)$, be an optimal solution to **(P)** according to Theorem 5.1. Then, define corresponding (optimal) primal and dual residuals as

$$\begin{aligned}\bar{r}_N^{\mathrm{pr}}(\varphi^{\mathcal{N}})&:=\langle f,\varphi^{\mathcal{N}}\rangle_{V',V}-a(\bar{u}_N,\varphi^{\mathcal{N}};\bar{\mu})&&\text{for }\varphi^{\mathcal{N}}\in V^{\mathcal{N}},\\\bar{r}_N^{\mathrm{du}}(\varphi^{\mathcal{N}})&:=\langle \ell,\varphi^{\mathcal{N}}\rangle_{V',V}+a(\varphi^{\mathcal{N}},\bar{z}_N;\bar{\mu})&&\text{for }\varphi^{\mathcal{N}}\in V^{\mathcal{N}}.\end{aligned}$$

We define the corresponding Riesz representations $\bar{\rho}_N^{\mathrm{pr}},\bar{\rho}_N^{\mathrm{du}}\in V^{\mathcal{N}}$, i.e.,

$$\begin{aligned}(\bar{\rho}_N^{\mathrm{pr}},\varphi^{\mathcal{N}})_V&=\bar{r}_N^{\mathrm{pr}}(\varphi^{\mathcal{N}})=\langle f,\varphi^{\mathcal{N}}\rangle_{V',V}-a(\bar{u}_N,\varphi^{\mathcal{N}};\bar{\mu})&&\text{for all }\varphi^{\mathcal{N}}\in V^{\mathcal{N}},\\(\bar{\rho}_N^{\mathrm{du}},\varphi^{\mathcal{N}})_V&=\bar{r}_N^{\mathrm{du}}(\varphi^{\mathcal{N}})=\langle \ell,\varphi^{\mathcal{N}}\rangle_{V',V}+a(\varphi^{\mathcal{N}},\bar{z}_N;\bar{\mu})&&\text{for all }\varphi^{\mathcal{N}}\in V^{\mathcal{N}}.\end{aligned}$$

This in particular implies that

$$(g,\bar{r}_N^{\mathrm{pr}})_{(V^{\mathcal{N}})'}=\langle g,\bar{\rho}_N^{\mathrm{pr}}\rangle_{(V^{\mathcal{N}})',V^{\mathcal{N}}}\quad\text{for all }g\in(V^{\mathcal{N}})',$$

which will be used later. It is noticable to mention that we have in general $\bar{\rho}_N^{\mathrm{pr}}\notin V_{N^{\mathrm{pr}}}$ and $\bar{\rho}_N^{\mathrm{du}}\notin\tilde{V}_{N^{\mathrm{du}}}$.

## 5.3 First-Order Necessary Optimality Conditions

First we write the equality constraints in **(P)** in a compact from. For that purpose we introduce the nonlinear mapping $e=(e_1,e_2):X_N\to Y_N'$ by

$$\langle e(x_N),\lambda_N\rangle_{Y_N',Y_N}=\langle e_1(x_N),\lambda_N^1\rangle_{V'_{N^{\mathrm{pr}}},V_{N^{\mathrm{pr}}}}+\langle e_2(x_N),\lambda_N^2\rangle_{\tilde{V}'_{N^{\mathrm{du}}},\tilde{V}_{N^{\mathrm{du}}}}$$

for $x_N = (u_N, z_N, \mu) \in X_N^{ad}$ and $\lambda_N = (\lambda_N^1, \lambda_N^2) \in Y_N$. Here, we identify the dual $Y_N'$ with $V_{N\mathrm{pr}}' \times \tilde{V}_{N\mathrm{du}}'$ and we put

$$\begin{aligned}
\langle e_1(x_N), \lambda_N^1 \rangle_{V_{N\mathrm{pr}}', V_{N\mathrm{pr}}} &= \langle f, \lambda_N^1 \rangle_{V_{N\mathrm{pr}}', V_{N\mathrm{pr}}} - a(u_N, \lambda_N^1; \mu), \\
\langle e_2(x_N), \lambda_N^2 \rangle_{\tilde{V}_{N\mathrm{du}}', \tilde{V}_{N\mathrm{du}}} &= \langle \ell, \lambda_N^2 \rangle_{\tilde{V}_{N\mathrm{du}}', \tilde{V}_{N\mathrm{du}}} + a(\lambda_N^2, z_N; \mu).
\end{aligned}$$

Using (5.2b) we infer that

$$\begin{aligned}
\|e(x_N)\|_{Y_N'} &= \sup_{\|\lambda_N\|_{Y_N}=1} \langle e(x_N), \lambda_N \rangle_{Y_N', Y_N} \\
&= \sup_{\|\lambda_N^1\|_V=1} \langle e_1(x_N), \lambda_N^1 \rangle_{V_{N\mathrm{pr}}', V_{N\mathrm{pr}}} + \sup_{\|\lambda_N^2\|_V=1} \langle e_2(x_N), \lambda_N^2 \rangle_{\tilde{V}_{N\mathrm{du}}', \tilde{V}_{N\mathrm{du}}} \\
&\leq C_e \left(1 + \|u_N\|_V + \|z_N\|_V\right)
\end{aligned}$$

with $C_e = \max(\|f\|_{V'} + \|\ell\|_{V'}, \gamma)$.

To derive first-order optimality conditions for **(P)** we have to ensure that the mapping $e$ is continuously (Fréchet) differentiable and satisfies a standard constraint qualification; see, e.g., [7, 16].

**Proposition 5.1** *Suppose that the bilinear form $a(\cdot,\cdot;\mu)$ satisfies* (5.2)*. Further, $f, \ell \in V'$ holds and the functions $\vartheta^q$ are continuously differentiable for $1 \leq q \leq Q$. Then, the mapping e is continuously (Fréchet) differentiable and its (Fréchet) derivative at $x_N = (y_N, \mu) \in X_N^{ad}$, $y_N = (u_N, z_N)$, is given by*

$$\langle e'(x_N)x_N^\delta, \lambda_N \rangle_{Y_N', Y_N} = \langle e_1'(x_N)x_N^\delta, \lambda_N^1 \rangle_{V_{Npr}', V_{Npr}} + \langle e_2'(x_N)x_N^\delta, \lambda_N^2 \rangle_{\tilde{V}_{Ndu}', \tilde{V}_{Ndu}}$$

*for any direction $x_N^\delta = (u_N^\delta, z_N^\delta, \mu^\delta) \in X_N$ and for $\lambda_N = (\lambda_N^1, \lambda_N^2) \in Y_N$, where*

$$\begin{aligned}
\langle e_1'(x_N)x_N^\delta, \lambda_N^1 \rangle_{\tilde{V}_{Npr}', \tilde{V}_{Npr}} &= -a(u_N^\delta, \lambda_N^1; \mu) - \sum_{q=1}^{Q} a^q(u_N, \lambda_N^1) \nabla \vartheta^q(\mu)^\top \mu^\delta, \\
\langle e_2'(x_N)x_N^\delta, \lambda_N^2 \rangle_{V_{Ndu}', V_{Ndu}} &= a(\lambda_N^2, z_N^\delta; \mu) + \sum_{q=1}^{Q} a^q(\lambda_N^2, z_N) \nabla \vartheta^q(\mu)^\top \mu^\delta
\end{aligned}$$

*with $\nabla \vartheta^q(\mu) = (\vartheta_{\mu_1}^q(\mu), \ldots, \vartheta_{\mu_P}^q(\mu))^\top \in \mathbb{R}^P$ and $\vartheta_{\mu_i}^q = \frac{\partial \vartheta^q}{\partial \mu_i}$. Furthermore, the (Fréchet) derivative $e'(x_N) : X_N \to Y_N'$ is a surjective operator for every $x_N \in X_N^{ad}$.*

*Proof* It follows by standard arguments that $e$ is (Fréchet) differentiable for every $x_N \in X_N^{ad}$. Therefore, we only prove that the linear operator $e'(x_N)$ is onto. Let $F_N = (F_N^1, F_N^2) \in Y_N'$ be chosen arbitrarily. Then, $e'(x_N)$ is surjective if there exists an element $x_N^\delta = (u_N^\delta, z_N^\delta, \mu^\delta) \in X_N$ satisfying

$$e'(x_N)x_N^\delta = F_N \quad \text{in } Y_N'. \tag{5.13}$$

Equation (5.13) is equivalent with

$$e_1'(x_N)x_N^\delta = F_N^1 \text{ in } V_{N\mathrm{pr}}' \quad \text{in} \quad e_2'(x_N)x_N^\delta = F_N^2 \quad \text{in } \tilde{V}_{N\mathrm{du}}'. \tag{5.14}$$

Choosing $\mu^\delta = 0$ we obtain from (5.14) that

$$\begin{aligned} a(u_N^\delta, \lambda_N^1; \mu) &= -\langle F_N^1, \lambda_N^1 \rangle_{V'_{N\mathrm{pr}}, V_{N\mathrm{pr}}} && \text{for all } \lambda_N^1 \in V_{N\mathrm{pr}}, \\ a(\lambda_N^2, z_N^\delta; \mu) &= \langle F_N^2, \lambda_N^2 \rangle_{\tilde{V}'_{N\mathrm{du}}, \tilde{V}_{N\mathrm{du}}} && \text{for all } \lambda_N^2 \in \tilde{V}_{N\mathrm{du}}. \end{aligned} \tag{5.15}$$

Since the bilinear form $a(\cdot\,,\cdot\,;\mu)$ is bounded and coercive, there exists a unique pair $y_N^\delta = (u_N^\delta, z_N^\delta) \in Y_N$ solving (5.15). Summarizing, $x_N^\delta = (y_N^\delta, 0)$ solves (5.13) which implies that $e'(x_N)$ is surjective.

Next let us introduce the Lagrange functional $\mathscr{L} : X_N \times Y_N \to \mathbb{R}$ for $x_N = (x_N^1, x_N^2, \mu) \in X_N$ and $\lambda_N = (\lambda_N^1, \lambda_N^2) \in Y_N$ as

$$\begin{aligned} \mathscr{L}(x_N, \lambda_N) &= J(x_N) + \langle e(x_N), \lambda_N \rangle_{Y'_N, Y_N} \\ &= -\frac{1}{2}\left( \|f - a(u_N, \cdot\,;\mu)\|^2_{(V^{\mathscr{N}})'} + \|a(\cdot\,, z_N;\mu) + \ell\|^2_{(V^{\mathscr{N}})'} \right) \\ &\quad + \langle (f, \ell), \lambda_N \rangle_{Y'_N, Y_N} - a(u_N, \lambda_N^1; \mu) + a(\lambda_N^2, z_N; \mu). \end{aligned}$$

We infer from Proposition 5.1 that first-order necessary optimality conditions are given as follows [7, 16]: Let $\bar{x}_N = (\bar{y}_N, \bar{\mu}) \in X_N^{ad}$, $\bar{y}_N = (\bar{u}_N, \bar{z}_N) \in Y_N$, be a local solution to **(P)**. Then, there exists a Lagrange multiplier $\bar{\lambda}_N = (\bar{\lambda}_N^1, \bar{\lambda}_N^2) \in Y_N$ solving the following system

$$\mathscr{L}_{u_N}(\bar{x}_N, \bar{\lambda}_N) u_N^\delta = 0 \qquad \text{for all } u_N^\delta \in V_{N\mathrm{pr}}, \tag{5.16a}$$

$$\mathscr{L}_{z_N}(\bar{x}_N, \bar{\lambda}_N) z_N^\delta = 0 \qquad \text{for all } z_N^\delta \in \tilde{V}_{N\mathrm{du}}, \tag{5.16b}$$

$$\mathscr{L}_{\mu}(\bar{x}_N, \bar{\lambda}_N)(\mu^\delta - \bar{\mu}) \geq 0 \qquad \text{for all } \mu^\delta \in \mathscr{D}, \tag{5.16c}$$

where, for instance, $\mathscr{L}_{u_N}$ denote the (Fréchet) derivative of the Lagrangian with respect to the argument $u_N$. First we study (5.16a). For $u_N^\delta \in V_{N\mathrm{pr}}$ we find

$$\mathscr{L}_{u_N}(\bar{x}_N, \bar{\lambda}_N) u_N^\delta = (f - a(\bar{u}_N, \cdot\,;\bar{\mu}), a(u_N^\delta, \cdot\,;\bar{\mu}))_{(V^{\mathscr{N}})'} - a(u_N^\delta, \bar{\lambda}_N^1; \bar{\mu}).$$

Using the Riesz representation $\bar{\rho}_N^{\mathrm{pr}} \in V^{\mathscr{N}}$ of $\bar{r}_N^{\mathrm{pr}} \in (V^{\mathscr{N}})'$, we get

$$\begin{aligned} \mathscr{L}_{u_N}(\bar{x}_N, \bar{\lambda}_N) u_N^\delta &= (\bar{r}_N^{\mathrm{pr}}, a(u_N^\delta, \cdot\,;\bar{\mu}))_{(V^{\mathscr{N}})'} - a(u_N^\delta, \bar{\lambda}_N^1; \bar{\mu}) \\ &= a(u_N^\delta, \bar{\rho}_N^{\mathrm{pr}}; \bar{\mu}) - a(u_N^\delta, \bar{\lambda}_N^1; \bar{\mu}) = a(u_N^\delta, \bar{\rho}_N^{\mathrm{pr}} - \bar{\lambda}_N^1; \bar{\mu}). \end{aligned} \tag{5.17}$$

From (5.16a) and (5.17) we infer the first adjoint equation:

$$a(u_N^\delta, \bar{\lambda}_N^1; \bar{\mu}) = a(u_N^\delta, \bar{\rho}_N^{\mathrm{pr}}; \bar{\mu}) \quad \text{for all } u_N^\delta \in V_{N\mathrm{pr}}. \tag{5.18}$$

*Remark 5.4* Since in general $\bar{\rho}_N^{\mathrm{pr}} \notin V_{N\mathrm{pr}}$ holds, we obtain in general $\bar{\lambda}_N^1 \neq \bar{\rho}_N^{\mathrm{pr}}$. Rather, $\bar{\lambda}_N^1$ is the $a$-orthogonal projection of $\bar{\rho}_N^{\mathrm{pr}} \in V$ onto $\bar{\lambda}_N^1 \in V_{N\mathrm{pr}}$. □

Further, we have

$$\mathscr{L}_{z_N}(\bar{x}_N, \bar{\lambda}_N) z_N^\delta = -(\ell + a(\cdot\,, \bar{z}_N; \bar{\mu}), a(\cdot\,, z_N^\delta; \bar{\mu}))_{(V^{\mathscr{N}})'} + a(\bar{\lambda}_N^2, z_N^\delta; \bar{\mu}) \tag{5.19}$$

for any direction $z_N^\delta \in \tilde{V}_{N^{\mathrm{du}}}$. Using the Riesz representation $\bar{\rho}_N^{\mathrm{du}} \in V^{\mathcal{N}}$ of $\bar{r}_N^{\mathrm{du}} \in (V^{\mathcal{N}})'$, combining (5.16b) and (5.19) we get

$$\mathcal{L}_{z_N}(\bar{x}_N, \bar{\lambda}_N) z_N^\delta = a(\bar{\lambda}_N^2 - \bar{\rho}_N^{\mathrm{du}}, z_N^\delta; \bar{\mu}) = 0 \quad \text{for all } z_N^\delta \in V_{N^{\mathrm{du}}}$$

which gives the second adjoint equation

$$a(\bar{\lambda}_N^2, z_N^\delta; \bar{\mu}) = a(\bar{\rho}_N^{\mathrm{du}}, z_N^\delta; \bar{\mu}) \quad \text{for all } z_N^\delta \in V_{N^{\mathrm{du}}}. \tag{5.20}$$

*Remark 5.5* Analogous to Remark 5.4 we infer that $\bar{\lambda}_N^2$ is the $a$-orthogonal decomposition of $\bar{\rho}_N^{\mathrm{du}}$ onto $\tilde{V}_{N^{\mathrm{du}}}$. □

Next we consider (5.16c). Using the Riesz representations $\bar{\rho}_N^{\mathrm{pr}}, \bar{\rho}_N^{\mathrm{du}} \in V^{\mathcal{N}}$ of $\bar{r}_N^{\mathrm{pr}}$, $\bar{r}_N^{\mathrm{du}} \in (V^{\mathcal{N}})'$, respectively, it follows that

$$\begin{aligned}
\mathcal{L}_\mu(\bar{x}_N, \bar{\lambda}_N)\mu^\delta &= \sum_{q=1}^{Q} \nabla\vartheta^q(\bar{\mu})^\top \mu^\delta \, (\bar{r}_N^{\mathrm{pr}}, a^q(\bar{u}_N, \cdot))_{(V^{\mathcal{N}})'} \\
&+ \sum_{q=1}^{Q} \nabla\vartheta^q(\bar{\mu})^\top \mu^\delta \left( (-\bar{r}_N^{\mathrm{du}}, a^q(\cdot, \bar{z}_N))_{V'} + a^q(\bar{\lambda}_N^2, \bar{z}_N) - a^q(\bar{u}_N, \bar{\lambda}_N^1) \right) \\
&= \sum_{q=1}^{Q} \left( a^q(\bar{u}_N, \bar{\rho}_N^{\mathrm{pr}} - \bar{\lambda}_N^1) + a^q(\bar{\lambda}_N^2 - \bar{\rho}_N^{\mathrm{du}}, \bar{z}_N) \right) \nabla\vartheta^q(\bar{\mu})^\top \mu^\delta
\end{aligned} \tag{5.21}$$

for any direction $\mu^\delta \in \mathbb{R}^P$. We define the Jacobi matrix

$$D\vartheta(\bar{\mu}) = \begin{pmatrix} \nabla\vartheta^1(\bar{\mu})^\top \\ \vdots \\ \nabla\vartheta^Q(\bar{\mu})^\top \end{pmatrix} \in \mathbb{R}^{Q\times P}$$

with $\nabla\vartheta^q(\mu) = (\vartheta^q_{\mu_1}(\mu), \ldots, \vartheta^q_{\mu_P}(\mu))^\top \in \mathbb{R}^P$ and $\vartheta^q_{\mu_i} = \frac{\partial\vartheta^q}{\partial\mu_i}$. Further, we set $\bar{\xi} = \bar{\xi}(\bar{x}_N, \bar{\lambda}_N) = (\bar{\xi}_1, \ldots, \bar{\xi}_Q)^\top \in \mathbb{R}^Q$ with

$$\bar{\xi}_q = a^q(\bar{u}_N, \bar{\rho}_N^{\mathrm{pr}} - \bar{\lambda}_N^1) + a^q(\bar{\lambda}_N^2 - \bar{\rho}_N^{\mathrm{du}}, \bar{z}_N) \quad \text{for } 1 \le q \le Q.$$

Then, we derive from (5.16c) and (5.21)

$$\left(D\vartheta(\bar{\mu})^\top \bar{\xi}\right)^\top (\mu^\delta - \bar{\mu}) \ge 0 \quad \text{for all } \mu^\delta \in \mathcal{D}. \tag{5.22}$$

Summarizing we have proved the following result.

**Theorem 5.2** *Suppose that the bilinear form $a(\cdot,\cdot;\mu)$ satisfies* (5.2). *Further, $f, \ell \in V'$ holds and the functions $\vartheta^q$ are continuously differentiable for $1 \le q \le Q$. Let $\bar{x}_N = (\bar{y}_N, \bar{\mu}) \in X_N^{ad}$, $\bar{y}_N = (\bar{u}_N, \bar{z}_N) \in Y_N$, be a local solution to* (**P**). *Then, there exists a unique associated Lagrange multiplier pair $\bar{\lambda}_N = (\bar{\lambda}_N^1, \bar{\lambda}_N^2) \in Y_N$ satisfying together with $\bar{x}_N$ the first-order necessary optimality conditions* (5.18), (5.20) *and* (5.22).

The gradient $\nabla\hat{J}$ of the reduced cost functional $\hat{J}$ at a point $\mu \in \mathscr{D}$ is given by the formula [7, 16]

$$\nabla\hat{J}(\mu) = D\vartheta(\mu)^\top \xi \in \mathbb{R}^P, \tag{5.23}$$

where the components of the vector $\xi \in \mathbb{R}^Q$ are

$$\xi_q = a^q(u_N, \bar{\rho}_N^{\mathrm{pr}} - \lambda_N^1) + a^q(\lambda_N^2 - \bar{\rho}_N^{\mathrm{du}}, z_N) \quad \text{for } 1 \le q \le Q,$$

$(u_N, z_N) = \mathscr{S}(\mu)$ holds and $\lambda_N = (\lambda_N^1, \lambda_N^2) \in Y_N$ solves the dual system

$$\begin{aligned} a(u_N^\delta, \lambda_N^1; \mu) &= a(u_N^\delta, \rho_N^{\mathrm{pr}}; \mu) && \text{for all } u_N^\delta \in V_{N^{\mathrm{pr}}},\\ a(\lambda_N^2, z_N^\delta; \mu) &= a(\rho_N^{\mathrm{du}}, z_N^\delta; \mu) && \text{for all } z_N^\delta \in \tilde{V}_{N^{\mathrm{du}}}. \end{aligned}$$

Here, $\rho_N^{\mathrm{pr}}, \rho_N^{\mathrm{du}} \in V^{\mathscr{N}}$ are the Riesz representants of the residuals $r_N^{\mathrm{pr}}(\cdot\,;\mu)$, $r_N^{\mathrm{du}}(\cdot\,;\mu) \in (V^{\mathscr{N}})'$, respectively.

*Remark 5.6* Suppose that the bilinear form is given by $a(\cdot,\cdot;\mu) = \vartheta^1(\mu)a^1(\cdot,\cdot)$ (i.e., $Q = 1$) and $\vartheta^1(\mu) \neq 0$ holds for all $\mu \in \mathscr{D}$. Then, solutions to different parameter values are linearly dependent; see Remark 5.1-2) and Remark 5.3-2). Then, it follows from $\vartheta^1(\mu) \neq 0$, (5.18) and (5.20) that

$$\begin{aligned} a^1(u_N^\delta, \bar{\lambda}_N^1) &= a^1(u_N^\delta, \bar{\rho}_N^{\mathrm{pr}}) && \text{for all } u_N^\delta \in V_{N^{\mathrm{pr}}},\\ a^1(\bar{\lambda}_N^2, z_N^\delta) &= a^1(\bar{\rho}_N^{\mathrm{du}}, z_N^\delta) && \text{for all } z_N^\delta \in \tilde{V}_{N^{\mathrm{du}}}. \end{aligned}$$

In particular, $a^1(\bar{u}_N, \bar{\rho}_N^{\mathrm{pr}} - \bar{\lambda}_N^1) = a^1(\bar{\lambda}_N^2 - \bar{\rho}_N^{\mathrm{du}}, \bar{z}_N) = 0$ holds true, which gives $\xi_1 = 0$. Therefore, $\nabla\hat{J}(\mu) = 0$ is satisfied. This coincides with the observation in Remark 5.3-2 that the mappings

$$\mu \mapsto \|a(\mathscr{S}_N^{\mathrm{pr}}(\mu), \cdot\,;\mu) - f\|_{(V^{\mathscr{N}})'} \quad \text{and} \quad \mu \mapsto \|a(\cdot, \mathscr{S}_N^{\mathrm{du}}(\mu);\mu) + \ell\|_{(V^{\mathscr{N}})'}$$

are constant. □

## 5.4 Second-Order Derivatives

To solve **(P)** in our numerical experiments we apply a globalized sequential quadratic programming (SQP) method which is makes use of second-order derivatives of the Lagrange functional; see [13], for example. For that reason we address second-order optimality conditions in this section. We restrict ourselves to simple bounds, i.e., we assume that the bounded and convex parameter set $\mathscr{D}$ is given by

$$\mathscr{D} = \underbrace{\left[\mu_{a,1}, \mu_{b,1}\right] \times \ldots \times \left[\mu_{a,P}, \mu_{b,P}\right]}_{P\text{-times}} \subset \mathbb{R}^P$$

with lower and upper bounds $\mu_{a,i} \le \mu_{b,i}$, $1 \le i \le P$. Let $\bar{x}_N = (\bar{y}_N, \bar{\mu}) \in X_N^{ad}$, $\bar{y}_N = (\bar{u}_N, \bar{z}_N) \in Y_N$, be a solution to the first-order necessary optimatity conditions for **(P)**;

see Theorem 5.2. Moreover, the pair $\bar{\lambda}_N = (\bar{\lambda}_N^1, \bar{\lambda}_N^2) \in Y_N$ denotes for the associated unique Lagrange multiplier. We suppose that the functions $\vartheta^q$ are twice continuously differentiable. For $u_N^\delta, \tilde{u}_N^\delta \in V_{N^{\mathrm{pr}}}$ we deduce

$$\mathscr{L}_{u_N u_N}(\bar{x}_N, \bar{\lambda}_N)(u_N^\delta, \tilde{u}_N^\delta) = -(a(\tilde{u}_N^\delta, \cdot; \bar{\mu}), a(u_N^\delta, \cdot; \bar{\mu}))_{(V^{\mathscr{N}})'}. \tag{5.24}$$

Analogously, we find for $z_N^\delta, \tilde{z}_N^\delta \in \tilde{V}_{N^{\mathrm{du}}}$

$$\mathscr{L}_{z_N z_N}(\bar{x}_N, \bar{\lambda}_N)(z_N^\delta, \tilde{z}_N^\delta) = -(a(\cdot, \tilde{z}_N^\delta; \bar{\mu}), a(\cdot, z_N^\delta; \bar{\mu}))_{(V^{\mathscr{N}})'}. \tag{5.25}$$

Further, it follows that

$$\mathscr{L}_{u_N z_N}(\bar{x}_N, \bar{\lambda}_N)(u_N^\delta, z_N^\delta) = \mathscr{L}_{z_N u_N}(\bar{x}_N, \bar{\lambda}_N)(z_N^\delta, u_N^\delta). \tag{5.26}$$

for $u_N^\delta \in V_{N^{\mathrm{pr}}}$ and $z_N^\delta \in \tilde{V}_{N^{\mathrm{du}}}$. Using $\bar{r}_N^{\mathrm{pr}} = f - a(\bar{u}_N, \cdot; \bar{\mu}) \in V'$ and the Riesz representant $\bar{\rho}_N^{\mathrm{pr}} \in V$ of $\bar{r}_N^{\mathrm{pr}}$ we observe that

$$\begin{aligned}
\mathscr{L}_{\mu u_N}(\bar{x}_N, \bar{\lambda}_N)(u_N^\delta, \mu^\delta) &= \mathscr{L}_{u_N \mu}(\bar{x}_N, \bar{\lambda}_N)(u_N^\delta, \mu^\delta) \\
&= \sum_{q=1}^{Q} \nabla \vartheta^q(\bar{\mu})^\top \mu^\delta \Big( a^q(u_N^\delta, \bar{\rho}_N^{\mathrm{pr}} - \bar{\lambda}_N^1) - (a^q(\bar{u}_N, \cdot), a(u_N^\delta, \cdot; \bar{\mu}))_{(V^{\mathscr{N}})'} \Big).
\end{aligned}$$

for $u_N^\delta \in V_{N^{\mathrm{pr}}}$ and $\mu^\delta \in \mathbb{R}^P$. Let $\bar{\zeta}_N^{\mathrm{pr},q} \in V^{\mathscr{N}}$, $1 \le q \le Q$, denote the Riesz representants of $a^q(\bar{u}_N, \cdot) \in (V^{\mathscr{N}})'$, i.e.

$$\langle \bar{\zeta}_N^{\mathrm{pr},q}, \varphi^{\mathscr{N}} \rangle_V = a^q(\bar{u}_N, \varphi^{\mathscr{N}}) \quad \text{for all } \varphi^{\mathscr{N}} \in V^{\mathscr{N}}.$$

Then, we derive that

$$\begin{aligned}
&\mathscr{L}_{u_N \mu}(\bar{x}_N, \bar{\lambda}_N)(u_N^\delta, \mu^\delta) \\
&\quad = \sum_{q=1}^{Q} \nabla \vartheta^q(\bar{\mu})^\top \mu^\delta (a^q(u_N^\delta, \bar{\rho}_N^{\mathrm{pr}} - \bar{\lambda}_N^1) - a(u_N^\delta, \bar{\zeta}_N^{\mathrm{pr},q}; \bar{\mu}))
\end{aligned} \tag{5.27}$$

for $u_N^\delta \in V_{N^{\mathrm{pr}}}$ and $\mu^\delta \in \mathbb{R}^P$. As above we apply $\bar{r}_N^{\mathrm{du}} = \ell + a(\cdot, \bar{z}_N; \bar{\mu}) \in (V^{\mathscr{N}})'$ and the Riesz representant $\bar{\rho}_N^{\mathrm{du}} \in V^{\mathscr{N}}$ of $\bar{r}_N^{\mathrm{du}}$ we observe that

$$\begin{aligned}
\mathscr{L}_{\mu z_N}(\bar{x}_N, \bar{\lambda}_N)(\mu^\delta, z_N^\delta) &= \mathscr{L}_{z_N \mu}(\bar{x}_N, \bar{\lambda}_N)(z_N^\delta, \mu^\delta) \\
&= \sum_{q=1}^{Q} \nabla \vartheta^q(\bar{\mu})^\top \mu^\delta \Big( a^q(\bar{\lambda}_N^2 - \rho_N^{\mathrm{du}}, z_N^\delta) - (a^q(\cdot, \bar{z}_N), a(\cdot, z_N^\delta; \bar{\mu}))_{(V^{\mathscr{N}})'} \Big)
\end{aligned}$$

for $z_N^\delta \in \tilde{V}_{N^{\mathrm{du}}}$ and $\mu^\delta \in \mathbb{R}^P$. Let $\bar{\omega}_N^{\mathrm{du},q} \in V^{\mathscr{N}}$, $1 \le q \le Q$, denote the Riesz representants of $a^q(\cdot, \bar{z}_N) \in (V^{\mathscr{N}})'$, i.e.

$$\langle \bar{\omega}_N^{\mathrm{du},q}, \varphi^{\mathscr{N}} \rangle_V = a^q(\varphi^{\mathscr{N}}, \bar{z}_N) \quad \text{for all } \varphi^{\mathscr{N}} \in V^{\mathscr{N}}.$$

Then, we conclude that

$$\begin{aligned}&\mathscr{L}_{z_N\mu}(\bar{x}_N,\bar{\lambda}_N)(z_N^\delta,\mu^\delta)\\&\quad=\sum_{q=1}^{Q}\nabla\vartheta^q(\bar{\mu})^\top\mu^\delta\big(a^q(\bar{\lambda}_N^2-\rho_N^{\mathrm{du}},z_N^\delta)-a(\bar{\omega}_N^{\mathrm{du},q},z_N^\delta;\bar{\mu})\big)\end{aligned}\tag{5.28}$$

for $z_N^\delta\in\tilde{V}_{N^{\mathrm{du}}}$ and $\mu^\delta\in\mathbb{R}^P$. Finally, we find for $\mu^\delta,\tilde{\mu}^\delta\in\mathbb{R}^P$

$$\begin{aligned}&\mathscr{L}_{\mu\mu}(\bar{x}_N,\bar{\lambda}_N)(\mu^\delta,\tilde{\mu}^\delta)\\&\quad=\tilde{\mu}^{\delta,\top}\Bigg(\Big(\sum_{q=1}^{Q}\big(a^q(\bar{u}_N,\bar{\rho}_N^{\mathrm{pr}})-a^q(\bar{\rho}_N^{\mathrm{du}},\bar{z}_N)\big)\Big)\nabla^2\vartheta^q(\bar{\mu})\Bigg)\mu^\delta\\&\qquad-\sum_{q=1}^{Q}\nabla\vartheta^q(\bar{\mu})^\top\mu^\delta\nabla\vartheta^q(\bar{\mu})^\top\tilde{\mu}^\delta\Big(\|\bar{\zeta}_N^{\mathrm{pr},q}\|_V^2+\|\bar{\omega}_N^{\mathrm{du},q}\|_V^2\Big)\end{aligned}\tag{5.29}$$

with $\tilde{\mu}^{\delta,\top}=(\tilde{\mu}^\delta)^\top$.

The convergence of the SQP method relies on second-order sufficient optimality conditions for (**P**). For an arbitrary $\tau\ge 0$ let us define the set of *strongly active constraints* for the parameter $\bar{\mu}$ by

$$\begin{aligned}\mathscr{A}_\tau(\bar{\mu})&=\big\{i\in\{1,\dots,P\}\,\big|\,|(\nabla\hat{J}(\bar{\mu}))_i|\ge\tau\big\}\\&=\big\{i\in\{1,\dots,P\}\,\big|\,|(\nabla D\vartheta(\bar{\mu})^\top\bar{\xi})_i|\ge\tau\big\},\end{aligned}$$

where $(\nabla\hat{J}(\bar{\mu}))_i$ denotes the $i$-th component of the vector $\nabla\hat{J}(\bar{\mu})\in\mathbb{R}^P$. *Second-order sufficient optimality conditions* for (**P**) are as follows [16]: Let $\bar{x}_N=(\bar{y}_N,\bar{\mu})\in X_N^{ad}$, $\bar{y}_N=(\bar{u}_N,\bar{z}_N)\in Y_N$, be a solution to the first-order necessary optimatity conditions for (**P**); see Theorem 5.2. Moreover, the pair $\bar{\lambda}_N=(\bar{\lambda}_N^1,\bar{\lambda}_N^2)\in Y_N$ are the associated Lagrange multiplier. If there exists a $\kappa>0$ such that

$$\mathscr{L}_{x_Nx_N}(\bar{x}_N,\bar{\lambda}_N)(x_N^\delta,x_N^\delta)\ge\kappa\Big(\|u_N^\delta\|_V^2+\|z_N^\delta\|_V^2+\|\mu^\delta\|_{\mathbb{R}^P}^2\Big)\tag{5.30}$$

for all $x_N^\delta=(y_N^\delta,\mu^\delta)\in X_N$, $y_N^\delta=(u_N^\delta,z_N^\delta)$, satisfying $y_N^\delta\in\ker e'(\bar{x}_N)$ and

$$(\mu^\delta)_i\begin{cases}=0 & \text{if } i\in\mathscr{A}_\tau(\bar{\mu}),\\ \ge 0 & \text{if } \bar{\mu}_i=\mu_{a,i}\text{ and } i\notin\mathscr{A}_\tau(\bar{\mu}),\\ \le 0 & \text{if } \bar{\mu}_i=\mu_{b,i}\text{ and } i\notin\mathscr{A}_\tau(\bar{\mu}).\end{cases}$$

then $\bar{x}_N$ is a strictly local solution to (**P**).

Suppose that $x_N^\delta=(y_N^\delta,\mu^\delta)\in\ker e'(\bar{x}_N)$ with $y_N^\delta=(u_N^\delta,z_N^\delta)\in Y_N$. Then we have

$$a(u_N^\delta,\psi;\bar{\mu})=-\sum_{q=1}^{Q}\nabla\vartheta^q(\bar{\mu})^\top\mu^\delta\,a^q(\bar{u}_N,\psi)\qquad\text{for all }\psi\in V_{N^{\mathrm{pr}}},\tag{5.31a}$$

$$a(\phi,z_N^\delta;\bar{\mu})=-\sum_{q=1}^{Q}\nabla\vartheta^q(\bar{\mu})^\top\mu^\delta\,a^q(\phi,\bar{z}_N)\qquad\text{for all }\phi\in\tilde{V}_{N^{\mathrm{du}}}.\tag{5.31b}$$

Utilizing (5.2a), (5.3) and (5.31a) we find

$$\alpha_0 \|u_N^\delta\|_V^2 \le a(u_N^\delta, u_N^\delta; \bar{\mu}) \le \gamma \|\bar{u}_N\|_V \sum_{q=1}^{Q} \|\nabla \vartheta^q(\bar{\mu})\|_{\mathbb{R}^P} \|u_N^\delta\|_V \|\mu^\delta\|_{\mathbb{R}^P}$$

which implies

$$\|u_N^\delta\|_V \le \bar{C_1} \|\mu^\delta\|_{\mathbb{R}^P} \quad \text{for all } x_N^\delta = (y_N^\delta, \mu^\delta) \in \ker e'(\bar{x}_N). \tag{5.32a}$$

with $\bar{C_1} = \gamma \|\bar{u}_N\|_V \sum_{q=1}^{Q} \|\nabla \vartheta^q(\bar{\mu})\|_{\mathbb{R}^P}$. Analogously, we derive from (5.2a), (5.3) and (5.31a)

$$\|z_N^\delta\|_V \le \bar{C_2} \|\mu^\delta\|_{\mathbb{R}^P} \quad \text{for all } x_N^\delta = (y_N^\delta, \mu^\delta) \in \ker e'(\bar{x}_N). \tag{5.32b}$$

with $\bar{C_2} = \gamma \|\bar{z}_N\|_V \sum_{q=1}^{Q} \|\nabla \vartheta^q(\bar{\mu})\|_{\mathbb{R}^P}$. From (5.2b), (5.32) and (5.32b) we infer that

$$\begin{aligned} -\|a(u_N^\delta, \cdot; \bar{\mu})\|_{(V^{\mathcal{N}})'}^2 - \|a(\cdot, z_N^\delta; \bar{\mu})\|_{(V^{\mathcal{N}})'}^2 &\ge -\gamma^2 \left(\|u_N^\delta\|_V^2 + \|z_N^\delta\|_V^2\right) \\ &\ge -\gamma^2 \left(\bar{C}_1^2 + \bar{C}_2^2\right) \|\mu\|_{\mathbb{R}^P}^2. \end{aligned} \tag{5.33}$$

We set $\bar{C_3} = \gamma^2(\bar{C}_1^2 + \bar{C}_2^2)$. Then, we derive from (5.24)–(5.29) and (5.33) that

$$\begin{aligned} &\mathcal{L}_{x_N x_N}(\bar{x}_N, \bar{\lambda}_N)(u_N^\delta, u_N^\delta) = \\ &\ge -\bar{C_3} \|\mu^\delta\|_{\mathbb{R}^P}^2 + 2 \sum_{q=1}^{Q} \nabla \vartheta^q(\bar{\mu})^\top \mu^\delta \left(a^q(u_N^\delta, \bar{\rho}_N^{\mathrm{pr}} - \bar{\lambda}_N^1) - a(u_N^\delta, \bar{\zeta}_N^{\mathrm{pr},q}; \bar{\mu})\right) \\ &\quad + 2 \sum_{q=1}^{Q} \nabla \vartheta^q(\bar{\mu})^\top \mu^\delta \left(a^q(\bar{\lambda}_N^2 - \rho_N^{\mathrm{du}}, z_N^\delta) - a(\bar{\omega}_N^{\mathrm{du},q}, z_N^\delta; \bar{\mu})\right) \\ &\quad + \mu^{\delta,\top} \left(\left(\sum_{q=1}^{Q} \left(a^q(\bar{u}_N, \bar{\rho}_N^{\mathrm{pr}}) - a^q(\bar{\rho}_N^{\mathrm{du}}, \bar{z}_N)\right)\right) \nabla^2 \vartheta^q(\bar{\mu})\right) \mu^\delta \\ &\quad - \sum_{q=1}^{Q} |\nabla \vartheta^q(\bar{\mu})^\top \mu^\delta|^2 \left(\|\bar{\zeta}_N^{\mathrm{pr},q}\|_V^2 + \|\bar{\omega}_N^{\mathrm{du},q}\|_V^2\right) \end{aligned}$$

for all $x_N^\delta \in \ker e'(\bar{x}_N)$. Since

$$-\bar{C_3} \|\mu^\delta\|_{\mathbb{R}^P}^2 - \sum_{q=1}^{Q} |\nabla \vartheta^q(\bar{\mu})^\top \mu^\delta|^2 \left(\|\bar{\zeta}_N^{\mathrm{pr},q}\|_V^2 + \|\bar{\omega}_N^{\mathrm{du},q}\|_V^2\right) \le 0$$

holds and the matrix

$$\left(\sum_{q=1}^{Q} \left(a^q(\bar{u}_N, \bar{\rho}_N^{\mathrm{pr}}) - a^q(\bar{\rho}_N^{\mathrm{du}}, \bar{z}_N)\right)\right) \nabla^2 \vartheta^q(\bar{\mu})$$

need not be positive definite, the second-order sufficient optimality condition (5.30) is not obvious in our case.

*Remark 5.7* If $\bar{\mu}$ is strongly active in all $P$ components, it follows that $\mathscr{A}_\tau(\bar{\mu}) = \{1,\ldots,P\}$. Thus, $\mu^\delta = 0$ is satisfied. From (5.32) and (5.32b) we conclude that $y_N^\delta = 0$ holds. This imply the second-order necessary optimality conditions at $\bar{x}_N$. □

## 5.5 Numerical Experiments

In this section we present some numerial results for the described theory. We use two versions of the well-known Thermal-Block-Model (see e.g. [15]) as a model example. Model 1 consists of two blocks while Model 2 consists of 12 blocks , see Fig. 5.1. The parameter domain is chosen as $\mathscr{D} = [0.2,2]^P$, where $P$ again denotes the number of parameters, i.e., $P = 2$ for Model 1 and $P = 12$ for Model 2, see Fig. 5.1. We choose $|\hat{J}(\mu)| \leq \varepsilon_{\text{stop}} = 1e-5$ as stopping criteria for the Greedy-algorithm. Since $P = 2$ for Model 1, we can easily visualize the reduced cost functional $\hat{J}(\mu)$ in that case, see Fig. 5.2.

As we can deduce from the shape of the cost functional, the appropriate choice for an initial value for the optimization scheme[1] is crucial in order to avoid determining a local minimum only. Let us clarify this in Fig. 5.2 (b): Choosing an initial parameter $\mu_{\text{init}}^N$ in the left half of the plane will lead to a local minimum whereas an initial value located in the right half of the plane will yield the global optimum $(0.2,2)$. In order to avoid the output of a local minimum, we have used four different strategies:

1. `euclidian_mu`: $\mu_{\text{init}}^N$ is chosen by maximizing the Euclidian distance to the barycenter of the previously determined parameter values $\mu_i$, $1 \leq i \leq N-1$.

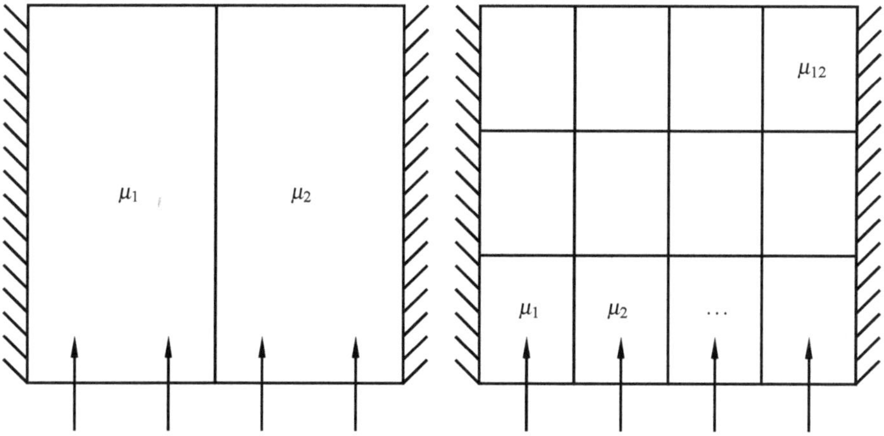


**Fig. 5.1.** Left: Model 1 (2 dimensions), right: Model 2 (12 dimensions)

[1] We used MATLAB's function `fmincon` for this. We set `options.TolCon=1e-6; options.TolFun=1e-6;` and `options.Algorithm='sqp'`, i.e., we used a MATLAB internal SQP algorithm.

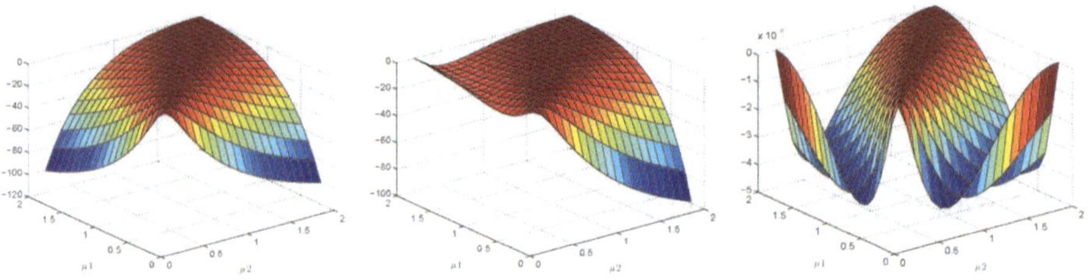

a $N=1$ b $N=2$ c $N=3$

**Fig. 5.2.** Reduced cost functional $\hat{J}(\mu)$ for Model 1. Note that the range for the $y$-axis for $N=3$ is $10^{-6}$

2. coarse_grid_mu: An equidistant coarse parameter-mesh consisting of $M=3^P=9$ points for Model 1 and $M=2^P=4096$ for Model 2 is used. We choose that parameter as initial value $\mu_{\text{init}}^N$ whose cost functional is minimal on that grid.
3. random_mu with "safety zone": $\mu_{\text{init}}^N$ is chosen randomly in $\mathscr{D}$, but ensuring a minimal distance (measured in the Euclidian norm) to all $\mu_i$, $1 \leq i \leq N-1$. This "safety zone" is chosen adaptively, i.e., the radius of the circular zone is decreased with increasing $N$. If we would not do that, we would get an $N^{\max}$, where no additional feasible points could be found.
4. multiple_random_mu: This is similar to coarse_grid_mu, except that we use a fixed number of $N_{\text{rand}}$ uniform randomly chosen parameters $\mu$ instead of a fixed coarse grid.

Especially in higher dimensions best results were obtained using multiple_random_mu and due to the curse of dimension coarse_grid_mu is not applicable properly in higher dimensions. We used an SQP algorithm as optimization scheme and compare the results to a classical training set strategy, using equidistant training sets consisting of $3^2=9$ respectively $10^2=100$ parameter values for Model 1 and $3^{12}=531.441$ parameter values for Model 2. Fig. 5.3 (left) shows the decay of the mean error estimator during the Greedy-process (i.e., with increasing $N$) for a randomly chosen test set of 10.000 parameters. We choose $\mu_0=(1,1)$ as initial snapshot-parameter. As expected (see [14]) the Greedy stops after two steps with $\mu_1=(\mu_{\min},\mu_{\max})$ and $\mu_2=(\mu_{\max},\mu_{\min})$. In this example there is no difference between using an optimization algorithm and using a training set strategy since the optimal parameter values $\mu_1$ and $\mu_2$ are contained in the training set. Hence, our optimization procedure is consistent with the known theory.

In Fig. 5.3 (right) the decay of the mean error estimator again for 10.000 randomly chosen parameters is shown for Model 2 with increasing basis size $N$. We observe the expected exponential decay and our optimization strategies perform as well as the classical training set strategy. This is remarkable since the number of reduced simulations needed in the offline phase is significantly smaller than for the classical training set Greedy approach (see Table 5.1).

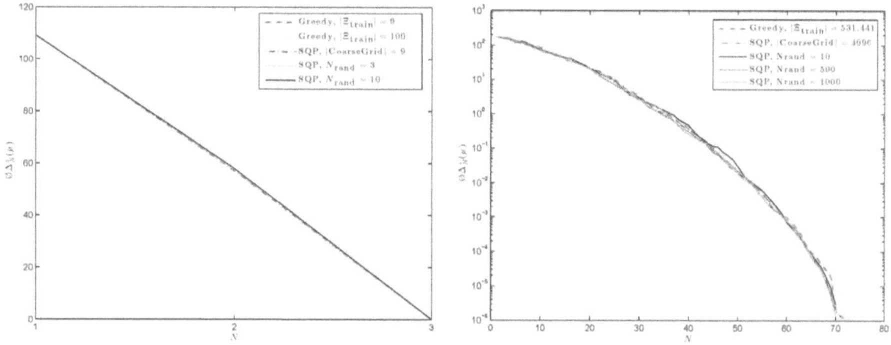


**Fig. 5.3.** Mean error estimators $\Delta_N^s(\mu)$ for different bases on a test set consisting of 10.000 randomly chosen parameters $\mu$ for Model 1 (left) and Model 2 (right). Note that on the left the scale is linear, while the right plot has a semilog scale

**Table 5.1.** Number of reduced simulations during the offline phase for Model 1 (left) and Model 2 (right)

| *Model 1* | *# RB simulations* |
|---|---|
| SQP, $N_{\text{rand}} = 3$ | 34 |
| SQP, $N_{\text{rand}} = 10$ | 98 |
| SQP, $\lvert\text{CoarseGrid}\rvert = 9$ | 101 |
| $\lvert\Xi_{\text{train}}\rvert = 9$ | 18 |
| $\lvert\Xi_{\text{train}}\rvert = 100$ | 200 |

| *Model 2* | *# RB simulations* |
|---|---|
| SQP, $N_{\text{rand}} = 10$ | 26.127 |
| SQP, $N_{\text{rand}} = 500$ | 55.536 |
| SQP, $N_{\text{rand}} = 1000$ | 91.983 |
| SQP, $\lvert\text{CoarseGrid}\rvert = 4096$ | 302.259 |
| $\lvert\Xi_{\text{train}}\rvert = 531.441$ | 37.732.311 |

In Table 5.1 we show the overall number of evaluations of $\hat{J}(\mu)$ - i.e., the number of reduced simulations – during the Greedy process in the offline phase for Model 1 and Model 2. Especially for Model 2 the Greedy algorithm combined with the optimization scheme needs much less function calls than the Greedy algorithm combined with a training set strategy. This can be an advantage in order to overcome the curse of dimension which prohibits to choose the training set arbitrarily large especially in high dimensions.

With our approach we were also able to generate bases for a $4 \times 4$-ThermalBlock consisting of 93 to 94 basis functions. The results are shown in Table 5.2. The coarse grid consisting of only 2 parameters per dimension already consists of $2^{16} = 65.536$ points which leads to a total number of reduced simulations of about 6.100.000. The classical Greedy, using a training set consisting of 3 parameters per dimension, would need $3^{16} = 43.046.721$ reduced simulations just for one new basis function. For a basis length of 90 this would result in approximately $3.87 \cdot 1e9$ reduced simulations during the offline phase, which nowadays clearly is out of scope.

**Table 5.2.** Number of reduced simulations during the offline phase for a $4\times 4$-Thermal Block

| *4×4-Thermal Block* | *# RB simulations* |
|---|---|
| SQP, $N_{\text{rand}} = 10$ | 43.265 |
| SQP, $N_{\text{rand}} = 500$ | 89.221 |
| SQP, $N_{\text{rand}} = 1000$ | 133.318 |
| SQP, $\lvert\text{CoarseGrid}\rvert = 65.536$ | 6.121.911 |
| $\lvert\Xi_{\text{train}}\rvert = 3^{16} = 43.046.721$ | — |

## References


1. Becker, R., Rannacher, R.: An optimal control approach to a posteriori error estimation in finite element methods. Acta Numerica **10**, 1–102 (2001)
2. Bui-Thanh, T.: Model-constrained optimization methods for reduction of parameterized systems. Pd.D. Thesis, MIT, USA (2007)
3. Bui-Thanh, T., Willcox, K., Ghattas, O.: Model reduction for large-scale systems with high-dimensional parametric input space. SIAM J. Sci. Comput. **30**, 3270–3288 (2008)
4. Drohmann, M., Haasdonk, B., Ohlberger, M.: Adaptive reduced basis methods for non-linear convection-diffusion equations. In: Fort, J., et al. (eds.) Finite Volumes for Complex Applications VI – Problems & Perspectives. Springer Proceedings in Mathematics 4, vol. 1, pp. 369–377. Springer-Verlag, Berlin Heidelberg (2011)
5. Haasdonk, B., Ohlberger, M.: Adaptive basis enrichment for the reduced basis method applied to finite volume schemes. In Proc. 5th International Symposium on Finite Volumes for Complex Applications, June 8–13, 2008, Aussois, France, pp. 471–478 (2008)
6. Hesthaven, J.S., Stamm, B., Zhang, S.: Efficient greedy algorithms for high-dimensional parameter spaces with applications to empirical interpolation and reduced basis methods. (Submitted 2012)
7. Hinze, M., Pinnau, R., Ulbrich, M., Ulbrich, S.: Optimization with PDE Constraints. Mathematical Modelling: Theory and Applications, vol. 23. Springer-Verlag, Berlin Heidelberg (2009)
8. Iapichino, L., Volkwein, S.: Greedy sampling of distributed parameters in the reduced-basis method by numerical optimization (Submitted 2013)
9. Kelley, C.T.: Iterative Methods for Optimization. Frontiers in Applied Mathematics, SIAM, Philadelphia, PA (1999)
10. Lass, O., Volkwein, S.: Adaptive POD basis computation for parametrized nonlinear systems using optimal snapshot location (Submitted 2012)
11. Maday, Y., Stamm, B.: Locally adaptive greedy approximations for anisotropic parameter reduced basis spaces. Submitted to SIAM J. Sci. Comput. (2012)
12. Nguyen, N.C., Veroy, K., Patera, A.T.: Certified real-time solution of parametrized partial differential equations. In: Yip, S. (ed.) Handbook of Materials Modeling, pp. 1523–1558. Springer, Netherlands (2005)
13. Nocedal, J., Wright, S.J.: Numerical Optimization, 2nd ed. Springer Series in Operation Research. Springer, New York (2006)
14. Patera, A.T., Rozza, G.: Reduced Basis Approximation and A Posteriori Error Estimation for Parametrized Partial Differential Equations. MIT Pappalardo Graduate

Monographs in Mechanical Engineering. Cambridge, MA (2007). Available from http://augustine.mit.edu/methodology/methodology_book.htm
15. Rozza, G., Huynh, D.B.P., Patera, A.T.: Reduced basis approximation and a posteriori error estimation for affinely parametrized elliptic coercive partial differential equations. Application to transport and continuum mechanics. Archives of Computational Methods in Engineering **15**, 229–275 (2008)
16. Tröltzsch, F.: Optimal Control of Partial Differential Equations. Theory, Methods and Applications. American Mathematical Society, Providence (2010)

# 6

# A Robust Algorithm for Parametric Model Order Reduction Based on Implicit Moment Matching

**Peter Benner and Lihong Feng**

**Abstract** Parametric model order reduction (PMOR) has received a tremendous amount of attention in recent years. Among the first approaches considered, mainly in system and control theory as well as computational electromagnetics and nanoelectronics, are methods based on multi-moment matching. Despite numerous other successful methods, including the reduced-basis method (RBM), other methods based on (rational, matrix, manifold) interpolation, or Kriging techniques, multi-moment matching methods remain a reliable, robust, and flexible method for model reduction of linear parametric systems. Here we propose a numerically stable algorithm for PMOR based on multi-moment matching. Given any number of parameters and any number of moments of the parametric system, the algorithm generates a projection matrix for model reduction by implicit moment matching. The implementation of the method based on a repeated modified Gram-Schmidt-like process renders the method numerically stable. The proposed method is simple yet efficient. Numerical experiments show that the proposed algorithm is very accurate.

## 6.1 Introduction

The modeling of many engineering and scientific applications leads to dynamical systems depending on parameters varying in different design stages or computer experiments. For example, in a thermal model [16], the film coefficient $k$ changes

P. Benner
Max Planck Institute for Dynamics of Complex Technical Systems, Sandtorstr. 1, 39106 Magdeburg, Germany
e-mail: benner@mpi-magdeburg.mpg.de

L. Feng (✉)
Max Planck Institute for Dynamics of Complex Technical Systems, Sandtorstr. 1, 39106 Magdeburg, Germany
e-mail: feng@mpi-magdeburg.mpg.de

A. Quarteroni, G. Rozza (eds.): *Reduced Order Methods for Modeling and Computational Reduction*, MS&A 9. DOI 10.1007/978-3-319-02090-7_6, 

with the temperature, this results in a parametric mathematical model

$$C\frac{dx(t)}{dt}+Gx(t)+kDx(t)=B, \quad y(t)=L^{\mathrm{T}}x(t). \tag{6.1}$$

In integrated circuits, due to process variations, the width of the interconnects is in fact a random variable, such that a non-parametric model

$$C\frac{dx(t)}{dt}=Gx(t)+Bu(t), \quad y(t)=L^{\mathrm{T}}x(t), \tag{6.2}$$

is not sufficient to describe the random variation. Therefore, in [22, 32], a linearized parametric system

$$\begin{aligned}(C_0+\lambda_1 C_1+\lambda_2 C_2)\tfrac{dx(t)}{dt} &= (G_0+\lambda_1 G_1+\lambda_2 G_2)x(t)+Bu(t),\\ y(t) &= L^{\mathrm{T}}x(t),\end{aligned} \tag{6.3}$$

is constructed. Here and below, the system matrices are $C, C_i, G, G_i \in \mathbb{R}^{n\times n}, i = 0,1,2$. $B \in \mathbb{R}^{n\times m_I}$ is the input matrix, $L \in \mathbb{R}^{n\times m_O}$ is the output matrix. $u(t) \in \mathbb{R}^{m_I}$ is the vector of input signals. $x(t) \in \mathbb{R}^n$ is the unknown vector. $y(t) \in \mathbb{R}^{m_O}$ is the vector of output responses. Many more examples for parametric systems can be found in the engineering literature, see, e.g., the benchmark examples in the recently published MOR wiki[1].

The above mentioned parametric systems are usually of very large dimensions as they often result from finite element discretizations of instationary partial differential equations (PDEs) defined on complex geometries. Solving the parametric systems by conventional simulation methods is often very time-consuming. On the one hand the parameters have to be provided as fixed values and these values cannot be changed during the simulation. On the other hand, if the dimension of the system is large, simulating such a system already once will be costly, and the cost of a design study requiring many runs with different parameter values ("many-query context") may be overwhelming.

Model order reduction (MOR) is an increasingly popular approach to overcome the obstacles posed by the computational demands in a many-query context. By MOR, a small dimensional approximate system can be derived, so that it can reliably replace the original system during the simulation. This can often save much simulation time and computer memory, see [2, 5, 6, 30] for some introductory texts on the topic and the presentation of the state-of-the-art.

The main goal of parametric model order reduction (PMOR) is to preserve parameters in the system as symbolic quantities in the reduced-order model. Thus, a change in parameters does not require to compute a new reduced-order model, but simply the evaluation of the reduced-order model for the new parameter values. If the error in the whole feasible parameter domain can be proven to satisfy an acceptable error tolerance, design and optimization of systems and devices can be significantly accelerated. First attempts at deriving MOR for linear parametric systems were based on extending the popular moment-matching methods (aka *Padé*

[1] See http://modelreduction.org.

*approximation, Krylov subspace-based MOR methods*) to parametric systems by multivariate power series expansions around appropriate interpolation points. Early references include [7, 8, 10, 16, 19, 21, 22, 27, 32, 34]. Later, other variants of (rational) interpolation techniques were derived, combining, e.g., balanced truncation and (sparse grid) interpolation [3], employing $\mathcal{H}_2$-optimal interpolation techniques [4], or using matrix and manifold interpolation techniques (e.g., [1,9,26]). Another large class of PMOR techniques is based on the Reduced Basis Method (RBM), originating in the fast approximation of parametric partial differential equations. The methods are also applicable in the context discussed here [20], but a dedicated comparison to the approaches mentioned here is deferred to future work. Therefore, we will not discuss this approach here any further and refer the reader to the survey [29] and other chapters in this volume.

In the following, we will discuss a robust implementation of the multi-moment matching methods first discussed in [7,8,19,34]. They have some advantages making them still the most popular approach used in practical applications:

- They are easy to implement and require almost no assumptions on system properties.
- Their cost is limited to a few (according to the number of employed expansion points) factorizations of sparse matrices and forward/backward solves using the computed factors. They do not require generation of trajectories and are therefore called "simulation-free" (in contrast to RBM and proper orthogonal decomposition (POD) methods). As a consequence, the "offline-phase" for computing the reduced-order model is cheap compared to RBM and POD, and it is often possible to achieve the goal encountered in practical industrial engineering design that the time for constructing the reduced-order model plus a simulation should be smaller than a single simulation of the full-order model.
- As they are simulation-free, no training inputs $u(t)$ need to be chosen so that the approximation quality is usually good for all feasible input signals, not only close to training inputs as in RB and POD methods.

Certainly, there are also some disadvantages: one has to first linearize parameter-dependencies (though polynomial forms are also possible, see, e.g., [10]), and the order of the reduced system may not be optimal. Nevertheless, improvements on these aspects are in progress, so that it is to be expected that multi-moment matching methods will remain competitive with other approaches in the future.

The MOR methods discussed here are based on projecting the unknown vector $x$ onto a small dimensional subspace. We use system (6.2) to briefly introduce the concept. If a projection matrix $V \in \mathbb{R}^{n \times q}$ has been determined, using $x \approx Vz$ we obtain the perturbed system

$$CV\frac{dz(t)}{dt} = GVz(t) + Bu(t) + e(t), \quad \hat{y}(t) = L^{\mathrm{T}}Vz(t),$$

with $e(t)$ the introduced residual. By Galerkin projection $V^{\mathrm{T}}e(t) = 0$, we get the reduced-order model:

$$V^{\mathrm{T}}CV\frac{dz(t)}{dt} = V^{\mathrm{T}}GVz(t) + V^{\mathrm{T}}Bu(t), \quad \hat{y}(t) = L^{\mathrm{T}}Vz(t),$$

where $z(t) \in \mathbb{R}^q$ is the unknown vector of the reduced model. The space dimension $q$ is often called the *order* of the reduced model. Therefore, the key step for MOR is how to get the projection matrix $V$, which determines the dimension and the accuracy of the reduced order model.

This paper is based on the ideas in [8], where the projection matrix $V$ is obtained by computing an orthonormal basis of the subspace spanned by the moment vectors. No detailed algorithm of computing the orthonormal basis is proposed in [8]. A simple way of generating $V$ is to first obtain the moment vectors by explicit matrix multiplications, and then all the columns of the computed moment vectors are orthonormalized to get the basis. However, this *explicit moment matching* procedure may lead to numerical instability, because higher order moment vectors usually become linearly dependent quickly as already observed in the non-parametric case, see [18]. We will demonstrate this effect in Sect. 6.2 for a practical example.

Our intention therefore is to develop an algorithm which computes the moment vectors implicitly rather than explicitly. In this way, good numerical stability can be preserved and an accurate orthonormal basis of the subspace spanned by the moment vectors can be obtained. The proposed algorithm can deal with both single-input and multiple-input systems without any limitation on the parameters in the system. It should be noted that this work dates back to first variants in 2007 [12, 13], and other comparable variants of implicit moment matching methods have been proposed [10, 21]. Here, we want to give a full account on the method discussed initially for only one parameter in [13].

In the following, we first review the method from [8] in Sect. 6.2 and explain the numerical instability resulting from explicit computation of the moments. In Sect. 6.3, we propose a numerical stable algorithm applicable to both single-input and multiple-input systems. The efficiency of the proposed algorithms is shown in Sect. 6.5 by simulating two examples from micro-electrical-mechanical systems (MEMS) and electrochemistry. Conclusions are given in the end.

## 6.2 Explicit Multi-Moment Matching PMOR

In this section, we give a short review of the method in [8] in order to explain the numerical instability of explicitly computing the moment vectors. A parametric system in time domain can be written as below,

$$\begin{aligned} C(s_1, s_2, \cdots, s_{p-1})\tfrac{dx}{dt}(t) &= G(s_1, s_2, \cdots, s_{p-1})x(t) + Bu(t), \\ y(t) &= L^{\mathrm{T}}x(t), \end{aligned} \tag{6.4}$$

where the system matrices $C(s_1, s_2, \cdots, s_{p-1})$, $G(s_1, s_2, \cdots, s_{p-1})$ are (maybe, non-linear, non-affine) functions of the parameters $s_1, s_2, \cdots, s_{p-1}$. A parametric system can also be stated in the frequency domain,

$$\begin{aligned} E(s_1,\ldots,s_p)x &= Bu(s_p), \\ y &= L^{\mathrm{T}}x, \end{aligned} \tag{6.5}$$

where the matrix $E \in \mathbb{R}^{n\times n}$ is parametrized. If the system in (6.5) is the Laplace transform of the system in (6.4), the new parameter $s_p$ is in fact the frequency parameter $s$, which corresponds to time $t$. The state $x$ is the Laplace transform of the unknown vector $x$ in (6.4).

### *6.2.1 Review*

The method in [8] is based on the representation of a parametric system in the frequency domain as in (6.5). In case of a nonlinear and/or non-affine dependence of the matrix $E$ on the parameters, the system in (6.5) is first transformed to an affine form

$$\begin{aligned} (E_0 + \tilde{s}_1 E_1 + \tilde{s}_2 E_2 + \ldots + \tilde{s}_p E_p)x &= Bu(s_p), \\ y &= L^{\mathrm{T}}x. \end{aligned} \tag{6.6}$$

Here the newly defined parameters $\tilde{s}_i, i = 1,\ldots,p$, might be some functions (rational, polynomial) of the original parameters $s_i$ in (6.5). To obtain the projection matrix $V$ for the reduced model, the state $x$ in (6.6) is expanded into a Taylor series at an expansion point $\tilde{s}_0 = (\tilde{s}_1^0,\ldots,\tilde{s}_p^0)^T$ as below,

$$\begin{aligned} x &= [I-(\sigma_1 M_1 + \ldots + \sigma_p M_p)]^{-1}\tilde{E}^{-1}Bu(s_p) \\ &= \sum_{m=0}^{\infty} [\sigma_1 M_1 + \ldots + \sigma_p M_p]^m \tilde{E}^{-1}Bu(s_p) \\ &= \sum_{m=0}^{\infty} \sum_{k_2=0}^{m-(k_3+\ldots+k_p)} \cdots \sum_{k_{p-1}=0}^{m-k_p} \sum_{k_p=0}^{m} [F^m_{k_2,\ldots,k_p}(M_1,\ldots,M_p)B_M u(s_p) \times \\ &\quad \sigma_1^{m-(k_2+\ldots+k_p)} \sigma_2^{k_2} \cdots \sigma_p^{k_p}, \end{aligned} \tag{6.7}$$

where $\sigma_i = \tilde{s}_i - \tilde{s}_i^0$, $\tilde{E} = E_0 + \tilde{s}_1^0 E_1 + \ldots + \tilde{s}_p^0 E_p$, $M_i = -\tilde{E}^{-1}E_i$, $i = 1,2,\ldots p$, and $B_M = \tilde{E}^{-1}B$. The $F^m_{k_2,\ldots,k_p}(M_1,\ldots,M_p)$ can be generated recursively as

$$F^m_{k_2,\ldots,k_p}(M_1,\ldots,M_p) = \begin{cases} 0, & \text{if } k_i \notin \{0,1,\ldots,m\}, i = 2,\ldots,p, \\ 0, & \text{if } k_2+\ldots+k_p \notin \{0,1,\ldots,m\}, \\ I, & \text{if } m = 0, \\ M_1 F^{m-1}_{k_2,\ldots,k_p}(M_1,\ldots,M_p) + M_2 F^{m-1}_{k_2-1,\ldots,k_p}(M_1,\ldots,M_p) + \ldots & \\ \quad \ldots + M_p F^{m-1}_{k_2,\ldots,k_p-1}(M_1,\ldots,M_p), & \text{else.} \end{cases}$$

For example, if there are two parameters $\tilde{s}_1, \tilde{s}_2$ in (6.6), $F^m_{k_2,\ldots,k_p}(M_1,\ldots,M_p) = F^m_{k_2}$ are:

$$
\begin{aligned}
&F_0^0 = I,\\
&F_0^1 = M_1F_0^0 = M_1,\; F_1^1 = M_2F_0^0 = M_2 \qquad (6.8)\\
&F_0^2 = M_1F_0^1 = (M_1)^2,\; F_1^2 = M_1F_1^1 + M_2F_0^1 = M_1M_2 + M_2M_1,\; F_2^2 = M_2F_1^1 = (M_2)^2,\\
&\ldots
\end{aligned}
$$

For the general case, the projection matrix $V$ is constructed as

$$
\begin{aligned}
&\text{range}\{V\}\\
&= \text{colspan}\{\bigcup_{m=0}^{m_q} \bigcup_{k_2=0}^{m-(k_p+\ldots+k_3)} \cdots \bigcup_{k_{p-1}=0}^{m-k_p} \bigcup_{k_p=0}^{m} F^m_{k_2,\ldots,k_p}(M_1,\ldots,M_p)B_M\}\\
&= \text{colspan}\{B_M, M_1B_M, M_2B_M, \ldots, M_pB_M, (M_1)^2B_M, (M_1M_2+M_2M_1)B_M, \ldots,\\
&\qquad (M_1M_p+M_pM_1)B_M, (M_2)^2B_M, (M_2M_3+M_3M_2)B_M, \ldots\}.
\end{aligned}
\qquad (6.9)
$$

We call the coefficients in the series expansion of the state $x$ in (6.7) the *moment vectors* of the parametric system. The corresponding moments of the transfer function are the moment vectors multiplied by $L^{\mathrm{T}}$ from the left. For example:

- $L^{\mathrm{T}}B_M$ is the 0th order moment; the columns in $B_M$ are the 0th order moment vectors.
- Similarly, $L^{\mathrm{T}}M_iB_M$, $i = 1,2,\ldots,p$, are the first order moments, and the columns in $M_iB_M$, $i = 1,2,\ldots,p$, are the first order moment vectors, which are the coefficients of $\tilde{s}_i$, $i = 1,\ldots,p$.
- The columns in $M_i^2B_M$, $i = 1,2,\ldots,p$, $(M_1M_i + M_iM_1)B_M$, $i = 2,\ldots,p$, $(M_2M_i + M_iM_2)B_M$, $i = 3,\ldots,p$, $\ldots$, $(M_{p-1}M_p + M_pM_{p-1})B_M$ are the second order moment vectors, which are the coefficients of $\tilde{s}_i^2$, $i = 1,2,\ldots,p$, $\tilde{s}_1\tilde{s}_i$, $i = 2,\ldots,p$, $\tilde{s}_2\tilde{s}_i$, $i = 3,\ldots,p$, $\ldots$, $\tilde{s}_{p-1}\tilde{s}_p$.

Since by moments we not only denote the Taylor coefficients corresponding to the Laplace variable $s = s_p$, but also those associated with the other parameters $s_i, i = 1,\ldots,p-1$, we consider them as multi-moments of the transfer function. To sum up, the set of coefficients corresponding to terms with powers summing up to $i$ is the set of the $i$-th order moment vectors. From the above construction of $V$, the subspace in (6.9) includes the 0-th order moment vectors till the $m_q$-th order moment vectors. The reduced model is computed as

$$
\begin{aligned}
(\hat{E}_0 + \tilde{s}_1\hat{E}_1 + \tilde{s}_2\hat{E}_2 + \ldots + \tilde{s}_p\hat{E}_p)z &= \hat{B}u,\\
\hat{y} &= \hat{L}^{\mathrm{T}}z,
\end{aligned}
\qquad (6.10)
$$

where $\hat{E}_i = V^{\mathrm{T}}E_iV$, $i = 0,1,2,\ldots,p$, $\hat{B} = V^{\mathrm{T}}B$, $\hat{L} = V^{\mathrm{T}}L$. Here we assume real expansion points, i.e., $\tilde{s}_i \in \mathbb{R}$ for all $i = 1,\ldots,p$. Otherwise, complex conjugate transposition might be needed to apply $V$ from the left. This results in a reduced model with complex system matrices, which is undesired in some applications. Alternatives to obtain a real reduced order model even for complex expansion points exist [14], but we leave out these technical details for clarity of presentation.

In time domain, the reduced system (6.4) is

$$\begin{aligned} V^T C(s_1,s_2,\ldots,s_{p-1})V\tfrac{dz}{dt} &= V^T G(s_1,s_2,\ldots,s_{p-1})Vz + V^T Bu(t), \\ \hat{y}(t) &= L^{\mathrm{T}} V z. \end{aligned} \tag{6.11}$$

Ideally, if the matrix $V$ forms an orthonormal basis of the subspace in (6.9), the multi-moments of the reduced model in (6.10) match the multi-moments of the original system in (6.6) up to $m_q$-th order [8]. However, if $V$ cannot be computed with sufficient numerical accuracy, the multi-moment matching property might be lost.

### 6.2.2 *Analysis*

Note that the subspace in (6.9) is not a Krylov subspace, therefore an orthonormal basis of the subspace spanned by the moment vectors cannot be computed by the standard Arnoldi algorithm. In [8], no algorithm for computation of the matrix $V$ is presented. If the moment vectors are computed explicitly by simple matrix-matrix/vector multiplication, the high order moments will become linearly dependent, so that it is difficult or even impossible to obtain an orthonormal basis for the subspace considered.

We employ the thermal model (6.1) with parameter $k \in [1, 10^9]$ (see Fig. 6.1) to illustrate this phenomenon. We observe the output of the system for $k = 10^9$. The moments vectors are first computed through explicit matrix multiplications (hence, explicit multi-moment matching)[2], then an orthogonalization process is applied to the moment vectors to get the final projection matrix $V$. Here we use the modified Gram-Schmidt process (with tolerance $10^{-11}$) to get a $V$ with orthonormal columns.

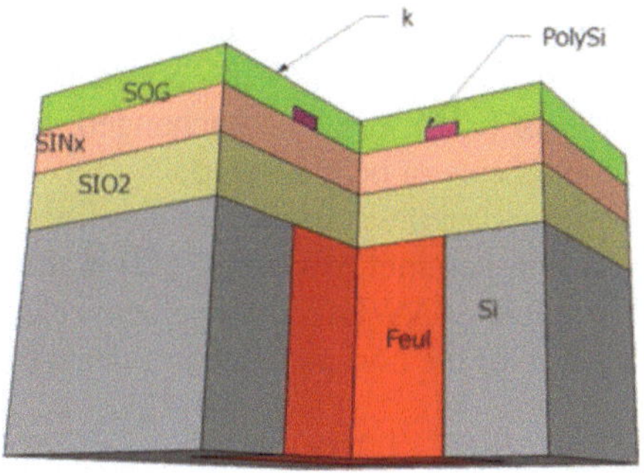


**Fig. 6.1.** Physical model of a microthruster unit for which a thermal MEMS model (6.1) is derived. Note that the film coefficient $k$ is applied at the top

[2] Here we use a nonzero expansion point for the Laplace variable $s$, $s_0 = 0.001$, a zero expansion point for $k$, $k_0 = 0$, to ensure that the matrix $\tilde{E}$ is nonsingular. For all the simulation results in Sect. 6.5.1, the same expansion points are taken for all the tested MOR methods: the non-parametric moment-matching MOR, the explicit multi-moment matching and the proposed Algorithm 6.1.

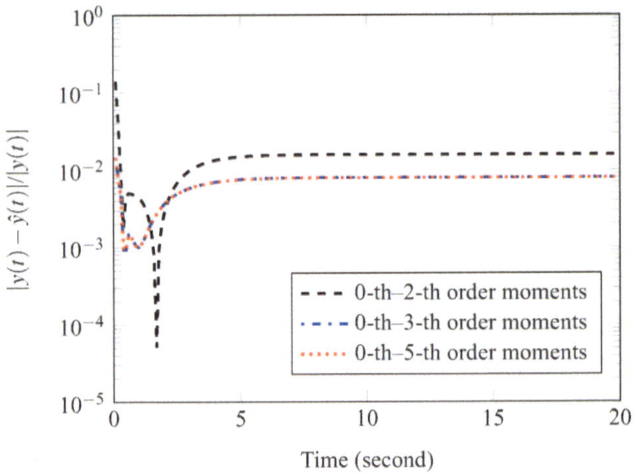


**Fig. 6.2.** Errors of the reduced models for the output responses of the thermal model (6.1), using explicit multi-moment matching (like in [28])

At first, we compute the moment vectors in (6.9) till the 2nd order to get the first reduced model. There are no vectors deleted during the orthogonalization process. The dashed line in Fig. 6.2 is the corresponding output error of the reduced model. If the moment vectors are computed up to order 3, the obtained second reduced model has smaller error. There is no deflation of the moment vectors either. If a more accurate model is to be derived, more moment vectors should be included. A third reduced model is obtained by computing the 0-th till the 5-th order moment vectors. This time, there are deflations during the modified Gram-Schmidt process. As a result, there is no increase in the number of the columns in the matrix $V$. The error of the reduced model is not further reduced, as can be seen from the dotted line in the figure. If the 6-th or higher order moment vectors are computed, the number of the columns in the matrix $V$ still remains unchanged, and the accuracy of the corresponding reduced model cannot be improved.

The work in [11] first points out the numerical instability of explicitly computing the moments of the linear non-parametric system (6.2) in the method AWE [28]. It explains the numerical problem of AWE from the eigenvector and eigenvalue point of view. The moment vectors of the non-parametric system (6.2) are

$$G^{-1}B, G^{-1}CG^{-1}B, (G^{-1}C)^2G^{-1}B, \ldots, (G^{-1}C)^qG^{-1}B, \ldots.$$

These vectors are used to construct the projection matrix $V$ and are computed explicitly in the method AWE. The computation of the $k$th moment vector $(G^{-1}C)^{k-1}G^{-1}B$ in fact corresponds to the power iteration $u_k = A^{k-1}b$, with $A = G^{-1}C$ and $b = G^{-1}B$ (we assume that $B$ is a vector for simplicity). This process converges rapidly to an eigenvector of $A$ associated to the eigenvalue of largest magnitude (assuming a simple eigenvalue). In the end, the computed vector $u_k$ contains only information of this "dominant" eigenvector, and the later computed vectors are all numerically linearly dependent to this eigenvector. This explanation also applies to the numerical instability of the explicit computation of the moment vectors in (6.9). Some part of the moment vectors in (6.9) are also of the power iteration form. For example,

$$M_iB_M, M_i^2B_M, M_i^3B_M, \ldots, M_i^qB_M, \ i = 1,2,\ldots,p.$$

If directly computed, they quickly converge to the respective dominant eigenvector of each matrix $M_i, \ \ i = 1,2,\ldots,p$.

In the next section, a numerically stable algorithm for implicitly computing the moment vectors is presented. The algorithm is applicable for both single-input, single-output systems and multiple-input, multiple-output systems. An orthonormal basis of the subspace spanned by the moment vectors can be obtained implicitly so that a more accurate reduced model can be derived.

## 6.3 A Robust Algorithm for Multi-Moment Matching PMOR

Taking a closer look at the power series expansion of $x$ in (6.7), we get the following equivalent, but different formulation,

$$\begin{aligned} x &= [I - (\sigma_1 M_1 + \ldots + \sigma_p M_p)]^{-1}\tilde{E}^{-1}Bu \\ &= \sum_{m=0}^{\infty} [\sigma_1 M_1 + \ldots + \sigma_p M_p]^m B_M u \\ &= B_M u + [\sigma_1 M_1 + \ldots + \sigma_p M_p] B_M u + [\sigma_1 M_1 + \ldots + \sigma_p M_p]^2 B_M u + \ldots \\ &\quad + [\sigma_1 M_1 + \ldots + \sigma_p M_p]^j B_M u + \ldots \end{aligned} \tag{6.12}$$

By defining

$$\begin{aligned} x_0 &= B_M, \\ x_1 &= [\sigma_1 M_1 + \ldots + \sigma_p M_p] B_M, \\ x_2 &= [\sigma_1 M_1 + \ldots + \sigma_p M_p]^2 B_M, \ldots, \\ x_j &= [\sigma_1 M_1 + \ldots + \sigma_p M_p]^j B_M, \ldots, \end{aligned}$$

we have $x = (x_0 + x_1 + x_2 + \cdots + x_j + \cdots)u$ and obtain the recursive relations

$$\begin{aligned} x_0 &= B_M, \\ x_1 &= [\sigma_1 M_1 + \ldots + \sigma_p M_p] x_0, \\ x_2 &= [\sigma_1 M_1 + \ldots + \sigma_p M_p] x_1, \ldots, \\ x_j &= [\sigma_1 M_1 + \ldots + \sigma_p M_p] x_{j-1}, \ldots. \end{aligned}$$

If we define a vector sequence based on the coefficient matrices of $x_j,\ j = 0, 1, \ldots$ as below,

$$\begin{aligned} R_0 &= B_M, \\ R_1 &= [M_1 R_0, M_2 R_0, \ldots, M_p R_0], \\ R_2 &= [M_1 R_1, M_2 R_1, \ldots, M_p R_1], \\ &\vdots \\ R_j &= [M_1 R_{j-1}, M_2 R_{j-1}, \ldots, M_p R_{j-1}], \\ &\vdots \end{aligned} \tag{6.13}$$

and let $R$ be the subspace spanned by the vectors in $R_j,\ j = 0, 1, \cdots, m$:

$$R = \text{colspan}\{R_0, \ldots, R_j, \ldots, R_m\},$$

we have $x \approx \hat{x} \in R$. We see that the terms in $R_j,\ j = 0, 1, \ldots, m$ are the coefficients of the parameters in the series expansion (6.12). They are also the $j$-th order moment vectors.

The next step is to construct an orthonormal basis $V$ of the subspace $R$ by taking use of the recursive relations between the $R_j$ in (6.13), such that the multi-moments of the original system are matched by those of the reduced model. A numerically stable algorithm for computing $V$ is given in Algorithm 6.1. All the vectors included in $R$ are orthogonalized to each other by the modified Gram-Schmidt (MGS) process once when constructed and then again after all $R_j$ have been computed. In this sense, the algorithm can be understood as a *repeated MGS process*. There is no limitation on the number of parameters, and the essential cost of applying $\tilde{E}^{-1}E_j$ only grows linearly in the number of parameters, while the cost for orthogonalization step essentially grows quadratically with $p$.

Some remarks on Algorithm 6.1 are in order.

*Remark 6.1* a) The application of $\tilde{E}^{-1}$ in Steps 2 and 18 is usually performed by computing once a (sparse) matrix factorization (Cholesky or LU, depending on the system structure) before the algorithm starts. Then each application of $\tilde{E}^{-1}$ means a forward/backward solve step. Hence, the whole algorithm requires only 1 matrix factorization, rendering it fairly cheap compared to other PMOR methods.

b) The application of $E_t$ in Step 18 is a (usually sparse) matrix-vector multiplication and precedes the forward/backward solve step, which is then applied to the resulting vector $E_t v_j$ using the precomputed factors of $\tilde{E}$.

c) For systems with multiple inputs, the input matrix $B$ has more than one column. All the columns in $R_0 = \tilde{E}^{-1}B$ are orthogonalized in Step 5 before the columns in $R_i, i > 0$ are computed. The variable *sum* counts the number of columns in $V$.

d) $m$ denotes the highest order of moments to be computed and is prescribed by the user.

**Algorithm 6.1** Compute $V = [v_1, v_2, \ldots, v_q]$ for a parametric system (6.6), where $B$ is generally considered as a matrix

Initialize $a_1 = 0$, $a_2 = 0$, $sum = 0$.
Compute $R_0 = \tilde{E}^{-1}B$.
**if** (multiple input) **then**
    Orthogonalize the columns in $R_0$ using MGS: $[v_1, v_2, \ldots, v_{q_1}] = \mathrm{orth}\{R_0\}$ with respect to a user given tolerance $\varepsilon > 0$ specifying the deflation criterion for numerically linearly dependent vectors.
    $sum = q_1$   % $q_1$ is the number of columns remaining after deflation w.r.t. $\varepsilon$.)
**else**
    Compute the first column in V: $v_1 = R_0/||R_0||_2$
    $sum = 1$
**end if**
% Compute the orthonormal columns in $R_1, R_2, \ldots, R_m$ iteratively as below
**for** $i = 1, 2, \ldots, m$ **do**
    $a_2 = sum$;
    **for** $t = 1, 2, \ldots, p$ **do**
        **if** $a_1 = a_2$ **then**
            $stop$
        **else**
            **for** $j = a_1 + 1, \ldots a_2$ **do**
                $w = \tilde{E}^{-1}E_t v_j$;
                $col = sum + 1$;
                **for** $k = 1, 2, \ldots, col - 1$ **do**
                    $h = v_k^T w$
                    $w = w - hv_k$
                **end for**
                **if** $\|w\|_2 > \varepsilon$ **then**
                    $v_{col} = \frac{w}{\|w\|_2}$;
                    $sum = col$;
                **end if**
            **end for**
        **end if**
    **end for**
    $a_1 = a_2$;
**end for**
Orthogonalize the columns in $V$ by MGS w.r.t. $\varepsilon$.

e) The index $t$ is used to refer to computations related to the $t$-th parameter $s_t$ corresponding to the coefficient $\tilde{E}^{-1}E_t$.

f) $a_2 - a_1$ is the number of columns added to $V$ corresponding to $R_{i-1}$.

g) $a_2 - a_1 = 0$ means that all the vectors corresponding to $R_{i-1}$ are deflated because they are linearly dependent (w.r.t. $\varepsilon$) to previous columns in $V$. In this case, there is no vector left which corresponds to $R_{i-1}$. As for a breakdown in a Krylov subspace method, we cannot continue to compute the columns in $V$ corresponding to $R_i$, hence the algorithm stops.

h) In Step 17, $j$ refers to the $j$-th column in $V$ and corresponds to a vector in $R_{i-1}$.

i) Steps 20–27 implement the MGS process. *col* is the subscript of the current column $v_{col}$ in $V$; it is orthogonalized to all the previous columns in $V$ by MGS.

j) In Step 24, $\|w\|_2 < \varepsilon$ is the criterion used to deflate vectors in $R_i$ that are linearly dependent (w.r.t. $\varepsilon$) to the previous vectors in $V$. It does not mean that all the vectors in $R_i$ are linearly dependent on the previous vectors in $V$. If linear dependence is determined by this criterion, we delete the vector $w$ and continue the algorithm till $a_1 = a_2$.

k) In Step 32, we orthonormalize all the columns in $V$ again using MGS to reduce $\|V^TV - I\|_2$ (where $I$ is the identity matrix of appropriate size) and to possibly further deflate columns. In this way, we perform a repeated MGS procedure. The final matrix $V$ has $q$ columns, which is equal to or less than the total number of vectors in $R_i$, $i = 0, 1, \cdots, m$.

l) When $p = 1$, the algorithm reduces to a block-Arnoldi-type process, with $R_0 = B_M$ being the starting block (the vectors in $R_0$ are the starting vectors), which can be used in moment-matching MOR for multiple input, non-parametric systems (see [17,25] for other variants of block Arnoldi processes used in moment-matching MOR).

It should be noted that analogously to moment-matching methods for non-parametric systems, a Petrov-Galerkin or oblique projection method can be constructed applying Algorithm 6.1 to $B$ replaced by $L$ and $\tilde{E}, E_t$ by $\tilde{E}^T, E_t^T$ (and not by complex conjugate transposition which would not yield the desired moment matching property). One would then obtain another orthogonal matrix $W$ whose columns form an orthogonal basis of a complementary subspace. The reduced-order model is then computed by oblique projection $\hat{E}_t = W^T E_i V$, $t = 0, \ldots, p$, etc., assuming the expansion point is chosen real. Technical issues as in standard oblique moment-matching methods will occur here even more pronounced, e.g., the number of computed columns for $V$ and $W$ may differ, the reduced-order model might loose stability, etc. We therefore restrict ourselves here to the presentation of the 1-sided (Galerkin/orthogonal) projection method to not obscure the presentation by too much technical details.

## 6.4 Multi-Moment Matching Property

In this section, we show that the reduced model obtained with the proposed Algorithm 6.1 has indeed the moment matching property derived in [8].

From the analysis in Sect. 6.3, the $R_i$ defined in (6.13) are composed of the coefficients in the series expansion of the state $x$ in frequency domain. The power series expansion of the transfer function of the original model (6.6) is, except for the left multiplication by $L^T$, the same due to the fact that for any feasible square-integrable function $u(\,.\,)$,

$$H(s_1,\ldots,s_p)u(s_p) = L^{\mathrm{T}}x(s_p).$$

(Note that $x$ depends implicitly on $s_1,\ldots,s_{p-1}$ which we omit for the ease of notation.) Hence, the $i$-th order multi-moments of the parametric transfer function $H$ are just the terms $L^{\mathrm{T}}R_i$, $i=0,1,2,\ldots$, where we recall that $R_i$ includes the set of the $i$-th order moment vectors of $x$. For the reduced model in (6.10), there are corresponding power series expansions of the state $z$ and the corresponding transfer function $\hat{H}$. We denote the coefficients in the series expansion of $z$ as

$$\begin{aligned}
\hat{R}_0 &= \hat{B}_M,\\
\hat{R}_1 &= [\hat{M}_1\hat{R}_0, \hat{M}_2\hat{R}_0, \ldots, \hat{M}_p\hat{R}_0],\\
\hat{R}_2 &= [\hat{M}_1\hat{R}_1, \hat{M}_2\hat{R}_1, \ldots, \hat{M}_p\hat{R}_1],\\
&\;\;\vdots\\
\hat{R}_j &= [\hat{M}_1\hat{R}_{j-1}, \hat{M}_2\hat{R}_{j-1}, \ldots, \hat{M}_p\hat{R}_{j-1}]\\
&\;\;\vdots
\end{aligned}$$

where $\hat{E} = \hat{E}_0 + s_1^0\hat{E}_1 + \ldots + s_p^0\hat{E}_p$, $\hat{B}_M = \hat{E}^{-1}\hat{B}$, and $\hat{M}_i = -\hat{E}^{-1}\hat{E}_i$, $i=1,\ldots,p$. The transfer function of the reduced model can be expressed by $z$ as

$$\hat{H}(s_1,\ldots,s_p)u(s_p) = \hat{L}^{\mathrm{T}}z(s_p).$$

Therefore, by the same variational argument as for the full-order system, the multi-moments of $\hat{H}$ are $\hat{L}^{\mathrm{T}}\hat{R}_i$, $i=0,1,2,\ldots$. Here, $\hat{E}_i$, $i=0,1,\ldots,p$, and $\hat{B},\hat{L}$ are defined in (6.10). Next we will prove that the multi-moments of $\hat{H}$ match the multi-moments of the original transfer function $H$. We summarize our analysis, using Lemma 6.1 and Lemma 6.2, in Theorem 6.1.

Suppose we construct the projection matrix $V$ by

$$\mathrm{range}(V) = \mathrm{colspan}\{R_0, R_1, R_2, \cdots, R_m\} =: \wp.$$

The following Lemma 6.1 is used to prove Lemma 6.2 (Lemma 6.1 recalls a known fact and appears in several papers, see e.g. [8]).

**Lemma 6.1** *If the column span of $V$ forms an orthonormal basis of $\wp$, then for any vector $\xi \in \wp$,*

$$\xi = VV^{\mathrm{T}}\xi. \tag{6.14}$$

**Lemma 6.2** *If the orthonormal projection matrix $V$ satisfies* $\mathrm{range}(V) = \wp$,*, then* $\hat{R}_i = V^{\mathrm{T}}R_i$, $i=0,1,\ldots,m$.

*Proof* Recall that $\hat{E} = V^T\tilde{E}V$. Thus, for $i=0$,

$$\hat{E}V^{\mathrm{T}}R_0 = V^{\mathrm{T}}\tilde{E}VV^{\mathrm{T}}R_0.$$

Since $\text{colspan}\{R_0\} \subseteq \wp$, we have $VV^{\mathrm{T}}R_0 = R_0$ by Lemma 6.1. Therefore, from the definition of $R_0$,

$$\hat{E}V^{\mathrm{T}}R_0 = V^{\mathrm{T}}\tilde{E}R_0 = V^{\mathrm{T}}\tilde{E}\tilde{E}^{-1}B = V^{\mathrm{T}}B = \hat{B}.$$

Hence, considering only the first and the last expression, we get,

$$V^{\mathrm{T}}R_0 = \hat{E}^{-1}\hat{B} = \hat{R}_0.$$

Thus, Lemma 6.2 is true for $i = 0$. Next, we assume that Lemma 6.2 is true for $i \leq j$, so that $\hat{R}_j = V^{\mathrm{T}}R_j$. We will prove that it is then also true for $i = j+1$. Since $\text{colspan}\{R_{j+1}\} \subseteq \wp$, by Lemma 6.1 and the definition of $R_{j+1}$, we get

$$\begin{aligned}\hat{E}V^{\mathrm{T}}R_{j+1} &= V^{\mathrm{T}}\tilde{E}VV^{\mathrm{T}}R_{j+1} \\ &= V^{\mathrm{T}}\tilde{E}R_{j+1} = V^{\mathrm{T}}\tilde{E}[-\tilde{E}^{-1}E_1R_j, -\tilde{E}^{-1}E_2R_j, \ldots, -\tilde{E}^{-1}E_pR_j] \\ &= V^{\mathrm{T}}[-E_1R_j, -E_2R_j, \ldots, -E_pR_j].\end{aligned} \tag{6.15}$$

Because $\text{colspan}\{R_j\} \subseteq \wp$, we know that $R_j = VV^{\mathrm{T}}R_j$ by Lemma 6.1. Hence, the last term of the above equation equals

$$V^{\mathrm{T}}[-E_1VV^{\mathrm{T}}R_j, -E_2VV^{\mathrm{T}}R_j, \cdots, -E_pVV^{\mathrm{T}}R_j]. \tag{6.16}$$

Therefore, by the definition of $\hat{E}_i$, $i = 1, \ldots, p$, and the assumption $\hat{R}_j = V^{\mathrm{T}}R_j$, (6.16) is equal to

$$[-\hat{E}_1\hat{R}_j, -\hat{E}_2\hat{R}_j, \cdots, -\hat{E}_p\hat{R}_j]. \tag{6.17}$$

Combining (6.15), (6.16) and (6.17), we obtain

$$\hat{E}V^{\mathrm{T}}R_{j+1} = [-\hat{E}_1\hat{R}_j, -\hat{E}_2\hat{R}_j, \cdots, -\hat{E}_p\hat{R}_j]. \tag{6.18}$$

Then from the definition of $\hat{R}_{j+1}$ we get

$$V^{\mathrm{T}}R_{j+1} = [-\hat{E}^{-1}\hat{E}_1\hat{R}_j, -\hat{E}^{-1}\hat{E}_2\hat{R}_j, \cdots, -\hat{E}^{-1}\hat{E}_p\hat{R}_j] = \hat{R}_{j+1}. \qquad \square$$

**Theorem 6.1** *If $V$ satisfies* $\text{range}(V) = \text{colspan}\{R_0, R_1, R_2, \cdots, R_m\}$, *then the multi-moments of the transfer function of the reduced model in* (6.10) *match those of the full system in* (6.6) *up to order $m$, i.e.* $L^{\mathrm{T}}R_i = \hat{L}^{\mathrm{T}}\hat{R}_i$, $i = 0, 1, \ldots, m$.

*Proof* From Lemma 6.2, and by the definition of $\hat{L}$, we have $\hat{L}^{\mathrm{T}}\hat{R}_i = L^{\mathrm{T}}VV^{\mathrm{T}}R_i$, $i = 0, 1, \ldots, m$. By Lemma 6.1, $VV^{\mathrm{T}}R_i = R_i$, therefore

$$\hat{L}^{\mathrm{T}}\hat{R}_i = L^{\mathrm{T}}R_i, \; i = 0, 1, \ldots, m. \qquad \square$$

## 6.5 Simulation Results

In this section, some simulation results are presented to show the efficiency of the proposed algorithm. We employ two examples, one being the thermal MEMS model considered before in Fig. 6.1 and the other one is from electrochemistry. Illustrated in Fig. 6.8 is the computational domain of the second model, where some chemical reactions take place.

## 6.5.1 Results for the thermal model

The thermal model is a generic example of a device with a single heat source, where the generated heat dissipates through the device to the surroundings. A heater is shown by the block made of PolySi. The exchange between surroundings and the device is modeled by convection boundary conditions with the film coefficient $k$ at the top. The corresponding mathematical parametric model is given in (6.1), where $k$ is the parameter. It is a single-input multiple-output system. For simplicity, we only observe a single output of the system, which is the temperature in the middle of the heater. As has been shown, the values of $k$ change significantly, $k \in [1, 10^9]$. The size of the system is $n = 4725$.

To implement Algorithm 6.1, we first need to transform the system into the frequency domain by the Laplace transformation assuming $x|_{t=0} = 0$ for all $k$. The corresponding system in the frequency domain is

$$\begin{aligned}(sC + G + kD)X(s) &= BU(s),\\ y &= L^{\mathrm{T}}X(s),\end{aligned}$$

where $s$ is considered as the second parameter of the system. Since $B$ is a vector, the projection matrix $V$ is constructed for the single input case in Algorithm 6.1.

### *Implicit vs. explicit moment vector computation*

In Sect. 6.2.2, we have analyzed the accuracy of the PMOR method from [8] if the moment vectors are explicitly computed. Here, we show the efficiency of the proposed Algorithm 6.1, and compare it with the explicit moment-matching described in Sect. 6.2.2.

In Fig. 6.3, the errors of three reduced models computed by Algorithm 6.1 are plotted. The dashed line is the error of the output produced by the reduced model by matching the multi-moments upto order 2. The dash-dotted line is the error by matching multi-moments up to order 3. The dotted line is the one obtained by matching the multi-moments up to order 5. Different from the errors in Fig. 6.2, the errors of the reduced models keep decreasing with the increasing number of matched multi-moments, whereas the errors of the reduced models in Fig. 6.2 do not change after matching up to 3rd order moments. In Fig. 6.4, the accuracy of the reduced models computed by explicit and implicit moment-matching is compared. The solid line and the dashed line represent the accuracy of the reduced model computed by explicit moment-matching. By matching multi-moments up to the same order, the implicit moment-matching method implemented in Algorithm 6.1 is more accurate than the explicit moment-matching.

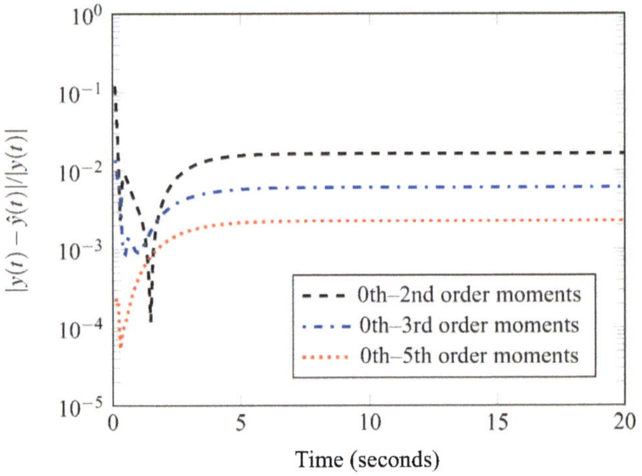


**Fig. 6.3.** Accuracy of Algorithm 6.1, the implicit multi-moment matching method

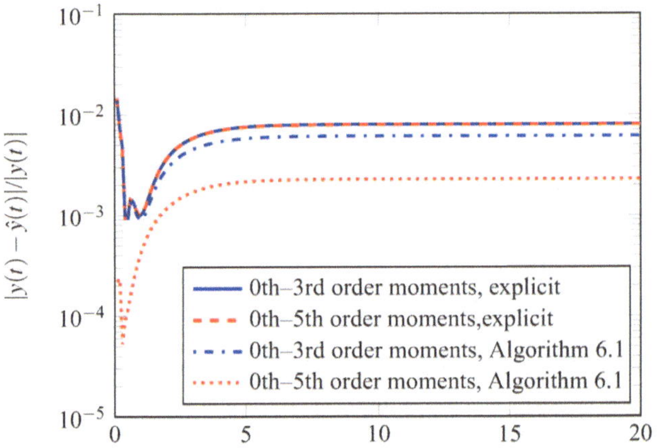


**Fig. 6.4.** Comparison between explicit and implicit multi-moment matching

### *PMOR vs. Non-Parametric MOR*

In order to show the importance and the advantage of PMOR, we compare the proposed PMOR Algorithm 6.1 with the standard non-parametric moment-matching MOR method (see e.g. [25]). For non-parametric MOR, all the parameters except for the Laplace variable $s$ must be fixed, such that the system becomes non-parametric. Hence, a standard non-parametric moment-matching method can be applied. Here the parameter $k$ is fixed to $k = 1$, and the moments are the simple moments associated with the Laplace variable $s$. The reduced models constructed by both methods are in the same form as in (6.11).

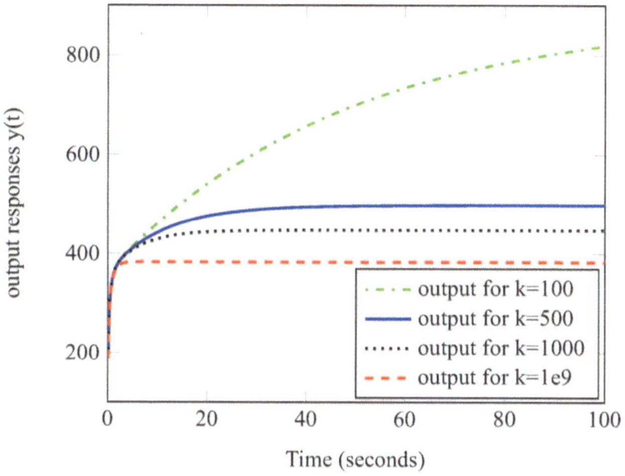


**Fig. 6.5.** Output responses of the original system (6.1) in the time-domain at different values of the parameter $k$

In Fig. 6.5, we plot the output responses corresponding to different values of $k$ by simulating the original parametric system (6.1) for several times. We see that the time-dependent output response varies much with $k$.

In Algorithm 6.1, the 0th order till the 8th order multi-moments are matched. That is, $\text{range}(V) = \text{colspan}\{R_0, R_1, \ldots, R_8\}$. The resulting reduced model is of order $q = 44$. For comparison, we could use the same order of moments associated with $s$ for the non-parametric MOR. However, the resulting reduced model is only of dimension $q = 9$. Instead, the two methods are compared with respect to the same order of the reduced model. To this end, the 0th order till the 43rd order moments are matched by the non-parametric MOR method, and the reduced model is of the same dimension $q = 44$.

In Fig. 6.6, the relative errors of each reduced model changing with different values of $k$ are plotted. Along the $x$-axis, the logarithm of the parameter $k$ is taken. Along the $y$ axis, the relative error defined as $||y(0,T;k) - \hat{y}(0,T;k)||_2/||y(0,T;k)||_2$ is plotted. Here $y(0,T,k) = (y(t_1;k),\ldots,y(t_N;k))^{\mathrm{T}}$ is a vector of the output responses at different time steps in the interesting time interval, $t_i \in [0,T], i = 1,\ldots,T_N$, for the current value of the parameter $k$, obtained by full simulation of the original system. The vector $\hat{y}(0,T;k)$ is obtained analogously from the output responses computed with the reduced model.

The solid line in the figure represents the errors produced by the reduced model with $q = 44$, obtained by non-parametric moment-matching MOR. It has good accuracy at the values of $k$ close to $k = 1$, the fixed value. However, when the value of $k$ grows, the error generally keeps increasing. As expected, the reduced model cannot catch the behavior of the output responses corresponding to values of $k$ far away from the fixed value. The accuracy of the reduced model computed with the proposed PMOR method is much higher, though there is still a very slow trend of

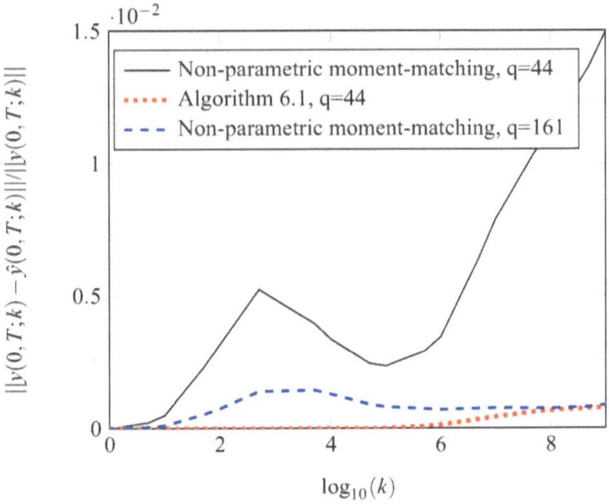


**Fig. 6.6.** Relative errors of the output responses of the thermal model in the time domain for different values of $k$, computed from the reduced models derived by non-parametric MOR and Algorithm 6.1, respectively. The orders of the three reduced models are $q = 44, 44, 161$, respectively

error increase with increasing value of $k$, see the dotted line. This is because a single expansion point for $k$, $k_0 = 0$, is used during the series expansion of the state vector $x$ (see (6.7) and (6.12)). Multiple point expansion can be used in combination with Algorithm 6.1 to further decrease the error of the reduced model for very large values of $k$, see [14].

To achieve the same level of accuracy as for the reduced model resulting from PMOR, a reduced model with dimension $q = 161$ must be constructed with the non-parametric moment-matching MOR, where the 0th till the 160th order moments are matched. The error of the reduced model is plotted using dashes. This shows that the PMOR method provides a more compact reduced model over the entire parameter domain.

### *Robustness of the Proposed Algorithm*

In Fig. 6.7, relative errors of three different reduced models constructed by Algorithm 6.1 are plotted. Each line represents the relative error between the output response of the reduced system and that of the original system according to different values of the parameter $k$. The definition of the relative error is the same as defined for Fig. 6.6. The line with the smallest error represents the error of the reduced system of order $q = 44$, for which the reduced system is obtained by $\text{range}(V) = \text{colspan}\{R_0, R_1, \ldots, R_8\}$. The line in the middle is the error of the reduced system with $q = 28$; it is derived from $\text{range}(V) = \text{colspan}\{R_0, R_1, \ldots, R_6\}$.

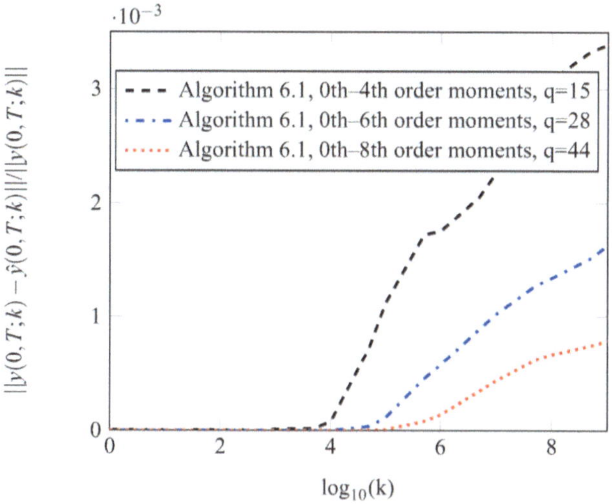


**Fig. 6.7.** Relative errors of the output responses of the thermal model in the time-domain computed from the reduced models with different order $q$, using Algorithm 6.1, for different values of $k$

The line on the top corresponds to the reduced system computed from $\text{range}(V) = \text{colspan}\{R_0, R_1, \ldots, R_4\}$. One can see that the error becomes smaller with increasing number of moment vectors used. All in all, the errors at all the values of the parameter $k$ are very small, and satisfy the accuracy requirement in real applications. Compared with the explicit multi-moment matching, and the non-parametric MOR, the proposed Algorithm 6.1 produces a much more accurate reduced model.

## 6.5.2 Results for the electrochemistry model

The detailed description and derivation of the model for the application depicted in Fig. 6.8 is available from the MORwiki[3]. The mathematical model after spatial discretization is

$$\begin{aligned} E\tfrac{dc}{dt} + Gc + s_1 D_1 c + s_2 D_2 c = F, \quad c(0) = c_0 \neq 0 \\ y = l^{\mathrm{T}} c, . \end{aligned} \tag{6.19}$$

The dimension of the system is $n = 16912$. Here, $E, G, D_1, D_2 \in \mathbb{R}^{n \times n}$ are system matrices. $l, F \in \mathbb{R}^n$ are constant vectors. $c(t) \in \mathbb{R}^n$ is the unknown vector. The two parameters $s_1 = e^{\beta u(t)}$, $s_2 = e^{-\beta u(t)}$ are functions of the voltage, where $\beta = 21.243036728240824$ is a constant. The voltage $u(t, \alpha)$ which is a function of

[3] http://morwiki.mpi-magdeburg.mpg.de/morwiki/index.php/Scanning_Electrochemical_Microscopy.

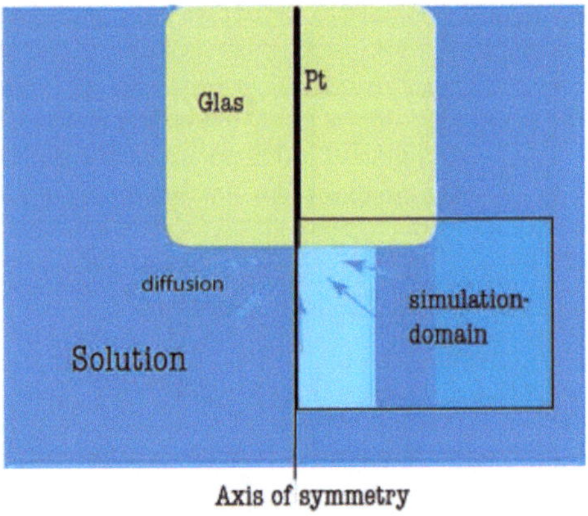


**Fig. 6.8.** Example from electrochemistry. The computational domain (indicated as simulation domain in the figure) under the 2D-axisymmetrical approximation includes the electrolyte under the electrode (the square at the middle top). Some chemical reactions take place in the computational domain. The interesting output is the total current over the electrode surface

time and $\alpha$, follows a symmetric, triangular waveform:

$$\begin{aligned} u(t,\alpha) &= u_0 + \alpha t, \quad 0 < t < t_\alpha, \\ u(t,\alpha) &= u_0 - \alpha t, \quad t_\alpha < t < 2t_\alpha. \end{aligned}$$

Here, the variable $\alpha$ takes four possible values $\alpha = 0.5, 0.05, 0.005, 0.0005$. The time point $t_\alpha$ actually varies with $\alpha$ by $t_\alpha = 4 \times 10^i$, when $\alpha = 0.5 \times 10^{-i}$, for $i = 0,1,2,3$.

The output $y(t)$ is the total current over the electrode surface, changing with the voltage $u(t,\alpha)$. The waveforms of the two parameters $s_1$ and $s_2$ as functions of time and the voltage $u(t,\alpha)$ are given in Fig. 6.9 and Fig. 6.10, respectively. Although both $s_1$ and $s_2$ are functions of the voltage $u$, hence are not independent, they are considered as two independent parameters in Algorithm 6.1. They can further be simply treated as two parameters independent of any argument, e.g. the time variable $t$, during the implementation of Algorithm 6.1, since the projection matrix $V$ is generated independently of the parameters.

To deal with the system with nonzero initial condition, we employ the transformation method in [15]. That is, we first transform the system into a system with zero initial condition by $\tilde{c} = c - c_0$. The resulting transformed system is

$$\begin{aligned} E\tfrac{d\tilde{c}}{dt} + G\tilde{c} + s_1 D_1 \tilde{c} + s_2 D_2 \tilde{c} = F - Gc_0 - s_1 D_1 c_0 - s_2 D_2 c_0, \\ y = l^{\mathrm{T}}(\tilde{c} + c_0), \quad \tilde{c}(0) = c(0) - c_0 = 0. \end{aligned} \tag{6.20}$$

By Laplace transform, the above system in frequency domain becomes

$$\begin{aligned} (sE + G + s_1 D_1 + s_2 D_2)x &= \tilde{F}u(s), \\ y &= l^{\mathrm{T}}(x + c_0 u(s)), \end{aligned} \tag{6.21}$$

where $x$ is the Laplace transform of the time-domain unknown vector $\tilde{c}$, $u(s) = 1/s$ is the Laplace transform of the constant 1, and $\tilde{F} = F - Gc_0 - s_1 D_1 c_0 - s_2 D_2 c_0$. As

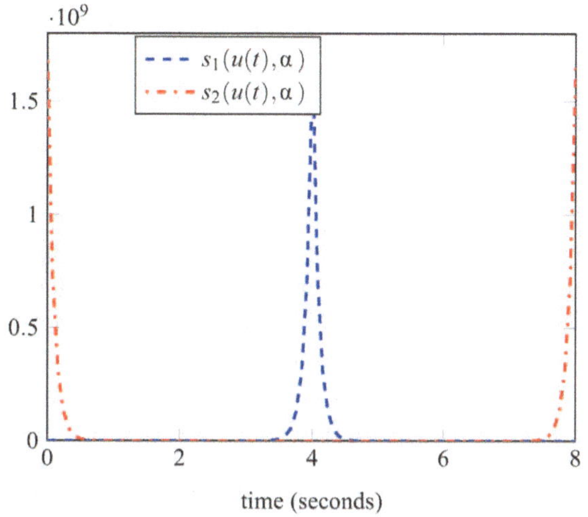


**Fig. 6.9.** $s_1(u(t,\alpha))$ and $s_2(u(t,\alpha))$ changing with time, $\alpha = 0.5$

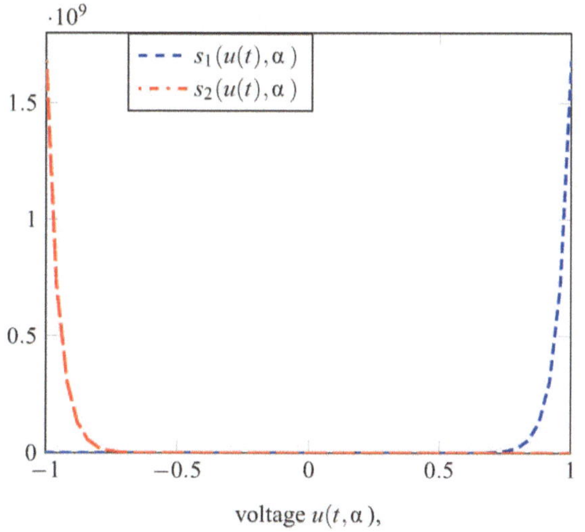


**Fig. 6.10.** $s_1(u(t,\alpha))$ and $s_2(u(t,\alpha))$ changing with the voltage $u(t,\alpha)$

explained above, $s_1(u(t))$ and $s_2(u(t))$ are treated as two constant parameters during the execution of Algorithm 6.1. As they are preserved in the reduced model, they can then again be varied with time according to their original definition when simulating the reduced model.

Note that the right-hand side of the system also depends on the two parameters $s_1, s_2$, which, however, is not a problem. Since the function $u(s)$ and the parameters

$s_1, s_2$ are both scalars, the right-hand side of the system (6.21) can be written as

$$(F - Gc_0 - s_1 D_1 c_0 - s_2 D_2 c_0)u(s) = [F - Gc_0, D_1 c_0, D_2 c_0] \begin{pmatrix} u(s) \\ -s_1 u(s) \\ -s_2 u(s) \end{pmatrix} = B\tilde{U},$$

where $B = [F - Gc_0, D_1 c_0, D_2 c_0]$, $\tilde{U} = [u(s), -s_1 u(s), -s_2 u(s)]^{\mathrm{T}}$. Therefore, the system in (6.21) can be considered as a multiple input system, so that the multiple input case in Algorithm 6.1 can be applied to construct the projection matrix $V$[4]. The time domain reduced model in the form of (6.11) is obtained by applying Galerkin projection, using $V$, to the transformed system in (6.20) [15].

Figures 6.11–6.15 show the simulation results of the original model (6.19) and the reduced model obtained by Algorithm 6.1. The figures display the currents as functions of the voltages $u(t, \alpha)$, which is the usual way to represent the so-called cyclic *voltammograms* of the electro-chemical reaction. The solid line is the result obtained by full simulation of the original large model, the dashed line is the result computed using the small reduced model. The results of the reduced model are accurate for a wide range of the dynamic behavior when the value of $\alpha$ changes by three orders of magnitude (0.005–0.5).

The dashed lines in the Figs. 6.11–6.13 show the simulation results of three different reduced models with $\alpha = 0.5$. As we have already seen, the projection matrix $V$ depends on the moment vectors of the system. If more moment vectors are used, the reduced model should become more accurate, at least in theory. The simulation

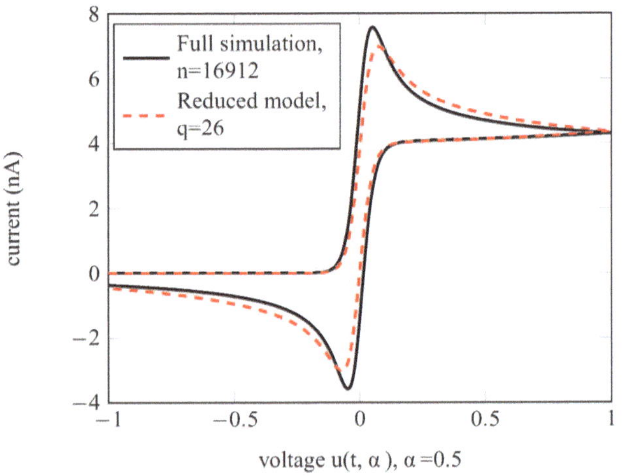


**Fig. 6.11.** The current $y$ as a function of the voltage $u(t, \alpha)$, for $\alpha = 0.5$, for both the full simulation and the PMOR method Algorithm 6.1 using the multiple input variant. The moments are matched up to 4th order, yielding a reduced model of dimension $q = 26$

[4] For this example, the zero expansion points $s^0 = 0$, $s_1^0 = 0$ and $s_2^0 = 0$ are used for all the three parameters $s, s_1, s_2$ in (6.21).

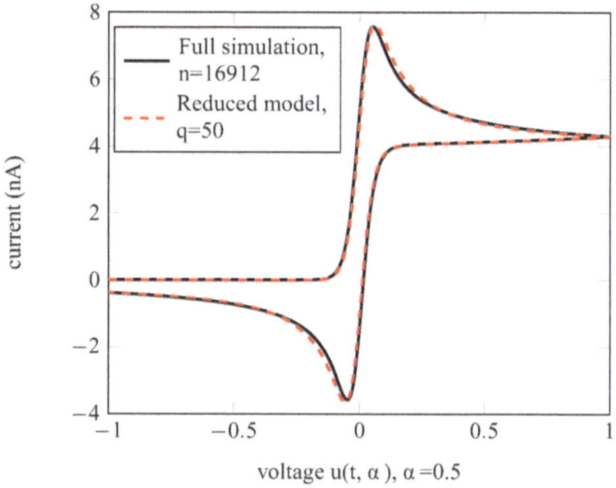


**Fig. 6.12.** The current $y$ as a function of the voltage $u(t,\alpha)$, for $\alpha = 0.5$, for both the full simulation and the PMOR method Algorithm 6.1 using the multiple input variant. The moments are matched up to 6th order, yielding a reduced space of dimension $q = 50$

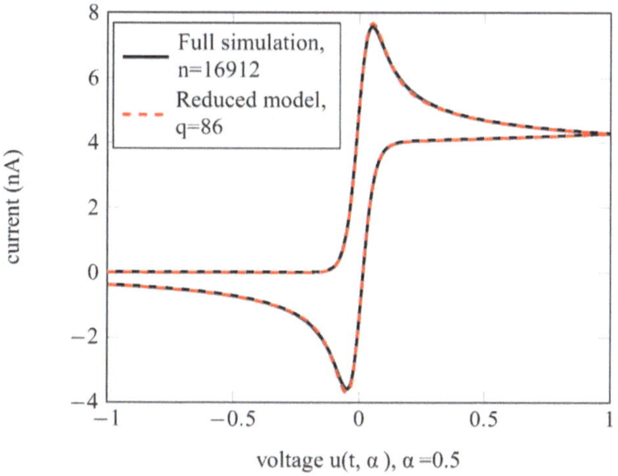


**Fig. 6.13.** The current $y$ as a function of the voltage $u(t,\alpha)$, for $\alpha = 0.5$, for both the full simulation and the PMOR method Algorithm 6.1 using the multiple input variant. The moments are matched up to 9th order, yielding a reduced space of dimension $q = 86$

results in Fig. 6.3 show this fact for the previous thermal MEMS problem. For the current problem, the simulation results in Figs. 6.11–6.13 further justify it. In contrast, if $V$ is computed by explicit matrix multiplications, the accuracy of the reduced model cannot be improved by using more moment vectors. In Fig. 6.11, the moment vectors from $R_0$ to $R_4$ are employed to compute $V$. In Fig. 6.12, $R_0$ till $R_6$ are used for the reduced model. The moment vectors from $R_0$ to $R_9$ are used in Fig. 6.13 to get $V$. The result in Fig. 6.13 is most accurate. The waveform of the current computed from the reduced model shows little difference from the solid line. In this case the order of the reduced model is $q = 86$. The relative error between the two currents is $\varepsilon|_{\alpha=0.5} = ||y-\hat{y}||_2/||y||_2 = 6.3 \times 10^{-4}$, where $y$ is the vector of the current at dense samples of the interesting time interval by full simulation, and $\hat{y}$ is the vector of the current at the same samples obtained by simulating the reduced model. The reduced model is good enough to replace the original model with space dimension $n = 16912$ in practical applications of the model.

Figures 6.14–6.15 show additional outcomes for other values of $\alpha$. Here we used the most accurate reduced model with range$(V) = \text{colspan}\{R_0, R_1, \ldots, R_9\}$ and study the effect when varying $\alpha$. The order of each reduced model is the same: $q = 86$.

The relative errors $\varepsilon$ are listed in Table 6.1 for a selection of different values of $\alpha$. All these simulation results show that accurate reduced models can be obtained with the proposed algorithm.

**Table 6.1.** $\varepsilon$ vs. $\alpha$

| $\alpha$ | $\alpha = 0.5$ | $\alpha = 0.05$ | $\alpha = 0.005$ | $\alpha = 0.0005$ |
|---|---|---|---|---|
| $\varepsilon$ | $6.3 \times 10^{-4}$ | $1.8 \times 10^{-5}$ | $1.64 \times 10^{-6}$ | $1.38 \times 10^{-6}$ |

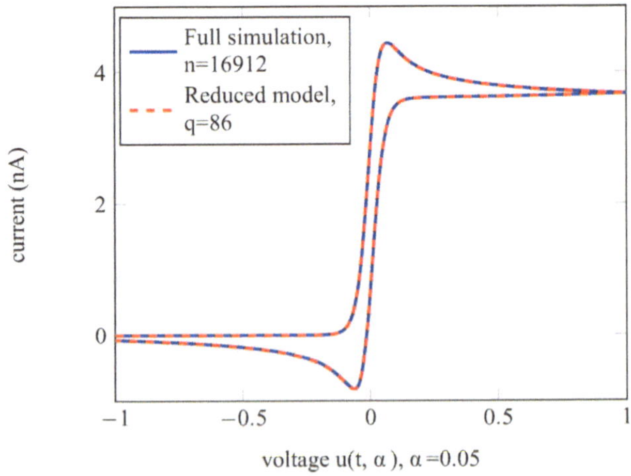


**Fig. 6.14.** The current $y$ as a function of the voltage $u(t,\alpha)$, for $\alpha = 0.05$, for both the full simulation and the PMOR method Algorithm 6.1 using the multiple input variant. The moments are matched up to 9th order, yielding a reduced space of dimension $q = 86$

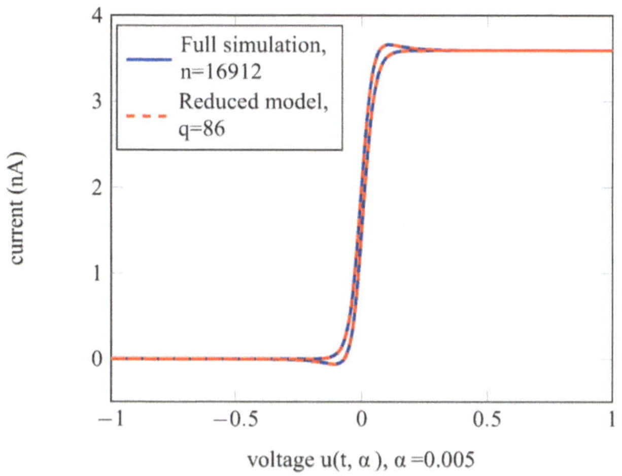


**Fig. 6.15.** The current $y$ as a function of the voltage $u(t,\alpha)$, for $\alpha = 0.005$, for both the full simulation and the PMOR method Algorithm 6.1 using the multiple input variant. The moments are matched up to 9th order, yielding a reduced space of dimension $q = 86$

## 6.6 Conclusions

A numerical stable algorithm for PMOR is explored in this paper. The algorithm is used to construct a projection matrix $V$ whose columns form an orthonormal basis of the subspace spanned by the moment vectors of the parametric system. Instead of explicit matrix-vector multiplications, a new moment vector is orthogonalized to all the previous ones during a (repeated) Modified Gram-Schmidt process. Numerical simulation results for both single input and multiple input parametric systems show that the proposed algorithm is very accurate and robust. Applications of the algorithm to parametric systems with more than three parameters can be found in [14].

The reduced parametric model can be used in optimization [35], in statistics [24], and in coupled simulations [23]. When used in statistics, it is important that quantities like mean and variance are well approximated. In applying PMOR for uncertainty quantification, one thus seeks to have a "statistics-preserving PMOR".

In some cases, the parameters may not be explicitly available. For instance, in modeling of electromagnetic problems, varying geometry may result in different meshes. For an approach to deal with this see [31].

Future research will focus on how to adaptively choose proper nonzero expansion points to attain a more compact model for systems with many (more than three) parameters. An error estimation for the state $x$ of the parametric system is proposed in [33] for an automatic sampling selection. For many applications, the output $y$ or the transfer function of the system is of interest, and an output-oriented error estimation for the proposed PMOR method is preferred, such that a more reliable reduced model can be obtained, automatically.

## References


1. Amsallem, D., Farhat, C.: An online method for interpolating linear parametric reduced-order models. SIAM J. Sci. Comput. **33**(5), 2169–2198 (2011). DOI 10.1137/100813051
2. Antoulas, A.: Approximation of Large-Scale Dynamical Systems. SIAM Publications, Philadelphia, PA (2005)
3. Baur, U., Benner, P.: Model reduction for parametric systems using balanced truncation and interpolation. at–Automatisierungstechnik **57**(8), 411–420 (2009)
4. Baur, U., Benner, P., Beattie, C., Gugercin, S.: Interpolatory projection methods for parameterized model reduction. SIAM J. Sci. Comput. **33**, 2489–2518 (2011)
5. Benner, P., Hinze, M., ter Maten, E.J.W. (eds.): Model Reduction for Circuit Simulation, Lecture Notes in Electrical Engineering, vol. 74. Springer-Verlag, Dordrecht, NL (2011)
6. Benner, P., Mehrmann, V., Sorensen, D. (eds.): Dimension Reduction of Large-Scale Systems, Lecture Notes in Computational Science and Engineering, vol. 45. Springer-Verlag, Berlin Heidelberg (2005)
7. Bond, B., Daniel, L.: Parameterized model order reduction of nonlinear dynamical systems. In Proc. International Conference on Computer-Aided Design pp. 487– 494 (2005)
8. Daniel, L., Siong, O., Chay, L., Lee, K., White, J.: A multiparameter moment-matching model-reduction approach for generating geometrically parameterized interconnect performance models. IEEE Trans. Comput.-Aided Des. Integr. Circuits Syst. **22**(5), 678–693 (2004)
9. Eid, R., Castañé-Selga, R., Panzer, H., Wolf, T., Lohmann, B.: Stability-preserving parametric model reduction by matrix interpolation. Math. Comp. Model. Dyn. Syst. **17**(4), 319–335 (2011)
10. Farle, O., Hill, V., Ingelström, P., Dyczij-Edlinger, R.: Multi-parameter polynomial order reduction of linear finite element models. Math. Comput. Model. Dyn. Syst. **14**(5), 421–434 (2008)
11. Feldmann, P., Freund, R.: Efficient linear circuit analysis by Padé approximation via the Lanczos process. IEEE Trans. Comput.-Aided Des. Integr. Circuits Syst. **14**, 639–649 (1995)
12. Feng, L., Benner, P.: Parametrische Modellreduktion durch impliziten Momentenabgleich (Parametric Model Reduction by Implicit Moment Matching). In: B. Lohmann, A. Kugi (eds.) Tagungsband GMA-FA 1.30 "Modellbildung, Identifizierung und Simulation in der Automatisierungstechnik", Workshop in Anif, 26.–28.9.2007, 34–47 (2007)
13. Feng, L., Benner, P.: A robust algorithm for parametric model order reduction based on implicit moment matching. Proc. Appl. Math. Mech. **7**, 1021.501–1021.502 (2008)
14. Feng, L., Benner, P., Korvink, J.G.: Subspace recycling accelerates the parametric macro-modeling of MEMS. Int. J. Numer. Meth. Engrg. **94**, 84–110 (2013)
15. Feng, L., Koziol, D., Rudnyi, E., Korvink, J.: Model order reduction for scanning electrochemical microscope: The treatment of nonzero initial condition. In Proc. Sensors pp. 1236–1239 (2004)
16. Feng, L., Rudnyi, E., Korvink, J.: Preserving the film coefficient as a parameter in the compact thermal model for fast electro-thermal simulation. IEEE Trans. Comput.-Aided Des. Integr. Circuits Syst. **24**(12), 1838–1847 (2005)
17. Freund, R.: Krylov subspace methods for reduced-order modeling in circuit simulation. J. Comput. Appl. Math. **123**, 395–421 (2000)
18. Gallivan, K., Grimme, E., Van Dooren, P.: Asymptotic waveform evaluation via a lanczos method. Appl. Math. Lett. **7**(5), 75–80 (1994)

19. Gunupudi, P., Khazaka, R., Nakhla, M., Smy, T., Celo, D.: Passive parameterized time-domain macromodels for high-speed transmission-line networks. IEEE Trans. Microwave Theory and Techniques **51**(12), 2347–2354 (2003)
20. Haasdonk, B., Ohlberger, M.: Efficient reduced models and a posteriori error estimation for parametrized dynamical systems by offline/online decomposition. Mathematical and Computer Modelling of Dynamical Systems **17**(2), 145–161 (2011). DOI 10.1080/13873954.2010.514703
21. Li, Y., Bai, Z., Su, Y.: A two-directional Arnoldi process and its application to parametric model order reduction. J. Comput. Appl. Math. **226**(1), 10–21 (2009)
22. Liu, Y., Pileggi, L., Strojwas, A.: Model order reduction of RC(L) interconnect including variational analysis. In Proc. Design Automation Conference pp. 201–206 (1999)
23. Lutowska, A.: Model order reduction for coupled systems using low-rank approximations. Ph.D. thesis, Eindhoven University of Technology (2012)
24. ter Maten, E., Pulch, R., Schilders, W., Janssen, H.: Efficient calculation of uncertainty quantification. CASA-Repprt 2012-38 (2012). http://www.win.tue.nl/analysis//reports/rana12-38.pdf
25. Odabasioglu, A., Celik, M., Pileggi, L.: PRIMA: passive reduced-order interconnect macromodeling algorithm. IEEE Trans. Comput.-Aided Des. Integr. Circuits Syst. **17**(8), 645–654 (1998)
26. Panzer, H., Mohring, J., Eid, R., Lohmann, B.: Parametric model order reduction by matrix interpolation. at–Automatisierungstechnik **58**(8), 475–484 (2010)
27. Phillips, J.: Variational interconnect analysis via PMTBR. In Proc. International Conference on Computer-Aided Design pp. 872–879 (2004)
28. Pillage, L., Rohrer, R.: Asymptotic waveform evaluation for timing analysis. IEEE Trans. Comput.-Aided Design **9**, 325–366 (1990)
29. Rozza, G., Huynh, D., Patera, A.: Reduced basis approximation and a posteriori error estimation for affinely parametrized elliptic coercive partial differential equations: application to transport and continuum mechanics. Arch. Comput. Methods Eng. **15**(3), 229–275 (2008). DOI 10.1007/s11831-008-9019-9
30. Schilders, W.H.A., van der Vorst, H.A., Rommes, J. (eds.): Model order reduction: Theory, research aspects and applications. Mathematics in Industry, vol. 13. Springer-Verlag, Berlin Heidelberg (2008)
31. Stavrakakis, K.: Model order reduction methods for parameterized systems in electromagnetic field simulations. Ph.D. thesis, TU Darmstadt (2012)
32. Tao, J., Zeng, X., Yang, F., Su, Y., Feng, L., Cai, W., Zhou, D., Chiang, C.: A one-shot projection method for interconnects with process variations. In Proc. International Symposium on Circuits and Systems pp. 333–336 (2006)
33. Villena, J.F., Silveira, L.M.: Multi-dimensional automatic sampling schemes for multipoint modeling methodologies. IEEE Trans. Comp-aid Design of Integrated Circuits and Systems **30**(8), 1141–1151 (2011). DOI 10.1080/13873954.2010.514703
34. Weile, D., Michielssen, E., Grimme, E., Gallivan, K.: A method for generating rational interpolant reduced order models of two-parameter linear systems. Appl. Math. Lett. **12**(5), 93–102 (1999)
35. Yue, Y.: The use of model order reduction in design optimization algorithms. Ph.D. thesis, KU Leuven (2012)

7

# On the Use of Reduced Basis Methods to Accelerate and Stabilize the Parareal Method

**Feng Chen, Jan S. Hesthaven and Xueyu Zhu**

**Abstract** We propose a modified parallel-in-time – parareal – multi-level time integration method that, in contrast to previously proposed methods, employs a coarse solver based on a reduced model, built from the information obtained from the fine solver at each iteration. This approach is demonstrated to offer two substantial advantages: it accelerates convergence of the original parareal method for similar problems and the reduced basis stabilizes the parareal method for purely advective problems where instabilities are known to arise. When combined with empirical interpolation methods (EIM), we develop this approach to solve both linear and nonlinear problems and highlight the minimal changes required to utilize this algorithm to accelerate existing implementations. We illustrate the advantages through algorithmic design, through analysis of stability, convergence, and computational complexity, and through several numerical examples.

## 7.1 Introduction

With the number of computational cores on large scale computing platforms increasing, the demands on scalability of computational methods likewise increase, due partly to an increasing imbalance between the cost of memory access, communication and arithmetic capabilities. Among other things, traditional domain decompo-

F. Chen
Brown University, 182 George Street, Providence, RI 02912, USA
e-mail: feng_chen_1@brown.edu

J.S. Hesthaven (✉)
EPFL-SB-MATHICSE, École Polytechnique Fédérale de Lausanne, 1007 Lausanne, Switzerland
e-mail: Jan.Hesthaven@epfl.ch

X. Zhu
Brown University, 182 George Street, Providence, RI 02912, USA
e-mail: xzhu@alumni.brown.edu

A. Quarteroni, G. Rozza (eds.): *Reduced Order Methods for Modeling and Computational Reduction*, MS&A 9. DOI 10.1007/978-3-319-02090-7_7, 

sition methods tend to stagnate in scaling as the number of cores increases and the computational cost is overwhelmed by other tasks. This suggests a need to consider the development of computational techniques that better balance these constraints and allow for the acceleration of large scale computational challenges.

A recent development in this direction is the parareal method, introduced in [16], that provide a strategy for 'parallel-in-time' computations and offers the potential for an increased level of parallelism. Relying on combining a computational inexpensive but inaccurate solver with an accurate and expensive but parallel solver, the parareal method utilizes an iterative, predictor-corrector procedure that allows the expensive solver to run across many processors in parallel. Under suitable conditions, the parareal iteration converges after a small number of iterations to the serial solution [3]. During the last decade, the parareal method has been applied successfully to a number of applications (cf. [17, 19]), demonstrating its potential, accuracy, and robustness.

As a central and serial component, the properties of the coarse solver can impact the efficiency and stability of the parareal algorithm, e.g., if an explicit scheme is used in both the coarse and the fine stage of the algorithm, the efficiency of the parareal algorithm is limited by the upper bound of the time step size [19]. One can naturally also consider a different temporal integration approach such as an implicit approach, although the cost of this can be considerable and often requires the development of a new solver. An attractive alternative is to use a simplified physics model as the coarse solver [2, 17, 18], thereby ignoring small scale phenomenon but potentially impacting the accuracy. The success of such an approach is typically problem specific.

While the choice of the coarse solver clearly impacts accuracy and overall efficiency, the stability of the parareal method is considerably more subtle. For parabolic and diffusion dominated problems, stability is well understood and observed in many applications [12]. However, for hyperbolic and convection dominated problems, the question of stability is considerably more complex and generally remains open [3, 8, 22]. In [8], the authors propose to regularly project the solution onto an energy manifold approximated by the fine solution. The performance of this projection method was demonstrated for the linear wave equation and the nonlinear Burgers' equation. As an alternative, the Krylov subspace parareal method builds a new coarse solver by reusing all information from the corresponding fine solver at previous iterations. The stability of this approach was demonstrated for linear problems in structural dynamics [10] and a linear 2-D acoustic-advection system [21]. However, the Krylov subspace parareal method appears to be limited to linear problems.

The approach of combining the reduced basis method [20] with the parareal method for parabolic equations was initiated in [13] in which it is demonstrated that a coarse solver based on an existing reduced model offers better accuracy and reduces the number of iterations in the examples considered. However, that work offers no discussion on the construction of the reduced model, nor was there any attempt to analyze the stability and convergence of the method.

Inspired by [13, 21], we propose a modified parareal method, referred to as the reduced basis parareal method in which the Krylov subspace is replaced by a subspace spanned by a set of reduced bases, constructed on-the-fly from the fine solver. This method inherits most advantages of the Krylov subspace parareal method and is observed to retain stability and convergence for linear wave problems. We demonstrate that this approach accelerates the convergence in situations where the original parareal already converges. However, it also overcomes several known challenges: (i) it deals with nonlinear problems by incorporating methodologies from the reduced basis methods; and (ii) the traditional coarse propagator is needed only once at the very beginning of the algorithm to generate an initial reduced basis. This allows for the time step restrictions to be relaxed as compared to the coarse solver of the original parareal method. The main difference between our method and [13] lies in the reduced approximation space and the construction of reduced bases. The reduced model, playing the role of the coarse solver, is updated for each iteration while the reduced model in [13] is built only once during an initial offline process. Among other advantages, this allows the proposed method to adapt the dimension of the reduced approximation space based on the regularity of the solution, while in [13] the reduced model remains fixed and must be developed using some other approach.

What remains of this paper is organized as follows. We first review the original parareal method in Sect. 7.2.1 and the Krylov subspace parareal method in Sect. 7.2.2. This sets the stage for Sect. 7.2.3 where we introduce the reduced basis parareal method and discuss different strategies to develop reduced models for problems with nonlinear terms. Section 7.3 offers some analysis of the stability, convergence, and complexity of the reduced basis parareal method and Sect. 7.4 demonstrates the feasibility and performance of the reduced basis parareal method through various linear and nonlinear numerical examples. We conclude the paper in Sect. 7.5.

## 7.2 Parareal Algorithms

To set the stage for the general discussion, let us first discuss the original and the Krylov subspace parareal methods in Sect. 7.2.1 and Sect. 7.2.2, respectively. We shall highlight issues related to stability and computational complexity to motivate the reduced basis parareal method, introduced in Sect. 7.2.3.

### *7.2.1 The original parareal method*

Consider the following initial value problem:

$$\begin{aligned} \mathbf{u}_t &= \mathbf{L}(\mathbf{u}) := \mathbf{A}\mathbf{u}(t) + \mathbf{N}(\mathbf{u}(t)), \quad t \in (0, T], \\ \mathbf{u}(0) &= \mathbf{u}_0, \end{aligned} \tag{7.1}$$

where $\mathbf{u} \in \mathbb{R}^N$ is the unknown solution, $\mathbf{L}$ is an operator, possibly arising from the spatial discretization of a PDE, with $\mathbf{A}$ being the linear part of $\mathbf{L}$, and $\mathbf{N}$ the nonlinear part.

In the following, we denote $F_{\delta t}$ as the accurate but expensive fine time integrator, using a constant time step size, $\delta t$. Furthermore, $G_{\Delta t}$ is the inaccurate but fast coarse time integrator using a larger time step size, $\Delta t$. Generally, it is assumed that $\Delta t \gg \delta t$.

The original parareal method is designed to solve (7.1) in a parallel-in-time fashion to accelerate the computation. First, $[0,T]$ is decomposed into $N_c$ coarse time intervals or elements:

$$0 = t_0 < \cdots < t_i < \cdots < t_{N_c} = T, \quad t_i = i\Delta T, \quad \Delta T = \frac{T}{N_c}. \tag{7.2}$$

Assume that

$$\Delta T = N_f \delta t, \quad N_f \in \mathbb{N}, \tag{7.3}$$

which implies that $T = N_c N_f \delta t$. Denote $F_{\delta t}(\mathbf{u}, t_{i+1}, t_i)$ as the accurate numerical solution integrated from $t_i$ to $t_{i+1}$ by using $F_{\delta t}$ with the initial condition $\mathbf{u}$ and the constant time step size $\delta t$. Similarly for $G_{\Delta t}(\mathbf{u}, t_{i+1}, t_i)$. Denote also $\mathbf{u}_n = F_{\delta t}(\mathbf{u}_0, T, 0)$ as the numerical solution generated using only the fine integrator. With the above notation, the original parareal method is shown below in Algorithm 7.1

Now assume that the $k$-th iterated approximation $\mathbf{u}_n^k$ is known. The parareal approach proceeds to the $k+1$-th iteration as

$$\mathbf{u}_{n+1}^{k+1} = G_{\Delta t}(\mathbf{u}_n^{k+1}, t_{n+1}, t_n) + F_{\delta t}(\mathbf{u}_n^k, t_{n+1}, t_n) - G_{\Delta t}(\mathbf{u}_n^k, t_{n+1}, t_n), \quad 0 \le k \le N_c - 1. \tag{7.4}$$

It is easy to see that $F_{\delta t}(\mathbf{u}_n^k, t_{n+1}, t_n)$ can be done in parallel across all temporal elements. If we take the limit of $k \to \infty$ and assume that the limit of $\{\mathbf{u}_n^k\}$ exists, we obtain [16]:

$$\mathbf{u}_{n+1}^{k+1} \to \mathbf{u}_{n+1} = F_{\delta t}(\mathbf{u}_n, t_{n+1}, t_n). \tag{7.5}$$

In order to achieve a reasonable efficiency, the number of iterations, $N_{it}$, should be much smaller than $N_c$.

To demonstrate the performance of the original parareal method, let us consider a few numerical examples, beginning with the viscous Burgers' equation:

$$\begin{aligned} &u_t + (\frac{u^2}{2})_x = \nu u_{xx}, \quad (x,t) \in (0, 2\pi) \times (0, T], \\ &u(x, 0) = \sin(x), \end{aligned} \tag{7.6}$$

where $T = 2$ and $\nu = 10^{-1}$. A $2\pi$-periodic boundary condition is used. The spatial discretization is a $P_1$ discontinuous Galerkin method (DG) with 100 elements [15] and the time integrator is a first-order forward Euler method. We use the following parameters in the parareal integration

$$N_c = 100, \quad N_{it} = 5, \quad \Delta t = 10^{-3}, \quad \delta t = 10^{-4}. \tag{7.7}$$

Figure 7.1 illustrates the $L_\infty$-error of the parareal solution at $T = 2$ against the number of iterations. Notice that for this nonlinear problem the algorithm converges after

**Algorithm 7.1** The original parareal method

Initialization:
$\mathbf{u}_0^0 = \mathbf{u}_0$;
**for** $i \leftarrow 0$ **to** $N_c - 1$ **do**
  $\mathbf{u}_{i+1}^0 = G_{\Delta t}(\mathbf{u}_i^0, t_{i+1}, t_i)$
**end**
Iterations:
$k = 0$;
**for** $k \leftarrow 0$ **to** $N_{it}$ **do**
  Parallel predictor step:
  **for** $i \leftarrow 0$ **to** $N_c - 1$ **do**
    $\mathbf{u}_{f}{}_{i+1}^{k} = F_{\delta t}(\mathbf{u}_i^k, t_{i+1}, t_i)$
  **end**
  Sequential correction step:
  **for** $i \leftarrow$ *0* **to** $N_c - 1$ **do**
    $\mathbf{u}_{i+1}^{k+1} = G_{\Delta t}(\mathbf{u}_i^{k+1}, t_{i+1}, t_i) - \mathbf{u}_{f}{}_{i+1}^{k} + G_{\Delta t}(\mathbf{u}_i^k, t_{i+1}, t_i)$
  **end**
**end**

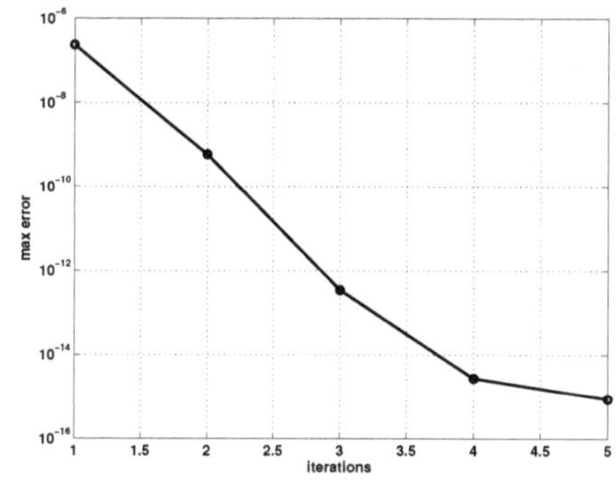


**Fig. 7.1.** The $L_\infty$-error at $T = 2$ against the number of iterations of the 1-D Burgers' equation using the original parareal method

only four iterations, illustrating the potential for an expected acceleration in a parallel environment.

As a second example, we consider the Kuramoto-Sivashinsky equation [25]:

$$\begin{aligned} \frac{\partial u}{\partial t} &= (\frac{u^2}{2})_x - u_{xx} - u_{xxxx}, \quad (x,t) \in (-8,8) \times (0,T], \\ u(x,0) &= \exp(-x^2) \end{aligned} \tag{7.8}$$

with final time $T = 40$ and periodic boundary conditions.

As a spatial discretization we use a Fourier collocation method with 128 points [14] and an IMEX scheme [1] as a time integrator, treating the linear terms implicitly

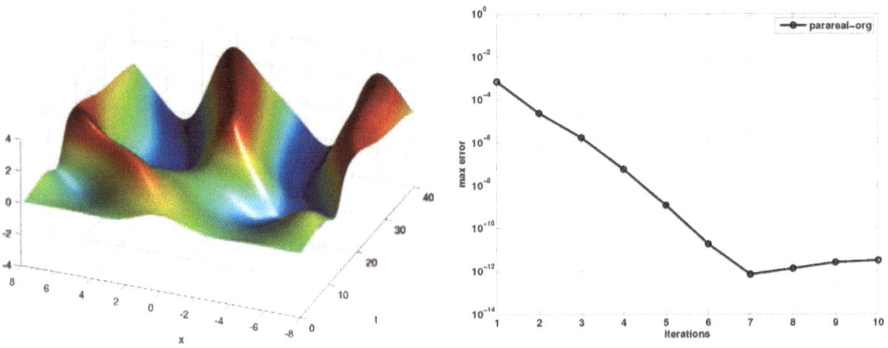


**Fig. 7.2.** The time evolution of the solution (left) and the $L_\infty$-error at $T = 40$ against the number of iterations (right) of the 1-D Kuramoto-Sivashinsky equation using the original parareal method

and the nonlinear term explicitly. The parameters in the parareal method are taken as

$$N_c = 100, \quad N_{it} = 5, \quad \Delta t = 10^{-2}, \quad \delta t = 10^{-4}. \tag{7.9}$$

Figure 7.2 (left) shows the time evolution of the chaotic solution to the Kuramoto-Sivashinsky equation with a Gaussian initial condition. In Fig. 7.2 (right), we show the $L_\infty$-error at $T = 40$ against the number of iterations. In this case, we take the solution computed by the fine solver as the exact solution. It is clear that the parareal solution converges, albeit at a slower rate. It should also be noted that $\Delta t/\delta t = 100$, indicating the potential for a substantial acceleration.

As a last and less encouraging example, we consider the 1-D advection equation

$$\begin{aligned} &u_t + a u_x = 0, \quad (x,t) \in (0,2\pi) \times (0,T], \\ &u(x,0) = \exp(\sin(x - at)), \end{aligned} \tag{7.10}$$

with a final time $T = 10$, $a = 2\pi$ and a $2\pi$-periodic boundary condition. We use a DG method of order 32 and 2 elements in space [15], a singly diagonal implicit fourth-order Runge-Kutta scheme in time (a five-stage fourth-order scheme, cf. S54b in [23]), and the parareal parameters:

$$N_c = 100, \quad N_{it} = 27, \quad \Delta t = 5 \times 10^{-2}, \quad \delta t = 10^{-4}. \tag{7.11}$$

Figure 7.3 shows the $L_\infty$-error at $T = 10$ against the number of iterations. The instability of the original parareal method is apparent, as has also been observed by others [3, 8, 22].

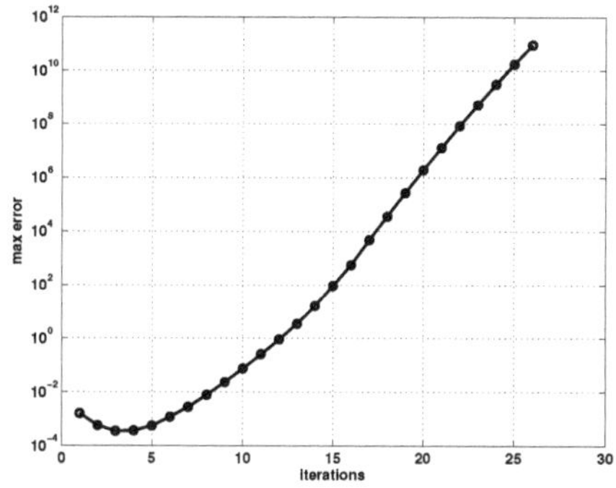


**Fig. 7.3.** The $L_\infty$-error at $T = 10$ against the number of iterations of the 1-D linear advection equation using the original parareal method

## *7.2.2 The Krylov Subspace Parareal Method*

We notice in Algorithm 7.1 that only $\{\mathbf{u}_{\mathbf{f}_{i+1}}^k\}_{i=0}^{N_c-1}$ is used in the advancement of the solution to $k+1$. To fix the stability issue, [10] proposed to improve the coarse solver by reusing information computed at all previous iterations and applied this idea to linear hyperbolic problems in structural dynamics. Recently, a similar idea was successfully applied to linear hyperbolic systems [21].

The basic idea of the Krylov subspace parareal method is to project $\mathbf{u}_i^{k+1}$ onto a subspace spanned by all numerical solutions integrated by the fine solver at previous iterations. Denote the subspace as

$$\mathbf{S}^k := \text{span}\{\mathbf{u}_{\mathbf{f}_i}^j,\ 1 \le i \le N_c,\ 1 \le j \le k\}. \tag{7.12}$$

The corresponding orthogonal basis set $\{\mathbf{s}_1, ..., \mathbf{s}_r\}$ is constructed through a QR factorization.

Denote $\mathbb{P}^k$ as the $L_2$-orthogonal projection onto $\mathbf{S}^k$. The previous coarse solver $G_{\Delta t}$ is now replaced by $K_{\Delta t}$ as:

$$K_{\Delta t}(\mathbf{u}, t_{i+1}, t_i) = G_{\Delta t}((\mathbb{I} - \mathbb{P}^k)\mathbf{u}, t_{i+1}, t_i) + F_{\delta t}(\mathbb{P}^k \mathbf{u}, t_{i+1}, t_i). \tag{7.13}$$

For a linear problem, $F_{\delta t}(\mathbb{P}^k \mathbf{u}, t_{i+1}, t_i)$ can be computed efficiently as

$$F_{\delta t}(\mathbb{P}^k \mathbf{u}, t_{i+1}, t_i) = F_{\delta t}(\sum_{j=1}^{N_c k} C_j \mathbf{s}_j, t_{i+1}, t_i) = \sum_{j=1}^{N_c k} C_j F_{\delta t}(\mathbf{s}_j, t_{i+1}, t_i), \tag{7.14}$$

where $F_{\delta t}(\mathbf{s}_j, t_{i+1}, t_i)$ are computed and stored once the $\mathbf{s}_j$'s are available. Since this approach essentially produces an approximation to the fine solver, the new coarse solver is expected to be more accurate than the old coarse solver. It was shown in [11] that as the dimension of $\mathbf{S}^k$ increases, $\mathbb{P}^k \to \mathbb{I}$ and $K_{\Delta t} \to F_{\delta t}$, thus achieving convergence. The algorithm outline is presented in Algorithm 7.2.

**Algorithm 7.2** The Krylov subspace parareal method

Initialization:
$\mathbf{u}_0^0 = \mathbf{u}_0$;
**for** $i \leftarrow 0$ **to** $N_c - 1$ **do**
  $\mathbf{u}_{i+1}^0 = G_{\Delta t}(\mathbf{u}_i^0, t_{i+1}, t_i)$
**end**
Iterations:
$k = 0$;
**for** $k \leftarrow 0$ **to** $N_{it}$ **do**
  Parallel predictor step:
  **for** $i \leftarrow 0$ **to** $N_c - 1$ **do**
    $\mathbf{u}_{f\,i+1}^{k} = F_{\delta t}(\mathbf{u}_i^k, t_{i+1}, t_i)$
  **end**
  Constructing reduced basis:
  Update $\mathbf{S}^{k-1}$ to $\mathbf{S}^k$ based on $\mathbf{u}_{i-1}^k, \mathbf{u_f}_i^k$
  Marching the basis:
  **for** $i \leftarrow 1$ **to** $N_r$ **do**
    $\mathbf{S}_{f_i} = F_{\delta t}(\mathbf{s}_i, 0, \Delta t)$;
  **end**
  Sequential correction step:
  **for** $i \leftarrow 0$ **to** $N_c - 1$ **do**
    $\mathbf{u}_{i+1}^{k+1} = K_{\Delta t}(\mathbf{u}_i^{k+1}, t_{i+1}, t_i) - \mathbf{u}_{f\,i+1}^{k} + K_{\Delta t}(\mathbf{u}_i^k, t_{i+1}, t_i)$
  **end**
**end**

To demonstrate the performance of the Krylov subspace parareal method, we use it to solve the linear advection equation, (7.10). In Fig. 7.4 (left) we show the $L_\infty$-error at $T = 10$ against the number of iterations. It is clear that the Krylov subspace parareal method stabilizes the parareal solver for this problem.

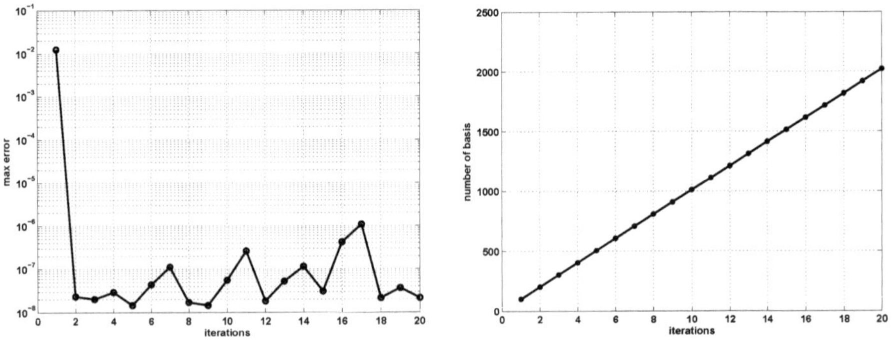


**Fig. 7.4.** The $L_\infty$-error at $T = 10$ against the number of iterations (left), and the number of bases (right) for solving the 1-D linear advection equation using the Krylov subspace parareal method

Two observations are worth making. First, the Krylov subspace parareal method needs to store all the values of $\mathbf{S}^k$ and $F(\mathbf{S}^k)$. As $k$ increases, this induces a memory requirement scaling $O(kN_cN)$ and this may be become a bottleneck as illustrated in Fig. 7.4 (right). Furthermore, the efficiency of the coarse solver depends critically on the assumption of linearity of the operator and it is not clear how to extend this framework to nonlinear problems. These constraints appear to limit the practicality of the method.

## 7.2.3 The reduced basis parareal method

Let us first recall a few properties of reduced basis methods that will subsequently serve as key elements of the proposed reduced basis parareal method.

### 7.2.3.1 Reduced Basis Methods

We are generally interested in solving the nonlinear ODE (7.1). As a system, the dimensionality of the problem can be very large, e.g., if the problem originates from a method-of-lines discretization of a nonlinear PDE, so to achieve a high accuracy, requiring a high number of degrees of freedom, $N$, and it is tempting to seek to identify an approximate model to enhance the computational efficiency without significantly impacting the accuracy.

A general representation of a reduced model in matrix-form is

$$\mathbf{u}(t) \approx \mathbf{V}_r\tilde{\mathbf{u}}(t), \tag{7.15}$$

where the $r$ columns of the matrix $\mathbf{V}_r$ represent a linear space - the reduced basis - and $\tilde{\mathbf{u}}(t) \in \mathbb{R}^r$ are the coefficients of the reduced model. Projecting the ODE system (7.1) onto $\mathbf{V}_r$, we recover the reduced system:

$$\mathbf{V}_r^T\mathbf{V}_r\frac{d\tilde{\mathbf{u}}(t)}{dt} = \mathbf{V}_r^T\mathbf{A}\mathbf{V}_r\tilde{\mathbf{u}}(t) + \mathbf{V}_r^T\mathbf{N}(\mathbf{V}_r\tilde{\mathbf{u}}(t)). \tag{7.16}$$

Assuming that $\mathbf{V}_r$ is orthonormal, this simplifies as

$$\frac{d\tilde{\mathbf{u}}(t)}{dt} = \mathbf{V}_r^T\mathbf{A}\mathbf{V}_r\tilde{\mathbf{u}}(t) + \mathbf{V}_r^T\mathbf{N}(\mathbf{V}_r\tilde{\mathbf{u}}(t)). \tag{7.17}$$

One is now left with specifying how to choose a good subspace, $\mathbf{V}_r$, to adequately represent the dynamic behavior of the solution and develop a strategy for how to recover the coefficients for the reduced model in an efficient manner. There are several ways to address this question, most often based on the construction of $\mathbf{V}_r$ through snapshots of the solution.

**Proper orthogonal decomposition.** The proper orthogonal decomposition (POD) [5, 6] is perhaps the most widely used approach to generate a reduced basis from a collection of snapshots. In this case, we assume we have a collection of $N_s$ snapshots

$$\mathbf{U} = [\mathbf{u}_1, ..., \mathbf{u}_{N_s}], \tag{7.18}$$

where each $\mathbf{u}_i$ is a vector of length $N$; this $N$ can be large as it reflects the number of degrees of freedom in system. The POD basis, denoted by $\{\phi_i\}_1^r \in \mathbb{R}^N$, is chosen as the orthonormal vectors that solve the minimization problem:

$$\min_{\phi_i \in \mathbb{R}^N} \sum_j^{N_s} ||\mathbf{u}_j - \sum_{i=1}^r (\mathbf{u}_j^T \phi_i)\phi_i||_2^2,$$
$$\text{subject to } \phi_i^T \phi_j = \delta_{ij} = \begin{cases} 1, \ i = j, \\ 0, \ \text{otherwise}. \end{cases} \tag{7.19}$$

The solution to this minimization problem is found through the singular value decomposition (SVD) of $\mathbf{U}$:

$$\mathbf{U} = \mathbf{V}\mathbf{\Sigma}\mathbf{W}^T, \tag{7.20}$$

where $\mathbf{V} \in \mathbb{R}^{N \times r}$ and $\mathbf{W} \in \mathbb{R}^{N_s \times r}$ are the left and right singular vectors, respectively, and $\mathbf{V}$ is the sought after basis. The entries of the diagonal matrix $\mathbf{\Sigma}$ provides a measure of the relative energy of each of the orthogonal vectors in the basis.

Once the basis is available, we can increase the computational efficiency for solving (7.17) by precomputing $\mathbf{V}_r^T \mathbf{A} \mathbf{V}_r$ of size $r \times r$. However, the computational complexity of the nonlinear term remains dependent on $N$ and, hence, potentially costly.

**Discrete Empirical Interpolation.** To address this, [7] proposed an approach, originating in previous work on empirical interpolation methods [4] but limited to the case of an existing discrete basis set. In this approach $\mathbf{N}(\mathbf{V_r}\tilde{\mathbf{u}}(t))$ is represented by $\tilde{\mathbf{N}}(t) \in \mathbb{R}^N$ which is subsequently approximated as

$$\mathbf{N}(\mathbf{V_r}\tilde{\mathbf{u}}(t)) \approx \tilde{\mathbf{N}}(t) \approx \mathbf{V_p}\mathbf{c}(t). \tag{7.21}$$

Here $\mathbf{V_p} = [\mathbf{v}_1, ..., \mathbf{v}_m]$ is an orthogonal POD basis set based on snapshots of $\mathbf{N}(t)$. To recover $\mathbf{c}(t)$, we seek a solution to an overdetermined system. However, rather than employing an expensive least square method, we extract $m$ equations from the original set of snapshots. Denote

$$\mathbf{P} = [\mathbf{e}_{p_1}, ...., \mathbf{e}_{p_m}] \in \mathbb{R}^{N \times m}, \tag{7.22}$$

where $\mathbf{e}_{p_1} = [0, ..., 0, 1, 0, ..., 0]^T \in \mathbb{R}^N$ (1 only appears on the $p_1$-th position of the vector). If $\mathbf{P}^T \mathbf{V_p}$ is nonsingular, $\mathbf{c}(t)$ can be uniquely determined by

$$\mathbf{P}^T \mathbf{N}(t) = \mathbf{P}^T \mathbf{V_p} \mathbf{c}(t),$$

resulting in a final approximation of $\tilde{\mathbf{N}}(t)$ as

$$\tilde{\mathbf{N}}(t) \approx \mathbf{V_p}(\mathbf{P}^T \mathbf{V_p})^{-1} \mathbf{P}^T \mathbf{N}(t).$$

The interpolation index $p_i$ is selected iteratively by minimizing the largest magnitude of the residual $\mathbf{r} = \mathbf{u}_k - \mathbf{V_{p,k}}\mathbf{c}$. The procedure, sometimes referred to as discrete empirical interpolation, is outlined in Algorithm 7.3.

**Algorithm 7.3** Empirical interpolation with a given discrete basis set

**input** : $\{\mathbf{v_k}\}_{k=1}^{m} \subset \mathbb{R}^N$ linearly independent POD bases of the nonlinear term
**output**: the interpolation operator $\mathbf{P}_m = [p_1, ..., p_m]$.

**begin**
  $\varepsilon = \max|\mathbf{u}_1|, p_1 = argmax|\mathbf{u}_1|$;
  $\mathbf{P} \leftarrow \{p_1\}$; $\mathbf{V_{p,1}} \leftarrow \{\mathbf{v}_1\}$;
  **for** $k \leftarrow 2$ **to** $M$ **do**
    Solve $\mathbf{P}^T\mathbf{v}_k = \mathbf{P}^T\mathbf{V_{p,k}c}(t)$ to obtain $\mathbf{c}(t)$;
    Compute the residual; $\mathbf{r} = \mathbf{v}_k - \mathbf{V_{p,k}c}$;
    $\varepsilon = \max|\mathbf{r}|, p_k = argmax|\mathbf{r}|$;
    $\mathbf{V}_{p,k} \leftarrow \mathbf{V}_{p,k-1} \cup \{\mathbf{v}_k\}$;
    $\mathbf{P}_k \leftarrow \mathbf{P}_{k-1} \cup \{p_k\}$;
  **end**
**end**

With the above approximation, we can now express the reduced system as

$$\frac{d\tilde{\mathbf{u}}(t)}{dt} = \mathbf{V}_r^T\mathbf{A}\mathbf{V}_r\tilde{\mathbf{u}}(t) + \mathbf{V}_r^T\mathbf{V_p}(\mathbf{P}^T\mathbf{V_p})^{-1}\mathbf{N}(\mathbf{P}^T\mathbf{V}_r\tilde{\mathbf{u}}(t)). \tag{7.23}$$

**Full Empirical Interpolation.** Pursuing the above approach further, one is left wondering if we can use a basis other than the computational expensive POD basis, and whether we can choose the interpolation position based on other guidelines. Addressing these questions leads us to propose a full empirical interpolation method.

It is well-known that the original empirical interpolation method is commonly used to separate the dependence of parameters and spatial variables [4], and that the method chooses 'optimal' interpolation points in a certain sense. We propose to consider time as a parameter, and use the empirical interpolation to construct the reduced bases $\mathbf{V_{E,k}}$ of $\mathbf{u}$ and the reduced bases $\mathbf{V_{pE,k}}$ of the nonlinear term, i.e.,

$$\mathbf{u}(t) \approx \mathbf{V_{E,k}c}(t), \quad \tilde{\mathbf{N}}(t) \approx \mathbf{V_{pE,k}c}(t). \tag{7.24}$$

The resulting reduced model can be written as

$$\frac{d\tilde{\mathbf{u}}(t)}{dt} = \mathbf{V_{E,k}}^T\mathbf{A}\mathbf{V_{E,k}}\tilde{\mathbf{u}}(t) + \mathbf{V_{E,k}}^T\mathbf{V_{pE,k}}(\mathbf{P}^T\mathbf{V_{pE,k}})^{-1}\mathbf{N}(\mathbf{P}^T\mathbf{V_{E,k}}\tilde{\mathbf{u}}(t)). \tag{7.25}$$

The essential difference between the models based on discrete empirical interpolation and the full empirical interpolation approach is found in the way in which one constructs the reduced basis set. In the former case, the importance of the basis elements is guided by the SVD and the relative size of the singular values, resulting in a potentially substantial cost. The latter case is based on the interpolation error and the basis in constructed in a full greedy fashion. A detailed comparative study of the performance between the two approaches is ongoing and will be presented in a forthcoming paper.

#### *7.2.3.2 The Reduced Basis Parareal Method*

Let us now introduce the new reduced basis parareal method. Our first observation is that the first term in (7.13) can be dropped under the assumption that the projection error vanishes asymptotically. Hence, for linear problems, we can replace $K_{\Delta t}$ by $\hat{K}_{\Delta t}$ as

$$\hat{K}_{\Delta t}(\mathbf{u}, t_{i+1}, t_i) = F_{\delta t}(\mathbb{P}^k \mathbf{u}, t_{i+1}, t_i) = \sum_{j=1}^{N_c k} C_j F_{\delta t}(\mathbf{s}_j, t_{i+1}, t_i). \tag{7.26}$$

This is essentially an approximation to the fine time integrator with an admissible truncation error. Keeping in mind that $F_{\delta t}$ is an expensive operation, we seek to reduce the dimension of $\mathbf{S}^k$ to achieve a better efficiency. If the solution to the ODE is sufficiently regular, it is reasonable to seek an $r$-dimensional subspace, $\mathbf{S}_r^k$ (the reduced basis space), of the original space $\mathbf{S}^k$. Now redefine $\mathbb{P}_r^k$ to be the orthogonal projection from $\mathbf{u}$ onto $\mathbf{S}_r^k$. Then (7.26) becomes

$$\hat{K}_{\Delta t}(\mathbf{u}, t_{i+1}, t_i) = F_{\delta t}(\mathbb{P}_r^k \mathbf{u}, t_{i+1}, t_i) = \sum_{j=1}^{r} C_j F_{\delta t}(\mathbf{s}_j, t_{i+1}, t_i), \tag{7.27}$$

which is essentially an approximation to the fine time integrator using the reduced model.

Consequently, our reduced basis parareal method for linear problems is as follows:

$$\mathbf{u}_{n+1}^{k+1} = F_{\delta t}(\mathbb{P}_r^k \mathbf{u}_n^{k+1}, t_{n+1}, t_n) + F_{\delta t}(\mathbf{u}_n^k, t_{n+1}, t_n) - F_{\delta t}(\mathbb{P}_r^k \mathbf{u}_n^k, t_{n+1}, t_n), \quad 0 \le k \le N_c - 1. \tag{7.28}$$

Depending on the construction of the reduced model, we refer to it as the POD parareal method or the EIM parareal method.

Algorithm 7.4 describes the basic steps of the reduced basis parareal method for linear problems. It follows a procedure similar to Algorithm 7.2, but requires less memory for storing the bases. Notice that for linear problems, the coarse solver is needed only for initializing the algorithm. After this first step, the fine solver produces all the information needed for the reduced model, and the algorithm no longer depends on the coarse solver.

For nonlinear problems, the relationship

$$F_{\delta t}(\mathbb{P}_r^k \mathbf{u}, t_{i+1}, t_i) = \sum_{j=1}^{r} C_j F_{\delta t}(\mathbf{s}_j, t_{i+1}, t_i) \tag{7.29}$$

does not generally hold, even if $\mathbb{P}^k \mathbf{u} \to \mathbf{u}$. Therefore, the Krylov subspace parareal method is not applicable. Fortunately, the knowledge of the development of reduced models using empirical interpolation offers insight into dealing with nonlinear problems, as mentioned in Sect. 7.2.3.1. We construct the coarse time integrator as follows:

$$\hat{K}_{\Delta t}(\mathbf{u}, t_{i+1}, t_i) = F_{\delta t}^r(\mathbb{P}_r^k \mathbf{u}, t_{i+1}, t_i), \tag{7.30}$$

**Algorithm 7.4** The reduced parareal method for a linear problem

Initialization:

$\mathbf{u}_0^0 = \mathbf{u}_0$;

**for** $i \leftarrow 0$ **to** $N_c - 1$ **do**

  $\mathbf{u}_{i+1}^0 = G_{\Delta t}(\mathbf{u}_i^0, t_{i+1}, t_i)$

**end**

Iterations:

$k = 0$;

**for** $k \leftarrow 0$ **to** $N_{it}$ **do**

  Parallel predictor step:

  **for** $i \leftarrow 0$ **to** $N_c - 1$ **do**

    $\mathbf{u}_{f_{i+1}}^k = F_{\delta t}(\mathbf{u}_i^k, t_{i+1}, t_i)$

  **end**

  Constructing reduced basis by POD or EIM:

  $U^k = \{\mathbf{u}_{f_{i+1}}^k, i = 0, ..., N_c, j = 0, ..., k\}$

  $\mathbf{S} = \text{POD}(U^k)$ or $\mathbf{S} = \text{EIM}(U^k)$ where $\mathbf{S} = \{\mathbf{s}_i, i = 1, ..., r\}$

  Marching the basis:

  **for** $i \leftarrow 1$ **to** $r$ **do**

    $\mathbf{S}_{f_i} = F_{\delta t}(\mathbf{s}_i, 0, \Delta t)$;

  **end**

  Sequential correction step:

  **for** $i \leftarrow 0$ **to** $N_c - 1$ **do**

    $\mathbb{P}^k \mathbf{u}_i^k = \Sigma_{j=1}^r C_j \mathbf{s}_j \leftarrow C_j$

    $\hat{K}_{\Delta t}(\mathbf{u}_i^{k+1}, t_{i+1}, t_i) = \Sigma_{j=1}^{N_r} C_j \mathbf{S}_{f_j}$

    $\mathbf{u}_{i+1}^{k+1} = \hat{K}_{\Delta t}(\mathbf{u}_i^{k+1}, t_{i+1}, t_i) - \mathbf{u}_{f_{i+1}}^k + \hat{K}_{\Delta t}(\mathbf{u}_i^k, t_{i+1}, t_i)$

  **end**

**end**

where $F_{\delta t}^r$ is the reduced model constructed by POD or EIM as we described in the previous section. Consequently, our reduced basis parareal method for nonlinear problems becomes

$$\mathbf{u}_{n+1}^{k+1} = F_{\delta t}^r(\mathbb{P}_r^k \mathbf{u}_n^{k+1}, t_{n+1}, t_n) + F_{\delta t}(\mathbf{u}_n^k, t_{n+1}, t_n) - F_{\delta t}^r(\mathbb{P}_r^k \mathbf{u}_n^k, t_{n+1}, t_n),$$
$$0 \leq k \leq N_c - 1. \tag{7.31}$$

As long as there exists a suitable reduced model for the problem, we can evaluate $\hat{K}_{\Delta t}$ efficiently while maintaining an accuracy commensurate with the fine solver. The reduced basis parareal method for nonlinear problems is outlined in Algorithm 7.5.

**Algorithm 7.5** The reduced parareal method for a nonlinear problem

Initialization:
$\mathbf{u}_0^0 = \mathbf{u}_0$;
**for** $i \leftarrow 0$ **to** $N_c - 1$ **do**
  $\mathbf{u}_{i+1}^0 = G_{\Delta t}(\mathbf{u}_i^0, t_{i+1}, t_i)$
**end**
Iterations:
$k = 0$;
**for** $k \leftarrow 0$ **to** $N_{it}$ **do**
  Parallel predictor step:
  **for** $i \leftarrow 0$ **to** $N_c - 1$ **do**
    $\mathbf{u}_{f\,i+1}^{\,k} = F_{\delta t}(\mathbf{u}_i^k, t_{i+1}, t_i)$
  **end**
  Constructing reduced basis:
  $U^k = \{\mathbf{u}_{f\,i+1}^{\,k}, i = 0, ..., N_c, j = 0, ..., k\}$
  $\mathbf{S} = \text{POD-DEIM}(U^k)$ or $\mathbf{S} = \text{EIM}(U^k)$ where $\mathbf{S} = \{\mathbf{s}_i, i = 1, ..., r\}$
  Sequential correction step:
  **for** $i \leftarrow 0$ **to** $N_c - 1$ **do**
    $\hat{K}_{\Delta t}(\mathbf{u}_i^k, t_{i+1}, t_i) = F_{\delta t}^r(\mathbb{P}_r^k \mathbf{u}_i^k, t_{i+1}, t_i)$
    $\hat{K}_{\Delta t}(\mathbf{u}_i^{k+1}, t_{i+1}, t_i) = F_{\delta t}^r(\mathbb{P}_r^k \mathbf{u}_i^{k+1}, t_{i+1}, t_i)$
    $\mathbf{u}_{i+1}^{k+1} = \hat{K}_{\Delta t}(\mathbf{u}_i^{k+1}, t_{i+1}, t_i) - \mathbf{u}_{f\,i+1}^{\,k} + \hat{K}_{\Delta t}(\mathbf{u}_i^k, t_{i+1}, t_i)$
  **end**
**end**

## 7.3 Analysis of the Reduced Basis Parareal Method

In the following we provide some analysis of the reduced basis parareal method to understand its stability, convergence and overall computational complexity. Throughout, we assume that there exists a reduced model for the continuous problem.

### *7.3.1 Stability analysis*

We first consider the linear case. Define the projection error:

$$g_j^k = ||(\mathbb{I} - \mathbb{P}_r^k)\mathbf{u}_j^k||_{L_2(0,T)}, \tag{7.32}$$

where $r$ is the dimension of the reduced space. We assume a projection error

$$g_j^k \leq \varepsilon, \quad \forall j, k, \tag{7.33}$$

and define:

$$C_{p,r} = \frac{\varepsilon}{\Delta T}, \quad \forall j, k. \tag{7.34}$$

It is reasonable to assume that the fine propagator is $L_2$ stable, i.e., there exists a nonnegative constant $C_F$ independent of the discretization parameters, such that,

$$||F_{\delta t}(\mathbf{v},t_{i+1},t_i)||_{L_2(0,T)} \leq (1+C_F\Delta T)||\mathbf{v}||_{L_2(0,T)}, \quad \forall \mathbf{v} \in L_2(0,T). \tag{7.35}$$

**Theorem 7.1 (Stability for the linear case)** *Under the assumption of* (7.33) *and* (7.35)*, the reduced basis parareal method is stable for* (7.1) *with* $\mathbf{N} \equiv \mathbf{0}$*, i.e., for each i and k,*

$$||\mathbf{u}_{i+1}^{k+1}||_{L_2(0,T)} \leq C_L e^{C_F(i+1)\Delta T}, \tag{7.36}$$

*where* $C_L$ *is a constant depending only on* $C_{p,r}$*,* $C_F$*, and* $\mathbf{u}_0$*.*

*Proof* Using the triangle inequality, linearity of the operator, and assumption (7.35), we obtain

$$\begin{aligned} ||\mathbf{u}_{i+1}^{k+1}||_{L_2(0,T)} \leq\, & ||F_{\delta t}(\mathbb{P}_r^k\mathbf{u}_i^{k+1},t_{i+1},t_i)||_{L_2(0,T)} + ||F_{\delta t}(\mathbf{u}_i^k,t_{i+1},t_i) \\ & - F_{\delta t}(\mathbb{P}_r^k\mathbf{u}_i^k,t_{i+1},t_i)||_{L_2(0,T)} \qquad (7.37) \\ \leq\, & (1+C_F\Delta T)||\mathbf{u}_i^{k+1}||_{L_2(0,T)} \\ & + (1+C_F\Delta T)||(\mathbb{I}-\mathbb{P}_r^k)\mathbf{u}_i^k||_{L_2(0,T)}. \qquad (7.38) \end{aligned}$$

Then, by the discrete Gronwall's lemma [9] and (7.33), we recover

$$\begin{aligned} ||\mathbf{u}_{i+1}^{k+1}||_{L_2(0,T)} \leq\, & (1+C_F\Delta T)^{i+1} \\ & \times (||\mathbf{u}_0^{k+1}||_{L_2(0,T)} + \Delta T\sum_{j=0}^{i}(1+C_F\Delta T)^{-j}C_{p,r}) \qquad (7.39) \\ =\, & (1+C_F\Delta T)^{i+1}||\mathbf{u}_0^{k+1}||_{L_2(0,T)} \\ & + \frac{1}{C_F}((1+C_F\Delta T)^{i+1}-1)C_{p,r} \qquad (7.40) \\ \leq\, & e^{C_F(i+1)\Delta T}||\mathbf{u}_0||_{L_2(0,T)} + \frac{1}{C_F}(e^{C_F(i+1)\Delta T}-1)C_{p,r}. \qquad (7.41) \end{aligned}$$

This completes the proof.

Note that if there exists an small integer $M$ (indicating a compact reduced approximation space) such that,

$$\lim_{r\to M} C_{p,r} = 0, \tag{7.42}$$

then we recover the same stability property as that of the fine solver:

$$||\mathbf{u}_{i+1}^{k+1}||_{L_2(0,T)} \leq e^{C_F(i+1)\Delta T}||\mathbf{u}_0||_{L_2(0,T)}.$$

For the nonlinear case, we further assume that there exists a nonnegative constant $C_r$, independent of the discretization parameters, such that,

$$||F_{\delta t}(\mathbf{v},t_{i+1},t_i) - F_{\delta t}^r(\mathbb{P}_r^k\mathbf{v},t_{i+1},t_i)||_{L_2(0,T)} \leq (1+C_r\Delta T)q_i^k, \quad \forall \mathbf{v} \in L_2(0,T), \tag{7.43}$$

where $q_i^k$ is the $L_2$-difference between the fine propagator and the reduced model using the same initial condition $\mathbf{v}$ at $t_i$. As before, we assume

$$q_j^k \le \varepsilon, \quad \forall j,k. \tag{7.44}$$

**Theorem 7.2 (Stability for the nonlinear case)** *Under assumptions* (7.35), (7.43) *and* (7.44), *the reduced basis parareal method is stable for* (7.1) *in the sense that for each i and k*

$$||\mathbf{u}_{i+1}^{k+1}||_{L_2(0,T)} \le C_N e^{C_\star (i+1)\Delta T}, \tag{7.45}$$

*where* $C_\star = \max\{C_F, C_r\}$ *and* $C_N$ *is a constant depending only on* $C_{p,r}$, $C_F$, $C_r$, *and* $\mathbf{u}_0$.

*Proof* Using the triangle inequality and assumptions (7.35) and (7.43), we have

$$\begin{aligned} ||\mathbf{u}_{i+1}^{k+1}||_{L_2(0,T)} \le\ & ||F_{\delta t}^r(\mathbb{P}_r^k \mathbf{u}_i^{k+1}, t_{i+1}, t_i)||_{L_2(0,T)} + ||F_{\delta t}(\mathbf{u}_i^k, t_{i+1}, t_i) \\ & - F_{\delta t}^r(\mathbb{P}_r^k \mathbf{u}_i^k, t_{i+1}, t_i)||_{L_2(0,T)} & (7.46)\\ \le\ & (1 + C_F \Delta T)||\mathbf{u}_i^{k+1}||_{L_2(0,T)} + (1 + C_r \Delta T) q_i^k. & (7.47) \end{aligned}$$

Next, by the discrete Gronwall's lemma and (7.44), we derive

$$\begin{aligned} ||\mathbf{u}_{i+1}^{k+1}||_{L_2(0,T)} \le\ & (1 + C_F \Delta T)^{i+1} \\ & \times (||\mathbf{u}_0^{k+1}||_{L_2(0,T)} + \Delta T \sum_{j=0}^{i} (1 + C_r \Delta T)^{-j} C_{p,r}) & (7.48)\\ =\ & (1 + C_F \Delta T)^{i+1} ||\mathbf{u}_0^{k+1}||_{L_2(0,T)} \\ & + \frac{1}{C_r}((1 + C_r \Delta T)^{i+1} - 1) C_{p,r} & (7.49)\\ \le\ & e^{C_F (i+1)\Delta T} ||\mathbf{u}_0||_{L_2(0,T)} + \frac{1}{C_r}(e^{C_r (i+1)\Delta T} - 1) C_{p,r}. & (7.50) \end{aligned}$$

This completes the proof.

### 7.3.2 Convergence analysis

To show convergence for the linear case, we first assume that there exists a nonnegative constant $C_F$, such that,

$$||F_{\delta t}(\mathbf{x}, t_{i+1}, t_i) - F_{\delta t}(\mathbf{y}, t_{i+1}, t_i)||_{L_2(0,T)} \le (1 + C_F \Delta T)||\mathbf{x} - \mathbf{y}||_{L_2(0,T)}, \quad \forall t_i > 0. \tag{7.51}$$

We define

$$w_j^k = ||(\mathbb{I} - \mathbb{P}_r^k)\mathbf{u}_j||_{L_2(0,T)}, \tag{7.52}$$

and assume that

$$w_j^k \le \varepsilon, \quad \forall j,k. \tag{7.53}$$

**Theorem 7.3 (Convergence for the linear case)** *Under assumption* (7.33), (7.42), (7.51), (7.53) *and* $\mathbf{N} \equiv \mathbf{0}$ *in* (7.1), *the reduced basis parareal solution converges to* $\mathbf{u}_{i+1}$ *for each i.*

*Proof* Using the reduced basis parareal formula and the linearity of the operator, we obtain

$$\begin{aligned}
\mathbf{u}_{i+1}^{k+1} - \mathbf{u}_{i+1} &= F_{\delta t}(\mathbb{P}_r^k \mathbf{u}_i^{k+1}, t_{i+1}, t_i) + F_{\delta t}(\mathbf{u}_i^k, t_{i+1}, t_i) \\
&\quad - F_{\delta t}(\mathbb{P}_r^k \mathbf{u}_i^k, t_{i+1}, t_i) - F_{\delta t}(\mathbf{u}_i, t_{i+1}, t_i) && (7.54) \\
&= F_{\delta t}(\mathbb{P}_r^k \mathbf{u}_i^{k+1}, t_{i+1}, t_i) - F_{\delta t}(\mathbb{P}_r^k \mathbf{u}_i, t_{i+1}, t_i) && (7.55) \\
&\quad + F_{\delta t}(\mathbf{u}_i^k, t_{i+1}, t_i) - F_{\delta t}(\mathbb{P}_r^k \mathbf{u}_i^k, t_{i+1}, t_i) && (7.56) \\
&\quad + F_{\delta t}(\mathbb{P}_r^k \mathbf{u}_i, t_{i+1}, t_i) - F_{\delta t}(\mathbf{u}_i, t_{i+1}, t_i). && (7.57)
\end{aligned}$$

By the triangular inequality and assumption (7.51), we recover

$$\begin{aligned}
||\mathbf{u}_{i+1}^{k+1} - \mathbf{u}_{i+1}||_{L_2(0,T)} &\leq (1 + C_F \Delta T)||\mathbf{u}_i^{k+1} - \mathbf{u}_i||_{L_2(0,T)} && (7.58) \\
&\quad + (1 + C_F \Delta T)||(\mathbb{I} - \mathbb{P}_r^k)\mathbf{u}_i^k||_{L_2(0,T)} && (7.59) \\
&\quad + (1 + C_F \Delta T)||(\mathbb{I} - \mathbb{P}_r^k)\mathbf{u}_i||_{L_2(0,T)}. && (7.60)
\end{aligned}$$

Finally by the discrete Gronwall's lemma, (7.33) and (7.53), we obtain

$$\begin{aligned}
||\mathbf{u}_{i+1}^{k+1} - \mathbf{u}_{i+1}||_{L_2(0,T)} &\leq (1 + C_F \Delta T)^{i+1}(||\mathbf{u}_0^{k+1} - \mathbf{u}_0||_{L_2(0,T)} && (7.61) \\
&\quad + \Delta T \sum_{j=0}^{i} (1 + C_F \Delta T)^{-j} C_{p,r} \\
&\quad + \Delta T \sum_{j=0}^{i} (1 + C_F \Delta T)^{-j} C_{p,r}) && (7.62) \\
&\leq 2\Delta T \sum_{j=0}^{i} (1 + C_F \Delta T)^{-j} C_{p,r} && (7.63) \\
&\leq \frac{2}{C_F}((1 + C_F \Delta T)^{i+1} - 1) C_{p,r} && (7.64) \\
&\leq \frac{2}{C_F}(e^{C_F (i+1) \Delta T} - 1) C_{p,r}, && (7.65)
\end{aligned}$$

which approaches zero as $r$ increases. This completes the proof.

For the nonlinear case, we must also assume that there exists a nonnegative constant $C_r$, such that,

$$\begin{aligned}
||F_{\delta t}(\mathbf{u}_i^k, t_{i+1}, t_i) - F_{\delta t}^r(\mathbb{P}_r^k \mathbf{u}_i^k, t_{i+1}, t_i)||_{L_2(0,T)} &\leq (1 + C_r \Delta T) q_i^k, \\
||F_{\delta t}(\mathbf{u}_i, t_{i+1}, t_i) - F_{\delta t}^r(\mathbb{P}_r^k \mathbf{u}_i, t_{i+1}, t_i)||_{L_2(0,T)} &\leq (1 + C_r \Delta T) p_i^k,
\end{aligned} \qquad (7.66)$$

where $q_i^k$ and $p_i^k$ represent the $L_2$-difference between the fine operator and the reduced solver using the same initial condition $\mathbf{u}_i^k$ and $\mathbf{u}_i$. As before, we assume that

$$p_j^k \leq \varepsilon, \quad \forall j,k. \tag{7.67}$$

**Theorem 7.4 (Convergence of the nonlinear case)** *Under assumptions* (7.42), (7.43), (7.44), (7.66) *and* (7.67), *the reduced basis parareal solution of* (7.1) *converges to* $\mathbf{u}_{i+1}$ *for each i.*

*Proof* Using the reduced basis parareal formula, we obtain

$$\begin{aligned}
\mathbf{u}_{i+1}^{k+1} - \mathbf{u}_{i+1} &= F_{\delta t}^r(\mathbb{P}_r^k \mathbf{u}_i^{k+1}, t_{i+1}, t_i) + F_{\delta t}(\mathbf{u}_i^k, t_{i+1}, t_i) \\
&\quad - F_{\delta t}^r(\mathbb{P}_r^k \mathbf{u}_i^k, t_{i+1}, t_i) - F_{\delta t}(\mathbf{u}_i, t_{i+1}, t_i) \qquad (7.68)\\
&= F_{\delta t}^r(\mathbb{P}_r^k \mathbf{u}_i^{k+1}, t_{i+1}, t_i) - F_{\delta t}^r(\mathbb{P}_r^k \mathbf{u}_i, t_{i+1}, t_i) \\
&\quad + F_{\delta t}(\mathbf{u}_i^k, t_{i+1}, t_i) - F_{\delta t}^r(\mathbb{P}_r^k \mathbf{u}_i^k, t_{i+1}, t_i) \\
&\quad + F_{\delta t}^r(\mathbb{P}_r^k \mathbf{u}_i, t_{i+1}, t_i) - F_{\delta t}(\mathbf{u}_i, t_{i+1}, t_i). \qquad (7.69)
\end{aligned}$$

By the triangular inequality and assumptions (7.66) and (7.43), we have

$$\begin{aligned}
||\mathbf{u}_{i+1}^{k+1} - \mathbf{u}_{i+1}||_{L_2(0,T)} &\leq (1 + C_F \Delta T)||\mathbf{u}_i^{k+1} - \mathbf{u}_i||_{L_2(0,T)} \\
&\quad + (1 + C_r \Delta T) q_i^k + (1 + C_r \Delta T) p_i^k. \qquad (7.70)
\end{aligned}$$

Then, by the discrete Gronwall's lemma, (7.44) and (7.67) we recover

$$\begin{aligned}
||\mathbf{u}_{i+1}^{k+1} - \mathbf{u}_{i+1}||_{L_2(0,T)} &\leq \frac{2}{C_r}((1 + C_r \Delta T)^{i+1} - 1) C_{p,r} \qquad (7.71)\\
&\leq \frac{2}{C_r}(e^{C_r (i+1) \Delta T} - 1) C_{p,r}, \qquad (7.72)
\end{aligned}$$

which approaches zero as $r$ increases under assumption (7.42).

For the above analysis it is worth emphasizing two points:

- The accuracy of the new parareal algorithm is $O(\varepsilon)$, since $C_{p,r}$ depends on $\varepsilon$ as a measure of the quality of the reduced model. We shall confirm this point by the numerical tests in Sect. 7.4.
- Theorem 7.3 and 7.4 indicate that if there exists a good reduced approximation space for the problem, the new parareal algorithm converges in one iteration.

### *7.3.3 Complexity Analysis*

Let us finally discuss the computational complexity of the reduced basis parareal method. Recall that the dimension of the reduced space is $r$ and that of the fine solution is $N$. This is assumed to be the same for the coarse and fine solvers although this may not be a requirement in general. The compression ratio is $R = r/N$. Following the notation of [21]: $\tau_{QR}(k), \tau_{RB}(k)$ (representing $\tau_{SVD}(k)$, $\tau_{EIM}(k)$, and $\tau_{DEIM}(k)$

in different scenarios) reflect computing times required by the corresponding operations at the $k$-th iteration. $\tau_c$ and $\tau_f$ is the time required by the coarse and fine solvers, respectively. $N_t = N_c N_f$ is the total number of time steps in one iteration with $N_c$ being the number of the coarse time intervals and $N_f$ the number of fine time steps on each coarse time interval. $N_p$ is the number of processors.

In [21], the speedup is estimated as

$$S(N_p) \approx \frac{N_t \tau_f}{N_c \tau_c + N_{it}(N_c \tau_c + N_t/N_p \tau_f) + N_{it} \tau_{QR}(it)} \tag{7.73}$$

$$= \frac{1}{(1+N_{it})(\frac{N_c}{N_t}\frac{\tau_c}{\tau_f}) + \frac{N_{it}\tau_{QR}(N_{it})}{N_t \tau_f} + \frac{N_{it}}{N_p}}. \tag{7.74}$$

In the reduced basis parareal method, $\tau_c = R^2 \tau_f$, since the complexity of the computation of the right hand side of system is $\mathbf{O}(r^2)$. In addition, $\tau_{QR}$ becomes $\tau_{SVD}$ or $\tau_{EIM}$. With this in mind, the speedup can be estimated as

$$S(N_p) = \frac{1}{(1+N_{it})(\frac{N_c}{N_t} R^2) + \frac{N_{it}\tau_{RB}(N_{it})}{N_t \tau_f} + \frac{N_{it}}{N_p}}. \tag{7.75}$$

Next, we examine the first two terms in the denominators of (7.74) and (7.75).

- In the first term, $\tau_c/\tau_f$ takes the role of $R^2$. Hence, we can achieve a comparable performance, if $R \approx \sqrt{\tau_c/\tau_f}$, i.e, if the underlying PDE solution can be represented by a reduced basis set of size $\mathbf{O}(\sqrt{\tau_c/\tau_f} N)$. Suppose that $\sqrt{\tau_c/\tau_f} = \sqrt{1/20} \approx 0.23$. This requires that $R < 1/4$, which is a reasonable compression ratio for many problems. In addition, it is possible to use a reduced basis approximation to achieve a better performance for cases where CFL conditions lead to restrictions for the coarse solver.
- For the second term, $\tau_{SVD} \approx \tau_{QR} \approx \mathbf{O}(N N_{it}^2 N_c^2)$, while $\tau_{EIM} \approx \mathbf{O}(r^3/2N_{it}N_c + rNN_{it}N_c)$. Therefore, $\tau_{SVD}/\tau_{EIM} \approx \mathbf{O}(2N_{it}N_c/Rr^2)$. As $N_c$ increases, $\tau_{EIM}$ becomes smaller. In addition, EIM has very good parallel efficiency and requires less memory during the computation.

Also note that $N_{it}$ would typically be different for the reduced basis parareal method and the original parareal method. If a reduced space exists, the modified algorithm usually converges within a few iterations, hence accelerating the overall convergence significantly.

## 7.4 Numerical Results

In the following, we demonstrate the feasibility and efficiency of the reduced basis parareal method for both linear and nonlinear problems. We generally use the solution obtained from the fine time integrator as the exact solution.

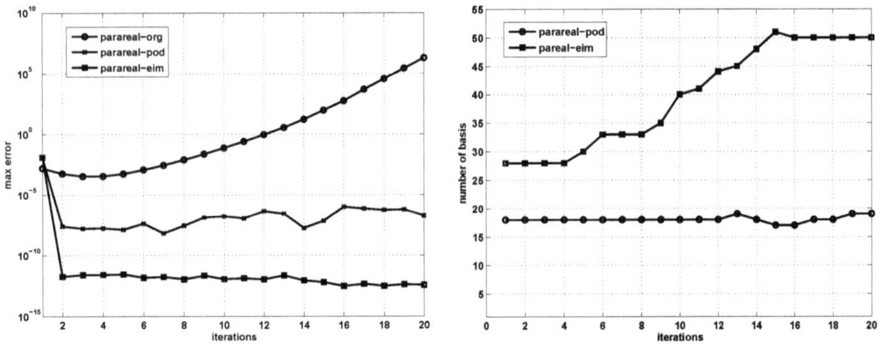


**Fig. 7.5.** POD parareal method, and the EIM parareal method for the 1-D advection equation. On the left we show the $L_\infty$-error at $T = 10$ against the number of iterations while the right shows the number of bases used for satisfying the tolerance of $\varepsilon(10^{-13})$ in the POD and EIM parareal methods across the iterations

## *7.4.1 The Linear Advection Equation*

We begin by considering the performance of the reduced basis parareal method and illustrate that it is stable for the 1-D linear advection equation (7.10). The spatial and temporal discretizations are the same as used in Sect. 7.2 and parameters in (7.11) are used.

In Fig. 7.5 (left), we show the $L_\infty$-error at $T = 10$ against the number of iterations for the original parareal method, the POD parareal method, and the EIM parareal method. The accuracy of the fine time integrator at $T = 10$ is $4 \times 10^{-13}$. The original parareal method is clearly unstable, while the other two remain stable. The very rapid convergence of the reduced basis parareal method reflects that the accuracy of reduced model is very high for this simple test case. As we will see for more complex nonlinear problems, this behavior does not carry over to general problems unless a high-accuracy reduced model is available.

In Fig. 7.5 (right), we show the number of bases used to satisfy the tolerance $\varepsilon$ in the POD parareal method and the EIM parareal method. Here $\varepsilon$ in the POD context is defined as the relative energy in the truncated mode and in the EIM context it is the interpolation error. In both cases, the tolerance in the basis selection using POD or EIM is set to $10^{-13}$. We note that the EIM parareal method achieves higher accuracy but requires more memory to store the bases. This suggests that one can explore a tradeoff between accuracy and efficiency for a particular application.

**Remark 7.1** It should be noted that if only snapshots from the previous iteration is used in the EIM basis construction, the scheme becomes unstable. However, when including all snapshots collected up to the previous iteration level, stability is restored.

Figure 7.6 (upper left) shows the convergence behavior of the EIM parareal algorithm with different tolerances ($\varepsilon = 10^{-k}, k = 2, 4, 6, 8, 10, 12$). The convergence

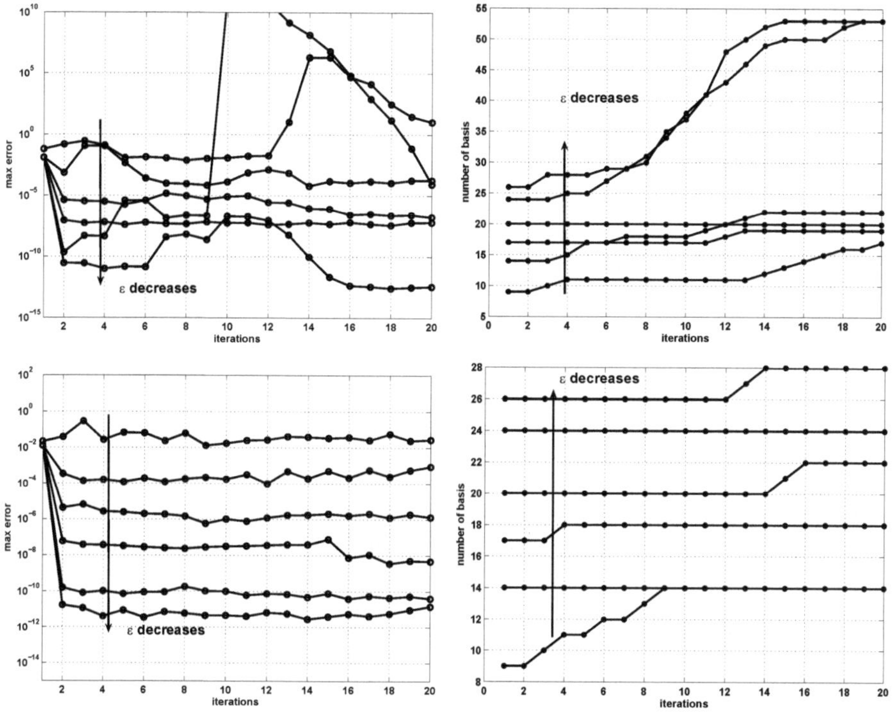


**Fig. 7.6.** The performance of the EIM parareal method for the 1-D advection equation against the tolerance used in the design of the reduced basis. On the upper left we show the $L_\infty$-error at $T = 10$ against the number of iterations as the tolerance $\varepsilon$ decreases and on the upper right the number of bases used for satisfying the tolerance as $\varepsilon$ decreases, where $\varepsilon = 10^{-k}, k = 2,4,6,8,10,12$; On the lower left and right, we show the corresponding convergence results and the number bases with the reorthogonalization procedure of the evolved basis

stagnates at a certain level and instability may set in after further iterations. There are two reasons for this: 1) as $\varepsilon$ becomes small, the reduced bases may become linear dependent, leading to a bad condition number of the related matrices that may impact stability; 2) the newly evolved reduced bases $\mathbf{S}_{f_i}$ for the fine solution may not be within $\mathbf{S}$ anymore. To resolve this problem, we first perform the reorthogonalization of the reduced bases to obtain a new space $\tilde{\mathbf{S}}$ and then project the newly evolved solution $\hat{K}_{\Delta t}(\mathbf{u}_i^{k+1}, t_{i+1}, t_i)$ back to $\tilde{\mathbf{S}}$. In Fig. 7.6 (lower left) we show the convergence results following this approach. Most importantly, stability is restored. Furthermore, the dependence of the final accuracy on $\varepsilon$ is clear. These results are consistent with Theorem 7.3, stating that the parareal solution converges to the serial solution integrated by the fine solver as long as the subspace $\mathbf{S}$ saturates in terms of accuracy. In practice, one can choose $\varepsilon$ such that the accuracy of the parareal solution and the serial fine solution are comparable.

### 7.4.2 The second order wave equation

To further evaluate the stability of the new parareal algorithm, we consider the second-order wave equation from [8]:

$$\begin{aligned} u_{tt} &= c^2 u_{xx}, \quad (x,t) \in (0,2\pi) \times (0,T], \\ u(x,0) &= f(x), \quad u_t(x,0) = g(x), \end{aligned} \tag{7.76}$$

where $T = 10$ and $c = 5$ and a $2\pi$-periodic boundary condition is used. The initial conditions are set as

$$f(x) = \sum_{l=-N}^{N} \hat{u}_l e^{ilx}, \quad g(x) = 0 \tag{7.77}$$

and

$$\hat{u}_l = \begin{cases} \frac{1}{|l|^p}, & l \neq 0, \\ 0 & l = 0. \end{cases}$$

and set $p = 4$. In the following we use a Fourier spectral discretization with 33 modes in space [14] and the velocity Verlet algorithm in time [24]. The following parameters are used in the parareal algorithm:

$$N_c = 100, \quad N_{it} = 10, \quad \Delta t = 10^{-3}, \quad \delta t = 10^{-4}. \tag{7.78}$$

The tolerance for POD is set to $10^{-11}$, respectively.

In Fig. 7.7 (left), we show the $L_\infty$-error at $T = 10$ against the number of iterations for the original parareal method and the POD parareal method. The original parareal method is clearly unstable, while the POD parareal remains stable and converges in one iteration. This confirms our analysis: if the reduced model is accurate enough, the reduced basis parareal should converge in one iteration. In Fig. 7.5 (right), we

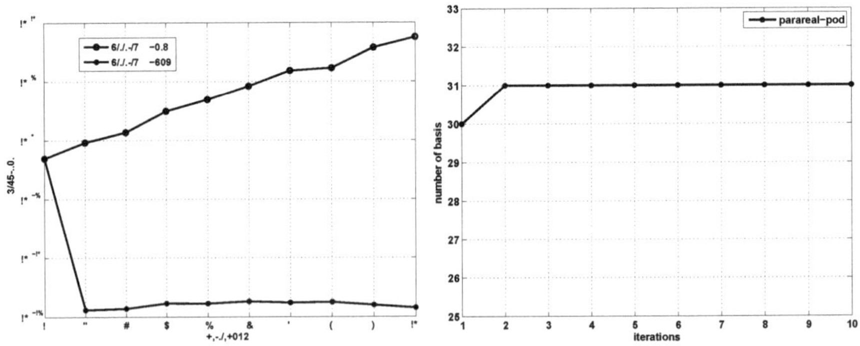


**Fig. 7.7.** Results obtained using the original parareal method, the POD parareal method for the 1-D second order wave equation. On the left we show the $L_\infty$-error at $T = 10$ against the number of iterations while the right shows the number of bases used for satisfying the tolerance of $\varepsilon(10^{-11})$ in the POD parareal method across the iterations

show the number of bases needed to satisfy the tolerance $\varepsilon$ in the POD parareal method.

### 7.4.3 Nonlinear Equations

Let us also apply the reduced basis parareal method to examples with nonlinear PDEs. We recall that the Krylov based approach is not applicable in this case.

#### 7.4.3.1 Viscous Burgers' Equation

We first consider the viscous Burgers' equation (7.6), with the same spatial and temporal discretization and the same parameters as in (7.7). To build the reduced basis, we set the tolerance for POD and EIM to be $10^{-15}$ and $10^{-10}$, respectively.

In Fig. 7.8 (left), we show the $L_\infty$-error at $T = 2$ against the number of iterations for the original parareal method, the POD parareal method, and the EIM parareal method. Note that in this case, the RB parareal performs worse than the original parareal does. It is a result of the reduced model not adequately capturing the information of the fine solver. Recall that in the nonlinear case, we have to deal with two approximations: one for the state variables and one for the nonlinear term. For the POD parareal algorithm, we choose the number of reduced bases based on the tolerance for the state variable $u$; alternatively, we can choose the dimension of the reduced approximation space based on the tolerance for the nonlinear term. The latter approach shows better convergence behavior in Fig. 7.8 (left, parareal-pod-modified). It is apparent that the quality of the reduced model directly impacts the convergence.

We emphasize that although the reduced basis parareal method converges slower than the original parareal, it is less expensive, as discussed in Sect. 7.2.3.1.

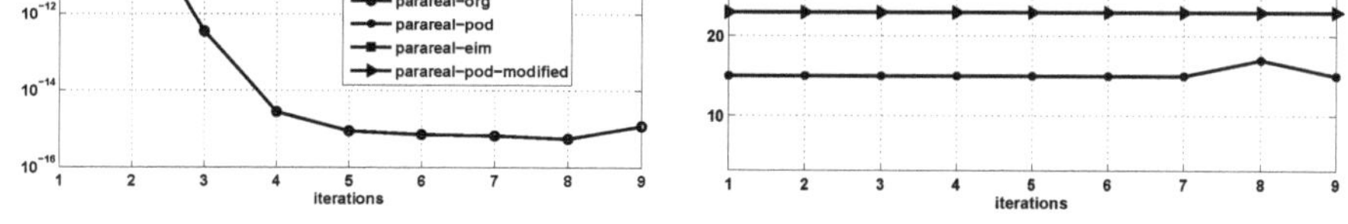


**Fig. 7.8.** We compare the performance of the original parareal method, the POD parareal method, the modified POD parareal and the EIM parareal method for the 1-D Burgers' equation. On the left we show the $L_\infty$-error at $T = 2$ against the number of iterations, while the right illustrates the number of bases

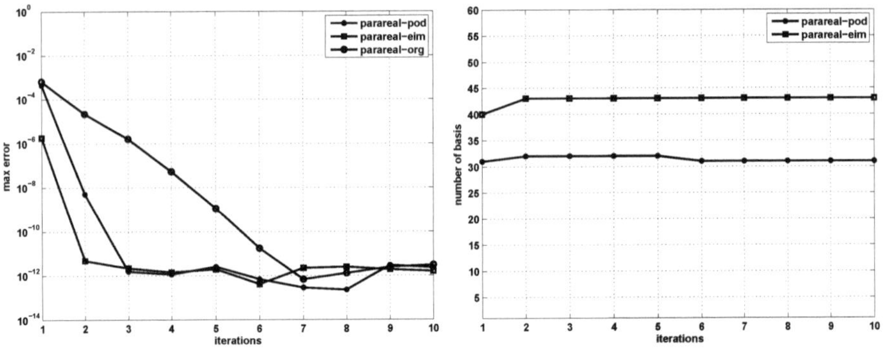


**Fig. 7.9.** We compare the performance of the original parareal method, the POD parareal method, and the EIM parareal method for the 1-D Kuramoto-Sivashinsky equation. On the left we show the $L_\infty$-error at $T = 40$ against the number of iterations, while the right shows the number of bases used against the number of iterations

#### *7.4.3.2 Kuramoto-Sivashinsky Equation*

Next we consider the Kuramoto-Sivashinsky equation (7.8). The same spatial and temporal discretization and the same parameters as in (7.9) are used. To build the reduced basis, we set the tolerance for POD and EIM to be $10^{-13}$ and $10^{-8}$, respectively.

In Fig. 7.9 we show the $L_\infty$-error at $T = 40$ against the number of iterations for the original parareal method, the POD parareal method, the modified POD parareal, and the EIM parareal method. It is clear that the reduced basis parareal method converges faster than the original parareal method. This is likely caused by the solution of the problem being smooth enough to ensure that there exists a compact reduced model. Moreover, to keep the corresponding tolerance, the number of degrees of freedom in the reduced basis parareal methods is roughly one-third that of the original parareal method.

#### *7.4.3.3 Allan-Cahn Equation: Nonlinear Source*

As a third nonlinear example we consider the 1-D Allan-Cahn equation:

$$\begin{aligned} \frac{\partial u}{\partial t} &= \nu u_{xx} + u - u^3, \quad (x,t) \in (0,2\pi) \times (0,T], \\ u(x,0) &= 0.25\sin(x), \end{aligned} \tag{7.79}$$

where $T = 2$ and $\nu = 2, 1, 10^{-1}, 10^{-2}$. A periodic boundary condition is assumed. We use a $P_1$ DG method with 100 elements in space [15] and a forward Euler scheme in time. The following parameters are used in the parareal algorithm

$$N_c = 200, \quad N_{it} = 5, \quad \Delta t = 1 \times 10^{-4}, \quad \delta t = 5 \times 10^{-6}. \tag{7.80}$$

We set the tolerance for POD and EIM to be $10^{-12}$ and $10^{-8}$, respectively.

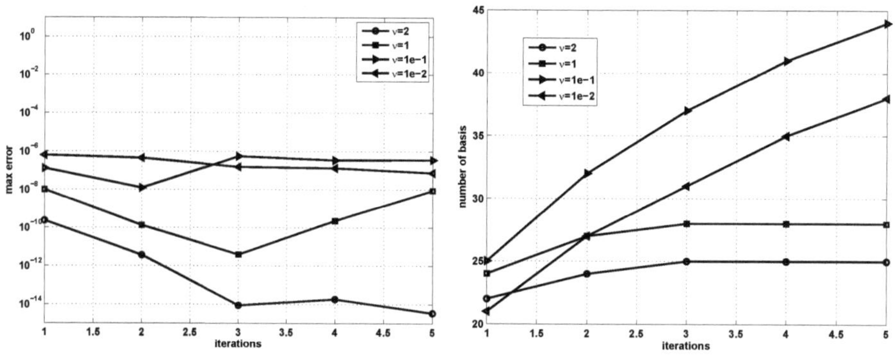


**Fig. 7.10.** The POD parareal method for the 1-D Allan-Cahn equation. On the left we show the $L_\infty$-error at $T=2$ against the number of iterations for different values of $\nu$ and on the right we show the number of bases

In Fig. 7.10 (left), we show the $L_\infty$-error at $T=2$ against the number of iterations for the POD parareal method for different values of $\nu$'s. It is clear that for larger values of $\nu$, the solution converges faster and less elements in the reduced basis is needed. This is expected since a larger $\nu$ indicates a smoother and more localized solution which is presumed to allow for an efficient representation in a lower dimensional space. Similar results are obtained by an EIM based parareal approach and are not reproduced here.

### *7.4.3.4 KdV Equation: Nonlinear Flux*

As a last example we consider the KdV equation (taken from [26]):

$$\begin{aligned} \frac{\partial u}{\partial t} &= -(\frac{u^2}{2})_x - \nu u_{xxx}, \quad (x,t) \in (-1,1) \times (0,T], \\ u(x,0) &= 1.5 + 0.5\sin(2\pi x), \end{aligned} \tag{7.81}$$

where $T=2$ and $\nu = 10^{-3}$ and we assume a periodic boundary condition. The equation conserves energy, much like the linear wave equation, but the nonlinearity induces a more complex behavior with the generation of propagating waves. In the parareal algorithm we use

$$N_c = 100, \quad N_{it} = 10, \quad \Delta t = 10^{-4}, \quad \delta t = 10^{-5}. \tag{7.82}$$

We use a first order local discontinuous Galerkin method (LDG) with 100 elements in space [15, 26] and an IMEX scheme in time [1], with the linear terms treated implicitly and the nonlinear term explicitly. We set the tolerance for POD and EIM to be $10^{-13}$ and $10^{-8}$, respectively.

In Fig. 7.11 (left) we show the $L_\infty$-error at $T=2$ against the number of iterations for the original parareal method, the POD parareal method, and the EIM parareal

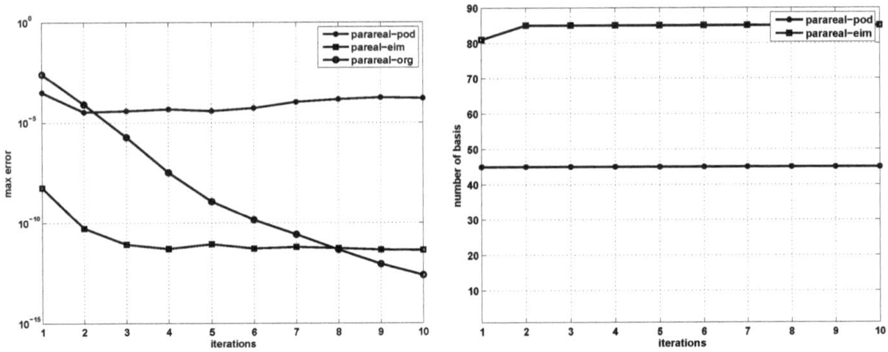


**Fig. 7.11.** We compare the performance of the original parareal method, the POD parareal method, and the EIM parareal method for the 1-D KdV equation. On the left we show the $L_\infty$-error at $T = 2$ against the number of iterations, while the right shows the number of bases used against the number of iterations

method. While the POD parareal method does not work well in this case, the EIM parareal method shows remarkable performance, i.e., it converges much faster than the original parareal method. Note that even if the tolerance for the POD is smaller than that of the EIM, it does not guarantee that the reduced model error based on the POD approach is smaller. There are two reasons: 1) the meaning of the tolerance in the context of the POD and the EIM are different. 2) in the convergence proof of (7.71), the constants $C_r, C_{p,r}$ depend on the details of the reduced approximation and the dimension of reduced approximation space, which impact the final approximation error.

## 7.5 Conclusions

In this paper, we propose an approach to produce and use a reduced basis method to replace the coarse solver in the parareal algorithm. We demonstrate that, as compared with the original parareal method, this new reduced basis parareal method has improved stability characteristics and efficiency, provided that the solution can be represented well by a reduced model. The analysis of the method is confirmed by the computational results, e.g., the accuracy of the parareal method is determined by the accuracy of the fine solver and the reduced model, used to replace the coarse solver. Unlike the Krylov subspace parareal method, this approach can be extended to include both linear problems and nonlinear problems, while requiring less storage and computing resources. The robustness and versatility of the method has been demonstrated through a number of different problems, setting the stage for the evaluation on more complex problems.

**Acknowledgements** The authors acknowledge partial support by OSD/AFOSR FA9550-09-1-0613 and AFOSR FA9550-12-1-0463.

## References

1. Ascher, U.M., Ruuth, S.J., Wetton, B.T.R.: Implicit-explicit methods for time-dependent partial differential equations. SIAM J. Numer. Anal. **32**(3), 797–823 (1995)
2. Baffico, L., Bernard, S., Maday, Y., Turinici, G., Zérah, G.: Parallel-in-time molecular-dynamics simulations. Physical Review E **66**(5) (2002)
3. Bal, G.: On the Convergence and the Stability of the Parareal Algorithm to Solve Partial Differential Equations. In: Domain decomposition methods in science and engineering, pp. 425–432. Lecture Notes in Computational Science and Engineering, Vol. 40. Springer-Verlag, Berlin Heidelberg (2005)
4. Barrault, D., Maday, Y., Nguyen, N., Patera, A.: An empirical interpolation method: application to efficient reduced-basis discretization of partial differential equations. Comptes Rendus Mathematique **339**(9), 667 (2004)
5. Cavar, D., Meyer, K.E.: LES of turbulent jet in cross flow: Part 2 POD analysis and identification of coherent structures. Inter. J. Heat Fluid Flow **36**, 35–46 (2012)
6. Chatterjee, A.: An introduction to the proper orthogonal decomposition. Current Science-Bangalore **78**(7), 808 (2000)
7. Chaturantabut, S., Sorensen, D.: Nonlinear model reduction via discrete empirical interpolation. SIAM Journal on Scientific Computing **32**(5), 2737 (2010)
8. Dai, X., Maday, Y.: Stable parareal in time method for first and second order hyperbolic system. arXiv preprint arXiv:1201.1064 (2012)
9. Emmerich, E.: Discrete versions of Gronwall's lemma and their application to the numerical analysis of parabolic problems, 1st ed.. TU, Fachbereich 3, Berlin (1999)
10. Farhat, C.: Cortial, J.: Dastillung, C., Bavestrello, H.: Time-parallel implicit integrators for the near-real-time prediction of linear structural dynamic responses.. International journal for numerical methods in engineering **67**(5), 697 (2006)
11. Gander, M., Petcu, M.: in ESAIM: Analysis of a Krylov subspace enhanced parareal algorithm for linear problems. Proceedings, vol. 25, pp. 114–129 (2008)
12. Gander, M., Vandewalle, S.: Analysis of the parareal time-parallel time-integration method. SIAM Journal on Scientific Computing **29**(2), 556 (2007)
13. He, L.: The reduced basis technique as a coarse solver for parareal in time simulations. J. Comput. Math **28**, 676 (2010)
14. Hesthaven, J.S., Gottlieb, S., Gottlieb, D.: Spectral Methods for Time-Dependent Problems. Cambridge University Press, Cambridge, UK (2007)
15. Hesthaven, J.S., Warburton, T.: Nodal Discontinuous Galerkin Methods: Algorithms, Analysis, and Applications. Springer-Verlag, New York (2008)
16. Lions, J., Maday, Y., Turinici, G.: A "parareal" in time discretization of pde's. Comptes Rendus de l'Academie des Sciences Series I Mathematics **332**(7), 661 (2001)
17. Maday, Y., Turinici, G.: Parallel in time algorithms for quantum control: Parareal time discretization scheme. International journal of quantum chemistry **93**(3), 223 (2003)
18. Maday, Y.: Parareal in time algorithm for kinetic systems based on model reduction. High-dimensional partial differential equations in science and engineering **41**, 183
19. Nielsen, A.S.: Feasibility study of the parareal algorithm. MSc thesis, Technical University of Denmark (2012)

20. Rozza, G., Huynh, D., Patera, A.T.: Reduced basis approximation and a posteriori error estimation for affinely parametrized elliptic coercive partial differential equations. Archives of Computational Methods in Engineering **15**(3), 229 (2008)
21. Ruprecht, D., Krause, R.: Explicit parallel-in-time integration of a linear acoustic-advection system. Computers & Fluids **59**, 72 (2012)
22. Staff, G.; Rønquist, E.: Stability of the parareal algorithm. Domain decomposition methods in science and engineering pp. 449–456 (2005)
23. Skvortsov, L.M.: Diagonally implicit Runge-Kutta methods for stiff problems. Computational Mathematics and Mathematical Physics **46**(12), 2110 (2006). DOI 10.1134/S0965542506120098. http://www.springerlink.com/index/10.1134/S0965542506120098
24. Verlet, L.: Computer "experiments" on classical fluids. i. thermodynamical properties of lennard-jones molecules. Physical review **159**(1), 98 (1967)
25. Xu, Y., Shu, C.: Local discontinuous galerkin methods for the Kuramoto-Sivashinsky equations and the Ito-type coupled KdV equations. Comp. Methods Appl. Mech. Engin. **195**(25), 3430-3447 (2006)
26. Yan, J., Shu, C.: A local discontinuous Galerkin method for KdV type equations. SIAM J. Num. Anal. **40**(2), 769–791 (2002)

# 8

# On the Stability of Reduced-Order Linearized Computational Fluid Dynamics Models Based on POD and Galerkin Projection: Descriptor vs Non-Descriptor Forms

**David Amsallem and Charbel Farhat**

**Abstract** The Galerkin projection method based on modes generated by the Proper Orthogonal Decomposition (POD) technique is very popular for the dimensional reduction of linearized Computational Fluid Dynamics (CFD) models, among many other typically high-dimensional models in computational engineering. This, despite the fact that it cannot guarantee neither the optimality nor the stability of the Reduced-Order Models (ROMs) it constructs. Short of proposing any variant of this model order reduction method that guarantees the stability of its outcome, this paper contributes a best practice to its application to the construction of linearized CFD ROMs. It begins by establishing that whereas the solution snapshots computed using the descriptor and non-descriptor forms of the discretized Euler or Navier-Stokes equations are identical, the ROMs obtained by reducing these two alternative forms of the governing equations of interest are different. Focusing next on compressible fluid-structure interaction problems associated with computational aeroelasticity, this paper shows numerically that the POD-based fluid ROMs originating from the non-descriptor form of the governing linearized CFD equations tend to be unstable, but their counterparts originating from the descriptor form of these equations are typically stable and reliable for aeroelastic applications. Therefore, this paper argues that whereas many computations are performed in CFD codes using the non-descriptor form of discretized Euler and/or Navier-Stokes equations, POD-based model reduction in these codes should be performed using the descriptor form of these equations.

D. Amsallem (✉)
Department of Aeronautics and Astronautics, Durand Building, 496 Lomita Mall, Stanford University, Stanford, 94305-4035, USA
e-mail: amsallem@stanford.edu

C. Farhat
Department of Aeronautics and Astronautics, Department of Mechanical Engineering, and Institute for Computational & Mathematical Engineering, Durand Building, 496 Lomita Mall, Stanford University, Stanford, 94305-4035, USA
e-mail: cfarhat@stanford.edu

A. Quarteroni, G. Rozza (eds.): *Reduced Order Methods for Modeling and Computational Reduction*, MS&A 9. DOI 10.1007/978-3-319-02090-7_8, 

## 8.1 Introduction

Linearized Computational Fluid Dynamics (CFD) models are ubiquitous in many applications pertaining to fluid dynamics. These include flow control, sensitivity analysis, shape optimization, flow stability analysis, and dynamic fluid-structure perturbation problems such as flutter, among others. In general, these computational models are less CPU intensive than their nonlinear counterparts. Nevertheless, because of the large dimensionality of these CFD models and the time-criticality of the aforementioned applications, there is a growing interest in developing Model Order Reduction (MOR) methods for constructing Reduced-Order Models (ROMs) that can capture the main characteristics of their high-dimensional counterparts at a fraction of the CPU cost they entail. A large class of such MOR methods is based on projection methods. These map a large number of degrees of freedom to a small number of generalized coordinates using a right Reduced-Order Basis (ROB). They also constrain the residual resulting from this approximation to be orthogonal to a left ROB.

The Proper Orthogonal Decomposition (POD) [27] – also known as the Singular Value Decomposition (SVD) – is a non-intrusive technique for generating a right ROB. Galerkin projections – that is, projections using identical left and right ROBs – with POD "modes" constitute a popular mean for constructing CFD-based linear ROMs [1,2,5,11,14,23,29]. This, despite the fact this approach for model reduction does not guarantee neither the optimality nor the stability of the ROMs it produces. To address the issue of ROMs constructed without a guaranteed stability, stabilization methods [4, 6] have been developed. In the specific context of CFD applications, more intrusive POD-based techniques have also been successfully developed for MOR. As it can be expected, each of these alternative approaches for restoring or guaranteeing stability has advantages and shortcomings.

Alternatively, this paper sheds some light on the behavior of the basic POD-based Galerkin projection method for CFD applications. It also proposes a best practice for reducing the occurrence of unstable POD-based linear ROMs that has proved to be effective for a large number of CFD problems. It conjectures that a large number of these occurrences is promoted by the application of the reduction process to the non-descriptor form [24] of the governing CFD equations. This form of an Ordinary Differential Equation (ODE) (or a set of them), which is also known as the "autonomous form of an ODE system," is characterized by the identity matrix as the coefficient of the term with the highest derivative. It is popular in many computational engineering applications including multibody dynamics [15, 17], molecular dynamics [26], and CFD [2,7,16,20,23]. This paper also shows numerically that, on the other hand, when MOR is applied to the descriptor form [24] of the governing equations, stable CFD ROMs are typically obtained.

To this effect, the remainder of this paper is organized as follows. Section 8.2 sets the stage for linearized Arbitrary Lagrangian Eulerian (ALE) CFD problems with moving boundaries and their semi-discretization by a finite volume method. The emphasis on moving boundaries is due to their predominant role in generating unsteady flows – even in the absence of turbulence – and their importance in dynamic fluid-structure applications. This section also introduces the descriptor and non-descriptor forms of a Linear Time-Invariant (LTI) system. Section 8.3 overviews the POD-based Galerkin projection model in the context of linearized CFD problems. More specifically, it shows that whereas the snapshot solutions computed using either the descriptor or non-descriptor form of a CFD-based LTI system are identical, the linear ROMs obtained by reducing both forms of this system using a Galerkin projection method are different. Section 8.4 focuses on realistic dynamic fluid-structure interaction problems to illustrate the formulated conjecture. It also highlights the robustness of Galerkin projections with POD modes when applied to the descriptor form of the governing fluid equations. Finally, Sect. 8.5 summarizes and this paper and concludes it.

## 8.2 Linearized CFD-Based Analysis

### *8.2.1 Governing Equations in Descriptor Form*

The semi-discretization of the ALE form of the Navier-Stokes equations with moving boundaries by a finite volume method leads to the following system of ODEs

$$\dot{\widehat{(\mathbf{V}(\mathbf{x})\mathbf{w})}} + \mathbf{F}(\mathbf{w}, \mathbf{x}, \dot{\mathbf{x}}) = \mathbf{0}, \tag{8.1}$$

where:

- a dot denotes the derivative with respect to time $t$;
- $\mathbf{V} \in \mathbb{R}^{N_f \times N_f}$ is a diagonal matrix storing the cell volumes and $N_f$ denotes the dimension of the semi-discrete fluid system;
- $\mathbf{w}(t) \in \mathbb{R}^{N_f}$ denotes the time-dependent conservative fluid state vector;
- $\mathbf{F} \in \mathbb{R}^{N_f}$ denotes the vector of numerical fluxes;
- $\mathbf{x}$ denotes the vector position of the CFD mesh nodes.

The linearization of (8.1) about an equilibrium state $(\mathbf{w}_0, \mathbf{x}_0, \dot{\mathbf{x}}_0)$ designated by the subscript 0 leads to the following set of ODEs [21]

$$\mathbf{V}_0 \delta\dot{\mathbf{w}} + \mathbf{H}_0 \delta\mathbf{w} + \mathbf{R}_0 \delta\dot{\mathbf{x}} + \mathbf{G}_0 \delta\mathbf{x} = \mathbf{0}, \tag{8.2}$$

where:

- $\delta$ designates a small perturbation of the quantity it is applied to;
- The subscript 0 designates the evaluation of a quantity at the equilibrium state $(\mathbf{w}_0, \mathbf{x}_0, \dot{\mathbf{x}}_0)$;
- $\mathbf{H}_0 = \left.\dfrac{\partial \mathbf{F}}{\partial \mathbf{w}}\right|_0 \in \mathbb{R}^{N_f \times N_f}$ denotes the Jacobian of the vector of numerical fluxes with respect to $\mathbf{w}$, at the equilibrium state $(\mathbf{w}_0, \mathbf{x}_0, \dot{\mathbf{x}}_0)$;
- $\mathbf{R} = \mathbf{E}_0 + \left.\frac{\partial \mathbf{F}}{\partial \dot{\mathbf{x}}}\right|_0 \in \mathbb{R}^{N_f \times N_f}$, where, using Einstein's notation, $E_{0_{ij}} = \left.\frac{\partial A_{il}}{\partial x_j}\right|_0 w_{0_l}$ denotes the Jacobian of the vector of numerical fluxes with respect to $\dot{\mathbf{x}}$, at the equilibrium state $(\mathbf{w}_0, \mathbf{x}_0, \dot{\mathbf{x}}_0)$;

- $\mathbf{G} = \frac{\partial \mathbf{F}}{\partial \mathbf{x}}\Big|_0 \in \mathbb{R}^{N_f \times N_f}$ denotes the Jacobian of the vector of numerical fluxes with respect to $\mathbf{x}$, at the equilibrium state $(\mathbf{w}_0, \mathbf{x}_0, \dot{\mathbf{x}}_0)$.

To keep the notation as simple as possible, the symbold $\delta$ and the subscript 0 are dropped throughout the remainder of this paper. Hence, Eq. (8.2) is re-written as

$$\mathbf{V}\dot{\mathbf{w}} + \mathbf{H}\mathbf{w} + \mathbf{R}\dot{\mathbf{x}} + \mathbf{G}\mathbf{x} = \mathbf{0}. \tag{8.3}$$

Equation (8.3) above is said to be in "descriptor" form because $\mathbf{V} \neq \mathbf{I}_{N_f}$, where $\mathbf{I}_{N_f}$ denotes the identity matrix of dimension $N_f$. Equation (8.3) is also referred to here as a LTI system because all matrices **V**, **H**, **R**, and **G** are time-independent.

The reader is reminded that the leading matrix **V** is diagonal and that its entries store the volumes of the cells of the CFD mesh. For external flow problems around rigid or flexible bodies, the cells are usually very small near the wall boundaries, and very large near the far-field artificial boundaries. Hence for such CFD problems, **V** is diagonal but ill-conditioned.

### *8.2.2 Governing Equations in Non-Descriptor Form*

The nonlinear equations (8.1) can also be written as

$$\mathbf{r}(\mathbf{w}, \mathbf{x}, \dot{\mathbf{x}}) = \mathbf{0}, \tag{8.4}$$

where

$$\mathbf{r}(\mathbf{w}, \mathbf{x}, \dot{\mathbf{x}}) = \mathbf{V}(\mathbf{x})\dot{\mathbf{w}} + \dot{\mathbf{V}}(\mathbf{x})\mathbf{w} + \mathbf{F}(\mathbf{w}, \mathbf{x}, \dot{\mathbf{x}}). \tag{8.5}$$

Hence, given an iterate fluid state vector $(\mathbf{w}^k, \mathbf{x}^k, \dot{\mathbf{x}}^k)$, $\mathbf{r}(\mathbf{w}^k, \mathbf{x}^k, \dot{\mathbf{x}}^k)$ designates the residual associated with it – that is, the residual associated with a $k$-th iteration applied to the solution of Eq. (8.4).

Consider next the scaled residual

$$\tilde{\mathbf{r}}(\mathbf{w}, \mathbf{x}, \dot{\mathbf{x}}) = \mathbf{V}^{-1}(\mathbf{x})\mathbf{r}(\mathbf{w}, \mathbf{x}, \dot{\mathbf{x}}). \tag{8.6}$$

From a purely numerical analysis viewpoint, it could be argued that scaling **r** by $\mathbf{V}^{-1}$ is a bad idea because it involves the multiplication of the governing nonlinear equations of equilibrium (8.1) by the inverse of a matrix. However, in both steady and unsteady CFD codes, it is common practice to work with the scaled residual introduced above for the following reasons:

- scaling the entries of **r** by the corresponding inverses of the cell volumes magnifies the residual in the small cells. In this case, given a stopping criterion and a convergence tolerance, the solution of Eq. (8.4) delivered by a finite number of iterations is most accurate in the flow regions where the cells are the smallest. This is highly desirable because the smallest cells are typically located in the flow regions where accuracy is most sought-after in the first place;
- after time-discretization, scaling the entries of **r** by the corresponding inverses of the cell volumes accelerates the convergence of an iterative process based on local time-stepping and applied to the steady-state solution of Eq. (8.3) [18];

- since $\mathbf{V}$ is diagonal, inverting this matrix is trivial.

The scaling (8.6) is associated with the following nonlinear semi-discrete fluid equations of equilibrium

$$\mathbf{I}_{N_f}\dot{\mathbf{w}} + \mathbf{V}^{-1}\dot{\mathbf{V}}(\mathbf{x})\mathbf{w} + \mathbf{V}^{-1}\mathbf{F}(\mathbf{w},\mathbf{x},\dot{\mathbf{x}}) = \mathbf{0}. \tag{8.7}$$

The linearization of these equations about the equilibrium state $(\mathbf{w}_0, \mathbf{x}_0, \dot{\mathbf{x}}_0)$ leads to

$$\mathbf{I}_{N_f}\dot{\mathbf{w}} + \mathbf{V}^{-1}\mathbf{H}\mathbf{w} + \mathbf{V}^{-1}\mathbf{R}\dot{\mathbf{x}} + \mathbf{V}^{-1}\mathbf{G}\mathbf{x} = \mathbf{0}. \tag{8.8}$$

Equation (8.8) above is said to be in "non-descriptor" form, because the matrix coefficient of its leading term (or term with the highest derivative) is the identity matrix $\mathbf{I}_{N_f}$. This LTI system is mathematically equivalent to its counterpart (8.3) which is written in descriptor form.

It is conjectured here that because the non-descriptor form (8.7) of the nonlinear semi-discrete fluid equations of equilibrium prevails in many CFD codes, POD-based model reduction is performed in some if not many of these codes using either inadvertently or purposely the non-descriptor form (8.8) of the governing linearized semi-discrete fluid equations of equilibrium. For example, this is the case for the POD-based model reductions performed in [20], [23], [2], [16], and [7]. For this reason, one objective of this paper is to analyze the differences, if any, between the linear ROMs constructed by reducing the descriptor form of the governing equations (8.3), and their counterparts constructed by reducing the non-descriptor form (8.8) of these equations.

## 8.3 Model Order Reduction via Galerkin Projection Based on POD Modes

Whether applied to the descriptor or non-descriptor form of a LTI fluid system, a projection-based MOR method generates another LTI fluid system of much smaller dimension $k_f << N_f$. In general, such a MOR method operates using two ROBs:

- a right (or trial) ROB $\Phi \in \mathbb{R}^{N_f \times k_f}$ which has full-column rank and is introduced to approximate the state vector $\mathbf{w}(t)$ as follows:

$$\mathbf{w}(t) \approx \Phi \mathbf{w}_r(t). \tag{8.9}$$

  In this case, the approximate state vector is uniquely defined by the vector of generalized coordinates $\mathbf{w}_r \in \mathbb{R}^{k_f}$. Substituting this approximation into the LTI fluid system of interest yields a non-zero residual $\mathbf{r}(t) \in \mathbb{R}^{N_f}$;
- a left (or test) ROB $\Psi \in \mathbb{R}^{N_f \times k_f}$ which also has full-column rank, and is introduced to limit the magnitude of the residual $\mathbf{r}(t)$ by constraining it to satisfy the orthogonality condition $\Psi^T \mathbf{r}(t) = \mathbf{0}$, where the superscript $T$ designates the transpose.

When $\Psi \neq \Phi$, a projection-based MOR method is also known as a Petrov-Galerkin approximation method. When $\Psi = \Phi$, it is known as a Galerkin approximation method.

In the remainder of this paper, the focus is set on the Galerkin projection method ($\Psi = \Phi$), and the POD technique for constructing the ROB $\Phi$.

### 8.3.1 Snapshot Collection

The POD technique based on numerical snapshots [27] computes a trial ROB $\Phi$ by compressing the information contained in solution snapshots of the system of interest. For LTI systems, these snapshots can be computed either in the time domain, or in the frequency domain. To simplify notation, only the case of a single forcing input function $\mathbf{x}(t)$ is presented below. However, it is noted that the extension to multiple inputs is straightforward (for example, see [23]).

In the time domain, solution snapshots are obtained by computing the dynamic response of the LTI system of interest to a given impulse forcing input and collecting samples of the time-dependent response $\{\mathbf{w}(t_i)\}_{i=1}^{N_w}$, where $N_w$ denotes the number of snapshots, in a matrix $\mathbf{W}$ as follows

$$\mathbf{W} = \mathbf{W}(t_1, \cdots, t_{N_w}) = \left[\mathbf{w}(t_1) \ \ldots \ \mathbf{w}(t_{N_w})\right]. \tag{8.10}$$

In the frequency domain, complex-valued snapshots [19, 22, 23, 28, 30] are obtained by formulating and solving the dynamic response problem in the frequency domain, and collecting samples of the frequency-dependent response in a similar matrix $\mathbf{W}$. For example, when working with the LTI system (8.3) written in descriptor form, the following frequency domain problems are formulated and solved

$$(j\omega_i \mathbf{V} + \mathbf{H})\mathbf{w}(\omega_i) = -(j\omega_i \mathbf{R} + \mathbf{G})\mathbf{x}, \qquad i = 1, 2, \cdots, N_w, \tag{8.11}$$

where $\omega_i$ denotes a sampled circular frequency of interest, $j$ denotes the pure imaginary complex number satisfying $j^2 = -1$, and $\mathbf{x}$ denotes the amplitude of a harmonic mesh motion driven by a harmonic displacement of the body around which the flow is computed. Then, the computed complex-valued samples $\mathbf{w}(\omega_i)$ are collected in a snapshot matrix $\mathbf{W}$ as follows

$$\mathbf{W} = \mathbf{W}(\omega_1, \cdots, \omega_{N_w}) = \left[\mathrm{Re}\left(\mathbf{w}(\omega_1)\right) \ldots \mathrm{Re}\left(\mathbf{w}(\omega_{N_w})\right) \mathrm{Im}\left(\mathbf{w}(\omega_1)\right) \ldots \mathrm{Im}\left(\mathbf{w}(\omega_{N_w})\right)\right]. \tag{8.12}$$

Similarly, when working with the LTI fluid system (8.8) written in non-descriptor form, the frequency domain problems are formulated as follows

$$(j\omega_i \mathbf{I}_{N_f} + \mathbf{V}^{-1}\mathbf{H})\mathbf{w}(\omega_i) = -(j\omega_i \mathbf{V}^{-1}\mathbf{R} + \mathbf{V}^{-1}\mathbf{G})\mathbf{x}, \qquad i = 1, 2, \cdots, N_w, \tag{8.13}$$

and the computed complex-valued samples $\mathbf{w}(\omega_i)$ are collected in a similar snapshot matrix $\mathbf{W}$ as above.

At this point, it is noted that whether collected in the time or frequency domain, and except for round-off effects, the snapshots are independent of the form in which the underlying LTI system is written. This is because both descriptor and non-descriptor forms of a LTI system are mathematically equivalent. However, as it will be shown below, the trial ROBs $\boldsymbol{\Phi}$ constructed using these snapshots differ.

### 8.3.2 Reduction of the Descriptor Form of the Governing Equations

Suppose that the LTI fluid system written in descriptor form (8.3) is chosen as the computational fluid model of interest. Note that the diagonal matrix $\mathbf{V}$ is also a symmetric positive definite matrix and therefore defines a norm. Hence in this case, after all solution snapshots are computed in either the time or frequency domain, the POD technique proceeds with performing the eigenvalue decomposition

$$\mathbf{W}^T\mathbf{V}\mathbf{W} = \hat{\mathbf{U}}^d\hat{\Lambda}^d\hat{\mathbf{U}}^{d^T}, \tag{8.14}$$

where $\mathbf{W}^T\mathbf{V}\mathbf{W} \in \mathbb{R}^{N_w \times N_w}$ is usually a small-size matrix, and the superscript $d$ designates the descriptor form of the underlying governing equations. Next, this decomposition is truncated to account only for the first $k_f$ eigenvalues of $\hat{\Lambda}^d$ and their corresponding eigenvectors, and the trial ROB $\boldsymbol{\Phi}^d$ is constructed as follows

$$\boldsymbol{\Phi}^d = \mathbf{W}\mathbf{U}^d\Lambda^{d^{-\frac{1}{2}}}, \tag{8.15}$$

where $\mathbf{U}^d$ and $\Lambda^d$ are the truncated counterparts of $\hat{\mathbf{U}}^d$ and $\hat{\Lambda}^d$, respectively.

Alternatively, $\boldsymbol{\Phi}^d$ can be constructed by first computing the SVD of the matrix $\mathbf{V}^{\frac{1}{2}}\mathbf{W}$, retaining the first $k_f$ left singular vectors $\mathbf{Y}$, and finally performing the following matrix-matrix multiplication

$$\boldsymbol{\Phi}^d = \mathbf{V}^{-\frac{1}{2}}\mathbf{Y}. \tag{8.16}$$

Finally, performing a Galerkin projection of the governing equations (8.3) using $\Psi = \boldsymbol{\Phi} = \boldsymbol{\Phi}^d$ leads to the reduced-order LTI fluid system

$$\dot{\mathbf{w}}_r + \left(\boldsymbol{\Phi}^{d^T}\mathbf{H}\boldsymbol{\Phi}^d\right)\mathbf{w}_r + \left(\boldsymbol{\Phi}^{d^T}\mathbf{R}\right)\dot{\mathbf{x}} + \left(\boldsymbol{\Phi}^{d^T}\mathbf{G}\right)\mathbf{x} = \mathbf{0}, \tag{8.17}$$

which is also referred to here as the linear fluid ROM based on the descriptor form of the governing equations.

### 8.3.3 Reduction of the Non-Descriptor Form of the Governing Equations

If on the other hand the non-descriptor form (8.8) of the LTI system of interest is chosen as the computational model of interest, the POD process proceeds with performing the following eigen decomposition instead

$$\mathbf{W}^T\mathbf{W} = \hat{\mathbf{U}}^{nd}\hat{\Lambda}^{nd}\hat{\boldsymbol{\Phi}}^{nd^T}, \tag{8.18}$$

where the superscript $nd$ designates the non-descriptor form of the underlying equations, and $\boldsymbol{\Phi}^{nd}$ is constructed as in (8.15). Then, performing a Galerkin projection of the governing equations (8.8) using this trial ROB leads to

$$\dot{\mathbf{w}}_r + \left(\Phi^{nd^T}\mathbf{V}^{-1}\mathbf{H}\Phi^{nd}\right)\mathbf{w}_r + \left(\Phi^{nd^T}\mathbf{V}^{-1}\mathbf{R}\right)\dot{\mathbf{x}} + \left(\Phi^{nd^T}\mathbf{V}^{-1}\mathbf{G}\right)\mathbf{x} = \mathbf{0}. \tag{8.19}$$

This reduced LTI fluid system is also referred to here as the linear fluid ROM based on the non-descriptor form of the governing equations.

### 8.3.4 Comparison of Alternative Reduced-Order Models

From (8.14) and (8.18), if follows that for $\mathbf{V} \neq \mathbf{I}_{N_f}$, $\mathbf{U}^d \neq \mathbf{U}^{nd}$, and therefore $\Phi^d \neq \Phi^{nd}$. Hence, the linear fluid ROM (8.17) based on the descriptor form of the governing equations is in general different from its counterpart (8.19) based on the non-descriptor form of the governing equations.

*Remark.* The reader can check that if the descriptor form of the LTI fluid system of interest is reduced by a Galerkin projection method, but its non descriptor form is reduced instead by a Petrov-Galerkin method, the choices

$$\Phi^{nd} = \Phi^d, \quad \Psi^d = \Phi^d, \quad \text{and} \quad \Psi^{nd} = \mathbf{V}\Phi^d \tag{8.20}$$

lead to two linear fluid ROMs that are identical. However, the focus of this work is set exclusively on the popular Galerkin projection method, and on the POD modes.

## 8.4 Applications to Dynamic Fluid-Structure Interaction Problems

Now that it has been established that the reductions of the descriptor and non-descriptor forms of a LTI system by a Galerkin projection method lead to two different ROMs, it remains to assess whether in the case where the ROBs are generated using the POD technique, the two alternative ROMs exhibit or not different accuracy and numerical stability properties for interesting applications. This is the objective of this section which, for this purpose, focuses on a special class of fluid-structure interaction problems known as aeroelasticity. Such problems are usually characterized by a linear, elastic structural subsystem, and a high-speed compressible fluid subsystem. The present focus on aeroelastic applications is motivated by the fact that linearized CFD is rapidly becoming a very competitive approach for modeling the fluid component of a perturbed aeroelastic system, primarily because it provides a relatively low-cost mean for capturing the effects of shocks in the transonic regime.

To this effect, each of the two fluid ROMs developed in Sectsions 8.3.2 and 8.3.3 is coupled here with a classical modal ROM of the structural subsystem past which the flow is computed, in order to obtain in each case a linear fluid-structure ROM. Then, two *inviscid* aeroelastic applications are considered: the flutter analysis in transonic air speeds of a wing-store-fuel configuration and of a F/A-18 fighter jet configuration, respectively. For each application, the numerical stability of the fluid ROM constructed using the descriptor or non-descriptor form of the governing fluid equations is assessed, and its effect on the behavior of the corresponding coupled fluid-structure ROM is highlighted.

### 8.4.1 Linearized Coupled Fluid-Structure Reduced-Order Models

CFD-based linearized computational fluid models are rapidly becoming the preferred computational models for representing the behavior of a compressible fluid subsystem in a coupled fluid-structure system. For example, they are very popular nowadays in aeronautics for the flutter analysis of modern aircraft in the transonic and other nonlinear regimes [10, 21, 25], and for loads analysis. In this and many other related contexts, the structural subsystem of interest is typically represented by a linearized finite element model that can be described by the following set of linear ODEs

$$\mathbf{M}\ddot{\mathbf{u}} + \mathbf{D}\dot{\mathbf{u}} + \mathbf{K}\mathbf{u} = \sqrt{p_\infty}\mathbf{P}\mathbf{w}, \tag{8.21}$$

where $\mathbf{u} \in \mathbb{R}^{N_s}$ denotes the vector of structural displacements of dimension $N_s$, $\mathbf{M} \in \mathbb{R}^{N_s \times N_s}$, $\mathbf{D} \in \mathbb{R}^{N_s \times N_s}$ and $\mathbf{K} \in \mathbb{R}^{N_s \times N_s}$ are the finite element structural mass, damping, and stiffness matrices, respectively, $p_\infty$ is the free-stream pressure, and $\mathbf{P} \in \mathbb{R}^{N_s \times N_f}$ denotes the Jacobian of the aerodynamic forces acting on the wet surface of the structure with respect to the fluid state vector $\mathbf{w}$. The reader is reminded that in this work, all vector quantities appearing in a linearized context are perturbations and that the symbold $\delta$ is dropped to keep the notation as simple as possible (see Sect. 8.2.1).

Modal decomposition is perhaps the most popular MOR method for a LTI structural subsystem such as that described in Eq. (8.21). In this case, the ROB $\mathbf{X} \in \mathbb{R}^{N_s \times k_s}$ is constructed using the first $k_s$ natural mode shapes of the structural subsystem, and Eq. (8.21) is reduced by a Galerkin projection onto the subspace of dimension $k_s$ spanned by the columns of $\mathbf{X}$. In other words, $\mathbf{u}(t)$ is approximated as follows

$$\mathbf{u}(t) \approx \mathbf{X}\mathbf{u}_r(t), \tag{8.22}$$

where $\mathbf{u}_r \in \mathbb{R}^{k_s}$ is the vector of generalized (modal) coordinates, and Eq. (8.21) is transformed into the following linear structural ROM

$$\ddot{\mathbf{u}}_r + \mathbf{D}_r\dot{\mathbf{u}}_r + \Omega_r^2\mathbf{u}_r = \sqrt{p_\infty}\mathbf{P}_r\mathbf{w}_r, \tag{8.23}$$

where

$$\mathbf{D}_r = \mathbf{X}^T\mathbf{D}\mathbf{X}, \quad \mathbf{P}_r = \mathbf{X}^T\mathbf{P}\boldsymbol{\Phi}, \tag{8.24}$$

$\boldsymbol{\Phi} = \boldsymbol{\Phi}^d$ if the descriptor form of the LTI fluid subsystem is reduced, or $\boldsymbol{\Phi} = \boldsymbol{\Phi}^{nd}$ if the non-descriptor form of the LTI fluid subsystem is reduced, and $\Omega_r \in \mathbb{R}^{k_s \times k_s}$ is the diagonal matrix of natural circular frequencies of the structure.

Assimilating the ALE fluid mesh with a quasi-static pseudo-structure [9, 12] and enforcing the compatibility of the displacements of the structural subsystem and the ALE fluid mesh across the wet surface of the structure results in a linear relationship between $\mathbf{x}$ and $\mathbf{u}$ that can be written as [21]

$$\mathbf{x} = \mathbf{K}^\star\mathbf{u}, \tag{8.25}$$

where $\mathbf{K}^\star$ is a time-independent operator described in [13, 21]. Hence, substituting the above relationship into the LTI fluid subsystem written in descriptor form (8.3) yields

$$\mathbf{V}\dot{\mathbf{w}} + \mathbf{H}\mathbf{w} + \mathbf{R}\mathbf{K}^\star\dot{\mathbf{u}} + \mathbf{G}\mathbf{K}^\star\mathbf{u} = \mathbf{0}, \tag{8.26}$$

and substituting it into the LTI fluid subsystem written in non-descriptor form (8.8) yields

$$\dot{\mathbf{w}} + \mathbf{V}^{-1}\mathbf{H}\mathbf{w} + \mathbf{V}^{-1}\mathbf{R}\mathbf{K}^\star\dot{\mathbf{u}} + \mathbf{V}^{-1}\mathbf{G}\mathbf{K}^\star\mathbf{u} = \mathbf{0}. \tag{8.27}$$

Let

$$\mathbf{H} = \sqrt{\frac{p_\infty}{\rho_\infty}}\overline{\mathbf{H}} \quad \text{and} \quad \mathbf{G} = \sqrt{\frac{p_\infty}{\rho_\infty}}\overline{\mathbf{G}}, \tag{8.28}$$

where $\rho_\infty$ denotes the free-stream density, and $\overline{\mathbf{H}}$ and $\overline{\mathbf{G}}$ do not depend neither on the free-stream pressure $p_\infty$ nor on $\rho_\infty$ [22, 23], but only on the free-stream Mach number $M_\infty$.

Substituting the above expressions of $\mathbf{H}$ and $\mathbf{G}$ into Eq. (8.26) and Eq. (8.27) leads in the descriptor case to

$$\mathbf{V}\dot{\mathbf{w}} + \sqrt{\frac{p_\infty}{\rho_\infty}}\overline{\mathbf{H}}\mathbf{w} + \mathbf{R}\mathbf{K}^\star\dot{\mathbf{u}} + \sqrt{\frac{p_\infty}{\rho_\infty}}\overline{\mathbf{G}}\mathbf{K}^\star\mathbf{u} = \mathbf{0}, \tag{8.29}$$

and in the non-descriptor case to

$$\dot{\mathbf{w}} + \sqrt{\frac{p_\infty}{\rho_\infty}}\mathbf{V}^{-1}\overline{\mathbf{H}}\mathbf{w} + \mathbf{V}^{-1}\mathbf{R}\mathbf{K}^\star\dot{\mathbf{u}} + \sqrt{\frac{p_\infty}{\rho_\infty}}\mathbf{V}^{-1}\overline{\mathbf{G}}\mathbf{K}^\star\mathbf{u} = \mathbf{0}. \tag{8.30}$$

Next, reducing the LTI fluid subsystems (8.29) and (8.30) by the Galerkin projection method based on $\boldsymbol{\Phi}^d$ and $\boldsymbol{\Phi}^{nd}$, respectively, leads to the following general expression of the linear fluid ROM

$$\dot{\mathbf{w}}_r + \sqrt{\frac{p_\infty}{\rho_\infty}}\mathbf{H}_r\mathbf{w}_r + \mathbf{R}_r\dot{\mathbf{u}}_r + \sqrt{\frac{p_\infty}{\rho_\infty}}\mathbf{G}_r\mathbf{u}_r = \mathbf{0} \tag{8.31}$$

where

$$\mathbf{H}_r = \boldsymbol{\Phi}^{d^T}\overline{\mathbf{H}}\boldsymbol{\Phi}^d, \quad \mathbf{R}_r = \boldsymbol{\Phi}^{d^T}\mathbf{R}\mathbf{K}^\star\mathbf{X}, \quad \mathbf{G}_r = \boldsymbol{\Phi}^{d^T}\overline{\mathbf{G}}\mathbf{K}^\star\mathbf{X} \tag{8.32}$$

in the descriptor case, and

$$\mathbf{H}_r = \boldsymbol{\Phi}^{nd^T}\mathbf{V}^{-1}\overline{\mathbf{H}}\boldsymbol{\Phi}^{nd}, \quad \mathbf{R}_r = \boldsymbol{\Phi}^{nd^T}\mathbf{V}^{-1}\mathbf{R}\mathbf{K}^\star\mathbf{X}, \quad \mathbf{G}_r = \boldsymbol{\Phi}^{nd^T}\mathbf{V}^{-1}\overline{\mathbf{G}}\mathbf{K}^\star\mathbf{X} \tag{8.33}$$

in the non-descriptor case.

Finally, re-writing Eq. (8.23) in first-order form and combining it with Eq. (8.31) leads to the following coupled fluid-structure linear ROM of dimension $k_f + 2k_s$

$$\begin{bmatrix}\dot{\mathbf{w}}_r \\ \ddot{\mathbf{u}}_r \\ \dot{\mathbf{u}}_r\end{bmatrix} = \begin{bmatrix} -\sqrt{\frac{p_\infty}{\rho_\infty}}\mathbf{H}_r & -\mathbf{R}_r & -\sqrt{\frac{p_\infty}{\rho_\infty}}\mathbf{G}_r \\ \sqrt{p_\infty}\mathbf{P}_r & -\mathbf{D}_r & -\boldsymbol{\Omega}_r^2 \\ \mathbf{0} & \mathbf{I}_{k_s} & \mathbf{0}\end{bmatrix}\begin{bmatrix}\mathbf{w}_r \\ \dot{\mathbf{u}}_r \\ \mathbf{u}_r\end{bmatrix}. \tag{8.34}$$

This ROM can be used for several fluid-structure applications ranging from real-time control to real-time flutter analysis. In the latter case, the onset of flutter at a given free-stream Mach number $M_\infty$ can be established by fixing the free-stream density $\rho_\infty$ and increasing the free-stream pressure $p_\infty$ until this coupled fluid-structure ROM becomes unstable. At this point, the free-stream pressure $p_\infty$ reaches a critical value denoted here by $p_\infty^{cr}$. This fast approach to flutter analysis requires however that the ROM (8.34) be stable outside the flutter point. In [4], it was shown that this in turn requires that the reduced fluid matrix $\mathbf{H}_r$ be stable. Hence, the application of the linear ROM (8.34) to the flutter analysis of a coupled fluid-structure system highlights the importance of requiring the chosen MOR method to preserve the stability of the LTI system or subsystem to which it is applied (for example, see [4]).

### 8.4.2 Flutter Analysis of a Wing-Store-Fuel Configuration

Consider first the wing-store-fuel aeroelastic configuration described in [8] and graphically depicted in Fig. 8.1. For a fixed altitude characterized by specific values of the free-stream pressure $p_\infty$ and density $\rho_\infty$, a flight condition for this configuration is defined here by an additional pair of data values corresponding to the free-stream Mach number $M_\infty$ and fuel fill level in the store (or tank). The hydroelastic effects due to the presence of fuel inside the tank modify the structural properties of the system and affect its aeroelastic characteristics. The High-Dimensional computational fluid and structural Models (HDMs) developed in [8] for this aeroelastic configuration have the dimensions $N_f = 689,485$ and $N_s = 6,834$, respectively.

For every flight condition of interest, 44 real-valued fluid snapshots are generated by exciting the wall boundary of the structural system by each of its first $k_s = 4$ structural mode shapes at each of six equispaced reduced frequencies in the interval $[0, 0.0125]$. Then, these snapshots are compressed by the POD technique to construct a suite of fluid ROBs of dimension $k_f \in \{1, \cdots, 40\}$. A corresponding suite of fluid ROMs of the same dimension $k_f$ is also constructed by performing Galerkin projections of both descriptor and non-descriptor forms of the LTI fluid subsystem onto these ROBs.

In all cases, the structural ROM is constructed as in (8.23) with $k_s = 4$ and rewritten in first-order form.

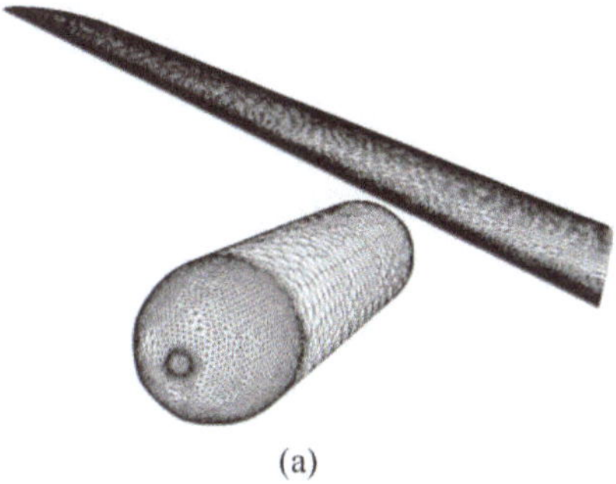

(a)

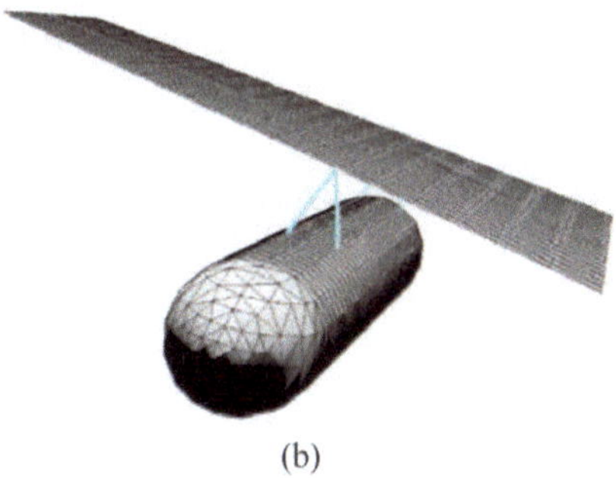

(b)

**Fig. 8.1.** High-dimensional aeroelastic model of a wing-store configuration. (a) CFD surface grid; (b) Finite element structural model

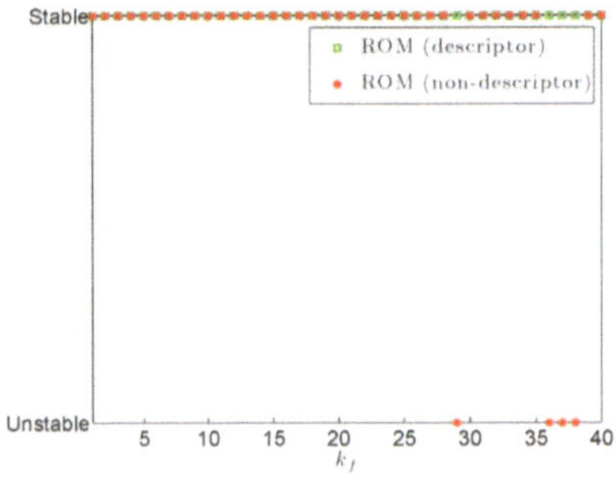


(a) Stability

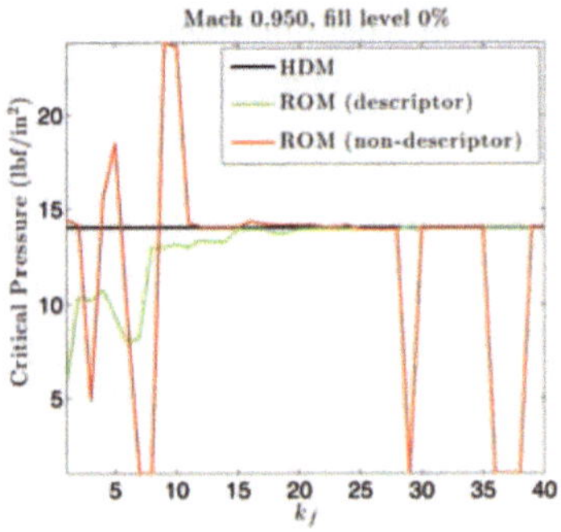


(b) Accuracy

**Fig. 8.2.** Wing-store-fuel configuration ($M_\infty = 0.95$ and empty tank): stability of the fluid ROM as a function of its dimension, and accuracy of the critical free-stream pressure predicted using the corresponding aeroelastic ROM

The first considered flight condition is defined by $M_\infty = 0.95$ and an empty tank. In this case, Figure 8.2(a) reports on the stability of the constructed fluid ROMs – that is, the stability of the matrices $\mathbf{H}_r$. Figure 8.2(b) reports on the accuracy they deliver for the prediction of the critical pressure. All fluid ROMs originating from the descriptor form of the LTI fluid subsystem are found to be stable. On the other hand, the fluid ROMs of dimension $k_f \in \{29, 36, 37, 38\}$ originating from the non-descriptor form of the LTI fluid subsystem are found to be unstable. Consequently, each unstable fluid ROM leads to an erroneous prediction of the critical pressure using the coupled fluid-structure ROM (8.34) (see Fig. 8.2(b)). In contrast, all fluid-structure ROMs of dimension $k_f \geq 15$ originating from the descriptor form of the LTI fluid subsystem deliver accurate predictions of the critical pressure. Similar results were also reported in [3] where a preliminary study of this problem was first performed.

The second and third considered flight conditions are defined by $M_\infty = 1.1$ and 31% fuel fill level in the tank, and $M_\infty = 0.75$ and 69% fuel fill level, respectively. Figures 8.3 and 8.4 report on the stability of the constructed fluid ROMs and accuracy of the corresponding aeroelastic ROMs for these two cases, respectively. These figures confirm the trends observed for the first flight condition and lead to similar conclusions.

### *8.4.3 Flutter Analysis of an F/A-18 Aircraft Configuration*

Next, an aeroelastic HDM of a full F/A-18 configuration with tip missiles is considered (see Fig. 8.5). Here, the dimension of the fluid HDM is $N_f = 1{,}054{,}500$, and that of the structural HDM is $N_s = 9{,}046$. The latter is reduced by Galerkin projection on a modal basis with $k_s = 10$ flexible structural mode shapes.

The free-stream condition is set to $M_\infty = 0.99$. Then, 210 fluid snapshots are computed in the frequency domain by exciting the wall boundary of the aircraft configuration using all 10 structural modal displacements individually, each at 21

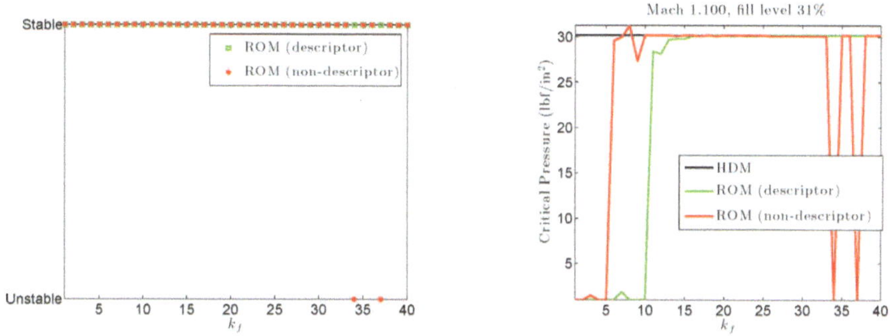


(a) Stability (b) Accuracy

**Fig. 8.3.** Wing-store-fuel configuration ($M_\infty = 1.1$ and 31% fuel fill level): stability of the fluid ROM as a function of its dimension, and accuracy of the critical free-stream pressure predicted using the corresponding aeroelastic ROM

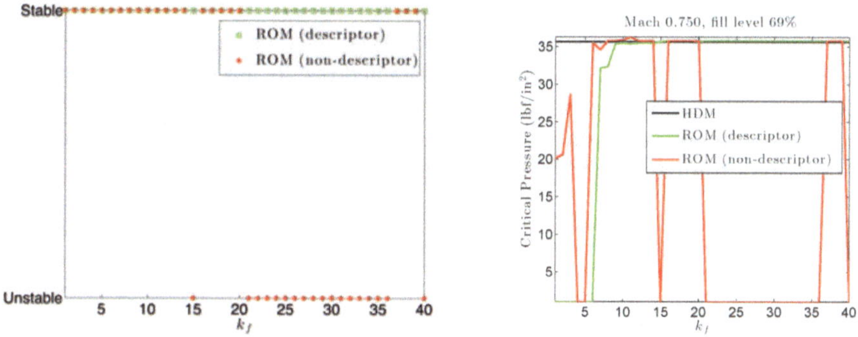


(a) Stability (b) Accuracy

**Fig. 8.4.** Wing-store-fuel configuration ($M_\infty = 0.75$ and 69% fuel fill level): stability of the fluid ROM as a function of its dimension, and accuracy of the critical free-stream pressure predicted using the corresponding aeroelastic ROM

equispaced reduced frequencies in the interval $[0, 0.04]$: 10 of the computed solution snapshots – more specifically, those associated with the zero reduced frequency – are real-valued, and all other 200 solution snapshots are complex-valued. In other words, the corresponding real-valued matrix $\mathbf{W}$ (8.12) has in this case 410 columns. These are compressed by the POD technique to construct a suite of ROBs and two associated suites of fluid ROMs of dimension $k_f \in \{1, \cdots, 400\}$. The first suite of fluid ROMs is obtained by Galerkin projection of the descriptor form of the LTI fluid subsystem on the computed suite of ROBs. The second one is computed by Galerkin projection of the non-descriptor form of the LTI fluid subsystem on the same suite of ROBs. Then, several instances of the coupled fluid-structure ROM (8.34) are constructed by coupling the modal structural ROM of dimension $k_s = 10$ with each of the computed fluid ROMs.

**Fig. 8.5.** F18/A configuration at $M_\infty = 0.99$: steady-state surface pressure

First, the stability of the constructed fluid ROMs, and more specifically that of the constructed matrices $\mathbf{H}_r$, is assessed. The obtained results are reported in Fig. 8.6. Once again, all fluid ROMs originating from the descriptor form of the LTI fluid subsystem are found to be stable, but more than half of those originating from its non-descriptor form are found to be unstable.

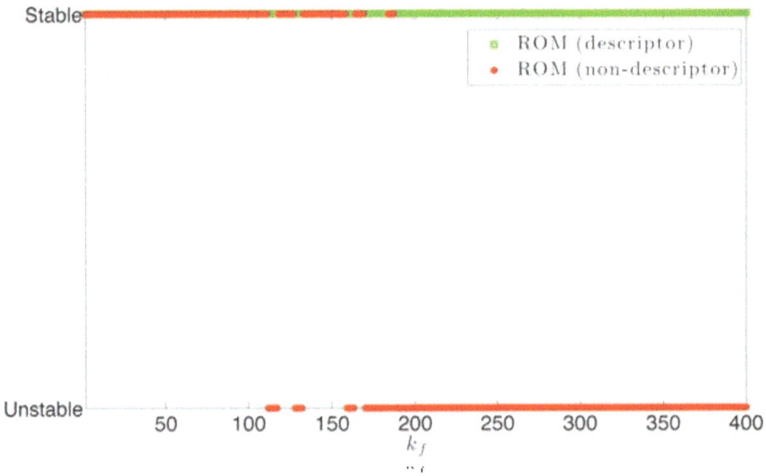


**Fig. 8.6.** F18/A aeroelastic configuration at $M_\infty = 0.99$: stability of the fluid ROM as a function of its dimension

Next, the accuracy of each constructed aeroelastic ROM is assessed by examining the critical pressure it predicts. The obtained results are reported in Fig. 8.7. From this figure and Fig. 8.6, the reader can observe that every unstable fluid ROM leads to an erroneous prediction of the critical pressure by the corresponding coupled fluid-

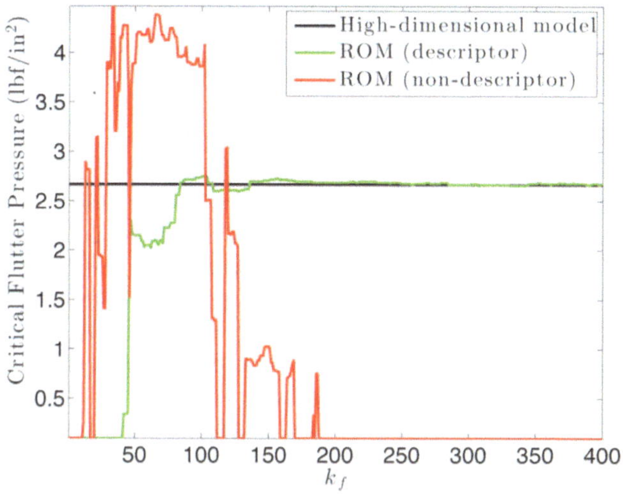


**Fig. 8.7.** F18/A aeroelastic configuration at $M_\infty = 0.99$: accuracy of the critical free-stream pressure predicted using the aeroelastic ROM

structure ROM. Furthermore, none of the aeroelastic ROMs originating from the non-descriptor form of the fluid LTI subsystem delivers an accurate prediction of the critical pressure. On the other hand, all aeroelastic ROMs originating from this descriptor form and of dimension $k_f \geq 100$ deliver critical pressure predictions that match their HDM counterparts.

## 8.5 Conclusions

In theory, the Galerkin projection method equipped with Proper Orthogonal Decomposition (POD) modes does not guarantee the stability of the Reduced-Order Models (ROMs) it is often used for constructing. In practice, it is reported in some forums to generate ROMs that are more frequently unstable than stable. Yet, the POD-based Galerkin projection method is among the most popular methods for the dimensional reduction of Linear Time-Invariant (LTI) systems arising from linearized Computational Fluid Dynamics (CFD).

In general, a LTI system can be written in either descriptor or non-descriptor form. The non-descriptor form is characterized by the identity matrix as the coefficient of the highest derivative term in the governing set of Ordinary Differential Equations (ODEs). On the other hand, the leading matrix coefficient of the descriptor form of the same LTI system is usually different from the identity. Therefore, transforming a descriptor form of a given LTI system into its non-descriptor form typically involves pre-multiplying all matrix coefficients of the descriptor form by the inverse of its leading matrix coefficient. Because of the usual numerical issues associated with computing the inverse of a matrix and/or the solution of potentially ill-conditioned systems of equation, such a transformation could be dismissed *a priori* as a "trouble maker". Nevertheless, such a transformation is routinely performed – and for good reasons – in many computational engineering applications. These range from multibody dynamics [15, 17], to molecular dynamics [26], to CFD [2, 7, 16, 20, 23]. For example, the *nonlinear* semi-discrete equations of dynamic equilibrium governing a flow problem are often transformed from their descriptor form to their non-descriptor form, in order to improve accuracy and accelerate convergence. Indeed, the leading matrix coefficient of the governing set of nonlinear ODEs governing a CFD problem is usually a diagonal matrix storing the volumes of the mesh cells when semi-discretization is performed using a finite volume method, or the volumes of the mesh elements when semi-discretization is performed using a finite element method. Hence, in this case, transforming the descriptor form of the governing set of nonlinear ODEs into its non-descriptor counterpart is a trivial task. It amounts to scaling each entry of the residual vector associated with these equations by the inverse of the corresponding volume of the mesh cell or element. Given that for external flow problems the cells or elements of the mesh are usually very small in the vicinity of the wall boundaries and very large near the far-field artificial boundaries, the non-descriptor form of the governing nonlinear CFD equations magnifies the residuals associated with the small mesh cells or elements. Therefore, the application of a finite number of steps of an iterative procedure to the solution of the non-descriptor form of the governing nonlinear CFD equations delivers a higher accuracy in the flow regions where the mesh cells or elements are the smallest – that is, in the flow regions that matter most – than the application of these same steps to the descriptor form of these equations. Furthermore, when local time-stepping is applied to the solution of a steady-state CFD problem, scaling the residual vector by the inverse of the volumes of the cells or elements of the mesh is often observed to accelerate convergence. For these reasons, many CFD codes effectively operate on the non-descriptor form of the Euler or Navier-Stokes equations. Therefore, it can be conjectured that at least for software legacy reasons, many linearized CFD codes or modules also operate on the non-descriptor forms of the linearized Euler and Navier-Stokes equations. Hence, when model order reduction is or will be implemented in such codes, it is likely to be applied, whether inadvertently or purposely for the reasons outlined above, to the non-descriptor form of the governing ODEs. This conjecture is supported by references such as [2, 7, 16, 20, 23] and others.

In this paper however, it was shown that whereas the snapshot solutions computed using either the descriptor or non-descriptor form of a CFD-based LTI system are identical, the ROMs obtained by reducing both forms of this system using a Galerkin projection method are different. More importantly, using as background the field of linearized computational aeroelasticity, it was also shown numerically that in general, the fluid ROMs constructed by applying the POD-based Galerkin projection method to the non-descriptor form of a CFD-based LTI subsystem of interest are more often unstable then stable. It was also shown that the stability of these ROMs is very sensitive to their dimension. This is consistent with the observations frequently reported in various forums about the inconsistent behavior of POD as far as stability is concerned. On the other hand, it was also shown numerically that for the same aeroelastic problems, the fluid ROMs constructed by applying the same POD-based Galerkin projection method to the descriptor form of the CFD-based LTI subsystem of interest are typically stable. Therefore, the findings reported in this paper suggest that when the objective is to construct a CFD-based linear fluid ROM using the POD-based Galerkin projection method, reducing the non-descriptor form of the linearized Euler or Navier-Stokes equations tends to promote the instability of the outcome ROM, whereas reducing the descriptor form of these equations tends to prevent it. Hence, a best practice in implementing the POD-based Galerkin projection method in a given CFD code for the purpose of constructing linear fluid ROMs is to apply this method to the descriptor form of the linearized Euler and Navier-Stokes equations, even when the nonlinear computational modules of this code operate on the non-descriptor form of these equations.

**Acknowledgements** The authors acknowledge partial support by the Army Research Laboratory through the Army High Performance Computing Research Center under Cooperative Agreement W911NF-07-2-0027, partial support by the Office of Naval Research under Grant N00014-11-1-0707, partial support by a research grant from King Abdulaziz City for Science and Technology (KACST), and partial support by The Boeing Company under Contract Sponsor Ref. 45047. The content of this publication does not necessarily reflect the position or policy of any of these supporters, and no official endorsement should be inferred.

## References

1. Amsallem, D., Carlberg, K., Cortial, J., Farhat, C.: A method for interpolating on manifolds structural dynamics reduced-order models. International Journal for Numerical Methods in Engineering **80**, 1241–1258 (2009)
2. Amsallem, D., Farhat, C.: Interpolation method for adapting reduced-order models and application to aeroelasticity. AIAA Journal **46**(7), 1803–1813 (2008)
3. Amsallem, D., Farhat, C.: On the stability of linearized reduced-order models: descriptor vs. non-descriptor form and application to fluid-structure interaction. AIAA Paper 2012-2687, 42nd AIAA Fluid Dynamics Conference and Exhibit. 25–28 June 2012, New Orleans, Louisiana pp. 1–12 (2012)
4. Amsallem, D., Farhat, C.: Stabilization of projection-based reduced-order models. International Journal for Numerical Methods in Engineering **91**, 358–377 (2012)
5. Berkooz, G., Holmes, P., Lumley, J.L.: The proper orthogonal decomposition in the analysis of turbulent flows. Annual Review of Fluid Mechanics **25**, 539–575 (1993)
6. Bond, B.N., Daniel, L.: Guaranteed stable projection-based model reduction for indefinite and unstable linear systems. In: 2008 IEEE/ACM International Conference on Computer-Aided Design (ICCAD), pp. 728–735. IEEE (2008)
7. Chen, G., Sun, J., Li, Y.-M.: Adaptive reduced-order-model-based control-law design for active flutter suppression. Journal of Aircraft **49**(4), 973–980 (2012)
8. Chiu, E., Farhat, C.: Effects of fuel slosh on flutter prediction. AIAA 2009-2682, 50th AIAA/ASME/ASCE/AHS/ASC Structures, Structural Dynamics, and Materials Conference (2009)

9. Degand, C., Farhat, C.: A three-dimensional torsional spring analogy method for unstructured dynamic meshes. Computers and Structures **80**, 305–316 (2002)
10. Dowell, E.H., Thomas, J.P., Hall, K.C.: Transonic limit cycle oscillation analysis using reduced order aerodynamic models. Journal of Fluids and Structures **18**, 17–27 (2004)
11. Epureanu, B.I.: A parametric analysis of reduced order models of viscous flows in turbomachinery. Journal of Fluids and Structures **17**, 971–982 (2003)
12. Farhat, C., Degand, C., Koobus, B., Lesoinne, M.: Torsional springs for two-dimensional dynamic unstructured fluid meshes. Computer Methods in Applied Mechanics and Engineering **163**, 231–245 (1998)
13. Farhat, C., van der Zee, K.G., Geuzaine, P.: Provably time-accurate loosely-coupled solution algorithms for transient nonlinear computational aeroelasticity. Computer Methods in Applied Mechanics and Engineering **195**, 1973–2001 (2006)
14. Hall, K.C., Thomas, J.P., Dowell, E.H.: Proper orthogonal decomposition technique for transonic unsteady aerodynamic flows. AIAA Journal **38**(2), 1853–1862 (2000)
15. Hardt, M., Oskar, V.S.: Dynamic modeling in the simulation, optimization, and control of bipedal and quadrupedal robots. ZAMM, Journal of Applied Mathematics and Mechanics/Zeitschrift für Angewandte Mathematik und Mechanik **83**, 648–662 (2003)
16. Hu, P., Bodson, M., Brenner, M.: Towards real-time simulation of aeroservoelastic dynamics for a flight vehicle from subsonic to hypersonic regime. AIAA 2008-6375, AIAA Atmospheric Flight Mechanics Conference and Exhibit (2008)
17. Jain, A.: Unified formulation of dynamics for serial rigid multibody systems. Journal of Guidance, Control, and Dynamics **14**, 531–542 (1991)
18. Kelley, C.T., Keyes, D.E.: Convergence Analysis of Pseudo-Transient Continuation. SIAM Journal of Numerical Analysis **35**(2), 508–523 (1998)
19. Kim, T.: Frequency-domain Karhunen-Loeve method and its application to linear dynamic systems. AIAA Journal **36**(11), 2117–2123 (1998)
20. Lesoinne, M., Farhat, C.: A CFD based method for solving aeroelastic eigenproblems in the subsonic, transonic, and supersonic regimes. AIAA Journal of Aircraft **38**, 628–635 (2001)
21. Lesoinne, M., Sarkis, M., Hetmaniuk, U., Farhat, C.: A linearized method for the frequency analysis of three-dimensional fluid/structure interaction problems in all flow regimes. Computer Methods in Applied Mechanics and Engineering **190**, 3121–3146 (2001)
22. Lieu, T., Farhat, C.: Adaptation of aeroelastic reduced-order models and application to an F-16 configuration. AIAA Journal **45**(6), 1244–1257 (2007)
23. Lieu, T., Farhat, C., Lesoinne, M.: Reduced-order fluid/structure modeling of a complete aircraft configuration. Computer Methods in Applied Mechanics and Engineering **195**(41–43), 5730–5742 (2009)
24. Luenberger, D.G.: Dynamic equations in descriptor form. IEEE Transactions on Automatic Control, pp. 312–321 (1977)
25. Mortchéléwicz, G.: Aircraft aeroelasticity computed with linearized RANS equations. 43rd Annual Conference on Aerospace Sciences, Tel Aviv, Israel (2003)
26. Nagarajan, V., Jain, A., Goddard, W.: Constant temperature constrained molecular dynamics: The Newton-Euler inverse mass operator method. The Journal of Physical Chemistry **100**, 10,508–10,517 (1996)
27. Sirovich, L.: Turbulence and the dynamics of coherent structures. Part I: Coherent structures. Quarterly of Applied Mathematics **45**(3), 561–571 (1987)
28. Thomas, J.P., Dowell, E.H., Hall, K.C.: Nonlinear inviscid aerodynamic effects on transonic divergence, flutter, and limit-cycle oscillations. AIAA Journal **40**, 638–646 (2002)

29. Thomas, J.P., Dowell, E.H., Hall, K.C.: Three-dimensional transonic aeroelasticity using proper orthogonal decomposition-based reduced order models. Journal of Aircraft **40**(3), 544–551 (2003)
30. Willcox, K., Peraire, J.: Balanced model reduction via the proper orthogonal decomposition. AIAA Journal **40**(11), 2323–2330 (2002)

9

# Model Order Reduction in Fluid Dynamics: Challenges and Perspectives

**Toni Lassila, Andrea Manzoni, Alfio Quarteroni and Gianluigi Rozza**

**Abstract** This chapter reviews techniques of model reduction of fluid dynamics systems. Fluid systems are known to be difficult to reduce efficiently due to several reasons. First of all, they exhibit strong nonlinearities – which are mainly related either to nonlinear convection terms and/or some geometric variability – that often cannot be treated by simple linearization. Additional difficulties arise when attempting model reduction of unsteady flows, especially when long-term transient behavior needs to be accurately predicted using reduced order models and more complex features, such as turbulence or multiphysics phenomena, have to be taken into consideration. We first discuss some general principles that apply to many parametric model order reduction problems, then we apply them on steady and unsteady viscous flows modelled by the incompressible Navier-Stokes equations. We address questions of *inf-sup* stability, certification through error estimation, computational issues and – in the unsteady case – long-time stability of the reduced model. Moreover, we provide an extensive list of literature references.

T. Lassila
MATHICSE-CMCS Modelling and Scientific Computing,
Ecole Polytechnique Fédérale de Lausanne, Station 8, CH-1015 Lausanne, Switzerland
e-mail: toni.lassila@epfl.ch

A. Manzoni
SISSA Mathlab - International School for Advanced Studies, Santorio A, Via Bonomea 265, I-34136 Trieste, Italy
e-mail: andrea.manzoni@sissa.it

A. Quarteroni
MATHICSE-CMCS Modelling and Scientific Computing, Ecole Polytechnique Fédérale de Lausanne, Station 8, CH-1015 Lausanne, Switzerland, and MOX – Dipartimento di Matematica, Politecnico di Milano, P.za Leonardo da Vinci 32, I-20133 Milano, Italy
e-mail: alfio.quarteroni@epfl.ch

Gianluigi Rozza (✉)
SISSA Mathlab - International School for Advanced Studies, Santorio A, Via Bonomea 265, I-34136 Trieste, Italy
e-mail: gianluigi.rozza@sissa.it

A. Quarteroni, G. Rozza (eds.): *Reduced Order Methods for Modeling and Computational Reduction*, MS&A 9. DOI 10.1007/978-3-319-02090-7_9, 

## 9.1 Introduction

Numerical methods for Computational Fluid Dynamics (CFD) are by now essential in engineering applications dealing with flow simulation and control, such as the ones arising in aerodynamics, hydrodynamics and, more recently, in physiological flows. In despite of a constant increase in available computational power, numerical simulations of turbulent flows, multiscale and multiphysics phenomena, flows separation and/or bifurcation phenomena are still very demanding, possibly requiring millions or tens of millions of degrees of freedom and several days of CPU time on powerful parallel hardware architectures. This effort is even more substantial whenever we are interested in the repeated solution of the fluid equations for different values of model parameters, such as in flow control or optimal design problems (*many-query* contexts), or in *real time* flow visualization and output evaluation.

These problems represent a remarkable challenge to classical numerical approximations techniques, such as Finite Elements (FE), Finite Volumes or spectral methods. In fact, these methods require huge computational efforts (and also data/memory management) if we are interested to provide accurate response, thus making both *real-time* and *many-query* simulations unaffordable. For this reason, we need to rely on suitable *Reduced-Order Models* (ROMs) – that can reduce both the amount of CPU time and storage capacity – in order to enhance the computational efficiency in these contexts.

This chapter reviews the current state-of-the art for the model reduction of parametrized fluid dynamics equations. In particular, we focus on the incompressible Navier-Stokes equations, because of their ubiquitous presence in fluid flow applications and the fact that they involve the most important features and challenges relevant to *nonlinear model reduction*. These equations are usually written in primitive variables as follows: find the velocity field $\mathbf{u} : \Omega \times [0,T) \to \mathbb{R}^d$ and pressure field $p : \Omega \times (0,T) \to \mathbb{R}$ such that

$$\begin{aligned} \frac{\partial \mathbf{u}}{\partial t} + (\mathbf{u}\cdot\nabla)\mathbf{u} + \nabla p - \frac{1}{\mathrm{Re}}\triangle \mathbf{u} &= 0, && \text{in } \Omega \times (0,T) \\ \nabla \cdot \mathbf{u} &= 0, && \text{in } \Omega \times (0,T) \\ \mathbf{u}(x,0) &= \mathbf{u}_0(x), && \end{aligned} \tag{9.1}$$

where $\Omega \subset \mathbb{R}^d$ denotes the fluid domain, $\mathrm{Re} = |\mathbf{u}_{\max}|L/\nu$ is the nondimensional Reynolds number, $L$ is a characteristic length, $\nu$ is the fluid kinematic viscosity and $|\mathbf{u}|_{\max} = \max_{\mathbf{x}\in\Omega} |\mathbf{u}|$. In addition, suitable boundary conditions need to be prescribed in order to solve problem (9.1), see e.g. [51, 80, 118].

The Navier-Stokes equations are the most accurate continuum-based approximation for viscous flows where both convective and diffusive effects contribute, and they are known to accurately reproduce many interesting physical phenomena observed in fluids, such as the onset of turbulence. Concerning the functional setting required to frame the analysis of problem (9.1), let us denote $(H_0^1(\Omega))^d \subset V \subset (H^1(\Omega))^d$ and $Q \subset L^2(\Omega)$. The solution of (9.1) is such that $(\mathbf{u},p) \in L^2(0,T;V) \times C^0(0,T;Q)$; see e.g. [102, 118] for the definition of Sobolev spaces and more details about this functional setting. Moreover, let us introduce a further functional space $W \subseteq V \times Q$, denote $\langle \cdot, \cdot \rangle_X$ the scalar product over a generic space $X$ and $\|\cdot\|_X$ its induced norm. When the subscript is omitted, $\langle \cdot, \cdot \rangle$ denotes in the following the $L^2$-scalar product and $\|\cdot\|$ the induced norm, respectively.

In many applications, the fluid problem can depend in addition on a number of parameters. In this case we deal with a parametric model reduction problem. We denote $\mu \in \mathscr{P} \subset \mathbb{R}^P$ a vector of $P$ parameters of interest for a given fluid dynamics problem, as in the case of the Reynolds number appearing in (9.1). Other typical examples deal with different physical parametrizations (e.g. by considering Grashof number, Prandtl number, inflow velocity peaks, etc. as parameters) or geometrical parametrization, i.e. when the fluid domain $\Omega = \Omega(\mu)$ depends on a set of parameters allowing to describe/modify its shape. For the sake of simplicity, in this chapter we will focus on physical parameters, whereas several details about flexible but efficient geometrical parametrizations can be found e.g. in [87].

Model reduction of the Navier-Stokes equations is a challenging task because their solutions tend to exhibit complex phenomena at multiple temporal and spatial scales, which means they are difficult to reduce to low-dimensional models without losing at least some of the scales. In the case of unsteady flows, application of the standard "method of lines" to the time-discretization of the unsteady Navier-Stokes equations leads in three dimensions to the lack of sharp long-time stability estimates. It is well known [68] that application of the discrete Grönwall lemma leads to excessive growth of error bounds in time, because standard linear stability analysis of the unsteady Navier-Stokes equations results in stability constants that can be of the order $C_s \sim \exp(\mathrm{Re}\, T)$. While turbulence has sometimes been offered as an explanation to this difficulty, the underlying situation is more delicate. The same type of problem is exhibited by the one-dimensional Burgers' equation, which does not possess turbulent solutions. This also makes hard to provide meaningful error bounds for the solutions of ROMs for the unsteady Navier-Stokes equations.

During the last three decades, several efforts in theoretical foundations, numerical investigations and methodological improvements have made possible to develop general ideas in reduced order modelling and to tackle several problems arising in fluid dynamics. Among a number of early contributions, we want to highlight the most important – in our opinion – that date back to the late 1980s (see e.g. [39, 98, 114]). These were mainly based on *ad hoc* selection of the basis functions, without the benefit of a formal algorithm. Indeed, model reduction has come into play as a truly invaluable tool in CFD applications only once systematic strategies for constructing quasi-optimal bases were made available.

For the sake of exposition, we limit ourselves to describe two main algorithms for choosing the basis on which to build ROMs, namely the *Proper Orthogonal Decomposition* (POD) and the *(greedy) Reduced basis* (RB) methods. They share several features but have been historically introduced and developed to address different types of problems – POD is typically applied to build bases for *time*-dependent problems, while the greedy RB method is usually applied to build bases for *parameter-*

dependent problems. Moreover, we provide detailed remarks and references about extensions of these techniques and alternative strategies. We do not address in this review the case of *combined time and parameter*-dependent problems; the interested reader can refer to some recent works concerning error estimates for ROMs in the case of acoustic Helmholtz and incompressible Navier-Stokes equations [63], the Boussinesq equations [70], and the viscous Burgers' equation using the method of lines [93] or in the space-time formulation [131].

### 9.1.1 Proper Orthogonal Decomposition

POD is the leading model reduction tool for the unsteady Navier-Stokes equations. It was first introduced in [83] in the context of fluid dynamics as a method for discerning and analyzing coherent structures in experimental turbulent flows, and more recently in direct numerical simulations of turbulent flows in [53, 126], where also the concept of space-time windowing of POD has been introduced, to identify turbulent effects in transitional flow that are highly localized both in space and time.

POD techniques reduce the dimensionality of a system by transforming the original unknowns onto a new set of $N_r$ variables (called POD modes, or principal components) such that the first few modes retain most of the energy present in all of the original unknowns. This allows to obtain a reduced, modal representation through a spectral decomposition which requires basic matrix computations (a singular value decomposition) also for nonlinear equations. For a deeper review on POD we recall here also the contribution of Bergman et al. and Grinberg et al. in this volume.

For the reader's convenience, we recall briefly the POD based on the *method of snapshots*, as presented in [114]. An approximation $\mathbf{u}_r(\mathbf{x},t)$ to the solution $\mathbf{u}(\mathbf{x},t)$ of (9.1) is sought as the sum of a base flow $\bar{\mathbf{u}}$ and a linear combination of some spatial modes $\Psi_i(\mathbf{x})$ through a set of temporal coefficients, as follows:

$$\mathbf{u}(\mathbf{x},t) \approx \mathbf{u}_r(\mathbf{x},t) := \bar{\mathbf{u}}(\mathbf{x}) + \sum_{i=1}^{N_r} a_i(t)\Psi_i(\mathbf{x}), \tag{9.2}$$

for a suitable $N_r \geq 1$, where $\bar{\mathbf{u}}(\mathbf{x}) := \int_0^T \mathbf{u}(\mathbf{x},\tau)\,d\tau$ is the time-averaged base flow. This *ansatz* is reasonable assuming that the flow field can be approximated by a stochastic process that is stationary in time and ergodic [60]. In Sect. 9.4.3 we will discuss some extensions in situations where such assumptions do not hold.

The spatial modes are assumed to satisfy the orthogonality relation $\langle \Psi_i, \Psi_j \rangle = 0$ if $i \neq j$, for $\langle \cdot, \cdot \rangle$ denoting a convenient scalar product, whereas the coefficients $a_i(t)$ satisfy the following system of ODEs

$$\begin{aligned} \frac{da_i(t)}{dt} &= F_i + \sum_{j=1}^{N_r} A_{ij} a_j(t) + \sum_{j=1}^{N_r} \sum_{k=1}^{N_r} C_{ijk} a_j(t) a_k(t), \quad t \geq 0 \\ a_i(0) &= \langle \Psi_i, \mathbf{u}_0 \rangle, \end{aligned} \tag{9.3}$$

for $i = 1, \ldots, N_r$, where the functional forms of the reduced system coefficient tensors

$$
\begin{aligned}
F_i &:= -\frac{1}{\mathrm{Re}}\langle \nabla \Psi_i, \nabla \bar{\mathbf{u}}\rangle - \langle \Psi_i, (\bar{\mathbf{u}}\cdot\nabla)\bar{\mathbf{u}}\rangle \\
A_{ij} &:= -\langle \Psi_i, (\bar{\mathbf{u}}\cdot\nabla)\Psi_j\rangle - \langle \Psi_i, (\Psi_j\cdot\nabla)\bar{\mathbf{u}}\rangle - \frac{1}{\mathrm{Re}}\langle \nabla\Psi_i, \nabla\Psi_j\rangle \\
C_{ijk} &:= -\langle \Psi_i, (\Psi_j\cdot\nabla)\Psi_k\rangle
\end{aligned}
\tag{9.4}
$$

are obtained by Galerkin projection of the original system (9.1) on the spatial modes $\Psi_1, \ldots, \Psi_{N_r}$ in (9.2). The resulting ROM is referred to as a *Galerkin ROM*.

We point out that the pressure terms do not appear in these equations, and our space is defined as $X \equiv V$. In fact, by construction the POD modes $\{\Psi_i\}_{i=1}^{N_r}$ are discretely divergence-free. However, for some flows we could be interested either in evaluating the pressure field through the ROM, or to explicitly enforce the divergence-free constraint in the ROM; we will go back to this point in Sect. 9.3.1.

From the structure of (9.2) we note immediately that trajectories of the reduced solution $\mathbf{u}_r$ live in an $N_r$-dimensional submanifold of the full space. Thus the accuracy of the ROM is implicitly dependent on the assumption that the trajectories of the full-order system (9.1) can reasonably be approximated on a much lower-dimensional submanifold. As we will see in the following sections, these two ingredients, namely *(i)* the expression of the approximate solution in a reduced-order model as a linear combination of properly selected snapshots and *(ii)* a projection onto the subspace spanned by the snapshot solutions in order to find the weights in the linear combination, are peculiar also to the *(greedy) reduced basis* methods.

We now focus on computation of the spatial modes $\{\Psi_i\}_{i=1}^{N_r}$. We start from a set of snapshot solutions $\mathbf{U}_n(\mathbf{x}) := \mathbf{u}(\mathbf{x}, t^n)$ of the trajectory $\mathbf{u}(\mathbf{x}, t)$ at some selected times $t^n$, for $n = 1, \ldots, N_s$. These solutions can be either obtained through accurate numerical simulations of the discretized Navier-Stokes equations (9.1), or by experimental measurements of the physical system. In the former case, a POD approach is premised upon a "truth approximation" space $X_h \subset X$ of (typically very large) dimension, for which the snapshot solutions $\mathbf{U}_n(\mathbf{x}) := \mathbf{u}_h(\mathbf{x}, t^n)$ of the (truth approximation of the) trajectory $\mathbf{u}_h(\mathbf{x}, t)$ at some selected times $t^n$, for $n = 1, \ldots, N_s$. Nonetheless, we omit the subscript $h$ wherever possible. The snapshots are typically equispaced in time along the entire period $T$ and obtained after discarding the initial transient of the flow until a stable regime is reached and the flow can be modelled as a stochastic process that is stationary in time[1].

The POD space $X_{N_r}^{\mathrm{POD}} := \mathrm{span}\{\Psi_i : i = 1, \ldots, N_r\}$ of dimension $1 \le N_r \le N_s$, for a suitable $N_s$, is defined as the subspace which minimizes the least-squares discrepancy between the snapshots $\{\mathbf{U}_i(\mathbf{x})\}_{i=1}^{N_r}$ and their best approximation in the $X$-norm:

$$
X_{N_r}^{\mathrm{POD}} := \underset{X_{N_r} \subset X_{N_s},\, \dim(X_{N_r}) = N_r}{\arg\inf} \frac{1}{N_r} \sum_{i=1}^{N_s} \left\| \mathbf{U}_i(\cdot) - \Pi_{X_{N_r}}(\mathbf{U}_i(\cdot)) \right\|_{(L^2(\Omega))^d}^2, \tag{9.5}
$$

[1] In practice, $N_r$ POD modes are required to resolve the first $N_r/2$ temporal harmonics, and these can be computed from $N_s = 2N_r$ snapshots [96].

where $\Pi_{X_{N_r}}$ denotes the $(L^2(\Omega))^d$ projection onto the subspace $X_{N_s}$; for incompressible fluid problems this means that the POD basis is the best approximation basis in the sense of capturing the kinetic energy contained in the snapshots.

From a practical point of view, we form the correlation matrix $\mathbb{C} \in \mathbb{R}^{N_s \times N_s}$, whose components are

$$\mathbb{C}_{nm} := \frac{1}{T} \int_{\Omega} \left[\mathbf{U}_n(\mathbf{x}) - \bar{\mathbf{U}}(\mathbf{x})\right] \cdot \left[\mathbf{U}_m(\mathbf{x}) - \bar{\mathbf{U}}(\mathbf{x})\right] \, d\mathbf{x}, \tag{9.6}$$

where $\bar{\mathbf{U}}(\mathbf{x}) := \frac{1}{N} \sum_{n=1}^{N} \mathbf{U}_n(\mathbf{x})$ is the ensemble average that approximates the base flow $\bar{\mathbf{u}}$. Then, we compute the eigenpairs $(\lambda_k, \boldsymbol{\psi}_k)$, $k = 1, \ldots, N_s$ (with positive eigenvalues ordered by decreasing size) of $\mathbb{C}$. The central result of POD states that the *optimal* subspace $X_{N_r}^{\text{POD}}$ of dimension $N_r$ satisfying (9.5) is such that

$$\Psi_i = \tilde{\Psi}_i / \|\tilde{\Psi}_i\|_W, \qquad \tilde{\Psi}_i = \sum_{n=1}^{N_s} \psi_{i,n}(\mathbf{U}_n(\mathbf{x}) - \bar{\mathbf{U}}(\mathbf{x})), \qquad 1 \leq k \leq N_s, \tag{9.7}$$

being $\psi_{in} = (\boldsymbol{\psi}_i)_n$ the $n$-th component of the $i$-th eigenvector. In this way, the basis functions $\{\Psi_i\}_{i=1}^{N_s}$ are $L^2$-orthonormal[2].

The POD can equally be applied to the reduction of parametric fluid flow problems (see e.g. the parametric studies in [33] for rotating transitional flow, in [54] for modeling the airflow in a large public building, and in [67] for the analysis of turbulent plane channel flow). In fact, if the system (9.1) depends in addition on a vector $\boldsymbol{\mu} \in \mathscr{P} \subset \mathbb{R}^P$ of $P$ parameters of interest, we can follow the same procedure, except that the snapshots are now sampled also in the parameter space. It should be noted, however, that even if the POD procedure is the same in both the time interval and the parameter space, the practical results will differ considerably, due to the *causal* nature of time as opposed to other types of physical parameters.

So far we have not mentioned the treatment of boundary conditions that need to be imposed on (9.1). In the case of homogeneous boundary conditions, the snapshots as well as their linear combinations will naturally satisfy the same boundary conditions so that nothing special needs to be done. If we have non-homogeneous Dirichlet boundary conditions, the linear combinations of snapshots will not in general satisfy them, and neither will the ROM solution. To remedy this problem we can either subtract the non-homogeneous boundary values from the snapshots before constructing the POD basis, or add an additional constraint equation to the ROM that enforces the boundary condition. These two methods can also be applied to parameter-dependent problems with multiple parameters in the boundary data. For a comparison of the two approaches we refer to [54], where both methods were found to produce similar results.

---

[2] For numerical stability reasons the POD eigenvalues are usually not computed from the correlation matrix itself, but rather as the squares of the singular values of the snapshot matrix obtained by collecting all the snapshots as column vectors.

More difficulties arise when the non-homogeneous boundary conditions depend on time. This is a very typical case when POD-based ROMs are used for boundary control applications on unsteady flows. In [112] the time-dependent velocity boundary condition was handled by augmenting the Galerkin system (9.3) with a penalty term, so that (9.3) can be written as

$$\frac{da_i(t)}{dt} = F_i + \sum_{j=1}^{N_r} A_{ij} a_j(t) + \sum_{j=1}^{N_r} \sum_{k=1}^{N_r} C_{ijk} a_j(t) a_k(t) + \tau \left[ U_i^{\text{in}}(t) - \sum_{j=1}^{N_r} M_{ij} a_j(t) \right],$$
$$a_i(0) = (\Psi_i, \mathbf{u}_0), \tag{9.8}$$

where the boundary tensors are $U^{\text{in}}$ and $M$ are defined as

$$U_i^{\text{in}}(t) := \int_{\Gamma_{\text{in}}} \Psi_i(\mathbf{x}) \cdot \mathbf{u}_{\text{in}}(\mathbf{x})\, ds, \quad M_{ij} := \int_{\Gamma_{\text{in}}} \Psi_i(\mathbf{x}) \cdot \Psi_j(\mathbf{x})\, ds \tag{9.9}$$

with the assumption that the time-averaged base flow is zero on the inflow section $\Gamma_{\text{in}}$, i.e. $\bar{\mathbf{u}}|_{\Gamma_{\text{in}}} \equiv 0$. The penalty term $\tau > 0$ was chosen such that the correct asymptotically stable solution was obtained. This can be understood as a weak imposition of the Dirichlet condition that approaches strong imposition as $\tau \longrightarrow \infty$.

### *9.1.2 Reduced Basis Construction by Greedy Algorithms*

A popular strategy for constructing ROMs in the case of parameter-dependent problems is that of using *greedy* algorithms, based on the idea of selecting at each step the locally optimal element. This option can be seen as an alternative to POD strategy of previous section, yet preferable in the context of parametrized problems for reasons that will be sketched later on.

Before describing the greedy algorithm, let us formulate a steady version of problem (9.1), depending on a set of parameters $\boldsymbol{\mu} \in \mathscr{P} \subset \mathbb{R}^P$, in a convenient way also for the following. Here we introduce the weak form, which was not the case in Sec. 9.1.1 to go from (9.2) to (9.3). The weak form of parametrized steady Navier-Stokes equations reads as follows: find $(\mathbf{u}, p) = (\mathbf{u}(\boldsymbol{\mu}), p(\boldsymbol{\mu})) \in V \times Q$ such that

$$\begin{cases} a(\mathbf{u}, \mathbf{w}; \boldsymbol{\mu}) + b(p, \mathbf{w}; \boldsymbol{\mu}) + c(\mathbf{u}, \mathbf{u}, \mathbf{w}; \boldsymbol{\mu}) = F(\mathbf{w}; \boldsymbol{\mu}), & \forall \mathbf{w} \in V \\ b(q, \mathbf{u}; \boldsymbol{\mu}) = G(q; \boldsymbol{\mu}), & \forall q \in Q, \end{cases} \tag{9.10}$$

where the parametrized bilinear and trilinear forms are defined as follows:

$$a(\mathbf{v}, \mathbf{w}; \boldsymbol{\mu}) = \int_\Omega \frac{\partial \mathbf{v}}{\partial x_i} \nu_{ij}(\cdot; \boldsymbol{\mu}) \frac{\partial \mathbf{w}}{\partial x_j} d\Omega, \quad b(q, \mathbf{w}; \boldsymbol{\mu}) = -\int_\Omega q \chi_{ij}(\cdot; \boldsymbol{\mu}) \frac{\partial w_j}{\partial x_i} d\Omega, \tag{9.11}$$

$$c(\mathbf{v}, \mathbf{w}, \mathbf{z}; \boldsymbol{\mu}) = \int_\Omega v_i \chi_{ji}(\cdot; \boldsymbol{\mu}) \frac{\partial w_m}{\partial x_j} z_m \, d\Omega. \tag{9.12}$$

In what follows, we consider the more general case including the pressure field, so that $X = V \times Q$. Here $\boldsymbol{\mu}$ may denote both physical and geometrical parameters, whose action on the problem is encoded by parametrized tensors $\nu(\cdot;\boldsymbol{\mu}),\chi(\cdot;\boldsymbol{\mu})$. We point out that tensors components might depend *a priori* on both parameter components and spatial coordinates; see e.g. [87, 100, 109] for their complete derivation. Furthermore, $F(\cdot;\boldsymbol{\mu})$ and $G(\cdot;\boldsymbol{\mu})$ are linear forms accounting for non-homogeneous boundary data and source terms. Until stated otherwise, summation over repeated indices is understood.

The goal of the Reduced Basis (RB) method is to compute a low-dimensional approximation $(\mathbf{u}_r(\boldsymbol{\mu}),p_r(\boldsymbol{\mu}))$ of the solution to problem (9.10) by seeking a linear combination of well-chosen solutions[3] $(\Psi_i,\xi_i)=(\mathbf{u}(\boldsymbol{\mu}^i),p(\boldsymbol{\mu}^i))$ of problem (9.10), corresponding to specific choices of the parameter values:

$$\mathbf{u}_r(x;\mu):=\sum_{i=1}^{N_r}u_i(\mu)\Psi_i(x),\qquad p_r(x;\mu):=\sum_{i=1}^{N_r}p_i(\mu)\xi_i(x),\tag{9.13}$$

where the coefficients $u_i(\boldsymbol{\mu}),p_i(\boldsymbol{\mu})$ are computed by solving the following nonlinear algebraic system:

$$\begin{cases}\displaystyle\sum_{j=1}^{N_r}A_{ij}(\boldsymbol{\mu})u_j(\boldsymbol{\mu})+\sum_{j=1}^{N_r}B_{il}(\boldsymbol{\mu})p_l(\boldsymbol{\mu})+\sum_{j=1}^{N_r}C_{ijk}(\boldsymbol{\mu})u_j(\boldsymbol{\mu})u_k(\boldsymbol{\mu})=F_i(\boldsymbol{\mu}),\\ \displaystyle\sum_{j=1}^{N_r}B^T u_j(\boldsymbol{\mu})=G_l(\boldsymbol{\mu}),\end{cases}\tag{9.14}$$

with $i=1,\dots,N_r$. Reduced spaces for pressure and velocity fields (denoted respectively $Q_{N_r}^{RB}$ and $V_{N_r}^{RB}$) have the same dimension in the case of physical parametrizations, whereas geometrical parametrizations require modifying the velocity space in order to manage the divergence-free constraint; see Sect. 9.3.1. As in the case of problem (9.1), the functional forms appearing in (9.14) are obtained by Galerkin projection of the original problem (9.10) onto the RB space $X_{N_r}^{RB}=V_{N_r}^{RB}\times Q_{N_r}^{RB}$, spanned by the solutions $(\Psi_i,\xi_i)$, so that, for $1\le i,j,k\le N_r$,

$$\begin{aligned}&A_{ij}(\boldsymbol{\mu}):=a(\Psi_i,\Psi_j;\boldsymbol{\mu}),\quad B_{kj}(\boldsymbol{\mu})=b(\xi_k,\Psi_j;\boldsymbol{\mu}),\quad C_{ijk}(\boldsymbol{\mu})=c(\Psi_i,\Psi_j,\Psi_k;\boldsymbol{\mu})\\&F_i(\boldsymbol{\mu}):=F(\Psi_i;\boldsymbol{\mu}),\quad G_l=G(\xi_l;\boldsymbol{\mu}),\end{aligned}\tag{9.15}$$

resulting again in a Galerkin ROM. In the parametrized setting the goal is to approximate uniformly well all the elements of the parametric manifold of solutions

$$M(\boldsymbol{\mu})=\{\mathbf{U}(\boldsymbol{\mu}):=(\mathbf{u}(\boldsymbol{\mu}),p(\boldsymbol{\mu}))\in X,\quad \boldsymbol{\mu}\in\mathscr{P}\}$$

using finite dimensional subspaces $X_{N_r}^{RB}$ generated from elements of $M(\boldsymbol{\mu})$.

From a practical point of view, this approach is premised upon a classical Finite Element (FE) method "truth approximation" space $X_h\subset X$ of (typically very large) dimension. The RB method thus consists in a low-order approximation of the "truth"

[3] Gram-Schmidt orthonormalization is required in order to ensure the algebraic stability of the reduced basis approximation. Furthermore, in case of parameter-dependent geometries, the velocity space has to be enriched, as detailed in Sect. 9.3.

manifold $M_h = \{\mathbf{U}_h(\boldsymbol{\mu}) := (\mathbf{u}_h(\boldsymbol{\mu}), p_h(\boldsymbol{\mu})) \in X_h : \boldsymbol{\mu} \in \mathscr{D}\}$. Nonetheless, we omit the subscript $h$ wherever possible.

Next we address the construction of these subspaces. The so-called *greedy* algorithm, first proposed in [125], provides a quasi-optimal procedure for sampling the parameter space $\mathscr{P}$ – and so the manifold $M(\boldsymbol{\mu})$.

Thus, we seek a set of snapshot functions $\{\mathbf{U}(\boldsymbol{\mu}^1), \mathbf{U}(\boldsymbol{\mu}^2), \dots, \mathbf{U}(\boldsymbol{\mu}^{N_r})\}$ such that each $\mathbf{U}(\boldsymbol{\mu}) \in M(\boldsymbol{\mu})$ is well approximated by the elements of the subspace $X_{N_r} = \mathrm{span}\{\mathbf{U}(\boldsymbol{\mu}^n), 1 \le n \le N_r\}$, according to the following algorithm:

```
S_1 = {μ^1}; compute U(μ^1); X_1^RB = span{U(μ^1)};
for n = 2 : N_r
    compute U(μ^n) = argmax_{W∈M(μ)} ||W − Π_{X_{n−1}} W||_X;
    set S_n = S_{n−1} ∪ {μ^n};
    set X_n^RB = X_{n−1}^RB ∪ span{U(μ^n)};
    if max_{W∈M(μ)} ||W − Π_{X_n^RB} W||_X ≤ ε*_tol
        setN_r = n − 1;
    end;
end.
```

where $\Pi_{X_n}$ is the orthogonal projection w.r.t. the scalar product induced by $\|\cdot\|_X$ onto $X_n^{RB}$. Thus, at each step $n = 1, \dots, N_r$, $\mathbf{U}(\boldsymbol{\mu}^n)$ is the *worst case* element, which maximizes the error in approximating the subspace $M(\boldsymbol{\mu})$ using the elements of $X_n^{RB}$. However, this procedure (sometimes called *strong greedy* algorithm) is computationally infeasible: finding the maximum of the error of best approximation $\|\mathbf{W} - \Pi_{X_n}\mathbf{W}\|_X$ in $X_n^{RB}$ would require a suitable maximization algorithm, which would also involve a large number of solutions of the full-order system (9.10). In a more feasible variant of this algorithm – sometimes called *weak greedy algorithm* – we replace the max over $\mathbf{W} \in M(\boldsymbol{\mu})$ with a max over a very fine sample $\Xi_{train} \subset \mathscr{D}$ of cardinality $|\Xi_{train}| = n_{train}$, and the true error $\|\mathbf{W} - \Pi_{X_n}\mathbf{W}\|_X$ with a suitable error estimate $\Delta_n(\boldsymbol{\mu})$, satisfying

$$c_\Delta \Delta_n(\boldsymbol{\mu}) \le \|\mathbf{W} - \Pi_{X_n}^{RB}\mathbf{W}\|_X \le C_\Delta \Delta_n(\boldsymbol{\mu}), \qquad \forall \mathbf{W} \in M(\boldsymbol{\mu}) \tag{9.16}$$

for some constants $C_\Delta > c_\Delta > 0$. In this way, $\mathbf{U}(\boldsymbol{\mu}^n) = \arg\max_{\mathbf{W}\in M(\boldsymbol{\mu})} \Delta_n(\boldsymbol{\mu})$ can be computed more effectively, under the assumption that the surrogate error $\Delta_n(\boldsymbol{\mu})$ is cheap to evaluate. In Sect. 9.3.2 we recall some a posteriori error estimates for reduced basis approximations for steady Navier-Stokes equations, and refer to [101] for their practical numerical implementation.

We point out that greedy-RB sampling methods are similar in objective to, but substantially different in approach from, the POD methods, which are more expensive from a computational standpoint. In fact, in the former we only need to compute the $N_r$ retained snapshots (or *winning candidates*), which are typically very few. Only the error estimate has to be evaluated over the whole train set $\Xi_{train}$, which is very large – this is the reason why we require that the surrogate error must be cheap to evaluate. Instead, in the latter we must compute all the $N_s$ candidate snapshots as well as compute the SVD of a large matrix.

We conclude this section by mentioning also some additional techniques quite close to POD for generating efficiently reduced spaces, the Centroidal Voronoi Tessellation (CVT) [25, 26, 46] and the Proper Generalized Decomposition (PGD) method [92], which has been recently applied to the solution of Navier-Stokes equations [47, 117]. Recent contributions are also contained in this book, see the chapter by Farhat and Amsallem dealing both with POD and Galerkin projection, and by Urban et al. A comparison on reduced representation approximations is provided instead by Bebendorf et al. in this volume.

The rest of the chapter is structured as follows: In Sect. 9.2 we lay out some general guidelines that should be considered before attempting to build a ROM for any specific fluid problem. In Sect. 9.3 we address some issues related to approximation stability and error estimation which occur in the *reduced basis* approximation of steady-state solutions of parametrized Navier-Stokes equations. Moreover, in Sect. 9.4 we discuss specific issues related to the POD/Galerkin -based ROMs, such as the need for stabilizing the ROM, and how to ensure that the proper long-term behavior is recovered by the ROM. Some final remarks and a quick glance on current developments in the field are given in Sect. 9.5.

## 9.2 Some Principles of Model Reduction of Fluid Systems

In this section we try to condense some fundamental principles to take into account when building ROMs that are known to most practitioners in the reduced-order modelling community but not always clearly communicated or established in literature. They are based both on our personal experience as well as on the general impression conveyed by state-of-the art literature on this subject. We have included motivating examples and several references to literature. Moreover, together with the description of these fundamental principles, we also sketch the basic ingredients of reduced-order models for the computational reduction of PDEs.

### *9.2.1 "Never try to reduce the irreducible"*

Once a full-order computational model for the fluid dynamics problem has been constructed, e.g. by means of finite elements or finite volumes discretizations, we may begin the process of constructing a suitable reduced-order model (ROM). The first step is to verify the assumption that the trajectories of the system live on a low-dimensional submanifold of the full space. From a practical point of view such a check is straightforward: it is sufficient to compute several trajectories of the full-order dynamical system, to collect snapshots into one matrix, and to perform a POD by using the singular value decomposition of this matrix. If the decay of the singular values is sufficiently rapid, then a limited number of modes will potentially suffice to represent the solution trajectories and an attempt at building a ROM can be performed.

It is easy to construct examples where slow decay or even no decay of the singular values of empirical snapshots is obtained. Consider for instance the one-dimensional linear transport equation

$$\begin{aligned} \partial_t u(x,t) + c\partial_x u(x,t) = 0, \qquad (x,t) \in \mathbb{R} \times (0,T) \\ u(x,0) = u_0(x), \quad x \in \mathbb{R} \end{aligned} \tag{9.17}$$

with solution $u(x,t) = u_0(x - ct)$. Take $N_s$ snapshots of this solution at times $t = 0, \Delta t, 2\Delta t, \ldots, (N_s - 2)\Delta t, T$. Assume that $u_0 \in L^2(\mathbb{R})$ and localized so that the measure of its support $\lambda(\mathrm{spt}(u_0)) < |c|\Delta t/2$. Thus, it follows that

$$\int_{\mathbb{R}} u_j(\xi)\, u_k(\xi)\, d\xi = \int_{\mathbb{R}} u_0(\xi - cj\Delta t)\, u_0(\xi - ck\Delta t)\, d\xi = \|u_0\|_{L^2(\mathbb{R})}\, \delta_{jk} \tag{9.18}$$

so that the correlation matrix of the snapshots (9.6) is diagonal with all eigenvalues equal. The singular values of the snapshot matrix do not decay at all, so that snapshot-based POD is not successful at representing traveling waves.

Using the empirical singular values to measure the feasibility of model reduction can also be theoretically justified. As already mentioned, the subset where solutions of the dynamical system live has typically the structure of a compact manifold $M(\boldsymbol{\mu})$ belonging to some larger function space $X$. To quantify how well such a manifold can be approximated by *Galerkin projection* onto a low-dimensional subspace, one can rely on the concept of Kolmogorov $n$-width, defined as

$$d_n(M;X) := \inf_{X_n \subset X} \sup_{\mathbf{u} \in M} \inf_{\tilde{\mathbf{u}} \in X_n} \|\mathbf{u} - \tilde{\mathbf{u}}\|_X \tag{9.19}$$

where the first infimum is taken over all linear subspaces $X_n \subset X$ of dimension $n$. The decay of $d_n \to 0$ as $n \to \infty$ can then be used as a measure of how many (POD or greedy-RB) modes need to be considered for the ROM (9.2) – the faster the decay, the smaller need to be the dimension of the linear subspace.

In the case that one is able to obtain exponential convergence in the $n$-width, that is to say $d_n(M;X) \leq C\exp(-\alpha n^{\beta})$ for some constants $C, \alpha, \beta > 0$, exponential convergence is also inherited (albeit at a reduced rate) by reduced-order approximations, and the same equivalence holds also for algebraic convergence rates, as was recently proved in [18]. Results regarding the connection between $n$-width decay rates and greedy algorithm converges rates can be found in [18, 22] for parametric problems, in [55] for time-dependent problems, and results regarding the $n$-width decay rates for parameter-dependent elliptic PDEs in [77, 86]. We stress that such results rely on a suitable sampling algorithm (such as the greedy algorithm, that selects proper time instances $t^n$ or parameter points $\boldsymbol{\mu}^n$) where to compute the snapshots $\mathbf{U}_n(\cdot, t^n)$ (respectively $\mathbf{U}_n(\cdot, \boldsymbol{\mu}^n)$) according to a reliable estimate of the error between the

ROM and the full-order model. This is in order to actually find a (quasi-)optimal approximation space. We will revisit this point in Sect. 9.2.2.

Exponentially fast convergence of numerical approximations is often linked to spectral approximations of smooth (analytic) functions. In the case of elliptic coercive PDEs with random coefficients it was shown (see e.g. [10], Lemma 3.2) that if an elliptic and uniformly coercive parametric bilinear form $a : X \times X \times \mathscr{P} \to \mathbb{R}$ (consider for instance the scalar equivalent of the one defined in (9.11)) is such that

$$a(w,w;\boldsymbol{\mu}) \geq \nu_{\min}\|w\|_X^2 \quad \text{for all } w \in X, \boldsymbol{\mu} \in \mathscr{P} \subset \mathbb{R} \tag{9.20}$$

and its dependence on $\boldsymbol{\mu}$ is analytic, then also the solutions $u(\boldsymbol{\mu})$ of

$$a(u(\boldsymbol{\mu}),w;\boldsymbol{\mu}) = f(w) \quad \text{for all } w \in X \tag{9.21}$$

for any $f \in X'$ are analytic functions of $\boldsymbol{\mu}$, provided that the parameter range $\mathscr{P} = [\boldsymbol{\mu}_{\min}, \boldsymbol{\mu}_{\max}]$ is bounded. The analyticity is then sufficient to prove exponential convergence of certain approximations to the solutions by expanding the solution as a power series. For example, when an approximation $u_{h,p}$ is obtained by using the FE method in space (with mesh size $h$) and the spectral collocation method in parameter (with polynomial order $p$), an exponential convergence result was obtained in [10] (Theorem 4.1): for any $\boldsymbol{\mu} \in \mathscr{D}$

$$\|u(\boldsymbol{\mu}) - u_{h,p}(\boldsymbol{\mu})\|_{L^2_P(\mathscr{P})\otimes X} \leq \frac{1}{\sqrt{\nu_{\min}}} \inf_{w\in L^2_P(\mathscr{P})\otimes X} \left( \frac{1}{|\mathscr{P}|} \int_{\mathscr{P}\times\Omega} \nu |\nabla(u(\boldsymbol{\mu}) - w)|^2 \right)^{1/2}$$
$$+ C\exp\left( -p \log\left[ \frac{2\tau}{|\mathscr{P}|} \left( \sqrt{1 + \frac{|\mathscr{P}|^2}{4\tau^2}} \right) \right] \right), \tag{9.22}$$

where the (sub)exponential convergence rate in $p$ depends on the distance $\tau > 0$ between $\mathscr{P}$ and the nearest singularity in the complex (parameter) plane. Unfortunately, theoretical results that give estimates on the regularity of Navier-Stokes solution with respect to parameters acting on boundary terms, external forces, or initial data require stringent assumptions of small data and small Reynolds number that are not usually fulfilled by realistic flows. Nevertheless, exponential convergence of ROM approximation is often recovered also in nonlinear fluid problems.

### 9.2.2 *"If it is not in the snapshots, it is not in the ROM"*

We now turn to the question of how to choose the dimension $N_r$ of the reduced space, so that we can take advantage of a substantial computational reduction but dealing with a reliable reduced-order model. In the case of the greedy-RB algorithm, reliable error estimates $\Delta_n(\boldsymbol{\mu})$ satisfying (9.16) can be used to assess the quality of the ROM, so that the sampling procedure stops when the error between the full-order model and the ROM is (estimated) lower than a give threshold, say $10^{-m}$ with $m \geq 2$, *uniformly* over the parameter space.

In the POD case, we can rely on the Relative Information Content (RIC) of the POD basis, which is defined as the ratio between the sum of the retained POD modes vs. the sum of the whole set of eigenvalues of the correlation matrix:

$$\mathrm{RIC} := \frac{\sum_{i=1}^{N_r} \lambda_i}{\sum_{i=1}^{N_s} \lambda_i}. \tag{9.23}$$

The RIC is usually chosen up to $100(1-\alpha)\%$ by retaining a limited number of the *most energetic* POD modes, being, say $\alpha \in [10^{-m}, 10^{-1}]$ for a suitable $m > 1$. Flow features that are not sufficiently energetic will be omitted in the POD and thus cannot be captured by the ROM. A possible way to check which features to retain is to use a spatially weighted $L^2$-norm in the computation of the POD that gives more weight to features located at particular sites of interest.

Individual snapshots in the ensemble can be weighted accordingly to their importance, as proposed in [33]. In a series of papers (see e.g. [31, 36, 37]) Navon et al proposed a dual-weighted POD method, where the weights assigned to each snapshot were derived from an adjoint related to the optimality system of a variational data assimilation problem in meteorology. It is also known that for compressible flows the choice of inner product and weighting of the different flow variables (velocity, pressure, speed of sound) in the snapshot matrix can have a large effect on the stability and accuracy of the ROM [14, 35]. Similarly, the $H^1$ inner product was recommended for the computation of POD modes for compressible Navier-Stokes equations in [66] for the purpose of enhancing stability.

On the other hand, in the case of parametrized problems, the approximation properties of the basis depend on the parameter points $\boldsymbol{\mu}^k$ where the snapshots are computed. It is known that, in general, a POD basis computed at a single parameter point is not a good approximation for solutions computed at different parameter points. This is another reason, in addition to computational efficiency pointed out in Sect. 9.1.2, why a greedy algorithm should be chosen in order to manage with a careful sampling of a parameter space. In summary, two typical improvements to the POD sampling process are adopted:

1. *Adaptivity*. In this case an initial POD basis is constructed and the resulting ROM used for simulations, but is later updated based on some problem-dependent criteria. This is a typical approach in ROM-based optimization and optimal control applications, such as those presented in [17, 103], where the snapshots and the POD are updated after every optimization step to improve the accuracy of the ROM near the optimal point. The price to be paid is that the cost of the optimization loop will increase due to the need of additional full-order simulations to update the ROM. The idea of a trust-region POD method was presented in [49]. In this case, the POD version of the optimality system is solved at each iteration within a trust-region radius $\Delta^{(k)}$ to obtain a quasi-optimal $c^{(k+1)}$ set of controls. Then the full-order Navier-Stokes equations are solved with the quasi-optimal controls to obtain $\mathbf{u}(c^{(k+1})$. The discrepancy between the ROM prediction and the full-order solution is then measured, and if its too large the step is rejected and the trust region radius decreased, $\Delta^{(k+1)} < \Delta^{(k)}$. Otherwise, the step is accepted and the trust region possibly increased, $\Delta^{(k+1)} \geq \Delta^{(k)}$. The ROM is updated after each iteration step to incorporate the newly computed snapshots.

2. *Optimality* (or *near-optimality*). A priori error estimates for POD approximations were introduced in [72] and can be used to gauge the total number of POD modes to retain to achieve a given representation accuracy at one single parameter point. In this case, snapshots are typically chosen iteratively by measuring the error of the current ROM at different trial points of the parameter space, then computing snapshots at the parameter point where the maximum error (estimate) is obtained and adding them to the ROM, like in the greedy-RB algorithm, first proposed in [52], and now standard in the parametric model reduction community. For sampling in time POD-greedy strategies have been proposed for linear evolution equations in [56], the viscous nonlinear Burgers' equation in [93], and the Navier-Stokes equations in [127].
   In [73] the authors derived sensitivity equations to measure the effect of adding new snapshots in the POD basis and use them to find optimal locations for new snapshots that minimize the error between the POD-solution and the trajectory of the full-order system. This can avoid the expensive computation of full-order trial solutions typically needed in a POD-greedy approach. Furthermore, in [24] the POD procedure was extended to incorporate goal-oriented quantities related to specific outputs of interest over the entire range of parameters.

### 9.2.3 "Exploit the known structure of the solutions"

Both POD and greedy-RB strategies use a set of full-order solutions to build a global basis for the approximation of the solution of a PDE problem for any given time $t \in (0, T)$ or parameter value $\boldsymbol{\mu} \in \mathscr{P}$. It is important to understand that the basis functions of a ROM do not really tell us much about the dynamical structure of a time-dependent problem. In the case of parameter-dependent problems, a parameter value $\boldsymbol{\mu}$ different from the snapshots $\boldsymbol{\mu}^1, \dots, \boldsymbol{\mu}^{N_r}$ may result in a flow regime that is qualitatively very different (for instance when the flow is parametrized with respect to the Reynolds number) than those exhibited at the snapshot parameters. Also parametrized geometrical features can greatly affect the qualitative behavior of the solutions. Thus, in order to make the ROM capable to represent the physics of the full model correctly, we need to let the equations play a role also at the evaluation level, for any new problem instance to solve. This is the reason why, in equations (9.3) or (9.14), we follow a *projection* approach, rather than an *interpolation*-based strategy. This makes more reliable also the evaluation of outputs derived from the solution, such as energy, stresses, vorticity, etc.

We still have to explain how to pursue a strong computational reduction when solving the problem obtained by plugging the reduced solution into the equations. Thus, we need to equip ROMs of previous sections with an efficient implementation aiming at decoupling the generation and projection stages. Let us focus, for the sake of clarity, on the case of parametrized problems. In particular, two ingredients need to come into play, in order to obtain the so-called *Offline/Online* splitting:

1. *Affine parameter dependence*. In order to speed up the evaluation of a reduced approximation when the differential operators depend on some parameters, the

key point is to isolate the contribute of parametrized quantities in the differential operators, so that expensive parameter-independent structures can be computed *Offline* and stored once, whereas inexpensive parameter-dependent quantities can be efficiently evaluated *Online* for each new value of the parameters.

To make the *Online* evaluation step efficient, we need to take the parametrized quantities out of the integrals appearing in (9.11)-(9.12). The usual assumption required in the reduced-basis methods is the so-called *affine parameter dependence*, i.e. we require that parametrized forms (9.11)-(9.12) can be expressed as linear combinations of parameter-independent operators:

$$a(\mathbf{v},\mathbf{w};\boldsymbol{\mu}) = \sum_{q=1}^{Q^a} \Theta_a^q(\boldsymbol{\mu}) a^q(\mathbf{v},\mathbf{w}), \qquad b(q,\mathbf{w};\boldsymbol{\mu}) = \sum_{q=1}^{Q^b} \Theta_b^q(\boldsymbol{\mu}) b^q(q,\mathbf{w}); \quad (9.24)$$

$$c(\mathbf{v},\mathbf{w},\mathbf{z};\boldsymbol{\mu}) = \sum_{q=1}^{Q^c} \Theta_c^q(\boldsymbol{\mu}) c^q(\mathbf{v},\mathbf{w},\mathbf{z}) \qquad (9.25)$$

for some integers $Q^a$, $Q^b$, $Q^c$, where $q$ is a condensed index of $i, j$ quantities. This is straightforward when dealing with common physical parametrizations (e.g. by considering Reynolds number, Grashof number, Prandtl number, inflow velocity peaks, etc. as parameters [42,44]) or simple affine geometrical parametrization – in all these cases, parametrized tensors entering in (9.11)-(9.12) depend only on parameter $\mu$. Instead, when parametrized tensors depend also on $x$, affinity assumptions (9.24)-(9.25) can only be recovered by suitable approximations, such as the ones based on the so-called Empirical Interpolation Method (EIM); see e.g. [15,85].

2. *Reduced matrix structures*. Once the parameters have been taken out of the operators by requiring the affine parameter dependence (9.24)-(9.25), the reduced operators (9.15) can be expressed (e.g. for the diffusion term) as

$$A_{ij}(\boldsymbol{\mu}) = a(\Psi_i,\Psi_j;\boldsymbol{\mu}) = \sum_{q=1}^{Q^a} \Theta_a^q(\boldsymbol{\mu}) a^q(\Psi_i,\Psi_j) = \sum_{q=1}^{Q^a} \Theta_a^q(\boldsymbol{\mu}) A_{ij}^q.$$

In order to make the *Online* evaluation independent of the dimension of the full-order space, structures $A^q$ and $C^q$ corresponding to parameter-independent operators must be constructed properly and stored during the *Offline* stage.

We remark that the basis functions are given by full-order approximations of (9.10) for selected values of the parameters, under the form

$$\Psi_i(x) = \sum_{m=1}^{N^{\mathbf{u}}} \Psi_i^m \phi_m^{\mathbf{u}}(x), \quad \xi_i(x) = \sum_{m=1}^{N^p} \xi_i^m \phi_m^p(x),$$

where $\{\phi_m^{\mathbf{u}}(x)\}_{m=1}^{N^{\mathbf{u}}}$, $\{\phi_m^p(x)\}_{m=1}^{N^p}$ are two bases of the full-order (velocity, resp. pressure) approximation spaces, of dimension $N^{\mathbf{u}}$, $N^p$, respectively. Thus, the assembling of reduced-order algebraic structures (9.15) can be efficiently performed by combining the matrices collecting the basis functions, given by

$$\mathbb{Z}_{\mathbf{u}} = [\Psi_1 | \dots | \Psi_{N_r}] \in \mathbb{R}^{N^{\mathbf{u}} \times N_r}, \qquad \mathbb{Z}_p = [\xi_1 | \dots | \xi_{N_r}] \in \mathbb{R}^{N^p \times N_r},$$

and the full-order algebraic structures. It is straightforward to check that, e.g.,

$$a^q(\Psi_i, \Psi_j) = \sum_{m=1}^{N^{\mathbf{u}}} \sum_{n=1}^{N^{\mathbf{u}}} \Psi_i^m a^q(\phi_m^{\mathbf{u}}, \phi_n^{\mathbf{u}}) \Psi_j^n, \qquad \text{i.e.} \qquad A^q(\boldsymbol{\mu}) = \mathbb{Z}_{\mathbf{u}}^T \tilde{A}^q(\boldsymbol{\mu})\, \mathbb{Z}_{\mathbf{u}},$$

where $\tilde{A}^q_{mn}(\mu) = a^q(\phi_m^{\mathbf{u}}, \phi_n^{\mathbf{u}})$ is the full-order stiffness matrix corresponding to the bilinear form $a^q(\cdot,\cdot)$. The same procedure can be applied to pressure and nonlinear terms as well; see e.g. [87, 100] for a detailed explanation.

Both these ingredients, together with the snapshot selection procedures and (wherever available) rigorous error estimates allow to successfully apply Galerkin ROMs to incompressible flows. However, some caveats should be mentioned.

For instance, the evaluation of the trilinear convective term – given by $C_{ijk}a_j(t)a_k(t)$ in (9.3) or $C_{ijk}(\boldsymbol{\mu})u_j(\boldsymbol{\mu})u_k(\boldsymbol{\mu})$ in (9.14) – even in the reduced-order formulation requires evaluating tensorial terms of relatively large sizes. This is even more involved when the size $Q^c$ of the affine expansion (9.25) is large. For more general nonpolynomial nonlinearities of the form

$$\left\langle \mathbf{f}\left( \bar{\mathbf{u}}(x) + \sum_{i=1}^{N_r} a_i(t)\Psi_i(x) \right), \bar{\mathbf{u}}(x) + \sum_{j=1}^{N_r} a_j(t)\Psi_j(x) \right\rangle \tag{9.26}$$

deflating the nonlinear terms to their full-order representations may be necessary in order to evaluate the nonlinear terms, negating many advantages of using a ROM in the first place. In order to reduce the online cost of evaluating the nonlinear term(s), several "hyper-reduction" techniques have been proposed, such as DEIM [30] (Discrete Empirical Interpolation Method), DBPIM [12] (Discrete Best Points Interpolation Method), MPE (Missing Point Estimation) [8] and GNAT [5] (Gauss-Newton with Approximated Tensor quantities). In general, most of these methods attempt to approximate the nonlinearity using linear combinations of the POD basis functions

$$\mathbf{f}\left( \bar{\mathbf{u}}(x) + \sum_{i=1}^{N_r} a_i(t)\Psi_i(x) \right) \approx f_0\, \bar{\mathbf{u}}(x) + \sum_{i=1}^{N_r} \tilde{f}_i(t)\, \Psi_i(x) \tag{9.27}$$

and differ mainly on the strategy of choosing the approximation coefficients $\tilde{f}_i(t)$. When the nonlinearity is treated using a Newton algorithm, a similar approximation can be applied to the Jacobian $J_f$, see e.g. [5, 30]. A contribution on discrete EIM (DEIM) is the chapter by Antil et al. in this book.

### 9.2.4 "Divide and conquer whenever possible"

The *rationale* behind the efficacy of the ROMs we have discussed so far is the regularity of the parameter dependence in the case of parametrized problems like (9.10)

– respectively, time continuity in the case of time-dependent problems like (9.1). In other words, solutions to these problems lie on a low-dimensional *manifold*, as already pointed out in Sect. 9.1.2. The more regular the manifold (and the parametric dependence), the more conveniently the solution can be approximated by a suitable combination of snapshots.

However, even laminar flow can experience strong qualitative changes (bifurcations) when critical parameters such as the Reynolds number is varied. For example, the flow behind a cylinder experiences first a transition from steady flow to a time-periodic flow, then a loss of periodicity in the vortex shedding, and finally transition to a chaotic turbulent regime as the Reynolds number is gradually increased. In order to make sure that a ROM approximates correctly the fluid flow in some range of the parameter(s), we require that the parameter space (or the time interval) are chosen such that the manifold is locally a branch of nonsingular solutions.

Although quite restrictive, this is a standard assumption also in the case of full-order approximations, based e.g. on the FE method (see e.g. [21,27]). Nevertheless, bases constructed using the greedy algorithm provide reliable approximations also in the case of bifurcation points included in the parameter space; for instance, ROMs have been used to track particular solution branches past the bifurcation point, see e.g. [97,119]. In case of parametrized flows, in order to minimize the required number of basis functions, a good ROM should be tailored so that different flow regimes can be captured in a reliable way. Since POD-based ROMs provide poor approximations away from the parameter values for which the snapshot solutions were computed, it rarely makes sense to try and develop one global approximation basis for the entire parameter space. Many works have been focused in these last years on possible strategies to rectify this aspect.

One possibility is to combine ROMs computed for different physical flow regimes. In [3,4] the ROMs computed at different parameter points were interpolated to obtain a new ROM that was valid also in the intermediate zone between the original parameter points. In [58,59] the parametric sensitivities of the POD modes were computed and added to the snapshot set, which improved the validity of the reduced solutions away from the parametric snapshots. However, in a more involved geometrical parametrization case the ROM failed completely, as it did not converge to the exact solution even when the number of POD modes was increased.

A "compact POD" approach based on goal-oriented Petrov-Galerkin projection was proposed in [28], in order to minimize the approximation error subject to a chosen output criteria, also including sensitivity information (with proper weighting coming from the Taylor-expansion) and including "mollification" of basis functions far away from the snapshot parameter. A further option, described in [5], exploits a *k-means* clustering procedure to construct local ROMs by grouping together nearby snapshots. In this way, once the snapshots have been computed, the reduced space is partitioned in subregions and a local reduced basis is assigned to each subregion. This can be seen as an adaptive version of a former strategy based on the so-called *Centroidal Voronoi Tessellation*, introduced in [46] and extended in [25,26].

Finally, we mention that local ROMs can be properly combined also in view of a further computational reduction for instance in the solution of parametrized problems featuring a repetitive geometrical structure – such as networks, or multi-domain configurations. The Reduced Basis Element (RBE) method combines domain decomposition with parametric ROMs, by exploiting nonconforming approaches – such as mortar methods or discontinuous Galerkin methods – between the subdomains and the greedy RB method within each subdomain. Recent application of the RBE method to fluid flows can be found e.g. in [43, 65, 81]. A more advanced variant exploits static condensation at the interdomain level [62] by connecting (at some interfaces, or ports, during the online stage) a library of reference, interchangeable components.

## 9.3 Model Reduction of Steady Viscous Flows

In this section we summarize those features which are peculiar to ROMs for parametrized steady viscous flows, such as *inf-sup* stability, correct treatment of pressure, suitable a posteriori error estimates. We also point out the analogies with the case of linear viscous flows modelled by Stokes equations. In particular, we exploit a greedy algorithm for the construction of the reduced space: at each step the basis of snapshots is augmented by the solution corresponding to the largest error estimate. The downside is that the method is completely reliant on the existence of computable a posteriori error bounds, which are not really available for the unsteady Navier-Stokes equations, as we mentioned. This is the main reason why, so far, this method has largely been limited to steady Navier-Stokes equations.

### *9.3.1 A Question of Stability: inf-sup Constants and Supremizers*

A feature of the standard POD-Galerkin ROM (9.3) is that the pressure term $-\nabla p$ has been completely eliminated. In fact, assuming that the POD modes $\Psi_i$ satisfy the strong incompressibility constraint by construction, $\nabla \cdot \Psi_i = 0$ pointwise, integration by parts of the pressure-gradient term evaluated on the POD modes gives

$$(\nabla p, \Psi_i) = \int_{\Omega} \nabla p \cdot \Psi_i \, d\mathbf{x} = -\int_{\Omega} p(\nabla \cdot \Psi_i) \, d\mathbf{x} + \int_{\partial\Omega} p(\Psi_i \cdot \boldsymbol{n}) \, ds, \qquad (9.28)$$

which demonstrates that the pressure only enters the ROM on the boundary and for enclosed flows ($\Psi_i \cdot \boldsymbol{n} \equiv 0$ on $\partial\Omega$) it vanishes completely from the equations. For instance, this is the case of a standard driven cavity problem. It should be noted, however, that the situation also depends on the choice of the adopted spatial discretization. For standard FE discretizations the incompressibility of solutions applies only elementwise, i.e.

$$\int_{K \in \mathscr{T}_h} \nabla \cdot \Psi_i \, d\mathbf{x} = 0 \quad \text{for all mesh elements } K \in \mathscr{T}_h \qquad (9.29)$$

so that unless piecewise constant functions in each mesh element $K$ are used for the pressure, the term $-\int_\Omega p(\nabla \cdot \Psi_i)\, d\mathbf{x}$ does not vanish identically. Nevertheless, this term is neglected for many flows as small and unnecessary to enforce the incompressibility of the ROM solutions. It is known that neglecting the pressure term for convectively unstable shear layers, especially ones with two-dimensional mixing layers, can result in large errors as was demonstrated in [96]. Pressure-extended ROMs include also the pressure in the equations, either by deriving the necessary terms in the expansion (9.3) to account for the pressure [96], or by performing a separate POD to construct another basis $\{\Phi_j\}_{j=1}^{N_r}$ for the pressure field [16, 79]. The benefit of the latter approach is that the pressure field is immediately recovered without any post-processing steps necessary.

We focus our analysis on pressure-extended ROMs, using a greedy algorithm to also build a basis for the pressure. In this way, for each selected parameter value, we compute both the (truth FE approximation of the) velocity and the pressure fields. Reduced velocity and pressure spaces result as follows:

$$V_{N_r} := \text{span}(\Psi_j \,:\, j = 1,\ldots,N_r), \quad Q_{N_r} := \text{span}(\Phi_j \,:\, j = 1,\ldots,N_r),$$

(we omit the superscript RB for the sake of brevity) where $\Psi_j \in V_h$ and $\Phi_j \in Q_h$ for any $j = 1,\ldots,N_r$, being $V_h$ and $Q_h$ the truth velocity and pressure approximation spaces. Mathematically, a necessary and sufficient condition ensuring the ROM stability is the *reduced (Brezzi) inf-sup condition*

$$\beta_r(\boldsymbol{\mu}) := \inf_{q \in Q_{N_r}} \sup_{\mathbf{v} \in V_{N_r}} \frac{b(q,\mathbf{v};\boldsymbol{\mu})}{\|q\|_Q \|\mathbf{v}\|_V} > 0, \tag{9.30}$$

which is obviously related to, but not implied by, the *full-order (Brezzi) inf-sup condition*

$$\beta_h(\boldsymbol{\mu}) := \inf_{q \in Q_h} \sup_{\mathbf{v} \in V_h} \frac{b(q,\mathbf{v};\boldsymbol{\mu})}{\|q\|_Q \|\mathbf{v}\|_V} > 0, \tag{9.31}$$

for velocity and pressure spaces $V_h \supset V_{N_r}$ and $Q_h \supset Q_{N_r}$. We recall that $b(q,\mathbf{v};\boldsymbol{\mu})$ denotes the pressure/divergence bilinear form, defined in (9.11). We also point out that now the stability factors such as $\beta_r(\boldsymbol{\mu})$, $\beta_h(\boldsymbol{\mu})$ are functions of the parameter vector $\boldsymbol{\mu}$, rather than *constants*, as in usual discretization techniques. We remind that (9.31) is ensured e.g. by choosing as $V_h \times Q_h$ the space of Taylor-Hood $\mathbb{P}_2 - \mathbb{P}_1$ elements (see [19, 20]); however, this choice is not restrictive, the whole construction keeps holding for other spaces combinations as well (e.g. [99]).

Instead, in order to fulfill the reduced inf-sup condition (9.30), we define for each pressure basis function $\Phi_j$ the corresponding inner supremizer velocity function [106, 109]

$$T^{\boldsymbol{\mu}} \Phi_j := \underset{\mathbf{v} \in V_h}{\text{argsup}} \frac{b(\Phi_j,\mathbf{v};\boldsymbol{\mu})}{\|\mathbf{v}\|_V}, \tag{9.32}$$

which can be obtained by solving the discrete elliptic problem

$$(T^{\boldsymbol{\mu}} \Phi_j, \mathbf{v})_V = b(\Phi_j,\mathbf{v};\boldsymbol{\mu}), \qquad \text{for all } \mathbf{v} \in V_h. \tag{9.33}$$

By applying (9.32) and enriching the RB velocity space $V_h$ to include the inner supremizers, we define a new extended velocity space as

$$V_{N_r}^* := V_{N_r} \oplus \text{span}(T^{\boldsymbol{\mu}} \Phi_j \, : \, j = 1, \ldots, N_r),$$

such that

$$\begin{aligned} 0 < \beta_h(\boldsymbol{\mu}) &= \inf_{q \in Q_h} \sup_{\mathbf{v} \in V_h} \frac{b(q, \mathbf{v}; \boldsymbol{\mu})}{\|q\|_Q \|\mathbf{v}\|_V} \leq \inf_{q \in Q_r} \sup_{\mathbf{v} \in V_h} \frac{b(q, \mathbf{v}; \boldsymbol{\mu})}{\|q\|_Q \|\mathbf{v}\|_V} \\ &= \inf_{q \in Q_r} \frac{b(q, T^{\boldsymbol{\mu}} q; \boldsymbol{\mu})}{\|q\|_Q \|T^{\boldsymbol{\mu}} q\|_V} \leq \inf_{q \in Q_r} \sup_{\mathbf{v} \in V_{N_r}^*} \frac{b(q, \mathbf{v}; \boldsymbol{\mu})}{\|q\|_Q \|\mathbf{v}\|_V} = \beta_r(\boldsymbol{\mu}). \end{aligned} \tag{9.34}$$

Thus, the inf-sup stability of the full-order space now implies the stability of the supremizer-enchanced reduced space, provided that this latter is enriched with the solutions of the *supremizer* equation (9.33).

We remark that, by enriching $V_{N_r}$ with the supremizers $\{T^{\boldsymbol{\mu}} \Phi_j\}_{j=1}^{N_r}$, the new RB velocity space $V_{N_r}^*$ has dimension $2N_r$, the double of the dimension $N_r$ of the RB pressure space.

The treatment of the (Brezzi) inf-sup stability through the supremizer operator is common to Stokes and Navier-Stokes equations, and more in general to any problem written under a *saddle-point* form. Further details about the efficient construction of the supremizer solutions and the Gram-Schmidt orthonormalization of the RB basis functions can be found for instance in [87, 106, 109], whereas a general context drawn for saddle-point problems has been developed in [50].

## *9.3.2 Certification of ROMs for the Steady Navier-Stokes Equations*

We now introduce the main aspects related with *a posteriori* error estimation in the RB context for parametrized steady Navier-Stokes equations. This approach is in analogy with the so-called *(Babuška) inf-sup stability theory* [11], which can be seen as a generalization to the Petrov-Galerkin case of the Lax-Milgram result for the Galerkin-type formulation. Its application to the Stokes problem is just a possible use, as shown in [106], where a general framework to compute error bounds for noncoercive problems solved by the RB method has been introduced. Within this framework, a joint residual-based estimation for velocity and pressure fields in the Stokes case can be easily obtained under the form

$$\|\mathbf{U}_h(\mu) - \mathbf{U}_r(\mu)\|_X \leq \frac{\|r(\boldsymbol{\mu})\|_{X'}}{\beta_{S,h}^{LB}(\boldsymbol{\mu})} =: \Delta_{N_r}(\boldsymbol{\mu}), \qquad \forall \mu \in \mathscr{D}, \tag{9.35}$$

where:

- $\mathbf{U}_h(\mu) = (\mathbf{u}_h(\boldsymbol{\mu}), p_h(\boldsymbol{\mu})) \in X_h = V_h \times Q_h$ and $\mathbf{U}_r(\boldsymbol{\mu}) = (\mathbf{u}_r(\boldsymbol{\mu}), p_r(\boldsymbol{\mu})) \in X_{N_r} = V_{N_r}^* \times Q_{N_r}$ denote the truth and the RB approximations of velocity and pressure;

- $\|r(\boldsymbol{\mu})\|_{X'} = \sup_{\mathbf{V}\in X_h} r(\mathbf{V};\boldsymbol{\mu})/\|\mathbf{V}\|_X$ is the dual norm of the global residual

$$r(\mathbf{V};\boldsymbol{\mu}) := r_{\mathbf{u}}^S(\mathbf{v};\boldsymbol{\mu}) + r_p(q;\boldsymbol{\mu}),$$

being

$$\begin{aligned} r_{\mathbf{u}}^S(\mathbf{v};\boldsymbol{\mu}) &:= F(\mathbf{v};\boldsymbol{\mu}) - a(\mathbf{u}_r(\mu),\mathbf{v};\mu) - b(p_r(\mu),\mathbf{v};\mu), \\ r_p(q;\boldsymbol{\mu}) &:= G(q;\boldsymbol{\mu}) - b(q,\mathbf{u}_r(\mu);\mu); \end{aligned} \tag{9.36}$$

- the bilinear form $A_S(\cdot,\cdot;\boldsymbol{\mu}) : X \times X \to \mathbb{R}$ denotes the global Stokes operator

$$A_S(\mathbf{U},\mathbf{V};\boldsymbol{\mu}) := a(\mathbf{u},\mathbf{v};\boldsymbol{\mu}) + b(p,\mathbf{v};\boldsymbol{\mu}) + b(q,\mathbf{u};\boldsymbol{\mu}); \tag{9.37}$$

- $\beta_{S,h}^{LB}(\boldsymbol{\mu})$ is a computable lower bound for the Babuška inf-sup stability factor $\beta_{S,h}(\boldsymbol{\mu})$, involving the global Stokes operator:

$$\exists\, \beta_{S,h}^{LB}(\boldsymbol{\mu}) > 0 : \; \beta_{S,h}(\boldsymbol{\mu}) = \inf_{\mathbf{U}\in X_h} \sup_{\mathbf{V}\in X_h} \frac{A_S(\mathbf{U},\mathbf{V};\boldsymbol{\mu})}{\|\mathbf{U}\|_X \|\mathbf{V}\|_X} \geq \beta_{S,h}^{LB}(\boldsymbol{\mu}), \quad \forall\, \boldsymbol{\mu} \in \mathscr{D}. \tag{9.38}$$

In this way, the stability of the reduced basis approximation is based on Brezzi's saddle point theory (and the introduction of a supremizer operator on the pressure terms), whereas a rigorous *a posteriori* error estimation procedure for velocity and pressure fields is based on Babuška's inf-sup constant.

Alternatively, we could rely on the Brezzi's theory also for the sake of error estimation, by deriving two distinct error bounds for velocity and pressure, as shown in [50]. However, despite their similar effectivity, these latter require the approximation of two stability factors (for stiffness and pressure/divergence terms) and of a continuity constant (for the stiffness term), which entail larger computational costs in a parametrized context.

In the Navier-Stokes case we can instead obtain a rigorous *a posteriori* error estimation by relying on the so-called Brezzi-Rappaz-Raviart (BRR) theory [21, 27], which is useful for the analysis of a wider class of nonlinear equations. We require some slight modifications with respect to the linear preliminaries, even if also for the Navier-Stokes problem the *a posteriori* error estimation takes advantage of the dual norm of residuals and of an effective lower bound of a suitable (parametric) stability factor, given in this case by the Babuška inf-sup constant referred not to the global Navier-Stokes operator

$$A(\mathbf{U},\mathbf{V};\boldsymbol{\mu}) = A_S(\mathbf{U},\mathbf{V};\boldsymbol{\mu}) + C(\mathbf{U},\mathbf{U},\mathbf{V};\boldsymbol{\mu}), \tag{9.39}$$

but to its *Fréchet* derivative (with respect to the first variable), defined as

$$dA(\mathbf{W};\boldsymbol{\mu})(\mathbf{U},\mathbf{V}) = A_S(\mathbf{U},\mathbf{V};\boldsymbol{\mu}) + C(\mathbf{W},\mathbf{U},\mathbf{V};\boldsymbol{\mu}) + C(\mathbf{U},\mathbf{W},\mathbf{V};\boldsymbol{\mu}), \tag{9.40}$$

when evaluated at $\mathbf{W} \in X$. In both cases, we denote by $C(\mathbf{U},\mathbf{U},\mathbf{V};\boldsymbol{\mu}) = c(\mathbf{u},\mathbf{u},\mathbf{v};\boldsymbol{\mu})$. In this framework, a joint residual-based estimation for velocity and pressure fields in the Navier-Stokes case takes the following form: for any $N_r \geq N^*(\boldsymbol{\mu})$,

$$\|\mathbf{U}_h(\boldsymbol{\mu}) - \mathbf{U}_r(\boldsymbol{\mu})\|_X \leq \frac{\beta_{NS,h}^{LB}(\boldsymbol{\mu})}{2\gamma(\rho;\boldsymbol{\mu})} \left(1 - \sqrt{1 - \tau_{N_r}(\boldsymbol{\mu})}\right) =: \Delta_{N_r}(\boldsymbol{\mu}), \quad \forall\, \boldsymbol{\mu} \in \mathscr{D} \tag{9.41}$$

provided that $\tau_{N_r}(\boldsymbol{\mu}) < 1$. In particular:

- $\tau_{N_r}(\boldsymbol{\mu})$ is a *non-dimensional* measure of the residual, defined as

$$\tau_{N_r}(\boldsymbol{\mu}) = \frac{4\gamma(\rho;\boldsymbol{\mu})\|r(\boldsymbol{\mu})\|_{X'}}{(\beta^{LB}_{NS,h}(\boldsymbol{\mu}))^2};$$

  moreover, we denote $N^*(\boldsymbol{\mu})$ the smallest $N_r$ such that $\tau_{N_r}(\boldsymbol{\mu}) < 1$, for all $N_r \geq N^*(\boldsymbol{\mu})$. Since $\|r(\boldsymbol{\mu})\|_{X'}$ – and thus $\tau_{N_r}(\boldsymbol{\mu})$ – undergoes a fast decrease when $N_r$ increases, usually $N^*(\boldsymbol{\mu}) < 10$, so that (9.41) holds for reasonable dimensions $N_r$;

- $\gamma(\rho_h;\boldsymbol{\mu})$ is the (discrete) continuity constant of the trilinear form $c(\cdot,\cdot,\cdot;\boldsymbol{\mu})$, depending on the Sobolev embedding constant $\rho_h$ defined as

$$\rho_h^2 = \sup_{v\in V_h} \frac{\|v\|^2_{L^4(\Omega)}}{(v,v)_H};$$

- the dual norm $\|r(\boldsymbol{\mu})\|_{X'}$ of the global residual, which is given in this case by

$$\begin{aligned} r(\mathbf{V};\boldsymbol{\mu}) &:= r_{\mathbf{u}}(\mathbf{v};\boldsymbol{\mu}) + r_p(q;\boldsymbol{\mu}), \\ r_{\mathbf{u}}(\mathbf{v};\boldsymbol{\mu}) &:= r^S_{\mathbf{u}}(\mathbf{v};\boldsymbol{\mu}) - c(\mathbf{u}_r(\mu),\mathbf{u}_r(\mu),\mathbf{v};\mu); \end{aligned} \tag{9.42}$$

- $\beta^{LB}_{NS,h}(\boldsymbol{\mu})$ is a computable lower bound for the Babuška inf-sup stability factor $\beta_{NS,h}(\boldsymbol{\mu})$, involving the *Fréchet* derivative of the global Navier-Stokes operator:

$$\exists\ \beta^{LB}_{NS,h}(\boldsymbol{\mu}) > 0:\ \beta_{NS,h}(\boldsymbol{\mu}) = \inf_{\mathbf{V}\in X_h}\sup_{\mathbf{W}\in X_h} \frac{dA(\mathbf{U}_h(\boldsymbol{\mu});\boldsymbol{\mu})(\mathbf{V},\mathbf{W})}{\|\mathbf{V}\|_X\|\mathbf{W}\|_X} \geq \beta^{LB}_{NS,h}(\boldsymbol{\mu}). \tag{9.43}$$

We remark that the framework described above is essential the nonlinear extension of the much simpler linear *a posteriori* error estimation (9.35), to which the nonlinear error estimation (9.41) reduces in the limit that $\|r(\boldsymbol{\mu})\|_{X'} \to 0$.

*A posteriori* error estimation for the Navier-Stokes problem poses, from a computational standpoint, more severe challenges than for Stokes problem. We do not provide any detail about the evaluation of these quantities; the interested reader can refer, for instance, to [76, 87, 94, 124].

### 9.3.3 Relevant Computational Issues

Finally we point out the most relevant computational difficulties encountered in developing/applying the methodology presented in this section. We focus, in particular, on the evaluation of the *a posteriori* error bounds, a crucial aspect when attempting to build a reduced space with the greedy RB method.

With respect to linear problems, where the computational speedup between a reduced basis method and a truth approximation is usually about $10^2$, reduction may be even larger (sometimes up to one order of magnitude) in nonlinear problems. In this case, nonlinear solvers might require several iterations to converge to the solution. Each iteration entails a large linear system to solve in the case of the truth approximation. Instead, a reduced-order model requires at each iteration of the nonlinear solver the solution of a small linear system, which can be assembled by exploiting the precomputed structures (9.14).

Nevertheless, we need to rely on a suitable Offline/Online splitting to speed up our computation. Such a strategy is also required to evaluate in a very small amount of time the error estimates (9.35) or (9.41), so that all the parametric-dependent quantities appearing in these formulas can exploit the affine parametric dependence.

Moreover, error estimates should to be uniformly effective across entire parameter range, to avoid the greedy algorithm skew towards particular locations in parameter space. In this case, the basis resulting from the selected snapshots could be inadequate to uniformly approximate the whole manifold of solutions, or result larger than required. Essentially, we pursue the following strategy:

1. *Stability factors*. If the (Navier)-Stokes operator is parameter-dependent, so is the lower bound of the stability factor (9.38) or (9.43). In this case, computing its lower bound according to a suitable Offline/Online splitting is not easy. We face it by using the so-called *Successive Constraint Method* (SCM)[4] which converts the eigenproblem corresponding to the computation of (9.38) or (9.43) on the successive solution of suitable linear optimization problems.
   This algorithm has been applied for the first time to saddle point Stokes problems in [106], while a first extension to the nonlinear Navier-Stokes case has been considered in [87]. In case of physical parametrizations (for instance, involving the Reynolds number) and large parametric variations, stability factors might undergo large variations and the SCM algorithm is able to capture this behavior. Instead, according to our own experience, in case of geometrical parametrizations arising from local shape changes or simple scaling (or affine) transformations, piecewise constant approximations of the stability factors can provide good result at a very lower cost. In more involved cases, alternative heuristic strategies to derive lower bounds of stability factors might take advantage of suitable interpolation techniques (see e.g. [87]).

2. *Residuals*. A suitable Offline/Online splitting can be used to evaluate the dual norms of residuals (9.36)–(9.42). Indeed, these quantities can be expressed as the sum of products of $\boldsymbol{\mu}$-dependent known functions and $\boldsymbol{\mu}$-independent inner products, formed of more complicated but precomputable quantities, involving the Riesz representations of $r_{\mathbf{u}}(\boldsymbol{\mu})$ and $r_p(\boldsymbol{\mu})$.
   As already remarked in Sect. 9.2.3, in the case of nonlinear convective terms tensorial terms of relatively large sizes are generated; they depend on both the dimension $N_r$ of the reduced spaces and the parametric complexity $Q^c$ of the trilinear convective term. Unfortunately, evaluating and storing these structures

[4] This algorithm has been first introduced in [64] for both coercive and noncoercive problems, analyzed in [107] in the coercive case and afterwards improved in [32]. A general version using the so-called "natural norm" [110] has been analyzed in [61], where it has been applied to noncoercive problems such as Helmholtz equations – the simpler coercive case can be seen as a particular instance where the stability factor is just the coercivity constant.

might become computationally infeasible, so that an Offline/Online splitting for evaluating the dual norms of residuals is not always practicable.

A Galerkin projection is well suited for symmetric and coercive PDEs, as in this case it provides the optimal approximation in the corresponding energy norm. In the case of convection-dominated flows, symmetry is broken and no a priori optimality can be ascertained. Indeed, a large gap between the magnitude of the observed nonlinear residuals and the true error between full and reduced solution may exist.

A remedy consists in using Petrov-Galerkin methods, with different spaces of test and trial functions. They are usually presented in the guise of stabilization methods, such as in the case of the Streamline-Upwind Petrov-Galerkin (SUPG) method. However, one is then left with the question of how to choose the test space. Recent works on optimal or near-optimal choice of Petrov-Galerkin test spaces were presented in [40, 41] and [38]. These options are "optimal" in the sense that they give the best possible ratio of continuity constant to stability constant in the energy norm estimates. In the finite element or discontinuous Galerkin context, optimal test spaces are usually avoided, as this would lead to using test functions with global support. However, in the ROM setting one does not care too much if the reduced order system is full, as it is typically small enough to be solved with direct solvers (the reduced dimension $N_r$ is typically in the range $10-10^2$).

In fact, the optimal test spaces are precisely equivalent to the method of supremizers used in [106, 109] to stabilize ROMs for the Stokes equations. Unfortunately, in the parametrized setting one has to face the fact that the optimal test spaces (and also the supremizers) usually depend explicitly on the parameters and thus suitable strategies to recover the Offline/Online splitting must be devised.

## 9.4 Model Reduction of Unsteady Viscous Flows

In this section we provide an overview of some reduction techniques available for unsteady viscous flows. We do not restrict ourselves to parametrized problems and RB methods; rather, we provide a quick survey of more general ROM techniques based on the study of the stability of the underlying dynamical system – arising for instance from *model order reduction for ODEs* – addressed in the following chapters of the book. We start by recalling that current approaches for constructing reduced basis approximations of time-dependent parametrized PDEs exploit a combined POD-greedy procedure – POD in time to capture the causality associated with the evolution equation, greedy procedure for sampling the parameter space and treat more efficiently extensive ranges of parameter variation [93].

Certified RB methods have been applied to parametrized (moderate Reynolds) unsteady viscous flows in [70], where a nonisothermal viscous flow is modelled by Boussinesq equations describing natural convection. Parameters are the Grashof number and the gravity direction. In [48] an improved *h-p* adaptive certified method is introduced to address the same natural convection problem, which has also been applied to a multiscale Stokes Fokker-Planck system modelling liquid crystals in

[71]. More recent contributions in the field adopt a *space-time* Petrov-Galerkin variational approach to improve the control of the exponentially growing energy estimates in the linear case [123] dealing with convection-conduction problems, for Burgers' equations [131], Boussinesq equations for moderate Grashof number flows exhibiting steady periodic responses [129] and even addressing interesting hydrodynamic stability problems for moderate Reynolds number flows in an eddy-promoted channel [130]. The chapter by Chen et al. in this monograph is focused on reduced approximations for the Parareal method.

### 9.4.1 Model reduction of linearized time-invariant systems

The POD modes discussed in Sect. 9.2.2 only represent the statistical information content of the set of snapshots without taking into account the underlying dynamical system. Many examples of fluid dynamics where a POD-Galerkin ROM described exactly the limit cycle of the system exist, however they completely miss the long-time dynamical behavior of its trajectories.

An example of a dynamical system whose POD-modes are able to exactly represent the stable limit cycle, but for which a Galerkin ROM gives incorrect dynamics was described in [95]. The quadratically nonlinear ODE system

$$\begin{cases} \dot{u}_1(t) = \mu u_1(t) - u_2(t) - u_1(t)u_3(t) \\ \dot{u}_2(t) = \mu u_2(t) + u_1(t) - u_2(t)u_3(t) \, , \\ \dot{u}_3(t) = -u_3(t) + u_1^2(t) + u_2^2(t) \end{cases}$$

has one fixed point at $\mathbf{u} = (0,0,0)$, which is unstable, and an asymptotically stable limit cycle $\mathbf{u}_{\mathrm{LS}}(t) = (\sqrt{\mu}\cos(t), \sqrt{\mu}\sin(t), \mu)$. All trajectories tend towards the limit cycle. Since

$$\frac{1}{t-t_0}\int_{t_0}^{t} \mathbf{u}(\tau)\, d\tau \quad \underset{t\to\infty}{\longrightarrow} \quad (0,0,\mu),$$

the POD basis of dimension 2 is given by

$$\bar{\mathbf{u}} := (0,0,\mu), \quad \Psi_1 = (1,0,0), \quad \Psi_2 = (0,1,0)$$

and is able to exactly represent the stable limit cycle:

$$\mathbf{u}_{\mathrm{LS}}(t) = \bar{\mathbf{u}} + \sqrt{\mu}\cos(t)\Psi_1 + \sqrt{\mu}\sin(t)\Psi_2.$$

However, Galerkin projection on the POD basis of dimension 2 using the Euclidean inner product leads to

$$\mathbf{u}_r(t) := \bar{\mathbf{u}} + a_1(t)\Psi_1 + a_2(t)\Psi_2.$$

The coefficients of the ROM are given by the dynamical system

$$\begin{cases} \dot{a}_1(t) = -a_2(t) \\ \dot{a}_2(t) = a_1(t) \end{cases},$$

which is only marginally stable and whose trajectories remain on a circle of radius $r = (a_1^2(t_0) + a_2^2(t_0))^{1/2}$ for all time without converging asymptotically towards the correct limit cycle.

In order to capture the correct temporal dynamics, the characteristics of the dynamical system (fixed points, periodic solutions, and their (in)stability) should be preserved by the ROM – such ROMs are built based on analyzing the stability of the underlying dynamical system. In this section we discuss some, namely, linearized time-invariant flows, which exhibit asymptotically stable periodic steady-states.

For *linear time-invariant systems* (LTIs), system-theoretical reduction methods such as *balanced truncation* [7,91] are more effective, in the sense that they provide a ROM that has nearly the best possible approximation error. A linearized input-output system in state-space form is

$$\begin{cases} \frac{d\mathbf{U}}{dt}(t) = A\mathbf{U}(t) + B\mathbf{S}(t) & \text{for } t \in (t_0, t_f) \\ \mathbf{Y}(t) = C\mathbf{U}(t) & \text{for } t \in (t_0, t_f) \\ \mathbf{U}(t_0) = \mathbf{U}_0 \end{cases} \quad (9.44)$$

with inputs (controls) $\mathbf{S}$ and outputs (observations) $\mathbf{Y}$. If the system (9.44) is stable, the controllability and observability Gramians are the matrices defined respectively as

$$W_c = \int_{t_0}^{t_f} e^{A\tau} BB^* e^{A^*\tau}\, d\tau, \quad W_o = \int_{t_0}^{t_f} e^{A^*\tau} C^* C e^{A\tau}\, d\tau \quad (9.45)$$

which can be computed from the Lyapunov equations:

$$AW_c + W_c A^* + BB^* = 0, \qquad A^* W_o + W_o A + C^* C = 0;$$

see e.g. [105] for further details. The controllability Gramian $W_c$ measures to what degree each state of the system (9.44) is excited by an input; in particular, $W_c$ is positive definite if and only if all states are reachable with some input $\mathbf{S}(t)$. Instead, the observability Gramian $W_o$ measures to what degree each state excites future outputs; in particular, $W_o$ is positive definite if and only if any initial state $\mathbf{U}(t_0) = \mathbf{U}_0$ can be uniquely determined from $\mathbf{Y}(t)$ on $(t_0, t_f)$.

A balancing transformation $\mathbb{T}$ is sought to transform the state variables of the LTI into equivalent "balanced state variables", $\hat{\mathbf{U}} = \mathbb{T}\mathbf{U}$, in a way that the transformed Gramians become equal and diagonal:

$$\mathbb{T}^{-1} W_c \mathbb{T}^{-*} = \mathbb{T}^* W_o \mathbb{T} = \Sigma = \mathrm{diag}(\hat{\sigma}_1, \ldots, \hat{\sigma}_N). \quad (9.46)$$

In the balanced coordinates, the states that are least influenced by the input also have the least influence on the output, and such a balancing transformation exists as long as the system is both controllable and observable (i.e., both $W_c$ and $W_o$ are positive definite). The $\{\hat{\sigma}_i\}$ are called the Hankel singular values; when sorted in descending order, we can split the balanced LTI system into two parts:

$$\begin{cases} \dfrac{d}{dt}\begin{bmatrix}\hat{\mathbf{U}}_1\\ \hat{\mathbf{U}}_2\end{bmatrix}(t) = \begin{bmatrix}\hat{A}_{11} & \hat{A}_{12}\\ \hat{A}_{21} & \hat{A}_{22}\end{bmatrix}\begin{bmatrix}\hat{\mathbf{U}}_1\\ \hat{\mathbf{U}}_2\end{bmatrix}(t) + \begin{bmatrix}\hat{B}_1\\ \hat{B}_2\end{bmatrix}\mathbf{S} & \text{for } t \in (t_0, t_f) \\ \mathbf{Y}(t) = \begin{bmatrix}\hat{C}_1 & \hat{C}_2\end{bmatrix}\begin{bmatrix}\hat{\mathbf{U}}_1\\ \hat{\mathbf{U}}_2\end{bmatrix}(t) & \text{for } t \in (t_0, t_f) \\ \hat{\mathbf{U}}(t_0) = \mathbb{T}\mathbf{U}_0, \end{cases} \tag{9.47}$$

where $\dim(\hat{\mathbf{U}}_1) = r$ and $\dim(\hat{\mathbf{U}}_2) = N - r$. A balanced truncation ROM is then obtained by retaining only the balanced state variables related to the first $r$ Hankel singular values:

$$\begin{cases} \dfrac{d}{dt}\hat{\mathbf{U}}_1(t) = \hat{A}_{11}\hat{\mathbf{U}}_1 + \hat{B}_1\mathbf{S} & \text{for } t \in (t_0, t_f) \\ \tilde{\mathbf{Y}}(t) = \hat{C}_1\hat{\mathbf{U}}_1(t) & \text{for } t \in (t_0, t_f) \cdot \\ \hat{\mathbf{U}}_1(t_0) = \mathbb{T}_1\mathbf{U}_0 \end{cases} \tag{9.48}$$

In other words, balanced truncation involves first changing the coordinates according to (9.46), and then truncating the least controllable/observable states, which have little effect on the input-output behavior.

When the exact transfer function $G(s) = C(sI - A)^{-1}B$ of the LTI system is compared with the one obtained after balanced truncation, $\hat{G}(s) = \hat{C}_1(sI - \hat{A}_{11})^{-1}\hat{B}_1$, we have the following results [7]:

- any ROM with $r$ states and transfer function $\tilde{G}_r(s)$ has operator norm error at least $\|G - \tilde{G}_r\|_\infty > \hat{\sigma}_{r+1}$, where $\hat{\sigma}_{r+1}$ is the $(r+1)$st Hankel singular value;
- the balanced truncation ROM with $r$ states and transfer function $\hat{G}_r(s)$ has operator norm error bounded by $\|G - \hat{G}_r\|_\infty < 2\sum_{i=r+1}^{N}\hat{\sigma}_i$;
- if the full-order system (9.44) is stable, so is the balanced truncation ROM (9.48).

The Hankel singular values $\{\hat{\sigma}_i\}$ characterize also the Kolmogorov $n$-width discussed in Sect. 9.2.1 of the range space of the Hankel operator, see [45]. As already discussed in Sect. 9.2.1, the main requirement for constructing efficient ROMs is that the associated singular values decay reasonably fast. Previously, we used the decay of the empirical POD singular values to measure this, whereas in the balanced truncation method one looks at the Hankel singular values. In fact, there exists an interesting connection between the Hankel singular values and the empirical POD singular values – it was pointed out in [105] that the POD modes are *equivalent* to the modes obtained by balanced truncation provided that the snapshots $\mathbf{U}_i$ are taken as the impulse responses of the system and the inner product equal to the one induced by the observability Gramian.

Balanced truncation methods based on explicitly computing the Gramians in (9.45) by solving Lyapunov equations are generally too expensive to apply to large linear systems with millions of state variables. A possible remedy is the *balanced truncation POD* method [74, 128], in which the exact Gramians (9.45) are approximated using a method of snapshots:

$$\begin{aligned} W_c^e &= \frac{1}{K}\sum_{k=1}^{K}\frac{1}{w_k^2}\int_{t_0}^{t_f}\left(\xi_k(\tau)-\bar{\xi}_k\right)\left(\xi_k(\tau)-\bar{\xi}_k\right)^* d\tau, \\ W_o^e &= \frac{1}{K}\sum_{k=1}^{K}\frac{1}{w_k^2}\int_{t_0}^{t_f} Q_k\left(\zeta_k(\tau)-\bar{\zeta}_k\right)^*\left(\zeta_k(\tau)-\bar{\zeta}_k\right) Q_k^*\, d\tau. \end{aligned} \tag{9.49}$$

Here the empirical trajectories $\xi_k(t)$ and empirical outputs $\zeta_k(t)$ are computed by solving the system (9.44) using generalized impulse controls $\mathbf{S}_k(t) = w_k Q_k \mathbf{e}_k \delta(t)$, where $w_k > 0$ are positive weights, $Q_k \in \mathbb{R}^{P\times P}$ are orthogonal matrices, $\mathbf{e}_k \in \mathbb{R}^P$ are Euclidean unit vectors, and $\delta(t)$ is the one-dimensional Dirac delta distribution:

$$\begin{cases} \dfrac{d\xi_k}{dt}(t) = A\xi_k(t) + B\mathbf{S}_k(t) & \text{for } t \in (t_0, t_f) \\ \zeta_k(t) = C\xi_k(t) & \text{for } t \in (t_0, t_f) \\ \xi(t_0) = \mathbf{U}_0 \end{cases} \tag{9.50}$$

for each $k = 1,\ldots,K$. In the case of LTI systems, the empirical Gramians (9.49) coincide with the exact Gramians (9.45) provided that $K \geq P$ empirical impulse responses are computed. It was proposed in [75] to use the same balanced truncation POD method for dealing also with nonlinear flows. In this case the empirical Gramians (9.49) – which are (approximate) finite Gramians – are obtained by solving the nonlinear system and taking snapshots of the trajectory. By using these finite Gramians to perform the balancing we obtain the following, empirical balancing transformation $\mathbb{T}_e = [\mathbb{T}_{e,1}\ \mathbb{T}_{e,2}]$:

$$\mathbb{T}_e^{-1} W_c^e \mathbb{T}_e^{-*} = \mathbb{T}_e^* W_o^e \mathbb{T} = \Sigma = \operatorname{diag}(\hat{\sigma}_1,\ldots,\hat{\sigma}_N). \tag{9.51}$$

The balanced POD modes were applied in [105] to a linearized flow in a plane channel and a comparison was made between POD, balanced truncation, and balanced POD methods. The conclusion was that the balanced POD modes produced nearly identical results with the balanced truncation modes, and both methods significantly outperformed the standard POD modes. Another comparison on a problem of designing closed-loop controllers for flow over a cavity was done in [13], where again the balanced POD modes achieved a stable closed-loop controller with fewer ROM degrees-of-freedom. A difficulty related to balanced truncation is that the linearized system must be stable. An extension to unstable linear systems was proposed in [132] by decoupling the dynamics on the stable and unstable subspaces, and then truncating the relatively uncontrollable and unobservable modal representations on each subspace (see e.g. [132] for further details). This strategy was used to propose reduced-order controllers around linearly unstable steady states for flow around a cylinder in [122] and for flow past a flat plate in [1].

### 9.4.2 Stabilization of ROMs for Unsteady Navier-Stokes Equations

As mentioned before, usually a standard Galerkin projection-based ROM does not produce satisfactory results when applied to nonlinear unsteady Navier-Stokes equations. There do exist exceptions – for nonautonomous problems with strong external sources, such as periodically driven inflow, long-time drifting from asymptotically stable states was not observed in [84, 113]. The drifting of ROM trajectories in the general case is however a well-known problem and many attempts have been made to remedy it.

First works on stabilization experimented in adding artificial viscosity [9] to the reduced equations. The idea was further developed by extending the spectral vanishing viscosity method of Tadmor [115] to the Navier-Stokes equations in [111]. In long-time simulation of convection dominated flows some type of closure model that takes into account the energy transfer between the ROM modes is needed. In [29] a driven cavity problem at Re = 20,000 was successfully stabilized by adding a linear damping term in the Galerkin ROM. The computation of correct limit cycles was done in [2] by applying a shooting method. For a review of various stabilization methods for Galerkin ROMS we refer to [16].

### 9.4.3 Dynamic Mean-Field Representations and Shift-Modes

In many fluid dynamics systems, the Reynolds decomposition (9.2) together with Galerkin projection leads to unstable ROMs because the interaction between the time-averaged mean flow $\bar{\mathbf{u}}$ and the oscillating part of the flow field represented by the POD modes is neglected. In [95] this problem was analyzed and identified moving from the consideration that a Galerkin model without dynamic mean-field correction is unable to represent correctly the unstable fixed point of the dynamical system, which leads to structurally unstable ROMs (small perturbations in the model can cause divergent trajectories). This was found to occur even in problems where theoretically the POD-Galerkin ROM was able to capture the stable attractor exactly. As a result the periodic limit cycle was correctly captured, but transient dynamics of the ROM were off by orders of magnitude.

The simplest method proposed in [95] to correct the mean-field approximation error of POD is the inclusion of a *shift-mode* $\Psi_\Delta$, which is added to the POD basis in order to represent the correct unstable fixed point of the full-order system, resulting in the extended POD *ansatz*

$$\mathbf{u}(x,t) \approx \mathbf{u}_r(x,t) := \bar{\mathbf{u}}(x) + a_0(t)\Psi_\Delta(x) + \sum_{i=1}^{N_r} a_i(t)\Psi_i(\mathbf{x}). \tag{9.52}$$

For instance, in the case of the unsteady cylinder wake flow, the unstable fixed point corresponds to the solution $\mathbf{u}_s$ of the steady Navier-Stokes flow. The shift-mode is obtained by applying a Gram-Schmidt process to the correction term $\mathbf{u}_\Delta := \bar{\mathbf{u}} - \mathbf{u}_s$:

$$\Psi_\Delta^* := \mathbf{u}_\Delta - \sum_{i=1}^{N_r} (\mathbf{u}_\Delta, \Psi_i)\Psi_i, \qquad \Psi_\Delta := \frac{\Psi_\Delta^*}{\|\Psi_\Delta^*\|_\Omega} \tag{9.53}$$

and applying the Galerkin ROM to the expanded POD basis of dimension $N_r + 1$. This allows the ROM to represent exactly the unstable fixed point of the system. A comprehensive discussion of the various other types of mean-field corrections and their effects on the ROM predictions can be found in [116].

### *9.4.4 Model Reduction of Periodic Steady-State Solutions*

In Sect. 9.4.3 we have discussed the difficulties of building ROMs that are capable of accurately representing the transient dynamics of unsteady flows. In many applications of fluid dynamics, for example in turbomachinery or in large "straight" arteries in the human circulatory system, the behavior of the flow is such that all trajectories approach a single stable periodic solution. One option is then to disregard the simulation of the transient, and concentrate only on approximating the periodic steady-state solution.

For linearized flows the frequency-domain POD technique was introduced in [69]. It replaces the time-domain representation of the Galerkin-projected equations with a Fourier-domain representation for each individual harmonic. For fully nonlinear flows the individual harmonics are coupled by the nonlinear terms and no term-by-term analysis of the harmonics can be performed. To solve this problem, the *Harmonic Balance* (HB) method used for the study of harmonic ODEs was adapted for the efficient solution of time-periodic flows in [57,89,90]. After suitable spatial discretization of (9.1) the system

$$\begin{bmatrix} \dot{\mathbf{u}}_h \\ 0 \end{bmatrix} = -\begin{bmatrix} -(\mathbf{u}_h \cdot \nabla)\mathbf{u}_h - \nabla p_h + \nu\Delta\mathbf{u}_h + \mathbf{f}(t) \\ -\nabla\cdot\mathbf{u}_h \end{bmatrix} = -\begin{bmatrix} \mathbf{S}_1(\mathbf{U}) \\ \mathbf{S}_2(\mathbf{U}) \end{bmatrix} = -\mathbf{S}(\mathbf{U}), \tag{9.54}$$

is obtained, where the spatial operator $\mathbf{S}(\mathbf{U})$ depends nonlinearly on the solution $\mathbf{U} := (\mathbf{u}_h, p_h) \in \mathbb{R}^{N_h^u + N_h^p}$, $N_h^u$ and $N_h^p$ being the number of degrees of freedom of the discrete velocity and pressure fields, respectively.

The method starts from the assumption that this system admits a periodic steady-state solution $\mathbf{U}_\infty(t)$ with known period $T$, so that $\mathbf{U}_\infty(t) = \mathbf{U}_\infty(t+T)$ for all $t$. If in addition the spatial operator is time periodic with the same period $T$, they can both be represented using Fourier series expansions as

$$\mathbf{U}_\infty(t) = \sum_{k=-\infty}^{\infty} \widehat{\mathbf{U}}_k \exp\left(\frac{2\pi ikt}{T}\right), \qquad \mathbf{S}(\mathbf{U}_\infty) = \sum_{k=-\infty}^{\infty} \widehat{\mathbf{S}}_k(\mathbf{U}_\infty) \exp\left(\frac{2\pi ikt}{T}\right), \tag{9.55}$$

where each $\widehat{\mathbf{U}}_k$ and $\widehat{\mathbf{S}}_k(\mathbf{U}_\infty)$ is a (discrete representation of a) complex-valued vector field over $\Omega$; by expressing $\widehat{\mathbf{S}}_k = \widehat{\mathbf{S}}_k(\mathbf{U}_\infty)$ we mean that each coefficient in the expansion of $\mathbf{S} = \mathbf{S}(\mathbf{U}_\infty)$ depends on (potentially all of) the spatial coefficients $\{\widehat{\mathbf{U}}_k\}_k$ of $\mathbf{U}_\infty(t)$. Since the periodic steady-state solution satisfies equation (9.54), its complex

Fourier coefficients $\widehat{\mathbf{U}}_k \in \mathbb{C}^{N_h^u+N_h^p}$ must satisfy

$$\frac{2\pi ik}{T}\widehat{\mathbf{U}}_k + \widehat{\mathbf{S}}_k(\mathbf{U}_\infty) = 0, \qquad \text{for all } k \in \mathbb{Z}. \tag{9.56}$$

The *harmonic balance* (HB) method starts by truncating the Fourier series to $2N+1$ terms and matching only those terms in (9.56), i.e.

$$\frac{2\pi ik}{T}\widehat{\mathbf{U}}_k + \widehat{\mathbf{S}}_k(\mathbf{U}_\infty) = 0, \qquad \text{for all } k = -N,\dots,N. \tag{9.57}$$

For real-valued fields $\widehat{\mathbf{U}}_{-k} = \overline{\widehat{\mathbf{U}}_k}$, so that only $N+1$ equations need to be solved.

If the flow is linear, all the harmonics decouple and we only need to solve $N+1$ uncoupled steady equations. For nonlinear flows, each $\widehat{\mathbf{S}}_k(\mathbf{U}_\infty)$ depends on all the $\widehat{\mathbf{U}}_k$ for $k = -N,\dots,N$ and thus the system (9.57) is a fully coupled nonlinear system of $(N+1)\times(N_h^u+N_h^p)$ complex-valued equations. Due to the nonlinearity of the spatial operator its Fourier series coefficients cannot be computed directly. This problem is solved either using the alternating frequency/time domain method, as was done in [57, 90], or by the asymptotic numerical method, as was done in [34]. Once the Fourier coefficients are known, the periodic steady state solution can be reconstructed with arbitrary temporary precision.

An advantage of HB compared to POD is that no full-order transient simulations need to be performed until the periodic steady-state is reached, nor is the ROM dependent on the initial condition of these simulations. For a comparison between POD and HB we refer to [82]. We remark that the HB method is very efficient in reducing the temporal complexity, as typically only $N < 10$ terms are needed to accurately represent the solution. However, it has no effect on the spatial complexity of the problem. Like many space-time formulations it requires the solution of a system that is several times larger than the one solved when using the more standard method of lines. So far the HB method has been applied mainly to industrial problems, such as the design and simulation of turbomachinery [57] and problems in aeroelasticity [120].

## 9.5 Conclusions

In this chapter we have presented an overview of model reduction methods for incompressible fluid dynamics, both in the steady and unsteady flow cases. The main focus was on Galerkin-projection based ROMs, and the main strategies for constructing the low-order projection basis have been discussed. Theoretical properties of ROMs for fluid problems are related to, e.g.: the possibility to reduce the dynamics of a fluid system to a low-dimensional submanifold, measured for instance by the very fast exponential convergence of empirical POD singular values or of the Kolmogorov $n$-width; the lack of long-time stability of Galerkin ROMs and the need for stabilization; the error estimation of the ROM in the case of steady flow problems; the gain of computational efficiency thanks to the online/offline paradigm that allows fast real-time ROM simulations as well as to the use of hyperreduction for treating the nonlinear terms in an efficient way.

*Ad hoc* reduced order modelling techniques have recently been proposed for optimal flow control problems [104, 108, 121], optimal shape design of devices related with fluid flows [6, 23, 58, 88], and the treatment of fluid-structure interaction problems [76, 78].

Far from having covered the subject exhaustively, we hope nonetheless that this chapter could offer the reader a contribution for understanding which type of ROM may be the best for his or her particular fluid dynamics application, having made extensive reference to available results in the literature.

## References

1. Ahuja, S., Rowley, C.: Feedback control of unstable steady states of flow past a flat plate using reduced-order estimators. J. Fluid Mech **645**, 447–478 (2010)
2. Akhtar, I., Nayfeh, A., Ribbens, C.: On the stability and extension of reduced-order Galerkin models in incompressible flows. Theor. Comp. Fluid Dyn. **23**(3), 213–237 (2009)
3. Amsallem, D., Cortial, J., Carlberg, K., Farhat, C.: A method for interpolating on manifolds structural dynamics reduced-order models. Int. J. Numer. Methods Engr. **80**(9), 1241–1258 (2009)
4. Amsallem, D., Farhat, C.: An online method for interpolating linear parametric reduced-order models. SIAM J. Sci. Comput. **33**(5), 2169–2198 (2011)
5. Amsallem, D., Zahr, M., Farhat, C.: Nonlinear model order reduction based on local reduced-order bases. Int. J. Numer. Methods Engr. **92**(10), 891–916 (2012)
6. Antil, H., Heinkenschloss, M., Hoppe, R.: Domain decomposition and balanced truncation model reduction for shape optimization of the Stokes system. Optim. Methods Softw. **26**(4–5), 643–669 (2011)
7. Antoulas, A.: Approximation of Large-Scale Dynamical Systems. SIAM (2005)
8. Astrid, P., Weiland, S., Willcox, K., Backx, T.: Missing point estimation in models described by proper orthogonal decomposition. IEEE. T. Automat. Contr. **53**(10), 2237–2251 (2008)
9. Aubry, N., Holmes, P., Lumley, J., Stone, E.: The dynamics of coherent structures in the wall region of a turbulent boundary layer. J. Fluid Mech. **192(115)**, 173,355 (1988)
10. Babuška, I., Nobile, F., Tempone, R.: A stochastic collocation method for elliptic partial differential equations with random input data. SIAM J. Numer. Anal. **45**(3), 1005–1034 (2007)
11. Babuška, I.: Error-bounds for finite element method. Numer. Math. **16**, 322–333 (1971).
12. Baiges, J., Codina, R., Idelsohn, S.: Explicit reduced order models for the stabilized finite element approximation of the incompressible Navier-Stokes equations. Int. J. Numer. Methods Fluids (2013). DOI 10.1002/fld.3777
13. Barbagallo, A., Sipp, D., Schmid, P.J.: Closed-loop control of an open cavity flow using reduced-order models. J. Fluid Mech. **641**(1), 1–50 (2009)
14. Barone, M.F., Kalashnikova, I., Segalman, D.J., Thornquist, H.K.: Stable Galerkin reduced order models for linearized compressible flow. J. Comp. Phys. **228**(6), 1932–1946 (2009)

15. Barrault, M., Maday, Y., Nguyen, N., Patera, A.: An 'empirical interpolation' method: application to efficient reduced-basis discretization of partial differential equations. C. R. Acad. Sci. Paris. Sér. I Math. **339**(9), 667–672 (2004)
16. Bergmann, M., Bruneau, C., Iollo, A.: Enablers for robust POD models. J. Comp. Phys. **228**(2), 516–538 (2009)
17. Bergmann, M., Cordier, L.: Optimal control of the cylinder wake in the laminar regime by trust-region methods and POD reduced-order models. J. Comp. Phys. **227**(16), 7813–7840 (2008)
18. Binev, P., Cohen, A., Dahmen, W., DeVore, R., Petrova, G., Wojtaszczyk, P.: Convergence rates for greedy algorithms in reduced basis methods. SIAM J. Math. Anal. **43**, 1457–1472 (2011)
19. Brezzi, F.: On the existence, uniqueness, and approximation of saddle point problems arising from Lagrangian multipliers. R.A.I.R.O., Anal. Numér. **2**, 129–151 (1974)
20. Brezzi, F., Fortin, M.: Mixed and Hybrid Finite Element Methods. Springer Series in Computational Mathematics, vol. 15. Springer-Verlag, New York (1991)
21. Brezzi, F., Rappaz, J., Raviart, P.: Finite dimensional approximation of nonlinear problems. Part I: Branches of nonsingular solutions. Numer. Math. **36**, 1–25 (1980)
22. Buffa, A., Maday, Y., Patera, A., Prud'homme, C., Turinici, G.: A priori convergence of the greedy algorithm for the parametrized reduced basis. ESAIM Math. Modelling Numer. Anal. **46**(3), 595–603 (2012)
23. Bui-Thanh, T., Willcox, K., Ghattas, O.: Parametric reduced-order models for probabilistic analysis of unsteady aerodynamics applications. AIAA J. **46**(10) (2008)
24. Bui-Thanh, T., Willcox, K., Ghattas, O., van Bloemen Waanders, B.: Goal-oriented, model-constrained optimization for reduction of large-scale systems. J. Comp. Phys. **224**(2), 880–896 (2007)
25. Burkardt, J., Gunzburger, M., Lee, H.: Centroidal Voronoi tessellation-based reduced-order modeling of complex systems. SIAM J. Sci. Comput. **28**(2), 459–484 (2006)
26. Burkardt, J., Gunzburger, M., Lee, H.: POD and CVT-based reduced-order modeling of Navier-Stokes flows. Comput. Meth. Appl. Mech. Engrg. **196**(1–3), 337–355 (2006)
27. Caloz, G., Rappaz, J.: Numerical analysis for nonlinear and bifurcation problems. In: P. Ciarlet, J. Lions (eds.) Handbook of Numerical Analysis, Vol. V, Techniques of Scientific Computing (Part 2), pp. 487–637. Elsevier Science B.V., Amsterdam (1997)
28. Carlberg, K., Farhat, C.: A low-cost, goal-oriented 'compact proper orthogonal decomposition' basis for model reduction of static systems. Int. J. Numer. Methods Engr. **86**(3), 381–402 (2011)
29. Cazemier, W., Verstappen, R., Veldman, A.: Proper orthogonal decomposition and low-dimensional models for driven cavity flows. Phys. Fluids **10**, 1685 (1998)
30. Chaturantabut, S., Sorensen, D.: Nonlinear model reduction via discrete empirical interpolation. SIAM J. Sci. Comput. **32**(5), 2737–2764 (2010)
31. Chen, X., Akella, S., Navon, I.: A dual-weighted trust-region adaptive POD 4-D Var applied to a finite-volume shallow water equations model on the sphere. Int. J. Numer. Methods Fluids **68**(3), 377–402 (2012)
32. Chen, Y., Hesthaven, J., Maday, Y., Rodriguez, J.: A monotonic evaluation of lower bounds for inf-sup stability constants in the frame of reduced basis approximations. C. R. Acad. Sci. Paris. Sér. I Math. **346**, 1295–1300 (2008)
33. Christensen, E., Brøns, M., Sørensen, J.: Evaluation of proper orthogonal decomposition–based decomposition techniques applied to parameter-dependent nonturbulent flows. SIAM J. Sci. Comput. **21**, 1419–1434 (2000)

34. Cochelin, B., Vergez, C.: A high order purely frequency-based harmonic balance formulation for continuation of periodic solutions. J. Sound Vibration **324**(1), 243–262 (2009)
35. Colonius, T., Rowley, C., Freund, J., Murray, R.: On the choice of norm for modeling compressible flow dynamics at reduced-order using the POD. In: Proc. 41st IEEE Conf. on Decision and Control, vol. 3, pp. 3273–3278. IEEE (2002)
36. Daescu, D., Navon, I.: Efficiency of a POD-based reduced second-order adjoint model in 4D-Var data assimilation. Int. J. Numer. Methods Fluids **53**(6), 985–1004 (2007)
37. Daescu, D., Navon, I.: A dual-weighted approach to order reduction in 4DVAR data assimilation. Monthly Weather Review **136**(3), 1026–1041 (2008)
38. Dahmen, W., Huang, C., Schwab, C., Welper, G.: Adaptive Petrov-Galerkin methods for first order transport equations. SIAM J. Numer. Anal. **50(5)** (2012)
39. Deane, A., Kevrekidis, I., Karniadakis, G., Orszag, S.: Low-dimensional models for complex geometry flows: Application to grooved channels and circular cylinders. Phys. Fluids **3**(10), 2337–2354 (1991)
40. Demkowicz, L., Gopalakrishnan, J.: A class of discontinuous Petrov-Galerkin methods. part I: The transport equation. Comput. Methods Appl. Mech. Engr. **199**(23-24), 1558–1572 (2010)
41. Demkowicz, L., Gopalakrishnan, J.: A class of discontinuous Petrov–Galerkin methods. II. optimal test functions. Numer. Methods Partial Differential Equations **27**(1), 70–105 (2011)
42. Deparis, S.: Reduced basis error bound computation of parameter-dependent Navier-Stokes equations by the natural norm approach. SIAM J. Num. Anal. **46**(4), 2039–2067 (2008)
43. Deparis, S., Løvgren, A.: Stabilized reduced basis approximation of incompressible three-dimensional Navier-Stokes equations in parametrized deformed domains. J. Sci. Comput. **50(1)**, 198–212 (2012)
44. Deparis, S., Rozza, G.: Reduced basis method for multi-parameter-dependent steady Navier-Stokes equations: Applications to natural convection in a cavity. J. Comp. Phys. **228**(12), 4359–4378 (2009)
45. Djouadi, S.: On the connection between balanced proper orthogonal decomposition, balanced truncation, and metric complexity theory for infinite dimensional systems. In: Proc. Am. Control Conf., June 30–July 2, Baltimore, MD, 2010 pp. 4911–4916 (2010)
46. Du, Q., Gunzburger, M.: Model reduction by proper orthogonal decomposition coupled with centroidal Voronoi tessellation. In: Proc. Fluids Engineering Division Summer Meeting, FEDSM2002-31051, ASME (2002)
47. Dumon, A., Allery, C., Ammar, A.: Proper general decomposition (PGD) for the resolution of Navier-Stokes equations. J. Comp. Phys. **230**, 1387–1407 (2011)
48. Eftang, J., Knezevic, D., Patera, A.: An "hp" certified reduced basis method for parametrized parabolic partial differential equations. Math. Comput. Model. Dynam. Syst. **17**(4), 395–422 (2011)
49. Fahl, M.: Trust-region methods for flow control based on reduced order modelling. Ph.D. thesis, Universität Trier (2001)
50. Gerner, A., Veroy, K.: Certified reduced basis methods for parametrized saddle point problems. SIAM J. Sci. Comp. **34**(5), A2812–A2836 (2012)
51. Girault, V., Raviart, P.A.: Finite element methods for Navier-Stokes equations: Theory and algorithms. Springer-Verlag, Berlin Heidelberg New York (1986)
52. Grepl, M., Patera, A.: A posteriori error bounds for reduced-basis approximations of parametrized parabolic partial differential equations. ESAIM Math. Modelling Numer. Anal. **39(1)**, 157–181 (2005)

53. Grinberg, L., Yakhot, A., Karniadakis, G.: Analyzing transient turbulence in a stenosed carotid artery by proper orthogonal decomposition. Ann. Biomed. Eng. **37**(11), 2200–2217 (2009)
54. Gunzburger, M., Peterson, J., Shadid, J.: Reducer-order modeling of time-dependent PDEs with multiple parameters in the boundary data. Comput. Methods Appl. Mech. Engrg. **196**, 1030–1047 (2007)
55. Haasdonk, B.: Convergence rates of the POD-greedy method. ESAIM Math. Modelling Numer. Anal. (2013). DOI 10.1051/m2an/2012045
56. Haasdonk, B., Ohlberger, M.: Reduced basis method for finite volume approximations of parametrized linear evolution equations. ESAIM Math. Modelling Numer. Anal. **42**(02), 277–302 (2008)
57. Hall, K., Thomas, J., Clark, W.: Computation of unsteady nonlinear flows in cascades using a harmonic balance technique. AIAA J. **40**(5), 879–886 (2002)
58. Hay, A., Borggaard, J., Akhtar, I., Pelletier, D.: Reduced-order models for parameter dependent geometries based on shape sensitivity analysis. J. Comp. Phys. **229**(4), 1327–1352 (2010)
59. Hay, A., Borggaard, J., Pelletier, D.: Local improvements to reduced-order models using sensitivity analysis of the proper orthogonal decomposition. J. Fluid Mech. **629**, 41–72 (2009)
60. Holmes, P., Lumley, J., Berkooz, G.: Turbulence, coherent structures, dynamical systems and symmetry. Cambridge University Press, Cambridge (1998)
61. Huynh, D., Knezevic, D., Chen, Y., Hesthaven, J., Patera, A.: A natural-norm successive constraint method for inf-sup lower bounds. Comput. Meth. Appl. Mech. Engrg. **199**(29–32), 1963–1975 (2010)
62. Huynh, D., Knezevic, D., Patera, A.: A static condensation reduced basis element method: approximation and a posteriori error estimation. ESAIM Math. Modelling Numer. Anal. **47**(1), 213–251 (2013)
63. Huynh, D., Knezevic, D., Peterson, J., Patera, A.: High-fidelity real-time simulation on deployed platforms. Comp. Fluids **43**(1), 74–81 (2011)
64. Huynh, D., Rozza, G., Sen, S., Patera, A.: A successive constraint linear optimization method for lower bounds of parametric coercivity and inf-sup stability costants. C. R. Acad. Sci. Paris. Sér. I Math. **345**, 473–478 (2007)
65. Iapichino, L., Quarteroni, A., Rozza, G.: A reduced basis hybrid method for the coupling of parametrized domains represented by fluidic networks. Comput. Methods Appl. Mech. Engrg. **221–222**, 63–82 (2012)
66. Iollo, A., Lanteri, S., Désidéri, J.: Stability properties of POD-Galerkin approximations for the compressible Navier-Stokes equations. Theor. Comp. Fluid Dyn. **13**(6), 377–396 (2000)
67. Johansson, P., Andersson, H., Rønquist, E.: Reduced-basis modeling of turbulent plane channel flow. Comput. Fluids **35**(2), 189–207 (2006)
68. Johnson, C., Rannacher, R., Boman, M.: Numerics and hydrodynamic stability: toward error control in computational fluid dynamics. SIAM J. Numer. Anal. **32(4)**, 1058–1079 (1995)
69. Kim, T.: Frequency-domain Karhunen-Loève method and its application to linear dynamic systems. AIAA J. **36**(11), 2117–2123 (1998)
70. Knezevic, D., Nguyen, N., Patera, A.: Reduced basis approximation and a posteriori error estimation for the parametrized unsteady Boussinesq equations. Math. Mod. and Meth. in Appl. Sc. **21(7)**, 1415–1442 (2011)

71. Knezevic, D.J.: Reduced basis approximation and a posteriori error estimates for a multiscale liquid crystal model. Math. Comput. Model. Dynam. Syst. **17**(4), 443–461 (2011)
72. Kunisch, K., Volkwein, S.: Galerkin proper orthogonal decomposition methods for a general equation in fluid dynamics. SIAM J. Numer. Anal. **40**(2), 492–515 (2003)
73. Kunisch, K., Volkwein, S.: Optimal snapshot location for computing POD basis functions. ESAIM Math. Modelling Numer. Anal. **44**(3), 509 (2010)
74. Lall, S., Marsden, J., Glavaški, S.: Empirical model reduction of controlled nonlinear systems. In: Proc. IFAC World Congress Vol. F. Int. Federation Automatic Control, Beijing, 1999 pp. 473–478 (1999)
75. Lall, S., Marsden, J., Glavaški, S.: A subspace approach to balanced truncation for model reduction of nonlinear control systems. Int. J. Robust Nonlinear Control **12**(6), 519–535 (2002)
76. Lassila, T., Manzoni, A., Quarteroni, A., Rozza, G.: A reduced computational and geometrical framework for inverse problems in haemodynamics. Int. J. Numer. Meth. Biomed. Engng. **29**(7), 741–776 (2013)
77. Lassila, T., Manzoni, A., Quarteroni, A., Rozza, G.: Generalized reduced basis methods and $n$-width estimates for the approximation of the solution manifold of parametric PDEs. In: F. Brezzi, P. Colli Franzone, U. Gianazza, G. Gilardi (eds.) Analysis and Numerics of Partial Differential Equations. INdAM Series, vol. 4. Springer (2013). Re-printed also on Bollettino Unione Matematica Italiana (UMI), under permission/agreement Springer-UMI
78. Lassila, T., Quarteroni, A., Rozza, G.: A reduced basis model with parametric coupling for fluid-structure interaction problem. SIAM J. Sci. Comput. **34(2)**, A1187–A1213 (2012)
79. Leblond, C., Allery, C., Inard, C.: An optimal projection method for the reduced-order modeling of incompressible flows. Comput. Methods Appl. Mech. Engrg. **200**, 2507–2527 (2011)
80. Lions, P.: Mathematical Topics in Fluid Mechanics. Oxford Lecture Series in Mathematics and Its Applications. Clarendon Press, Oxford, UK (1996)
81. Løvgren, A., Maday, Y., Rønquist, E.: A reduced basis element method for the steady Stokes problem. ESAIM Math. Modelling Numer. Anal. **40(3)**, 529–552 (2006)
82. Lucia, D., Beran, P., Silva, W.: Reduced-order modeling: new approaches for computational physics. Prog. Aerosp. Sci. **40**(1-2), 51–117 (2004)
83. Lumley, J.: The structure of inhomogeneous turbulent flows. In: Yaglom, A.M., Tatarski, (eds.) Atmospheric turbulence and radio wave propagation pp. 166–178 (1967)
84. Ma, X., Karniadakis, G., Park, H., Gharib, M.: DPIV-driven flow simulation: a new computational paradigm. Proc. R. Soc. A **459**(2031), 547–565 (2003)
85. Maday, Y., Nguyen, N., Patera, A., Pau, G.: A general multipurpose interpolation procedure: the magic points. Commun. Pure Appl. Anal. **8(1)** (2009)
86. Maday, Y., Patera, A., Turinici, G.: Global a priori convergence theory for reduced-basis approximation of single-parameter symmetric coercive elliptic partial differential equations. C.R. Acad. Sci. Paris. Sér. I Math. **335**, 1–6 (2002)
87. Manzoni, A.: Reduced models for optimal control, shape optimization and inverse problems in haemodynamics. Ph.D. thesis, École Polytechnique Fédérale de Lausanne, Lausanne (2012)
88. Manzoni, A., Quarteroni, A., Rozza, G.: Shape optimization of cardiovascular geometries by reduced basis methods and free-form deformation techniques. Int. J. Numer. Methods Fluids **70**(5), 646–670 (2012)

89. Maple, R., King, P., Orkwis, P., Wolff, J.: Adaptive harmonic balance method for non-linear time-periodic flows. J. Comp. Phys. **193**(2), 620–641 (2004)
90. McMullen, M.: The application of non-linear frequency domain methods to the Euler and Navier-Stokes equations. Ph.D. thesis, Stanford University, Stanford, CA, USA (2003)
91. Moore, B.: Principal component analysis in linear systems: Controllability, observability, and model reduction. IEEE Trans. Automat. Contr. **26(1)** (1981)
92. Néron, D., Ladevèze, P.: Proper generalized decomposition for multiscale and multi-physics problems. Arch. Comput. Methods Engrg. **17**(4), 351–372 (2010)
93. Nguyen, N., Rozza, G., Patera, A.: Reduced basis approximation and a posteriori error estimation for the time-dependent viscous Burgers equation. Calcolo **46**(3), 157–185 (2009)
94. Nguyen, N., Veroy, K., Patera, A.: Certified real-time solution of parametrized partial differential equations. In: Yip, S. (Ed.). Handbook of Materials Modeling pp. 1523–1558 (2005)
95. Noack, B., Afanasiev, K., Morzynski, M., Tadmor, G., Thiele, F.: A hierarchy of low-dimensional models for the transient and post-transient cylinder wake. J. Fluid Mech. **497**(1), 335–363 (2003)
96. Noack, B., Papas, P., Monkewitz, P.: The need for a pressure-term representation in empirical Galerkin models of incompressible shear flows. J. Fluid Mech. **523**(1), 339–365 (2005)
97. Noor, A., Peters, J.: Reduced basis technique for nonlinear analysis of structures. AIAA J. **18(4)**, 455–462 (1980)
98. Peterson, J.: The reduced basis method for incompressible viscous flow calculations. SIAM J. Sci. Stat. Comput. **10**, 777–786 (1989)
99. Quarteroni, A.: Numerical Models for Differential Problems, 2nd ed. Modeling, Simulation and Applications (MS&A), Vol. 8. Springer-Verlag Italia, Milano (2014)
100. Quarteroni, A., Rozza, G.: Numerical solution of parametrized Navier-Stokes equations by reduced basis methods. Numer. Methods Partial Differential Equations **23(4)**, 923–948 (2007)
101. Quarteroni, A., Rozza, G., Manzoni, A.: Certified reduced basis approximation for parametrized partial differential equations in industrial applications. J. Math. Ind. **1**(3) (2011)
102. Quarteroni, A., Valli, A.: Numerical Approximation of Partial Differential Equations. Springer-Verlag, Berlin Heidelberg (1994)
103. Ravindran, S.: Reduced-order adaptive controllers for fluid flows using POD. J. Sci. Comput. **15**(4), 457–478 (2000)
104. Ravindran, S.: A reduced-order approach for optimal control of fluids using proper orthogonal decomposition. Int. J. Numer. Meth. Fluids **34**, 425–448 (2000)
105. Rowley, C.: Model reduction for fluids, using balanced proper orthogonal decomposition. Int. J. Bifur. Chaos Appl. Sci. Engrg. **15**(3), 997–1014 (2005)
106. Rozza, G., Huynh, D., Manzoni, A.: Reduced basis approximation and a posteriori error estimation for Stokes flows in parametrized geometries: roles of the inf-sup stability constants. Numer. Math. **125**(1), 741–776 (2013). DOI 10.1007/s00211-013-0534-8
107. Rozza, G., Huynh, D., Patera, A.: Reduced basis approximation and a posteriori error estimation for affinely parametrized elliptic coercive partial differential equations. Arch. Comput. Methods Engrg. **15**, 229–275 (2008)

108. Rozza, G., Manzoni, A., Negri, F.: Reduction strategies for PDE-constrained optimization problems in haemodynamics. In: J. Eberhardsteiner et.al. (ed.) Proceedings of the European Congress on Computational Methods in Applied Sciences and Engineering (ECCOMAS 2012), Vienna, Austria, September 10–14, 2012 (2012)
109. Rozza, G., Veroy, K.: On the stability of reduced basis methods for Stokes equations in parametrized domains. Comput. Methods Appl. Mech. Engrg. **196**(7), 1244–1260 (2007)
110. Sen, S., Veroy, K., Huynh, P., Deparis, S., Nguyen, N., Patera, A.: "natural norm" a posteriori error estimators for reduced basis approximations. J. Comp. Phys. **217**(1), 37–62 (2006)
111. Sirisup, S., Karniadakis, G.: A spectral viscosity method for correcting the long-term behavior of POD models. J. Comp. Phys. **194**(1), 92–116 (2004)
112. Sirisup, S., Karniadakis, G.: Stability and accuracy of periodic flow solutions obtained by a POD-penalty method. J. Phys. D **202**(3), 218–237 (2005)
113. Sirisup, S., Karniadakis, G., Yang, Y., Rockwell, D.: Wave–structure interaction: simulation driven by quantitative imaging. Proc. R. Soc. A **460**(2043), 729–755 (2004)
114. Sirovich, L.: Turbulence and the dynamics of coherent structures. Part I: Coherent structures. Q. Appl. Math. **45**(3), 561–571 (1987)
115. Tadmor, E.: Convergence of spectral methods for nonlinear conservation laws. SIAM J. Numer. Anal. **26(1)**, 30–44 (1989)
116. Tadmor, G., Lehmann, O., Noack, B., Morzyński, M.: Mean field representation of the natural and actuated cylinder wake. Phys. Fluids **22**, 034,102 (2010)
117. Tamellini, L., Le Maître, O., Nouy, A.: Model reduction based on proper generalized decomposition for the stochastic steady incompressible Navier-Stokes equations. Tech. Rep. 26, MOX - Modellistica e calcolo scientifico, Politecnico di Milano (2012)
118. Temam, R.: Navier-Stokes Equations. AMS Chelsea, Providence, Rhode Island (2001)
119. Terragni, F., Vega, J.M.: On the use of POD-based ROMs to analyze bifurcations in some dissipative systems. Physica D: Nonlinear Phenomena 241(17), 1393–1405 (2012). DOI 10.1016/j.physd.2012.04.009
120. Thomas, J., Hall, K., Dowell, E.: A harmonic balance approach for modeling nonlinear aeroelastic behavior of wings in transonic viscous flow. In: 44 th AIAA/ASME/ASCE/AHS/ASC Structures, Structural Dynamics, and Materials Conference (2003)
121. Tonn, T., Urban, K., Volkwein, S.: Optimal control of parameter-dependent convection-diffusion problems around rigid bodies. SIAM J. Sci. Comput. **32**(3), 1237–1260 (2010)
122. Tu, J., Rowley, C.: An improved algorithm for balanced POD through an analytic treatment of impulse response tails. J. Comp. Phys. **231(16)** (2012)
123. Urban, K., Patera, A.T.: An improved error bound for reduced basis approximation of linear parabolic problems. Math. Comp. (2013). In press
124. Veroy, K., Patera, A.: Certified real-time solution of the parametrized steady incompressible Navier-Stokes equations: rigorous reduced-basis a posteriori error bounds. Int. J. Numer. Meth. Fluids **47(8-9)**, 773–788 (2005)
125. Veroy, K., Prud'homme, C., Rovas, D.V., Patera, A.T.: *A posteriori* error bounds for reduced basis approximation of parametrized noncoercive and nonlinear elliptic partial differential equations. In: Proceedings of the 16th AIAA Computational Fluid Dynamics Conference (2003). Paper 2003–3847
126. Wang, Z., Akhtar, I., Borggaard, J., Iliescu, T.: Proper orthogonal decomposition closure models for turbulent flows: A numerical comparison. Comput. Meth. Appl. Mech. Engrg. **237–240**, 10–26 (2012)

127. Weller, J., Lombardi, E., Bergmann, M., Iollo, A.: Numerical methods for low-order modeling of fluid flows based on POD. Int. J. Numer. Methods Fluids **63**(2), 249–268 (2010)
128. Willcox, K., Peraire, J.: Balanced model reduction via the proper orthogonal decomposition. AIAA J. **40**(11), 2323–2330 (2002)
129. Yano, M.: A space-time Petrov-Galerkin certified reduced basis method: application to the Boussinesq equations. Submitted to SIAM J. Scientific Computing (revised, 2013). Preprint available at augustine.mit.edu
130. Yano, M., Patera, A.T.: A space-time variational approach to hydrodynamic stability theory. Proceedings of Royal Society A, **469**(2155), article 2013 0036 (2013)
131. Yano, M., Patera, A.T., Urban, K.: A space-time certified reduced basis method for Burgers' equation (2013). Preprint available at augustine.mit.edu
132. Zhou, K., Salomon, G., Wu, E.: Balanced realization and model reduction for unstable systems. Int. J. Robust Nonlinear Control **9**(3), 183–198 (1999)

10

# Window Proper Orthogonal Decomposition: Application to Continuum and Atomistic Data

**Leopold Grinberg, Mingge Deng, Alexander Yakhot and George Em Karniadakis**

## 10.1 Introduction

Proper Orthogonal Decomposition (POD) is a powerful tool for analyzing multi-dimensional data, especially of vector fields in large-scale simulations. In this article we review the Window Proper Orthogonal Decomposition (WPOD) proposed in [7] for analysis of continuum data and in [5] for analysis of atomistic fields.

WPOD seeks for correlation of data obtained over certain time intervals (windows). In that sense it can be compared to the window Fourier analysis or the wavelet decomposition. However, WPOD imposes no requirement for the data to exhibit periodicity and does not require use of predefined spatial modes.

Here we are not seeking for data and dimensionality reduction only, but rather we attempt to partition the data with respect to different physical events reflected by changes in the POD eigenspectra. For example, in application of WPOD to atomistic fields, the ensemble solution can be effectively separated from the thermal fluctuations. Our basic measure for partitioning the POD expansion is the *rate of convergence* of POD eigenvalues, while in most implementations of POD in low-dimensional modeling the basic metric is the total energy associated with the low POD modes.

L. Grinberg (✉)
BM T.J. Watson Research Center, Cambridge, MA, 02142 USA, and
Division of Applied Mathematics, Brown University, Providence, RI 02912, USA
e-mail: leopoldgrinberg@us.ibm.com

M. Deng
Division of Applied Mathematics, Brown University, Providence, RI 02912, USA

A. Yakhot
Department of Mechanical Engineering, Ben-Gurion University, Beersheva 84105, Israel

G.E. Karniadakis
Division of Applied Mathematics, Brown University, Providence, RI 02912, USA

A. Quarteroni, G. Rozza (eds.): *Reduced Order Methods for Modeling and Computational Reduction*, MS&A 9. DOI 10.1007/978-3-319-02090-7_10, 

Several visualization packages such as VMD [21] exist for visualizing data from particle based simulations, however, these packages do not incorporate quantitative analyzis of the particle motion. In visualizations of multiscale atomistic data, often represented by millions or even billions of discrete particles, the large-scale features can be depicted by projecting the atomistic field on a continuum involving also a substantial data compression. The *correlated motion* of the media is then visualized using the continuum fields, while the motion of a small collection of particles (e.g., modeling polymers, red blood cells) can be described by tracking their trajectories and superimposing it onto the continuum data images.

In this chapter we provide the formulation of WPOD and present its utility in several disciplines. We start with a brief motivation and a mathematical formulation of POD and WPOD for continuum and atomistic data. Then, we present an application of WPOD for quantitative analysis of intermittent laminar-turbulent flow. We continue with application of WPOD for analysis of atomistic data, specifically, computing the ensemble solution and the PDF of thermal fluctuations. We discuss coupled atomistic-continuum multiscale simulations where WPOD is applied to co-process interface data. We demonstrate the utility of WPOD for multiscale visualization and conclude with a summary.

## 10.2 Motivation

There are several reasons for developing windowed spectral analysis tools as a post- and co-processing procedure. In the following, we name a few areas where WPOD can be successfully implemented.

i) *Molecular dynamics simulations* generate information at the microscopic level, e.g., positions and velocities of particles. The conversion of this microscopic information to macroscopic observables requires statistical averaging, i.e., computing ensemble averages. An ensemble is defined as a collection of all possible systems, which have different microscopic states but have an identical macroscopic or thermodynamic state. Traditional statistical averaging requires an extremely large number of samplings of microscopic information, which is often a major computational expense. Even with very high sampling rate, it is typical that the accuracy of the final results may still be quite low, which makes physical interpretations difficult.

In *stationary* flow simulations, the average solution $\bar{u}(\mathbf{x})$ is typically computed by sampling and averaging the trajectories of the particles over a subdomain $\Omega_p$ and over a very large time interval, i.e., $\bar{u}(\mathbf{X}_p) = \frac{1}{N_p}\sum_{n,j} u(t^n, \mathbf{x}_j)$, where $\mathbf{X}_p$ corresponds to the center of $\Omega_p$ and $N_p$ is the total number of particles $\mathbf{x}_j \in \Omega_p$ at any given time-step $t^n$. To obtain an *accurate* approximation to $\bar{u}(\mathbf{x})$, it is required that both the time and space intervals be sufficiently large. With respect to time, this can be achieved by integrating over thousands of time steps.

In *non-equilibrium* atomistic simulations, however, such as in simulations of unsteady flows, propagation of cracks in materials, simulations of polymers and red blood cells, etc., the standard statistical tools employed are even more problematic.

In *non-stationary flow* simulations, an ensemble average $\bar{u}(t,\mathbf{x})$ is required, but it is not obvious how to define a time interval $T \gg \Delta t$ (where $\Delta t$ is the time step) over which the solution can be averaged. It is possible to perform phase averaging, if the flow exhibits a limit cycle and integrate the solution over a large number of cycles, but for general cases phase averaging may not be a suitable technique. In some simulations, constructing the ensemble based on many realizations is the only choice, but even then the accuracy can be improved only proportionally to $\sqrt{N_r}$, where $N_r$ is the number of realizations. For simulation of small problems on a moderate number of computer processors, increasing $N_r$ is a reasonable approach. However, in simulations requiring over $O(10^5)$ computer processors, increasing $N_r$ by an order of magnitude is very inefficient giving the limited accuracy gain.

*The dynamics of deformable vesicles and cells* subject to stresses and thermal noise is characterized by a variety of phenomena as shown in recent biologically motivated experiments [2, 9, 10, 16, 20]. The vesicles and cells are typically simulated using particles linked by non-rigid bonds, e.g., polymers can be represented as a chain of particles, a membrane of blood cell can be represented as a network of particles with fixed connectivity. As an example of dynamically deformable object let us consider a vesicle which has three quasi-steady dynamical states in shear flow, namely, tank treading, tumbling and trembling. These dynamical states can be described in a framework of low-dimensional modes (or the dynamical process can be described by one or two degrees of freedom) [11, 12]. However, in relative strong flows where the stresses acting on the soft objects might be extremely big, the cell dynamics is much more complicated due to the nonlinear response to the stress and stochastic noise. The response to thermal noise becomes a central issue for such dynamic system, in a sense that it may mask the deterministic component of time-space fluctuations. Wrinkling patterns of membranes and semi-flexible polymer appear in elongational flow and shear flow [2,9,10,16,20]. To capture the mechanisms behind the nonlinear dynamics, one has to analyze the wrinkling patterns which are related to the excitement of high-order deformation modes. Since the wrinkles during the vesicle dynamics are not stationary and are excited only for a limited time, the often-used Fourier decomposition and spherical harmonics decomposition [2, 10, 16, 20] are actually not suitable because of the boundaries and unknown nonuniform tensions on the vesicle surface along the polymer chain. Instead, the Euler-Lagrange equation of the energy functional with respect to its configuration and nonuniform tension defines a set of orthogonal eigenfunctions (and eigenvalues) with satisfyied boundary conditions [9]. However, it is impossible, even numerically, to obtain the eigenfunctions when the tension function is unknown and unsteady. Thus, WPOD may become an enabling tool for analysis of the time-space nonlinear dynamical modes and their evolution process by constructing orthogonal eigenfunctions from dynamical trajectories and decoupling the correlated motion of particles from thermal fluctuations.

ii) *Coupled atomistic-continuum simulations* [22] is another area where robust techniques for computing the ensemble average are crucial for establishing interface conditions. In Fig. 10.1 we present a coupled atomistic-continuum simulation of platelet aggregation in an aneurysm [6]. The *unsteady* flow in the domain of brain

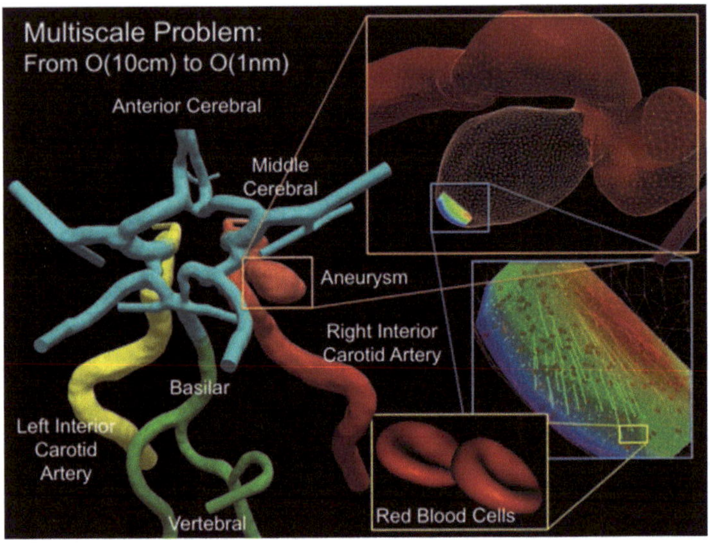


**Fig. 10.1.** Multiscale atomistic-continuum coupled simulation of a brain blood flow. Shown on the left is the computational domain of major brain arteries with an aneurysm (macrodomain) where the Navier-Stokes equations are solved; different colors correspond to different domain patches. Shown in the inset (right) is the microdomain ($3.93mm^3$) where dissipative particle dynamics is applied. The WPOD method is applied in this microdomain. Inside the microdomain: streamlines show instantaneous flow directions; red objects represent a fraction of red blood cells; dots represent a fraction of plasma particles; colors correspond to the ensemble average velocity. Courtesy of Argonne National Laboratory. Visualization by J. Insley

arteries was computed using a continuum approach, while the interactions between blood cells, plasma and arterial walls were simulated using the dissipative particle dynamics (DPD) method. The flow was simulated on up to 294,912 computer processors, with over 90% of computing power dedicated to the atomistic solver. Continuity in the continuum and atomistic fields were achieved by imposing proper interface conditions [6]. The interface conditions required extracting large-scale flow features from the atomistic simulations, i.e., filtering out thermal fluctuations from the atomistic data. The WPOD method was applied to compute the ensemble average of the stochastic fields. We note that an alternative approach based on concurrent multiple realizations would require millions of computer processors.

Coupled atomistic-continuum algorithms become more efficient and conservative if fluxes instead of state variables are exchanged in the coupled scheme [13]. This, in turn, requires smooth representations of interface fluxes for faster convergence and for preventing erroneous propagation of numerical artifacts into the domain. As we will demonstrate in Sect. 10.5, the gradients of the state variables – obtained in atomistic simulations of unsteady flow and processed in the standard way – are far from smooth and hence inappropriate for such multiscale simulations. The smoother

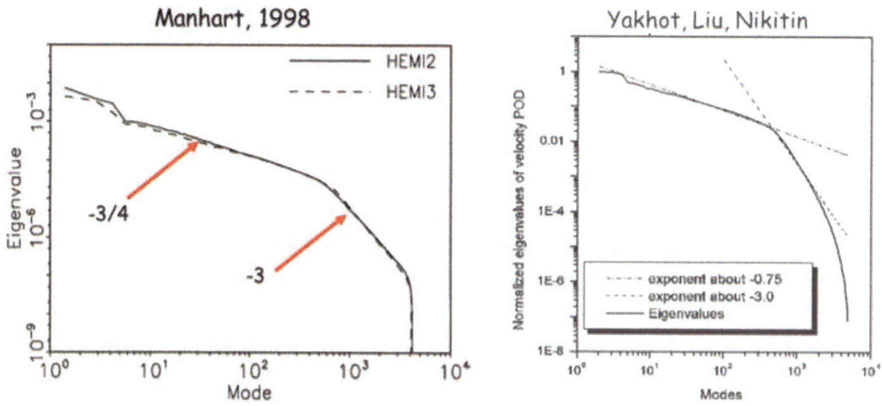


**Fig. 10.2.** Turbulent flow simulations: POD eigenspectra. *Left* - flow around a wall-mounted hemi-sphere (reproduction from [17]); *right* - flow around a wall-mounted cube (unpublished)

velocity field reconstructed with the WPOD allows better accuracy in predicting the field gradients.

iii) *WPOD for continuum* data was first applied for *quantification* of transitional blood flow, i.e., capturing the spatio-temporal transition from laminar to turbulent flow regimes [7]. Laminar flows exhibit a high degree of spatio-temporal correlation while turbulent flows show significantly lower. This is reflected in the dimensionality of the flow field data and correspondingly in the POD eigenvalue spectrum, where high order ("turbulent") modes converge slow and often at a constant rate. If the region of turbulent flow is very small compared to the size of the entire domain, analysis of POD spectrum or inspection of temporal modes performed on the entire domain is insufficient. To this end, WPOD can quantify precisely the distribution of the kinetic energy at different spatio-temporal windows.

Manhart [17] applied POD analysis of high-Reynolds number turbulent flows to analyze flow around a wall-mounted hemisphere; Yakhot *et. al.* (unpublished) applied POD to analyze turbulent flow around a wall-mounted cube. Both studies clearly showed the *power law decay* rate $\lambda_k \sim k^{-3/4}$ of the POD high eigenvalues (see Fig. 10.2). The fact that the same power law was obtained in two different POD studies is intriguing. To the best of our knowledge, it has not been shown that the value of the exponent is universal. Fourier analysis applied to turbulent fields shows that the energy spectrum follows a power law $\lambda_k \sim k^{-s}$ with $1 < s < 3$ depending on the turbulence nature. For homogeneous isotropic turbulence, the POD eigenmodes are simply the Fourier modes.

Flow in complex geometries is neither homogeneous nor isotropic but a power law energy decay provides additional evidence for the quatitative characterization of turbulence. In [7] a study of flow in a stenosed carotid artery was performed and high-frequency oscillations were detected downstream of the stenosed region. The presence of fluctuations is not sufficient evidence for the presence of turbulence, which is characterized by specific statistical properties, and hence the distribution of energy between different scales is a commonly used criterion. The spatio-temporal WPOD analysis of flow in the stenosed carotid artery presented in [7] reveals that during the systolic peak of the cardiac cycle the exponent $s$ is in the range of $s = 0.8$ to $s = 1.1$; this result was later confirmed in experiments by Kefayati & Poepping [14].

iv) *Multiscale visualizations of atomistic data.* The number of particles in MD or DPD simulations can be extremely large; this taxes not only the computational cost of simulations but also adversely affects data post-processing including visualization. In particle-based simulations, data for every 100M particles require about 5.2 GB of disk space for one snapshot, even by saving only the particle IDs, coordinates and velocity vector.

Typical particle-based simulations (MD or DPD) include solvent particles, e.g., particles representing the blood plasma, and non-solvent particles, e.g., proteins, red blood cells (RBCs), platelets, glycocalyx, etc. Proteins and blood cells are represented by a collection of bonded particles with fixed connectivity. Fortunately, the majority of the particles (at least in simulations considered here) represent the solvent, and visualization of discrete particle data can be substituted by presenting their average properties projected on a fixed grid. When the projection is combined with WPOD, the data also represents the collective and correlated in time and in space solution field. Representing the data for the plasma particles by its continuum analog typically reduces the storage requirements by 80 to 95 percent. The output data can be written on the disk or passed directly to the visualization software allowing real-time visualization. This real-time data compression and fast data commute from simulation to visualization also opens up new possibilities for effective computational steering.

## 10.3 Methodology

### *10.3.1 Proper Orthogonal Decomposition (POD)*

Proper Orthogonal Decomposition is a spectral analysis tool often employed for data compression and low-dimensional modeling. The method is also known as principal component analysis (PCA), singular value decomposition (SVD) and Karhunen Loève decomposition. Our formulation of WPOD is based on the method of *snapshots*, introduced by Sirovich [18] who applied POD in fluid dynamics for identification of coherent structures in a turbulent velocity field.

POD decomposes the time-space field $u(t,\mathbf{x})$ into an expansion of orthogonal temporal and spatial modes, i.e.,

$$u(t,\mathbf{x}) = \sum_{k=1}^{N_{pod}} a_k(t)\phi_k(\mathbf{x}). \tag{10.1}$$

The basis $\phi_k(\mathbf{x})$ is sought so that the approximation onto the first $K$ functions: $\hat{u}(\mathbf{x},t) = \sum_{k=1}^{K} a_k(t)\phi^k(\mathbf{x}),\ \ K \leq N_{pod}$ has the largest mean square projection.

To compute the space- time- POD modes over a time interval $T$ and space $\mathbf{x} \in \Omega$, a temporal auto-correlation covariance matrix $\mathbf{C}$ is constructed from the inner products between pairs of fields (snapshots) collected at times $t^i$ and $t^j$, $\quad i,j = 1,...,N_{pod}$:

$$C_{i,j} = \int u(t^i,\mathbf{x})u(t^j,\mathbf{x})d\Omega. \tag{10.2}$$

The temporal modes $a_k(t)$ are the eigenvectors of $\mathbf{C}$. Using orthogonality, the POD spatial modes $\phi_k(\mathbf{x})$ are calculated from

$$\phi_k(\mathbf{x}) = \int_T a_k(t)u(t,\mathbf{x})dt. \tag{10.3}$$

The eigenvalues $\lambda_k, \lambda_1 > \lambda_2 > ... > \lambda_{N_{pod}}, \ \forall k = 1,...,N_{pod}$ of the autocorrelation matrix $\mathbf{C}$ represent the energy level associated with the POD mode $k$.

### *10.3.2 Window Proper Orthogonal Decomposition (WPOD)*

In WPOD we consider splitting the time interval $T$ into overlapping or non-overlapping sub-intervals (windows) $\Delta T_m \in T, m = 1,2,...$, i.e., the POD analysis is performed over a sub-set of snapshots. Such approach is more adequate for analysis of fields where certain events exist over a relatively short (finite) time. Also, such window approach may be the only choice for data co-processing, where computer memory is sufficient for storing only a certain number of the most resent snapshots. Eigenspectrum analysis of $\mathbf{C}(\Delta T)$ constructed over shorter time intervals leads to better capturing of transitional phenomena, as will be illustrated in Sect. 10.4 where POD analysis of intermittent laminar-turbulent flow is presented. Implementation of POD as a co-processing tool naturally leads to performing the analysis over time windows, where the autocorrelation matrix is computed from the most recent $N_{pod}$ snapshots.

The concept of local reduced-order bases (ROB) developed in [1] uses POD applied on the snapshots of the system taken at various locations of the state space. As in WPOD, the solution space is partitioned into subdomains, when a local ROB is constructed and assigned to each subdomain. The difference is that the ROB approach uses for nonlinear system model order reduction, while the WPOD employs for analysis of nonlinear system evolving data.

POD requires spatial integration of fields represented at fixed grid (mapping from moving frame to fixed can be also considered). In atomistic simulations, field data is associated with Lagrangian particles moving in the computational domain and even exiting and entering the domain. Accordingly, WPOD in the atomistic simulations is performed in two steps. At the *first step*, a spatio-temporal averaging over relatively short time intervals is applied to project atomistic data on a set of fixed in time grid points $\mathbf{x}_p$. In our simulations the size of these time intervals is typically in the range of 50 to 500 time steps ($N_{ts} = [50\ 500]$):

$$u(\tau,\mathbf{x}_p) = \frac{1}{N_p} \sum_{n=n_1}^{n_1+N_{ts}} \sum_j u(t^n,\mathbf{x}_j), \ \tau = t^{n_1} + N_{ts}\Delta t/2.$$

At the *second step*, the method of snapshots is applied to a subset of $u(\tau^s, \mathbf{x})$, simi-

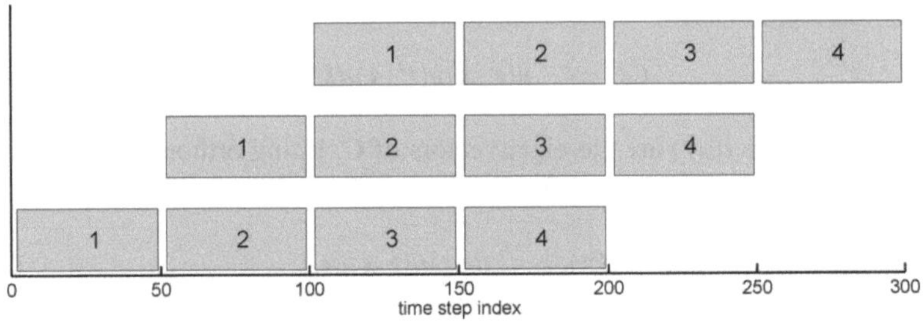


**Fig. 10.3.** Illustration of POD windows in processing molecular dynamics data. Each POD window is composed of $N_{pod} = 4$ snapshots. Each snapshot is obtained by spatio-temporal averaging of data over $N_{ts} = 50$ timesteps. Three POD windows are shown

larly to WPOD for continuum data. In Fig. 10.3 we provide a schematic illustration of the WPOD method for atomistic fields. WPOD is performed on data computed within a time interval $T_{pod} = N_{pod}N_{ts}\Delta t$. The correlation matrix **C** is based on the $N_{pod}$ most recent snapshots; it is updated every $N_{ts}$ time steps and its eigenvalues and corresponding eigenmodes are computed.

The WPOD method we present here transforms the velocity field to orthogonal modes, which can then be employed to approximate the ensemble solution $\bar{u}(t, \mathbf{x})$ and thermal fluctuations $u\prime(t, \mathbf{x})$. Unlike the simple spatio-temporal averaging, WPOD is effective in capturing the *correlations* in atomistic properties (e.g. velocity filed) over time intervals, and not just to smear off any temporal fluctuations in the solution. In addition, WPOD allows for a fast construction of the probability density function (PDF) of the fluctuating part (higher order modes) $u\prime(t, \mathbf{x})$, which can be utilized to gain better physical insight.

To separate the velocity field into $\bar{\mathbf{u}}(t, \mathbf{x})$ and $\mathbf{u}\prime(t, \mathbf{x})$, e.g., into the ensemble average solution and thermal fluctuations, several criteria based on adaptive analysis of the POD eigenvalues and eigenvectors can be employed. We denote the WPOD method as "adaptive" since the POD eigenproperties and the criteria for selecting the proper number of POD modes to compute $\bar{\mathbf{u}}(t, \mathbf{x})$ and $\mathbf{u}\prime(t, \mathbf{x})$ are re-examined every $N_{ts}$ time steps. To this end, the *first* criterion is based on the rate of convergence of POD eigenvalues. The *second* criterion is based on the level of noise present in temporal modes. The *third* criterion is based on the analysis of standard deviation of POD eigenvectors (the temporal modes).

To illustrate the two distinct rates of convergence of POD eigenspactra, we plot in Fig. 10.4 the POD eigenvalues and selected temporal eigenmodes obtained by a DPD simulation of pulsatile flow in a pipe; this will be discussed further the Sect. 10.5. The eigenvalues can be separated into two clearly distinct subsets according to the rate of convergence. The POD modes corresponding to $\lambda_k$ with very small rate of convergence ($k \geq 3$ in Fig. 10.4 (left) represent data with very short correlation length (i.e., thermal fluctuations), while the low POD modes represent fields characterized

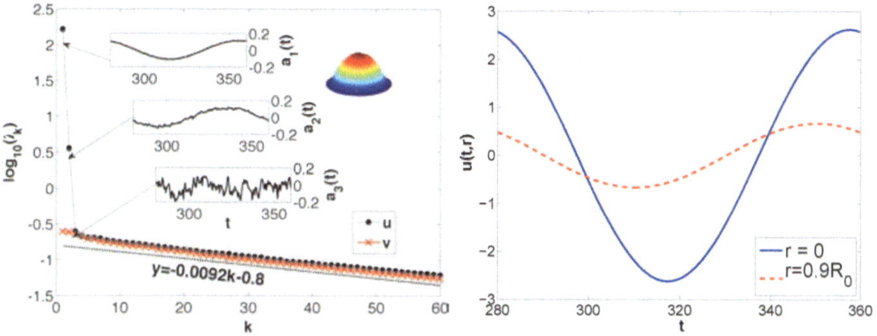


**Fig. 10.4.** DPD-simulation: 3D pipe flow driven by a time-periodic force. *Left*: Eigenspectra of velocity in $x-$ (stream-wise velocity component, black dots) and $y-$direction (red crosses), and three POD temporal modes. The first two eigenvalues of the WPOD analysis of analytical solution performed over the same time-window as in computation are $log_{10}(\lambda_1) = 2.208$ and $log_{10}(\lambda_2) = 0.525$. Top right corner: velocity profile (streamwise component) reconstructed with the first two POD modes at $t = 360$. *Right*: velocity trace at the center line ($r = 0$) and close to the wall ($r = 0.9R_0$; here $R_0$ is the pipe radius). (Adapted from [5])

by a long correlation length, i.e., the ensemble average. The WPOD analysis of the simulation data agrees very well with the WPOD analysis of the analytical solution performed over the same time-window as in computation. The first two POD eigenvalues obtained from the analytical solution are $log_{10}(\lambda_1) = 2.208$ and $log_{10}(\lambda_2) = 0.525$, while for $k > 2$ we have $log_{10}(\lambda_k) = 0$. The first two POD eigenvalues of the DPD simulation are $log_{10}(\lambda_1) = 2.217$ and $log_{10}(\lambda_2) = 0.54$, while for $k > 2$ we have $log_{10}(\lambda_k) < -0.6$. Considering the very small deviation between the eigenvalues of the analytical solution and numerical solution with DPD, we can expect that the error in evaluating $\bar{\mathbf{u}}$ will be of the same order. In fact, the error in simulation results processed with WPOD and compared to the analytical solution was of order $10^{-2}$, which is about two orders of magnitude smaller that the mean velocity.

To employ the *first* criterion to separate the velocity field into $\bar{\mathbf{u}}(t,\mathbf{x})$ and $\mathbf{u}\prime(t,\mathbf{x})$, we investigate the rate of convergence of the POD eigenvalues $\lambda_k$ as a function of mode number $k$. The high modes represent small-scale features with very short correlation time, i.e., the thermal fluctuations. In contrast to the $\lambda_k$ for the low modes, the convergence of $\lambda_k$ associated with the higher modes is very slow. We can then compute the ensemble average from the low, most energetic modes, characterized by the fast convergence rate of $\lambda_k$, while the fluctuations ($u\prime(t,\mathbf{x})$) are computed from the high slowly decaying modes.

*The first criterion* is based on fitting the curve $log_{10}(\lambda_k) = f(k)$ with piecewise $C^0$ continuous linear function; the method is illustrated in Fig. 10.5. Two least-square approximations are performed for a range of $\tilde{\tilde{k}}$ values to obtain coefficients $c_1(\tilde{\tilde{k}})$ and $c_2(\tilde{\tilde{k}})$ of $log(\lambda_k) - log(\lambda_{\tilde{\tilde{k}}}) = c_1(k - \tilde{\tilde{k}}), k \leq \tilde{\tilde{k}}$ and $log(\lambda_{\tilde{\tilde{k}}}) - log(\lambda_k) = c_2(k - \tilde{\tilde{k}}), k \geq \tilde{\tilde{k}}$. Second, the residual of the least-square approximations for each $\tilde{\tilde{k}}$ is computed

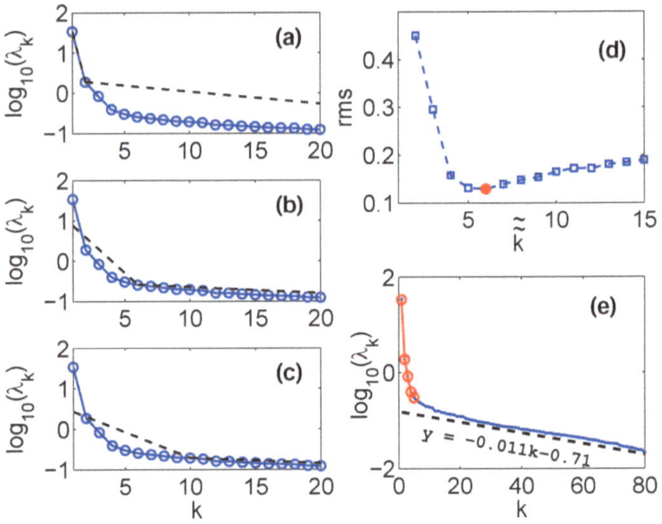


**Fig. 10.5.** MD simulation with DPD thermostatting of unsteady cylinder flow between two plates: eigenvalue analysis of velocity in $x-$ direction. (a)-(c) – POD eigenspectra and piecewise linear approximation with $\tilde{\tilde{k}} = 2$, 6 and 10. (d) – root mean square (rms) of the piecewise approximation error, red dot at $\tilde{\tilde{k}} = 6$ marks the smallest rms. (e) – five POD modes (highlighted in red) are selected to reconstruct the flow in $x$- direction. $N_r = 3, N_{ts} = 250, N_{pod} = 80$. (Adapted from [5])

and we set $\tilde{k} = \tilde{\tilde{k}} - 1$, where $\tilde{\tilde{k}}$ corresponds to the minimum residuals over all $\tilde{\tilde{k}}$. The ensemble average and the thermal fluctuations are then computed as $\bar{u}(\tau,\mathbf{x}) = \sum_{k=1}^{k=\tilde{k}} a_k(\tau)\phi_k(\mathbf{x})$ and $u\prime(\tau,\mathbf{x}) = u(\tau,\mathbf{x}) - \bar{u}(\tau,\mathbf{x})$. In flow simulations where the ensemble average does not depend on time, the number of modes required to compute $\bar{u}(\mathbf{x})$ is strictly one. Moreover, $\tilde{k} = 0$ corresponds to the field consisting of thermal fluctuations only; an example for such a field is presented in Fig. 10.4, where all eigenvalues corresponding to the cross-flow velocity (in $y-$direction) have practically the same rate of convergence. In general, $\tilde{k}$ increases with the complexity of a flow as we will see in the following examples.

To employ the *second criterion* to separate the velocity field into $\bar{\mathbf{u}}(t,\mathbf{x})$ and $\mathbf{u}\prime(t,\mathbf{x})$ we take into account the smoothness of temporal modes. For example, in Fig. 10.4 we observe that the first two temporal modes can be accurately approximated with a smooth function, e.g., high-order polynomials. The $L_2$-norm of the error in approximating the temporal modes $a_1(t)$ and $a_2(t)$ with fourth-order polynomial through a least squares method is about three orders of magnitude less than its maximum value, which means that the presence of noise in the first and second modes is negligible. This finding is also consistent with the accuracy in computing the $\lambda_1$ and $\lambda_2$. The high-frequency oscillations present in the third POD mode which, according to the first criterion was attributed to the thermal fluctuations, suggest that it cannot be accurately approximated with a relatively smooth function and according to the second criterion it should not be used to reconstruct $\bar{\mathbf{u}}(t,\mathbf{x})$. To adaptively select the POD modes representing $\bar{\mathbf{u}}(t,\mathbf{x})$, one needs to define a threshold relating the $L_2$-error in approximating $a_k(\tau)$ with respect to (for example) $L_\infty$-norm of $a_k(\tau)$. A good approximation to the energy of noise present in the low-order "smooth" POD modes can be derived by measuring the energy of oscillations present in high-order modes representing the thermal fluctuations.

Although the sample problem used to illustrate the behavior of the POD eigenspectra in atomistic simulations considers unsteady flow, the task of separating the eigemodes by the rate of convergence was relatively simple. It can be seen that the curve $y(k) = log_{10}(\lambda_k)$ can be approximated very well by a $C^0$ piecewise linear function, however, in more general cases, e.g., considering complex geometry or different types of particles, the transition from the very fast to very slow converging $\lambda_k$ can be smoother. In Fig. 10.5 we plot the eigenspectra obtained by MD simulation of unsteady flow past two cylinders [5]. It is clearly seen that the eigenvalues in the range of $k = 2$ to 6 smoothly change the rate of convergence from very fast to very slow. However, even for this case the method of finding $\tilde{k}$ and consequently separating the velocity fields into $\bar{\mathbf{u}}(t,\mathbf{x})$ and $\mathbf{u}\prime(t,\mathbf{x})$ using the first criterion only was quite robust.

In most of our particle-based simulations of *complex fluids* the transition from fast convergence (k O(1)) to steady slow convergence (k»1) is not sharp, and accordingly, approximating the curve of $\lambda(k)$ with piece-wise polynomial function defined on two elements may not result in approximation of that curve with close to zero error. However, such approximation still serves its goal very well, as the only purpose of it is to detect (even approximately) the range of POD modes associated with slow (and fixed-rate) converging eigenvalues. The sensitivity of $\bar{\mathbf{u}}(t,\mathbf{x})$ to errors in $\tilde{k}$ is very low and can be estimated from comparing the value of $\lambda_{\tilde{k}+1}$ to $\sum_1^{\tilde{k}} \lambda_k$ (see Fig. 10.5e).

The *third criterion* for detecting whether a certain POD mode should be used to reconstruct $\bar{\mathbf{u}}(t,\mathbf{x})$ or $\mathbf{u}\prime(t,\mathbf{x})$ is based on the analysis of the temporal modes $a_k(t)$, specifically, their mean value $< a_k >$ or standard deviation $STD(a_k)$:

$$STD(a_k) = \sqrt{\frac{\sum_i^{N_{pod}}(a_k(\tau^i) - < a_k >)^2}{N_{pod}-1}}, \quad < a_k > = N_{pod}^{-1}\sum_i^{N_{pod}} a_k(\tau^i).$$

Assuming that the flow simulated with the atomistic solver is not turbulent, the energy associated with higher POD modes will converge very fast, which means that only the fluctuating component resulting from the thermal fluctuations contributes to the slow converging modes. The random force introduced as a DPD thermostat is a function of independent identically distributed random variable with zero mean, hence it is expected that the thermal fluctuations due to the added random force will also have zero mean. Consequently, the temporal POD modes representing the thermal fluctuations will also have zero mean, which implies that their standard deviation will be $STD(a_k) = 1/(\sqrt{N_{pod}-1})$, since $a_k \cdot a_k \equiv 1$. The effects of applying the third criterion in simulations of complex flows will be discussed in Sect. 10.5, where the WPOD is applied to DPD simulation of complex flow including red blood cells and plasma. The three criteria presented above can be also combined to evaluate the optimal $\tilde{k}$. In our simulations we search for $\tilde{k}$ by combining either the first and the second or the first and the third criteria.

## 10.4 WPOD for Quantative Analysis of Intermittent Laminar-Turbulent Flow

We have already discussed in Sect. 10.2 the application of POD to turbulent flow. A POD eigenspectrum showing the peculiar power law decay rate $s = -3/4$ obtained in two independent studies has been presented. Here we apply WPOD for analysis of *intermittent* laminar-turbulent flow in a complex geometry domain. Specifically, high-resolution three-dimensional simulations were employed to study transient turbulent flow in a carotid arterial bifurcation with a stenosed Internal Carotid artery (ICA). The detailed description of the problem set-up and simulation can be found in [7]; here we only briefly describe the procedure and methodology and focus on the outcome of the WPOD analysis.

The geometrical model of the arteries was reconstructed from MRI images, and *in vivo* velocity measurements were incorporated in the simulations to provide time-varying inlet and outlet boundary conditions (see Fig. 10.6). Due to high degree of the ICA occlusion and variable flowrate, a transitional and intermittent flow between laminar and turbulent states is established.

The complex geometry, specifically the curvature of the ICA in the downstream region with respect to the stenosis, facilitates a Coanda-type jet formation and turbulent transition exhibiting hysteretic behavior with respect to changes in Reynolds

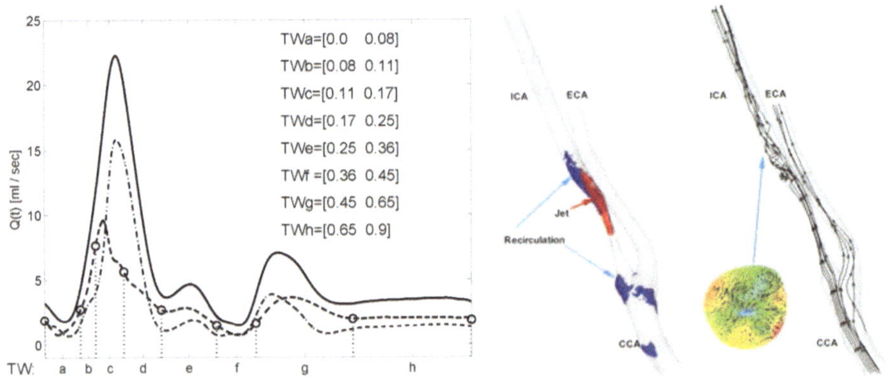


**Fig. 10.6.** Intermittent laminar-turbulent flow simulations in stenosed carotid artery. Left - waveform flow rates imposed in the inlet of the common carotid artery (solid), and outlets of the internal (dash) and external (dash - dot) carotid arteries and Time-Windows selected for WPOD data analysis. Right - flow patterns: iso-surfaces in a high-speed region (jet, red), blue iso-surfaces - back-flow regions; and instantaneous path-lines of swirling flow and cross-stream secondary flows. (Adapted from [7])

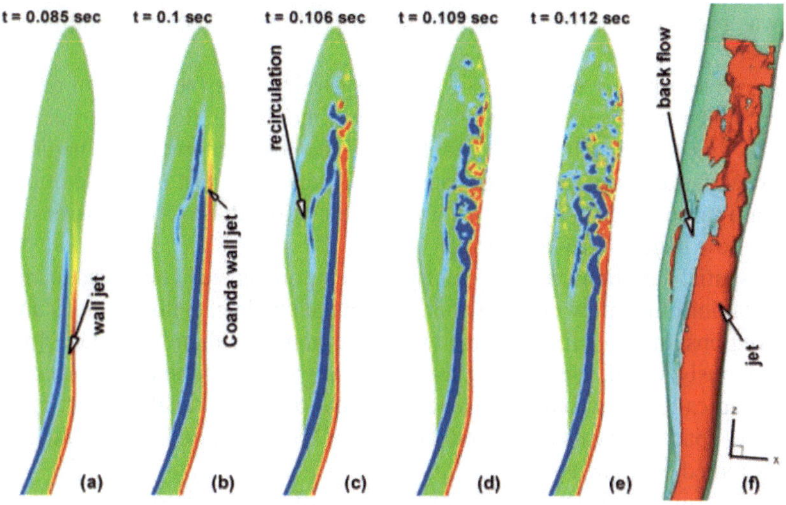


**Fig. 10.7.** (in color) Unsteady flow in carotid artery: transition to turbulence. (a-e) cross-flow vorticity contours $\Omega_y$ extracted along $y = -1.2$ in ICA. (f) region of ICA where flow becomes unstable, colors represent iso-surfaces of $w$-velocity (streamwise, along $z$-direction), $Re = 350$, $Ws = 4.375$. (Adapted from [7])

number. The main global flow features of the flow are presented in Fig. 10.6; while in common and external carotid arteries (CCA and ECA) pathlines are quite organized, those in ICA exhibit disorder in the poststenotic region featuring a swirling pattern. The blood flow along the curved wall is accompanied with decrease of the pressure on the wall, dropping below the surrounding pressure and resulting in the attachment of the fluid flow to the wall. The wall jet consists of an inner region, which is similar to a boundary layer, and an outer region wherein the flow resembles a free shear layer. These layers interact strongly and form a complex flow pattern.

Figure 10.7 demonstrates the jet-like effect created by the stenosis as predicted in our simulation. In Fig. 10.7 we show the contours of the $\Omega_y$-vorticity (transverse) that can be linked to the rolling up along the wall (see coordinate axes in Fig. 10.7f), for different stages of transition that show the wall jet breakdown. These results illustrate the onset of turbulence due to shear layer type instabilities of the Coanda wall jet in the post-stenotic region. Specifically, in Fig. 10.7a, the laminar state of the incoming flow is confirmed by the straight path traces. The jet outer region (marked by a blue trace) and the adjacent recirculation back-flow region (marked by a light-blue trace) form a free shear layer. In Fig. 10.7b, the jet moved downstream along the wall showing the early stage of interaction with the adjacent recirculation region. Figure 10.7c shows the perturbed shear layer at the leading edge of the jet. The tilted vorticity trace in Fig. 10.7d provides evidence of the stage of vortices coalescence in the outer region and the rolling up of the inner shear layer. The tilted wall jet rapidly breaks down leading to dispersion of the organized flow pattern (Figs. 10.7d,e). It should be noted that the breakdown gradually propagates upstream, a phenomenon that was predicted by Sherwin & Blackburn [19] using DNS in a simplified geometry.

The waveform curves consist of a brief systolic phase (acceleration and deceleration) and a longer diastolic phase with some increase in flow rate around t=0.55 (see Fig. 10.6). The early turbulent activity in the post-stenotic region begins at the mid-acceleration phase of the cardiac cycle. In the early part of deceleration there is intense turbulent activity; past the mid-deceleration phase, the intensities die out and the flow begins to re-laminarize.

The spatial variations in the geometry and temporal unsteadiness lead to intermittent behavior of the flow creating a jet, pockets of stable backflow regions and high shear regions, and *localized* in time and space transitional and turbulent flows. As we previously stated, we consider the power-law decay of the energy spectrum as a clear indication of turbulence. Hence, the goal of WPOD analysis here is to capture time- space intervals where convergence of high order POD eigenvalues can be approximated by a power-law.

To perform WPOD analysis we divide the cardiac cycle into eight *time-window* intervals denoted by $a \div h$ as illustrated in Fig. 10.6 (left); the time-windows have been chosen to represent different stages of the transient regime. We will refer to the time-windows as TWa÷TWh. In Fig. 10.8 we plot the POD eigenspectra computed over six consecutive time-windows; the eigenvalue spectrum slope provides an indication of a turbulent or laminar regime. The eigenvalue spectra distinguish clearly the presence of transition from laminar to turbulent regime shown in Fig. 10.8a and

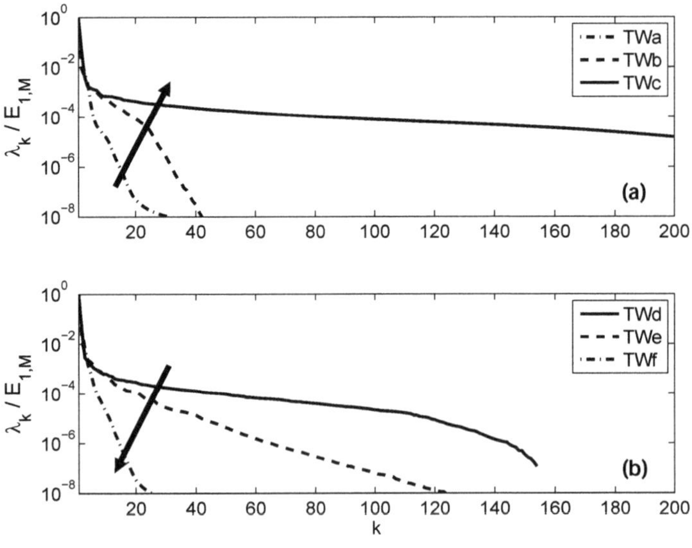


**Fig. 10.8.** WPOD: Eigenspectra obtained over different time-windows (see Fig. 10.6). The arrows denote the spectrum evolution in time: transition to turbulence is denoted by an upward-directed arrow, the downward-directed arrow refers to re-laminarization. (Adapted from [7])

followed by re-laminarization in Fig. 10.8b. The spectra obtained over the TWc and TWd time-windows, covering the transitional regime, display a slow and steady decay.

Time-window POD may not be sufficient for detection of spatially intermittent distribution of kinetic (turbulent) energy. Visualization of solution reconstructed from higher-order POD modes only does lead to capturing regions with high turbulent energy, however, it is not sufficient to *quantify* turbulence in each region. To this end, a *space-window* POD to detect regions with high kinetic energy can be employed. We analyze the eigenspectrum in two sub-domains of the ICA shown in Fig. 10.9: 1) the stenosis throat (sub-domain *AB*); 2) the post-stenotic region, from 12 to 22 mm downstream of the stenosis throat (sub domain *CD*).

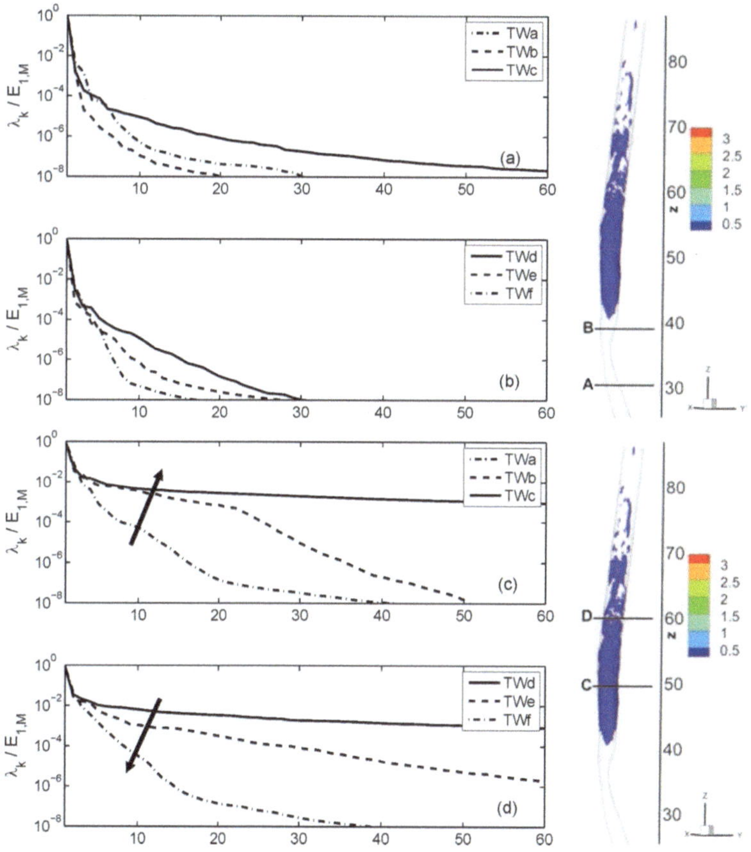


**Fig. 10.9.** POD eigenspectra obtained over different time-windows in *sub-domains AB* and *CD* (right): (a,c,e) - time-windows TWa, TWb and TWc (flow acceleration and transition to turbulence); (b,d,f) - time-windows TWd, TWe and TWf (flow deceleration and laminarization); (arrows show the time growth, color represents the *w*-iso-surface reconstructed from POD modes 20 to 50 at $t = 0.13$. (Adapted from [7])

Figure 10.9a,b shows fast decay of the POD eigenspectra computed in sub-domain *AB* where no turbulence was detected. In Fig. 10.9c,d we plot the spectra computed in sub-domain *CD*. The energy spectra in Fig. 10.9c reveal onset of turbulence and subsequent flow re-laminarization. The spectra obtained over TWc and TWd depict slow decay, typical for turbulent flow. The POD spectra in Fig. 10.9e,f show the same scenario of transition/re-laminarization although the turbulence here is very weak because it experiences a decay and eventually re-laminarization downstream of the stenosis (compare TWc and TWd curves in plots (e,f) with those in plots (c,d)).

Processing 3D data sets to determine if the energy spectra of a flow exhibits a power law decay may still carry excessive computational cost, particularly when simulation data is analyzed at the run time. In the next example we show that a 2D space-time WPOD can also be successfully employed for that purpose. To this end, the velocity field computed in ICA is extracted along 2-dimensional (2D) slices (space-windows), and then time-window POD analysis is performed. In Fig. 10.10 we show spectra computed over different time-windows using velocity fields obtained from a transverse to the main flow 2D slice at $z = 60$. Similarly to 3D POD analysis (see Figs. 10.9c,d), these POD spectra reveal transient flow regimes shown by arrows. POD spectra were also computed along longitudinal cross-sections provides the same information

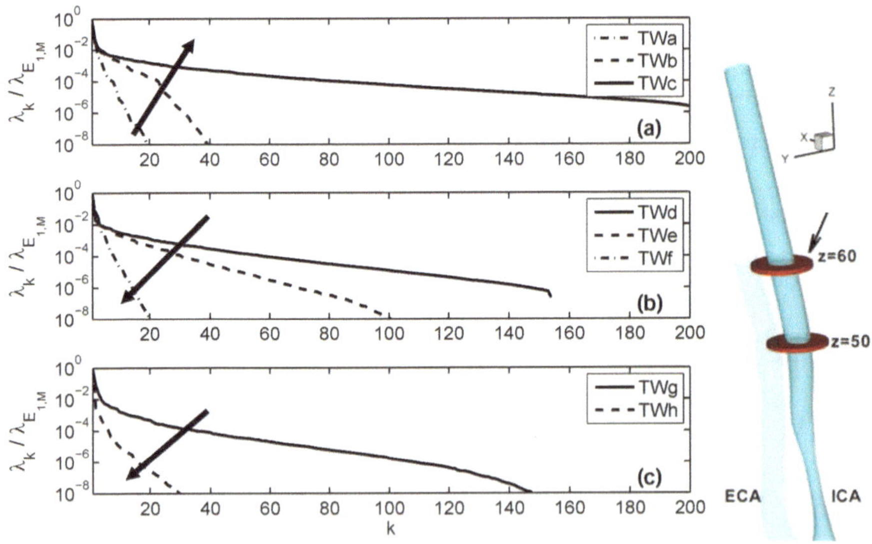


**Fig. 10.10.** 2D POD: eigenspectra obtained over different time intervals (see Fig. 10.6); velocity field is extracted at $z = 60$; (a) time-windows TWa, TWb and TWc (flow acceleration, and transition to turbulence); (b) time-windows TWd, TWe and TWf (flow deceleration, and laminarization); (c) time-windows TWg and TWh (diastole phase); arrows show the time growth

To quantify the *quasi-instantaneous* decay of the POD eigenvalues the following procedure can be implemented. Recall that turbulence is associated with existence of the high POD modes that exhibit a power law energy decay, namely

$$\lambda_k \sim k^{-s(t)}. \tag{10.4}$$

Let us extract flow field data from different planes, as shown in Fig. 10.10. For each time instant $t$ over the cardiac cycle, the POD analysis is performed over a relatively short time-window $t - \Delta t'/2 < t < t + \Delta t'/2$, where $\Delta t'$ is approximately 0.01 second at the systolic peak and approximately 0.1 second during the diastolic phase. Taking advantage of the high time resolution of simulations, each time interval $\Delta t'$ here is covered by 80 snapshots. The exponent $s(t)$ is then computed by approximating the convergence rate of modes $k = [2\ 20]$ with the power-law. In Fig. 10.11 we plot the exponent $s(t)$ of the POD eigenvalues. The double hump curves clearly indicate the transient nature of the flow. The low values of the slope $(0.8 < s < 1.1)$ correspond to the turbulent regime that occurs during the systolic phase. The secondary turbulence regime at $t \approx 0.55$ mentioned above is also captured by the low slope values in Fig. 10.11. The eigenspectra decay rate $0.8 < s < 1.1$ obtained in the simulation setup has been later confirmed in experimental setting by Kefayati and Poepping [14].

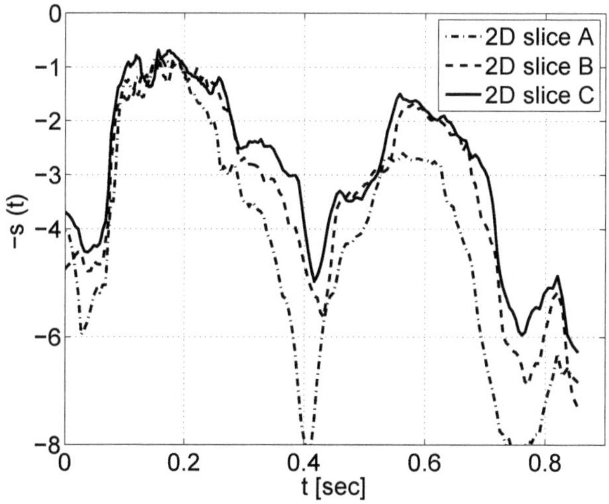


**Fig. 10.11.** 2D POD: decay rate of POD eigenspectra. Data are extracted along: slice $z = 50$ (2D slice A) and slice $z = 60$ (2D slice B) depicted in Fig. 10.10 (right); longitudinal slice with $x = const$ and located between $z = 50$ and $z = 60$. $s(t)$ is computed for the modes $k = 2 \div 10$

## 10.5 WPOD in Atomistic Simulations

In this section we focus on applications of WPOD in particle-based simulations. While the method is suitable for processing molecular dynamics and coarse-grained molecular dynamics data, we limit our discussion on the latter. Specifically, we consider application of WPOD to the data generated in dissipative particle dymanics simulations (DPD). Our focus is on the accuracy and computational efficiency in separating the ensemble solution from the thermal fluctuations. To exemplify the strength of the WPOD, particularly in analysis of non-stationary data, we selected two classes of problems: 1) simulation of unsteady flow where DPD particles have the same properties; and 2) simulation with four types of DPD particles representing healthy and diseased red blood cells, blood plasma and walls. The various particle types may have distinct cutoff radius, strength of dissipative force or other parameters defining the pairwise interactions as described in the following brief overview of the DPD method.

DPD is a mesoscopic particle method with each particle representing a *molecular cluster* rather than an individual molecule [3, 8, 15]. The DPD system consists of $N$ point particles interacting through conservative, dissipative and random forces, i.e., $\mathbf{F}_j = \mathbf{F}_j^C + \mathbf{F}_j^D + \mathbf{F}_j^R$ and all forces are truncated beyond the cutoff radius $r_c$, which defines the length scale in the DPD system. The motion of DPD particles is governed by the second Newton's law:

$$\frac{d\mathbf{r}_i}{dt} = \mathbf{u}_i \quad m_i \frac{d\mathbf{u}_i}{dt} = \mathbf{f}_i \quad \mathbf{f}_i = \sum_{j, i \neq j} \left( \mathbf{F}_{ij}^C + \mathbf{F}_{ij}^D + \mathbf{F}_{ij}^R \right). \tag{10.5}$$

The forces between particles $i$ and $j$ are computed from:

$$\mathbf{F}_{ij}^C = a_{ij} \left( 1 - \frac{r_{ij}}{r_c} \right) \hat{\mathbf{r}}_{ij}, \quad \mathbf{F}_{ij}^D = -\gamma \omega^D(r_{ij}) (\mathbf{u}_{ij} \cdot \hat{\mathbf{r}}_{ij}) \hat{\mathbf{r}}_{ij}, \quad \mathbf{F}_{ij}^R = \sigma \omega^R(r_{ij}) \frac{\xi_{ij}}{\sqrt{\Delta t}} \hat{\mathbf{r}}_{ij},$$

where $\mathbf{u}_{ij} = \mathbf{u}_i - \mathbf{u}_j$, $\Delta t$ is the time step, and $r_c$ is the cutoff radius beyond which all forces vanish. The coefficients $a_{ij}$, $\gamma$, and $\sigma$ define the strength of conservative, dissipative, and random forces, respectively; $\omega^D$ and $\omega^R(r_{ij}) = (1 - \frac{r_{ij}}{r_c})^k$ with the exponent $k$ are weight functions, and $\xi_{ij} = \xi_{ji}$ is a normally distributed random variable. The random and dissipative forces form a thermostat and must satisfy the fluctuation-dissipation theorem [3] leading to the two conditions: $\omega^D(r_{ij}) = \left[\omega^R(r_{ij})\right]^2$ and $\sigma^2 = 2\gamma k_B T$ with $T$ being the equilibrium temperature. The velocity fluctuations due to $F^R$ dominate in regions, where the ensemble velocity is expected to be small, for example next to the walls, which makes accurate extraction of $\overline{\mathbf{u}}(t, \mathbf{x})$ and its gradients a very difficult task.

Our *first example* is a DPD simulation of pipe flow driven by a time-periodic force, for which an analytical expression (Womersley velocity profile) for the ensemble average is known:

$$u(t, r) = -Im \left[ \frac{i \Delta P}{\omega \rho} \left( 1 - \frac{J_0(r\sqrt{-i\omega/\nu}}{J_0(D/2\sqrt{-i\omega/\nu})} \right) \exp^{i\omega t} \right], \tag{10.6}$$

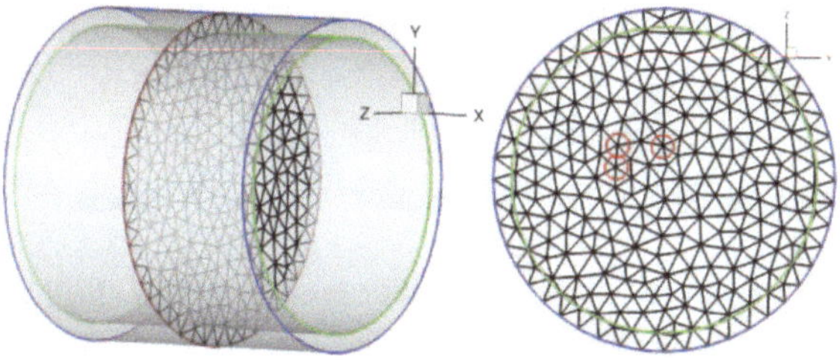


**Fig. 10.12.** (in color) Computational domain for an unsteady pipe flow problem. WPOD is performed over points $\mathbf{X}_p$ – vertices of the cross-sectional plane. The walls are represented by static particles located between the surfaces marked by green and blue edges. Red spheres centered at the vertices represent sampling region $\Omega_p$

where $\Delta P$ is the predefined pressure gradient, $\omega$ is the wave number, $\rho$ is the density, $D$ is the pipe diameter and $i = \sqrt{-1}$. The expected mean flow is one-dimensional.

The snapshots are collected on a set of grid points $\mathbf{X}_p$ distributed along cross-sectional plane in the middle of the computational domain, as presented in Fig. 10.12. The main flow characteristics are: Womersley and peak Reynolds numbers: $Ws = R\sqrt{\omega/\nu} = 3.81$, $Re_{peak} = U_{max}2R/\nu = 97.1$. Here $R = 10$ is the pipe radius, $\omega = 2\pi 0.0125$, $\nu = 0.54$ is the kinematic viscosity and $U_{max}$ is the maximum velocity at the centerline. The DPD coefficients employed in this study are: $a = 25$, $\gamma = 4.5$, $\sigma = 3.0$, $r_c = 1.0$, $k = 0.25$. The flow is due to the force acting in the direction of the centerline: $Fx = \Delta P \sin(\omega t)$, $\Delta P = 0.5$.

In Fig. 10.13a–f we plot the instantaneous ensemble average velocity profile along the streamwise ($x-$)direction and its derivative taken in the radial direction. The results are compared to those obtained with the standard spatio-temporal averaging technique (i.e., $N_{pod} = 1$), and to the exact analytical solution. In Fig. 10.13g,h we plot the PDF of the thermal fluctuations computed with the WPOD method for the streamwise and one of the cross-flow velocity components. In both cases a Gaussian PDF seems to be the best fit, consistent with the fact that the random force term in the DPD governing equations depends on a random Gaussian variable. The data in Fig. 10.13 show instantaneous solutions, however, the accuracy over the entire simulation remains about the same. In Fig. 10.14 (left) we show that WPOD provides significant computational savings: same accuracy can be achieved by averaging results of 24 simulations or by employing the WPOD method within a single simulation. The POD eigenspectra have been presented in Fig. 10.4 in Sect. 10.3.2. The accuracy in predicting the low-order eigenvalues, also discussed in Sect. 10.3.2, is also comparable to the data of Fig. 10.14 (left).

To achieve a certain accuracy, the WPOD approach allows balancing the number of POD modes and the length of time interval $N_{ts}\Delta t$ for obtaining a single snapshot. Figure 10.14 (right) shows that for $N_{pod}N_{ts} = const$ practically identical accuracy in reconstruction of the deterministic component of an unsteady flow can be achieved. We have observed that increasing the WPOD time interval $T_{pod} = N_{pod}N_{ts}$ also in-

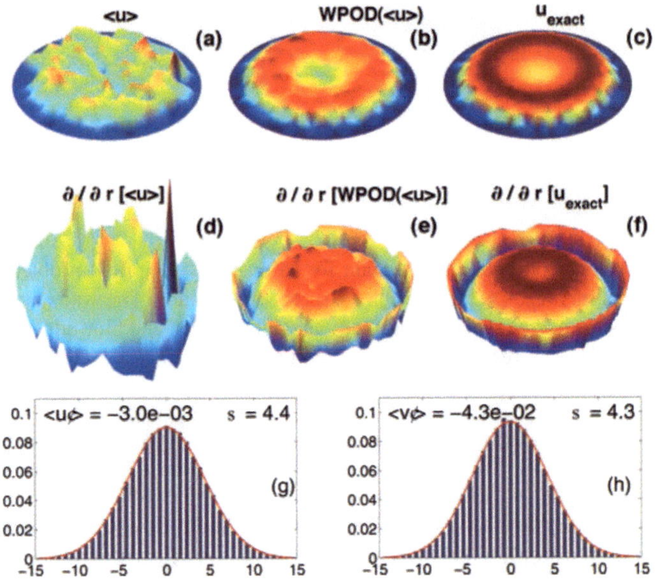


**Fig. 10.13.** (in color) DPD-simulation: 3D non-stationary pipe flow driven by a time-periodic pressure gradient. Ensemble average velocity (a-c) and its derivatives (d-f); $z-$axis and colors represent the velocity (derivative) magnitude. In all three cases the derivatives have been computed numericaly on first-order finite element grid. (g,h) - a histogram of probability density function (PDF) ($y-$axis) of thermal fluctuations. $N_r = 1$, $N_{ts} = 50$, $N_{pod} = 1$ and $N_{pod} = 160$, $\tilde{k} = 2$. Velocity and derivative profile correspond to simulation time $t = 340$ (see Fig. 10.4 (right) for reference). (Adapted from [5])

creases the accuracy; however, we note that parameter $N_{ts}$ cannot be very large if the process is expected to be non-stationary.

The WPOD time interval in this analysis spans exactly one cycle; the alternatives for the WPOD method for evaluation of timevarying ensemble solution are either performing phase averaging or averaging data over short time intervals across numerous realizations. Both alternative methods are computationally demanding. Moreover, for solutions with non-periodic in time manifolds phase averaging cannot be performed.

Our *second example* is blood flow simulation with healthy and malaria-infected red blood cells (RBC). The RBCs, solvent and walls are modeled by DPD particles. The properties of DPD particles of the red blood cells and plasma can be found in [4]. The total number of red blood cells is 42, and the membrane of each red blood cell is modeled by 500 DPD particles. The hematocrit level is: 30%. The total number of DPD particles including the solvent, the RBC particles and the wall particles is 60,067. The main characteristics of the flow simulations are: Reynolds number $Re = DU_{mean}/\nu = 0.24$; here $D = 20$ is the pipe diameter, $U_{mean} = 0.27$ is the average velocity in the streamwise direction and $\nu = 22.33$ is the kinematic viscosity (all in DPD units). The computational domain is a pipe with length $L = 40$ and radius

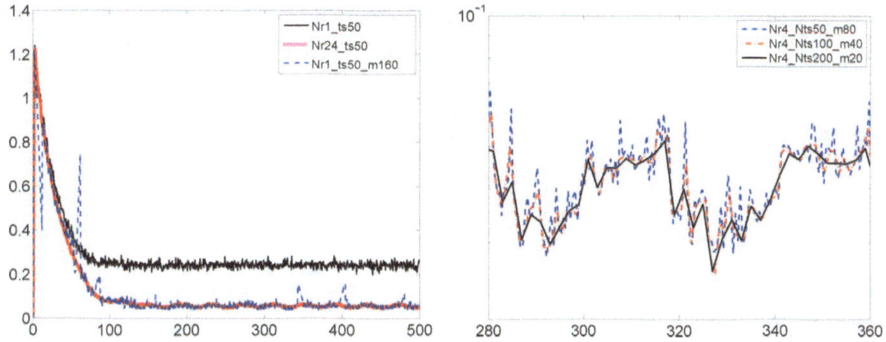


**Fig. 10.14.** (in color) Computational efficiency: DPD-simulation of 3D non-stationary pipe flow driven by a time-periodic pressure gradient. X-axis - time; Y-axis - $L_2$-error. *Left*: Comparable improvement in accuracy accuracy is achieved by either increasing the computational cost by a factor of 24 or by employing the WPOD method requiring negligible computational overhead. *Right*: The time window ($N_{pod}N_{ts} = 4000$) over which the data is processed is kept constant. Data is averaged over four concurrent realizations ($N_r = 4$); blue dash curve - $N_{ts} = 50$, $N_{pod} = 80$; red dot-dash curve - $N_{ts} = 100$, $N_{pod} = 40$; black solid curve - $N_{ts} = 200$, $N_{pod} = 20$. (Adapted from [5])

$R = 10$. The no-slip boundary condition on the wall is imposed by fixed particles. The points $\mathbf{X}_p$ are spread almost uniformly in the entire volume of the computational domain. In all three cases no assumptions of periodicity or axi-symmetry in sampling the data at the $\mathbf{X}_p$ points have been imposed, and the radius of spherical region $\Omega_p$ was $r_p = 0.75$. The flow is driven by a constant force, however, due to RBCs interactions, weak secondary flows are expected to develop. As a result of these secondary flows, a non-stationary, non-periodic and asymmetric (with respect to the pipe central axis) ensemble solution is expected.

Results of this simulation are presented in Fig. 10.15. As expected, considerably smoother ensemble average flow profiles correspond to simulation with WPOD. The POD eigenspectra of the three velocity components point to the dominance of the streamwise flow, which has about four orders of magnitude higher kinetic energy than the cross-flow components. The fast convergence of the first six eigenvalues of the crossflow velocity components suggests developing secondary flows due to interactions among the blood cells and between the blood cells and the plasma.

WPOD was applied at simulation run-time, and the number of POD modes required for ensemble solution reconstruction was computed dynamically according to the criteria discussed in Sect. 10.3.2. To select the number of POD modes ($\tilde{k}$) for reconstruction of $\hat{\mathbf{u}}(t,\mathbf{x})$ we first applied the first criterion (simulation I), e.g., $\tilde{k}$ was computed by minimizing the error in the piecewise linear approximation of the eigenspectrum for each velocity component separately. Second (simulation II), we repeated the simulation and applied the first and the third criteria for selecting the number of modes composing $\hat{\mathbf{u}}(t,\mathbf{x})$ in the following manner: The number of POD modes ($\tilde{k}$) were selected adaptively for each velocity component using the first cri-

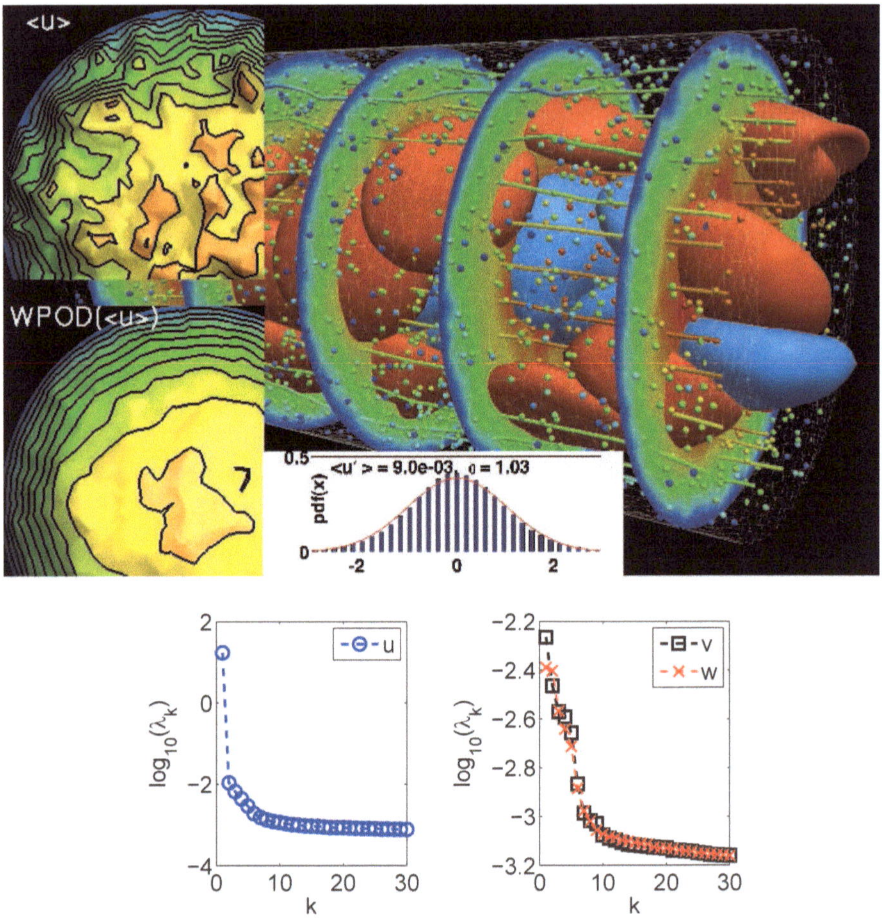


**Fig. 10.15.** (in color) DPD simulation of suspension of RBCs in a pipe flow, driven by a constant force. Upper: instantaneous position and deformation of the RBCs superimposed on the solvent flow processed with the WPOD method, and streamwise velocity profile (left inset) reconstructed with WPOD and standard averaging ($N_{pod} = 1$); PDF of the streamwise velocity fluctuations, the red curve depicts the fitted normal Gaussian PDF. Lower: eigenvalue spectra for three velocity components (only the first 30 eigenvalues are shown). $N_r = 1$, $N_{ts} = 250$, $N_{pod} = 160$. (Adapted from [5].Visualization with help of J. Insley Argonne National Laboratory, USA)

teria. Then, starting from the second mode we compared the standard deviation of temporal modes to $(1/\sqrt{(N_{pod}-1)}$; once standard deviation of $(a_k)$ was within 1% of the reference value the procedure was terminated and all lower order modes have been kept. In simulation I the average number of POD modes forming $\hat{\mathbf{u}}(t,\mathbf{x})$ was about 6 to 7 for the streamwise velocity and 8 to 9 for the cross-flow components, while in simulation II only the first POD mode was selected for the streamwise component and 3 to 5 modes for the cross-flow components. Although, the difference in parameter $\tilde{k}$ in simulations I and II may appear significant, the energy associated with the POD modes rejected by the third criterion is very low comparing to the energy contained in the preserved modes.

### *10.5.1 Analysis of Deformability in Cells*

Here we apply the POD to analyze red blood cells (RBC) deformations. The data is obtained by DPD simulations. Each RBC is represented by a cluster of 500 DPD particles, as illustrated in Fig. 10.16, for a detailed description on the RBC modeling we refer to [4]. The particles have fixed connectivity, and their position can be described as a function of two variables - time and particle index $PID$: $X = X(t, PID)$, $Y = Y(t, PID)$, $Z = Z(t, PID)$. The correlation matrix required for POD analysis is constructed from:

$$C_{ij} = \frac{1}{N_{particles}} \sum_{k=1}^{3} \left( \sum_{PID=1}^{500} X_k(t^i, PID) X_k(t^j, PID) \right), \quad X_1 = X, \; X_2 = Y, \; X_3 = Z.$$

To illustrate the POD eigenspectra of deformation of object represented by a collection of bonded particles, we present here POD analysis of three different types of a single red blood cell (RBC) dynamics: (i) RBC suspended in a solvent, (ii) RBC drifting in a weak tube flow resulting in a small cell deformation (bullet shape), and (iii) RBC rotating in a strong shear flow with large unsteady deformation. In all simulations the surrounding fluid (blood plasma) is represented by non-bonded DPD particles. The goal of POD is to separate the correlated motion of the cell membrane from thermal fluctuations.

The POD eigenspectra for different dynamics are shown in Fig. 10.17. We observe a typical power-law decay of high-order POD eigenvalues in all simulations.

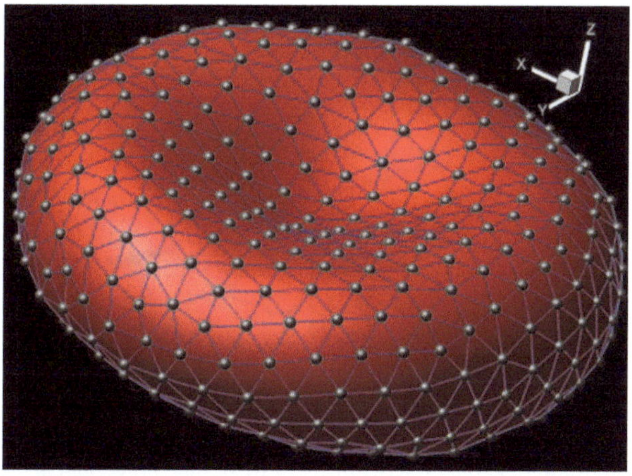


**Fig. 10.16.** DPD simulation of red blood cell. The cell membrane is constructed from 500 DPD particles, linked by non-rigid bonds

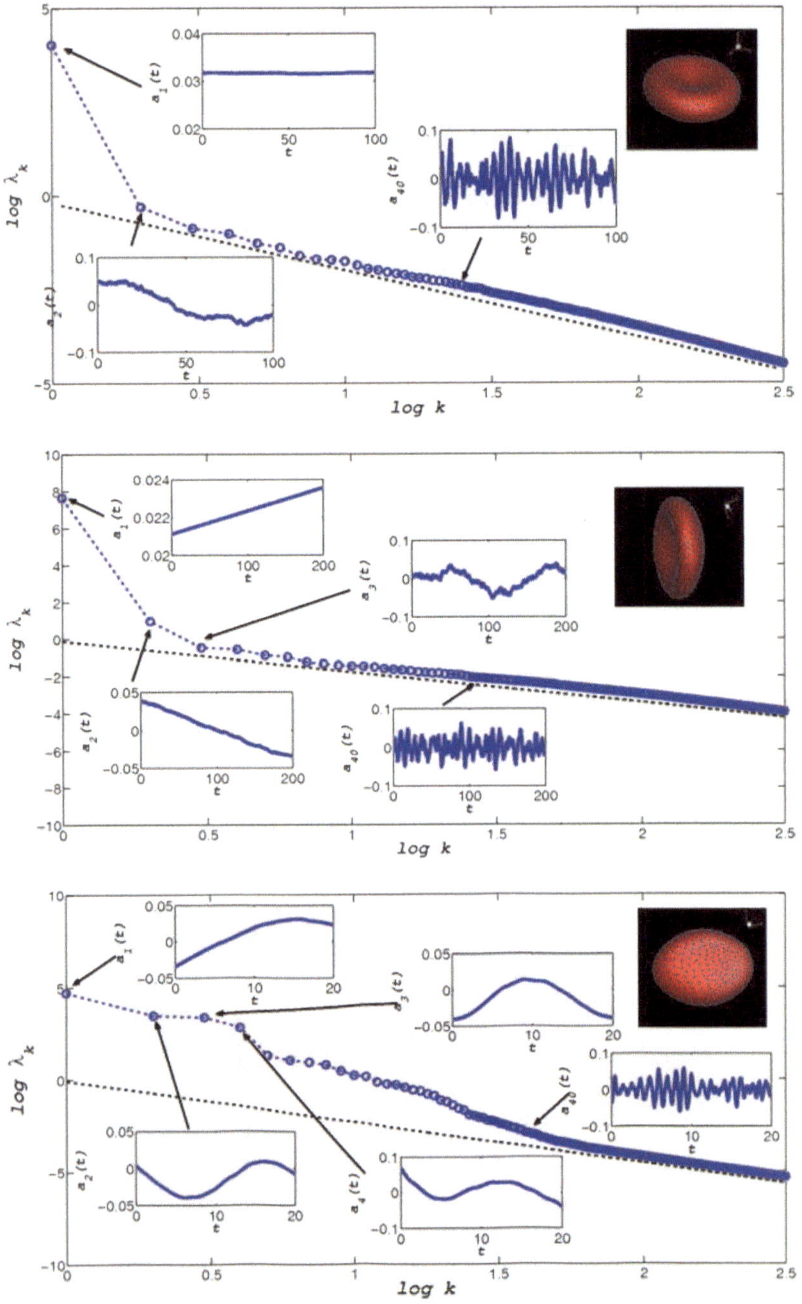


**Fig. 10.17.** POD analysis of red blood cell (RBC): eigenvalue spectra. *Top* – RBC fluctuates due to DPD thermostating; *middle* – RBC drifts in weak tube flow; *bottom* – RBC in strong shear flow

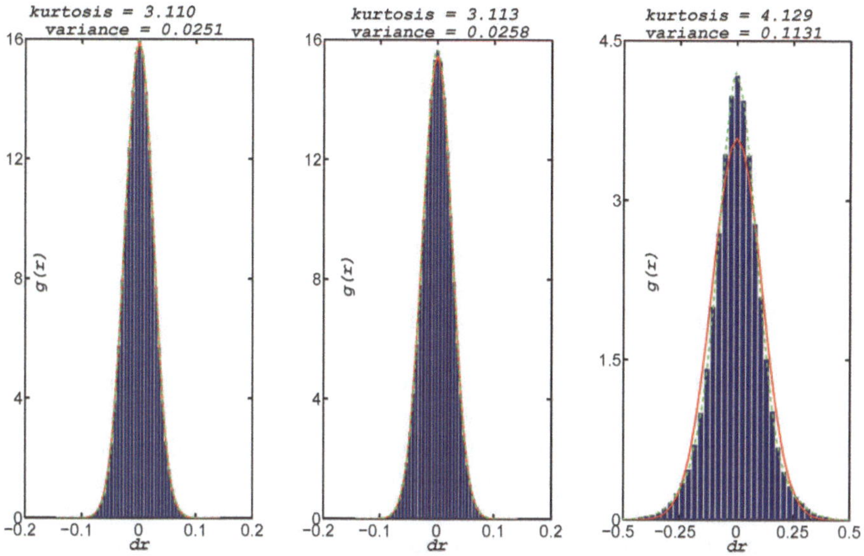


**Fig. 10.18.** POD analysis of red blood cell (RBC): probability density function of RBC conformation fluctuations (|**dr**|). Same is in Fig. 10.17. *Left* – RBC fluctuates due to DPD thermostating; *middle* - RBC drifts in weak tube flow; *right* – RBC in strong shear flow

As expected the number of POD modes characterizing the correlated motion of the cell is increasing with the complexity of the flow. The probability density function (PDF) of thermal fluctuations is computed by analysis of positions of RBC's vertices reconstructed from high-order POD modes whose eigenvalues converge according to the power-low.

To remove the thermal fluctuation from the data, we can reconstruct the position of RBC membrane particles from the low-order modes, while truncating all the modes for which convergence of eigenvalues can be approximated with the power-low:

$$\bar{\mathbf{X}}(t,PID) = \sum_{k=1}^{\tilde{k}} a_k(t)\phi_k(PID), \quad \mathbf{X} = [X\ Y\ Z].$$

To analyze the fluctuating component we subtract the reconstructed with low-order modes data from the original data:

$$\mathbf{dr}_f(t,PID) = \mathbf{X}(t,PID) - \bar{\mathbf{X}}(t,PID),$$

where **dr** denotes deformation due to thermal fuctuations.

As can be seen in Fig. 10.17, there are only one and two modes dominating the RBC dynamics in case (i) and (ii), respectively. As shown in Fig. 10.18, the conformation fluctuations in simulations (i) and (ii) are small and almost close to thermostat input stochastic noise, i.e., Gaussian random noise in our DPD simulation.

However, in strong shear flow the RBC dynamics is accompanied by the excitement of high-order membrane deformation modes, and more degrees of freedom are needed to describe such motions. The eigenspectrum decays much slower than in the first two cases, and it asymptotically converges to power-law decay. The conformation fluctuation is much bigger and deviates from Gaussian distribution (with kurtosis higher than 3.0) due to the interaction between the nonlinearity and thermal noise.

## 10.6 WPOD in Multiscale Visualization

Here we consider DPD simulations of RBCs in blood plasma. The particle data is projected on a fixed grid composing vertices of tetrahedral elements. The projection is performed by time-space sampling of particle velocities over short time intervals (50 to 500 time steps each), and WPOD is then applied for extracting the correlated motion of the flow while removing thermal fluctuations. The data is processed at the run-time and the number of POD modes for reconstructing the velocity field is computed adaptively using the criteria described in Sect. 10.3.2. As a result, the large-scale flow patterns are visualized using the projected and filtered with WPOD data, while the small scales are visualized using a small subset of particle data.

In Fig. 10.15 multiscale data from a DPD simulation of blood flow are visualized. The large-scale flow features are presented using continuum data computed by a projection of atomistic data onto a specified grid $\mathbf{X}_p$. The small scale features, such as RBC location, their interactions and membrane folding are visualized by presenting the DPD particles forming the blood cells membrane. In Fig. 10.19 we plot a snapshot from multiscale visualization of flow around an RBC adhering to the wall. The RBC also flips and deforms while it sporadically moves in the flow direction. The configuration of RBC is presented by plotting instantaneous position of 500 DPD particles representing the cell membrane, while the global flow field is presented by data projected onto a 3D grid and separated from the thermal fluctuations with the WPOD. Simultaneous animation of the RBC motion and the flow helps to better understand the interactions between the RBC motion and deformations and developing secondary flows.

## 10.7 Summary and Outlook

High performance computing has already exceeded the petaflop barrier and it now is soaring into the exascale, hence allowing great advances in biomedical simulations. Consequently, increasing the size and volume of such simulations leads inevitably to generation of massive volumes of multiscale data which must be analyzed, stored and visualized. Hence, it is clear that future research will focus not only on how to analyze existing vast volumes of data, but also how to move this data analysis into run-time in simulations as co-processing. This is particularly true for cardiovascular

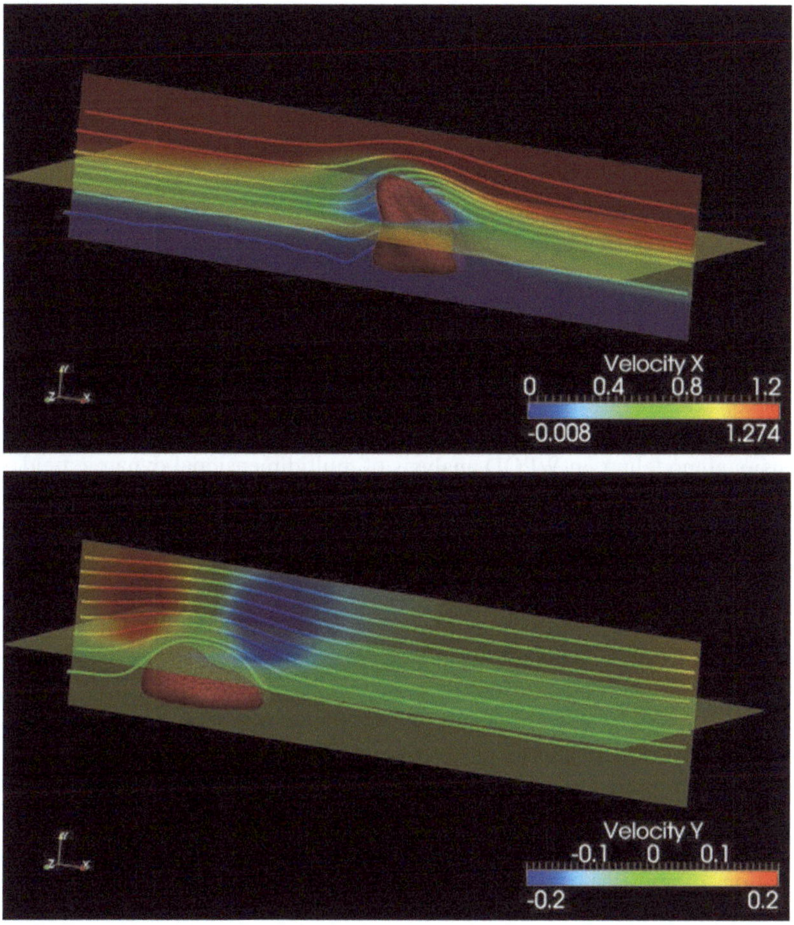


**Fig. 10.19.** Multiscale visualization of a flow around an adhesion cite. The membrane of the RBC is represented by 500 DPD bonded particles with a fixed connectivity. Blood plasma is also simulated with DPD particles. The atomistic data is projected on a 3D grid with WPOD forming a continuum data. The RBC is visualized by tracing the DPD particles while the global flow features are visualized using the continuum 3D data. Courtesy of Argonne National Laboratory. Visualization created by J. Insley

flows which are pulsatile with data over many cardiac cycles required for detailed analysis. Such shift will, in fact, allow users to reduce the cost of data processing and also to obtain more accurate conclusions based on substantially larger data samples than analyzing fields stored only sporadically.

Multiscale biological flow modeling involves both continuum and atomistic based methods. Computational methods for processing data at the continuum level are quite mature, although, there is still a need for exploring scalable approaches for run-time data compression and analysis, and, in particular, for quantitative analysis of and transitional and turbulent fields. At the same time, quantitative analysis and visualization tools for data of mesoscale and molecular simulations are non-existent! Quantifying deviations from equilibrium, computing ensemble solution and performing fluctuation analysis is one of the most important but expensive part of many particle-based simulations.

In this chapter we described our first attempts to develop data analysis tool based on window proper orthogonal decomposition (WPOD) with focus on cardiovascular flows for large arteries but also at the capillary level. First, WPOD was applied to analyze intermittent laminar-turbulent flow in a carotid bifurcation to capture the onset of turbulence and subsequent re-laminarization within one cardiac cycle. Subsequently, WPOD was applied for analysis of molecular and coarse grained molecular dynamics (DPD) simulation data in smaller arterioles and capillaries, including an analysis of deformability and fluctuations of a single red blood cell.

We have demonstrated that unlike the often-used energy based criterion in low-dimensional modeling, in WPOD analysis of multidimensional dynamic field, such as cardiovascular flows, it is *the rate of convergence* of the POD eigenvalues that reflects the underlying physical process.

A wider use of WPOD in diverse multiscale biological phenomena, from protein and cell dynamics, to flow-structure interactions in large arteries but also in arterioles and capillaries will help to further enhance its effectiveness while at the same time will provide new quantitative information of the dynamic multiscale processes analyzed. Specific new advances are required on developing sharper criteria for partitioning data based on POD eigenspectra, especially in cases where there is no clear separation of the eigenvalues. In a broader context, developing similar tools like WPOD would greatly facilitate effective analysis of multiscale biological phenomena involving heterogeneous but coupled continuum-atomistic simulations. Finally, such specialized computational methods for interactively exploring spatio-temporal correlations of multiscale phenomena should found their way into open source visualization software.

**Acknowledgements** Computations were performed on the IBM BlueGene/P computers of Argonne National Laboratory and Jülich Supercomputing Centre; and CRAY XT5 of National Institute for Computational Sciences and Oak Ridge National Laboratory. Visualization was performed on Eureka at the Argonne Leadership Computing Facility. Support was provided by DOE INCITE grant DE-AC02-06CH11357, NSF grant OCI-0904190. G. Em Karniadakis would like to acknowledge support by the DOE's Collaboratory on Mathematics for Mesoscopic Modeling of Materials (CM4) at PNNL.

## References

1. Amsallem, D., Zahr, M.J., Farhat, C.: Nonlinear model order reduction based on local reduced-order bases. International Journal for Numerical Methods in Engineering **92**(10), 891 (2012)

2. Deschamps, J., Kantsler, V., Segre, E., Steinberg, V.: Dynamics of a vesicle in general flow. Proc. Natl. Acad. Sci. U.S.A. **106**, 11444 (2009)
3. Espanol, P., Warren, P.: Statistical-mechanics of dissipative particle dynamics. Europhysics Letters **30**(4), 191 (1995)
4. Fedosov, D.A., Caswell, B., Karniadakis, G.E.: A multiscale red blood cell model with accurate mechanics, rheology, and dynamics. Biophys. J. **98**(10), 2215 (2010)
5. Grinberg, L.: Proper orthogonal decomposition of atomistic flow simulations. Journal of Computational Physics **231**(16), 5542–5556 (2012)
6. Grinberg, L., Fedosov, D.A., Karniadakis, G.E.: Proper orthogonal decomposition of atomistic flow simulations. Journal of Computational Physics (2012). doi 10.1016/j.jcp.2012.08.023
7. Grinberg, L., Yakhot, A., Karniadakis, G.E.: Analyzing transient turbulence in a stenosed carotid artery by proper orthogonal decomposition. Annals of Biomedical Engineering **37**(11), 2200 (2009)
8. Hoogerburgge, P.J., Koelman, J.M.V.A.: Simulating microscopic hydrodynamic phenomena with dissipative particle dynamics. Europhysics Letters **19**(3), 155 (1992)
9. Kantsler, V., Goldstein, R.E.: Fluctuations, dynamics, and the stretch-coil transition of single actin filaments in extensional flows. Phys. Rev. Lett. **108**, 038103 (2012)
10. Kantsler, V., Segre, E., Steinberg, V.: Vesicle dynamics in time-dependent elongation flow: Wrinkling instability. Phys. Rev. Lett. **99**, 178102 (2007)
11. Kantsler, V., Steinberg, V.: Orientation and dynamics of a vesicle in tank-treading motion in shear flow. Phys. Rev. Lett. **95**, 258101 (2005)
12. Kantsler, V., Steinberg, V.: Transition to tumbling and two regimes of tumbling motion of a vesicle in shear flow. Phys. Rev. Lett. **96**, 036001 (2006)
13. Karniadakis, G.E., Beskok, A., Aluru, N.: Microflows and Nanoflows: Fundamentals and Simulation, 2nd ed. Springer, New York (2005)
14. Kefayati, S., Poepping, T.L.: Transitional flow analysis in the carotid artery bifurcation by proper orthogonal decomposition and particle image velocimetry. Medical Engineering and Physics (2012). DOI 10.1016/j.medengphy.2012.08.020
15. Lei, H., Caswell, B., Karniadakis, G.E.: Direct construction of mesoscopic models from microscopic simulations. Physical Review E **81**, 026704 (2010)
16. Levant, M., Steinberg, V.: Amplification of thermal noise by vesicle dynamics. Phys. Rev. Lett. **109**, 268103 (2012)
17. Manhart, M.: Vortex shedding from a hemisphere in a turbulent boundary layer. Theoretical and Computational Fluid Dynamics **12**, 1 (1998)
18. Sirovich, L.: Turbulence and dynamics of coherent structures: I-iii. Quarterly of Applied Mathematics **45**, 561 (1987)
19. Sherwin, S.J., Blackburn, H.M.: Three-dimensional instabilities of steady and pulsatile axisymmetric stenotic flows. Journal of Fluid Mechanics **533**, 297 (2005)
20. Turitsyn, K., Vergeles, S.S.: Wrinkling of vesicles during transient dynamics in elongational flow. Phys. Rev. Lett. **100**, 028103 (2008)
21. Visual molecular dynamics. http://www.ks.uiuc.edu/Research/vmd
22. Werder, T., Walther, J., Koumoutsakos, P.: Hybrid atomistic-continuum method for the simulation of dense fluid flows. Journal of Computational Physics **205**, 373 (2005)

# 11

# Reduced Order Models at Work in Aeronautics and Medicine

**Michel Bergmann, Thierry Colin, Angelo Iollo, Damiano Lombardi, Olivier Saut and Haysam Telib**

**Abstract** We review a few applications of reduced-order modeling in aeronautics and medicine. The common idea is to determine an empirical approximation space for a model described by partial differential equations. The empirical approximation space is usually spanned by a small number of global modes. In case of time-periodic or mainly diffusive phenomena it is shown that this approach can lead to accurate fast simulations of complex problems. In other cases, models based on definition of transport modes significantly improve the accuracy of the reduced model.

M. Bergmann
Inria, F-33400 Talence, France
e-mail: michel.bergman@inria.fr

T. Colin
Université de Bordeaux and Inria, F-33400 Talence, France
e-mail: thierry.colin@inria.fr

A. Iollo (✉)
Université de Bordeaux and Inria, F-33400 Talence, France
e-mail: angelo.iollo@inria.fr

D. Lombardi
Inria, F-78153 Rocquencourt, France
e-mail: damiano.lombardi@inria.fr

O. Saut
CNRS and Inria, F-33400 Talence, France
e-mail: olivier.saut@inria.fr

Haysam Telib
Optimad Engineering, I-10143 Torino, Italy
e-mail: haysam.telib@optimad.it

A. Quarteroni, G. Rozza (eds.): *Reduced Order Methods for Modeling and Computational Reduction*, MS&A 9. DOI 10.1007/978-3-319-02090-7_11, 

## 11.1 Introduction

Progress in numerical simulation of partial differential equations (PDEs) allows accurate and reliable predictions of some complex phenomena in solid and fluid mechanics, solid state physics, geophysics, etc., at the price of significant code developments, difficult computational set ups and large high-performance computing infrastructures. Using reduced-order models (ROMs) one trades accuracy for speed and scalability, and counteracts the curse of dimension by significantly reducing the computational complexity. Thus ROMs represent an ideal building block of systems with real-time requirements, like interactive decision support systems that offer the possibility to explore various alternatives. In complex cases, the real-time requirements would not be met by standard numerical methods.

The construction of ROMs for design, optimization, control and data-driven systems is a non-trivial task and various alternative ways can be followed often without any guarantee that the ROM will effectively model the physical phenomenon in the application. Focusing for example on flows or environmental phenomena, different states can often be characterized by the presence or absence of qualitative flow features, by the structure of feature patterns and by the strength of such features. Proper orthogonal decomposition (POD) [7, 11] is a mean to extract such features from existing solution snapshots under the form of global modes. However, ROMs based on such POD modes are numerically unstable in unsteady, advection dominated models. Stabilization can be obtained by various ad hoc techniques (see [2, 5, 14] for example), but a general framework to determine accurate and robust unsteady ROMs is still lacking. Still, ROMs can be useful to model far-field conditions coupled to a complete model, or to regularize the solution of an inverse problem. We give in the following two examples in these directions.

Another central issue for ROMs is the quality of the approximation obtained thanks to a reduced number of empirical modes. These modes are determined from a set of snapshots that are relative to a particular configuration: geometry, physical parameters, boundary conditions. When the configuration varies there is no guarantee that the reduced basis will adequately approximate the solution. On the other hand, if the snapshot set from which the basis is obtained includes a large number of different configurations, by construction the reduced basis will enjoy better approximation properties when the configuration varies. Given the computational costs relative to a systematic exploration of the configuration space, optimal sampling strategies must be introduced. In the following, we present one strategy based on an estimation of the approximation error of the reduced base.

Nevertheless, there is a fundamental difficulty in approximating with global (for example POD) modes the displacement of, say, a flow feature in time or across the parameter space. Global modes are not optimal for advection. In particular, POD modes reduce to Fourier modes for translation invariant signals. An alternative idea is to define advection modes as the solution of an optimal transportation problem. An application to interpolate the solution of a PDE system across the parameter space based on the definition of advection modes is presented in the following.

## 11.2 Systematic sampling for ROM

We have considered an oscillating NACA0012 airfoil in a compressible flow as in the CT1 test case from AGARD-R702 report. This case corresponds to a Mach 0.6 flow at infinity past an oscillating NACA0012 airfoil. In the following the computations are inviscid; in the actual experiments the Reynolds number is $4.8 \times 10^6$. The parameter space is two dimensional: the oscillating frequency varies between $f^1$ =30Hz and $f^2$ =70Hz (CT1 case: 50Hz) whereas the amplitude of the oscillation varies between $\alpha_0^1$=1.6deg and $\alpha_0^2$=3.6deg (CT1: 2.5deg) with an average pitch of $\alpha_m$=3.0deg. We have implemented an algorithm to sample the parameter space in order to enrich the database of the POD basis functions. The objective of this procedure is to determine a set of POD modes that minimizes the approximation error across the parameter space $S = [\alpha_0^1, \alpha_0^2] \times [f^1, f^2]$.

The main idea is to build a recursive Voronoi diagram and the corresponding Delaunay triangulation based on the projection error of the POD representation. This is an extension of what it was proposed in a one-dimensional setting in [10]. Let $\mathscr{P}_n$ be the set of points $P_1, \cdots, P_n$ in the parameter space corresponding to actual high-fidelity simulations and $\mathscr{T}_n$ the corresponding Delaunay triangulation. For given number $M$ of POD modes (the size of the basis) we build a set of POD basis functions $\phi_i$, $i = 1, \cdots, M$ using the high-fidelity simulations corresponding to points $P_1, \cdots, P_n$. The number of POD modes $M$ is arbitrary fixed and is kept constant during the sampling process. Then we determine the representation error $E(P_k)$, $k = 1, \cdots, n$, corresponding to the residual in the $L^2$ norm of the projection of high fidelity solutions at $P_k$ on $\phi_i$, $i = 1, \cdots, M$. Let us denote $V(T_s)$ the set of vertexes of $T_s \in \mathscr{T}_n$. We select the triangle $T_{\max} \in \mathscr{T}_n$ for which the product of its area and the sum of $E(P_k)$, $P_k \in V(T_s)$, is maximum. The next point of the triangulation is the barycenter of $T_{\max}$. This new point is used to compute a new Delaunay triangulation. A Delaunay triangulation has thus to be performed at each sampling iteration.

As an example consider Fig. 11.1. The parameter space $S = [\alpha_0^1, \alpha_0^2] \times [f^1, f^2]$ is mapped to the unit square ($\alpha_0 = [\alpha_0^1, \alpha_0^2] \mapsto \overline{A} = [0,1]$ and $f = [f^1, f^2] \mapsto \overline{F} = [0,1]$) and is partitioned in 8 triangles relative to 7 simulation points that were obtained by iterating the method starting from points $P_1, P_2, P_3, P_4$. Both Delaunay triangulation (red) and Voronoi tessellation (blue) are presented. The new high-fidelity simulation point $P_8$ is added at the barycenter of the triangle relative to points $P_2, P_4, P_5$. For this triangle the product of the area times the sum of the representation errors at the vertexes is the highest.

The procedure implies the computation of $n$ space correlations of high-fidelity solutions for each new simulation point $P_{n+1}$. These operations are particularly efficient in the hybrid domain-decomposition ROM as the spatial extension of the snapshots and of the POD modes is reduced to a region close to the airfoil. The same procedure can be extended to higher-dimensional parameter spaces.

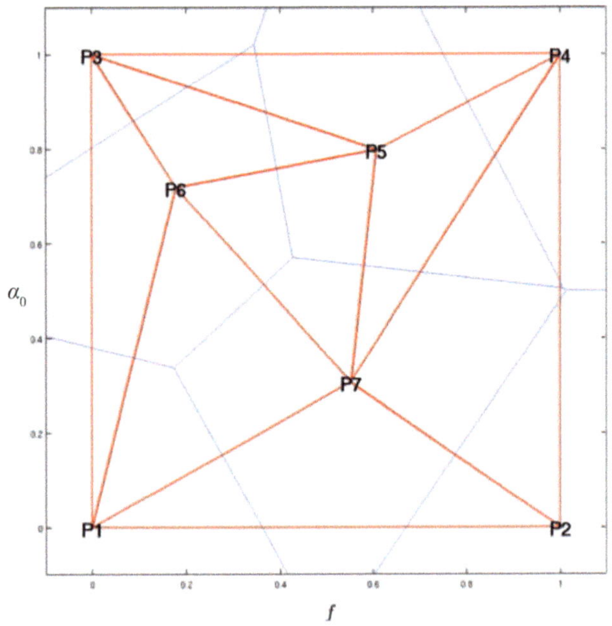


(a)

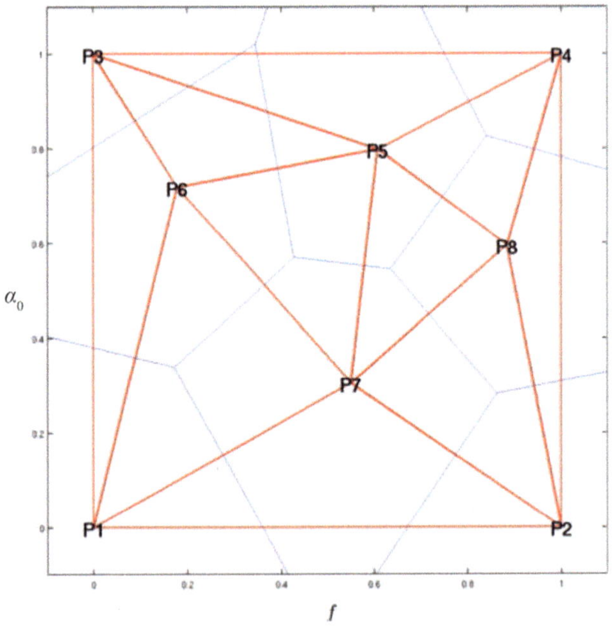


(b)

**Fig. 11.1.** Example of one iteration of the Voronoi tessellation algorithm. The parameter space subset $S$ is represented. $\alpha_0$ is on the ordinates and $f$ on the abscissa. (a) typical iteration (iteration 3); (b) next point is added (P8) and the triangulation updated

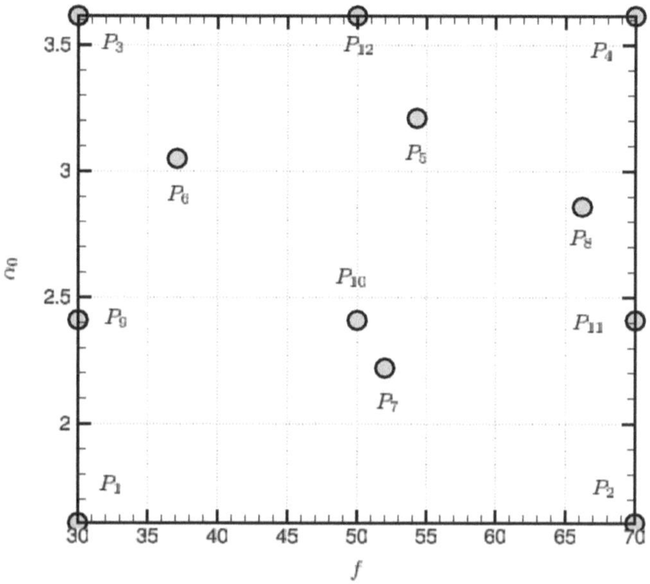


**Fig. 11.2.** Sampling of the parameter space

## 11.2.1 Results

We start with a POD basis, called $B_{Initial}$, computed from snapshots taken at 4 points $P_1$, $P_2$, $P_3$ and $P_4$ (see Fig. 11.2). 20 time snapshots are uniformly taken over one period for each point $P_i$, $1 \leq i \leq 4$. Starting from these points in parameter space, 4 additional points, denoted by $P_5$, $P_6$, $P_7$ and $P_8$ are determined using the method described above (Voronoi tessellation). A suboptimal POD basis, called $B_{Subopt}$ is then computed from these 8 points: $P_1$ to $P_8$. We want to compare the suboptimal basis performance to another basis composed with the same number of sampling points, but chosen without any specific critria. For instance, we consider an uniform-like basis, $B_{Uniform}$, computed using from $P_1$ to $P_4$ and $P_9$ to $P_{12}$. The points $P_9$ to $P_{12}$ are relative to already existing simulations that we exploit now for building a basis. No special criteria were used to specify these points. However, $M$ is the same for $B_{Uniform}$ and $B_{Subopt}$. . A summary of the high-fidelity simulation employed for each POD basis is represented in Table 11.1.

**Table 11.1.** POD basis summary

| *POD basis* | $P_1$ | $P_2$ | $P_3$ | $P_4$ | $P_5$ | $P_6$ | $P_7$ | $P_8$ | $P_9$ | $P_{10}$ | $P_{11}$ | $P_{12}$ |
|---|---|---|---|---|---|---|---|---|---|---|---|---|
| $B_{Initial}$ uniform | $X$ | $X$ | $X$ | $X$ | | | | | | | | |
| $B_{Uniform}$ uniform | $X$ | $X$ | $X$ | $X$ | | | | | $X$ | $X$ | $X$ | $X$ |
| $B_{Subopt}$ suboptimal | $X$ | $X$ | $X$ | $X$ | $X$ | $X$ | $X$ | $X$ | | | | |

**Table 11.2.** POD Basis $L^2$ projection errors $\times 10^4$. $P_T$ denotes the average error over the 12 points $P_i$. $B_{Uniform}$ and $B_{Subopt}$ are computed with 160 snapshots and $P_S$ the standard deviation. $B_{Initial}$ is computed with 80 snapshots

| E | $P_1$ | $P_2$ | $P_3$ | $P_4$ | $P_9$ | $P_{10}$ | $P_{11}$ | $P_{12}$ | $P_5$ | $P_6$ | $P_7$ | $P_8$ | $P_T$ | $P_S$ |
|---|---|---|---|---|---|---|---|---|---|---|---|---|---|---|
| $B_{Initial}$ | 3.71 | 3.75 | 7.36 | 4.80 | 6.20 | 5.25 | 5.58 | 3.80 | 4.69 | 4.53 | 3.75 | 4.63 | 4.82 | 1.12 |
| $B_{Uniform}$ | 3.85 | 4.07 | 6.70 | 5.29 | 4.91 | 4.20 | 4.87 | 4.18 | 4.38 | 4.29 | 3.89 | 4.45 | 4.60 | 0.79 |
| $B_{Subopt}$ | 3.24 | 3.23 | 5.42 | 5.41 | 5.11 | 4.62 | 4.99 | 4.20 | 3.74 | 3.37 | 3.01 | 3.06 | 4.08 | 0.95 |

The accuracy of the 3 POD basis is evaluated by computing the $L^2$ projection error of the whole snapshot set $P_1$ to $P_{12}$ onto each POD basis, see Table 11.2. In particular, we consider the average $L^2$ norm of the error on each variable: density, velocity components and speed of sound. This error norm can be biased by the normalization of the different physical quantities. However, in our case, the normalizations are such that all the variables have comparable absolute values and hence the average error over the different physical quantities is a reasonable measure of accuracy. The error $P_T$ denotes the average error evaluated over the whole set of points $P_1$ to $P_{12}$. The basis $B_{Subopt}$ shows the best average errors of about 15% compared to $B_{Uniform}$. Even for the extra uniform sampling points $P_9$ to $P_{12}$ that are not included in the $B_{Subopt}$ database, the errors obtained with $B_{Subopt}$ are close to those obtained with $B_{Uniform}$.

## 11.3 ROM by Domain Decomposition

Let $\Omega_a(t)$ denote the two-dimensional region enclosed by the airfoil at time $t$ and let $\Omega$ be such that $\Omega_a(t) \subset \Omega \subset \mathbb{R}^2$. The compressible Euler equations are defined on the domain $\Omega_c(t) := \Omega \backslash \Omega_a(t)$. Let us also define two rectangles $\mathscr{R}_e$ and $\mathscr{R}_i$ such that $\Omega_a(t) \subset \mathscr{R}_i \subset \mathscr{R}_e \subset \Omega$. The inner rectangle $\mathscr{R}_i$ always includes the airfoil during its oscillation about a point of the chord (see Fig. 11.3).

In $\Omega_c(t)$, we solve the unsteady compressible Euler equations on a fixed cartesian mesh to second order accuracy in space and time, as explained in [6]. We collect an appropriate solution database of $N$ flow snapshots.

Let $U^{(k)}$ be one solution snapshot in $\Omega_c(t_k)$, $1 \leq k \leq N$, restricted to $\mathscr{R}_e \backslash \mathscr{R}_i$ and defined in terms of primitive flow variables. We compute a Galerkin base of the form $\phi_i = \sum_{k=1}^{N} b_{ik}(U^{(k)} - \overline{U})$, with $1 \leq i \leq M$, $\overline{U} = 1/N \sum_{k=1}^{N} U^{(k)}$ and where the coefficients $b_{ik}$ are found as in [8], [1]. This decomposition is performed individually for each primitive variable, i.e. the flow velocity, the pressure and the speed of sound. Consequently each expansion gives an optimal representation of the original dataset relative to each physical variable.

Let us define $\widehat{U} = \overline{U} + \sum_{i=1}^{M} a_i \phi_i$. The number of global global modes $M$ is very small compared to the size of the computational grid in $\Omega_c(t)$.

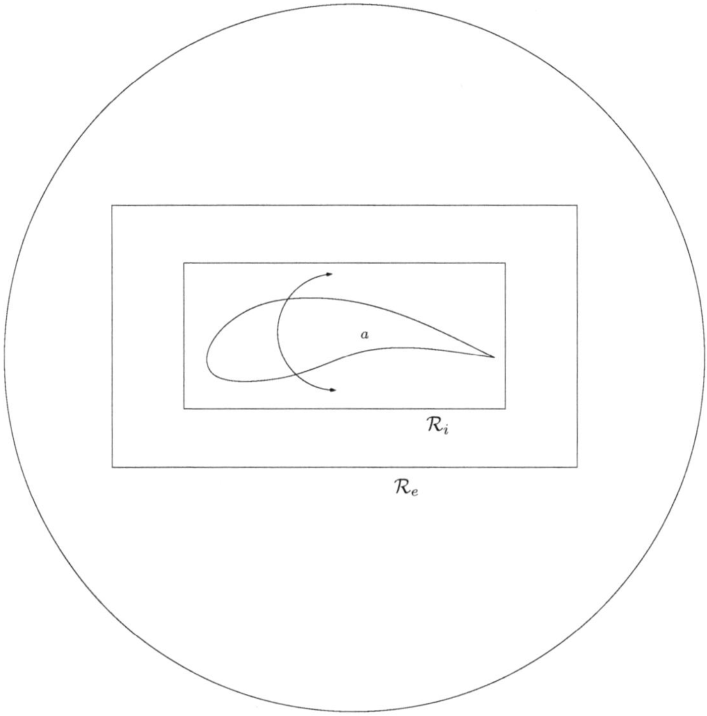


**Fig. 11.3.** Illustration of computational domain and subdomain definitions

The hybrid computational model is obtained by coupling the cartesian grid solver in $\mathscr{R}_e \backslash \Omega_a(t)$ and the Galerkin representation defined in $\mathscr{R}_e \backslash \mathscr{R}_i$. To this end, we follow the steps below:

- integrate the governing equations in $\mathscr{R}_e \backslash \Omega_a(t)$ by the cartesian solver, with given initial conditions $U^{(n)}$ in $\mathscr{R}_e \backslash \Omega_a(t)$ and boundary conditions on $\partial \mathscr{R}_e$;
- project the restriction to $\mathscr{R}_e \backslash \mathscr{R}_i$ of the updated solution $U^{(n+1)}$ on the subspace spanned by the POD modes $\phi_i$ and hence determine $\widehat{U}^{(n+1)}$;
- recover the boundary conditions to be imposed at the next time step on $\partial \mathscr{R}_e$ as the trace of $\widehat{U}^{(n+1)}$ on $\partial \mathscr{R}_e$;
- goto (1) until convergence is attained.

This algorithm is fully detailed in [3] for several idealized internal flows. The ratio between the computational cost to solve this hybrid scheme and the cost to solve the flow on the full domain is of the order of the ratio between the area of $\mathscr{R}_e \backslash \Omega_a(t)$ and that of $\Omega_c(t)$.

### 11.3.1 Oscillating Airfoil in Transonic Flow

We consider a two-dimensional flow past an oscillating NACA0012 airfoil. The airfoil oscillates about a point fixed at 25% of its chord according to a sinusoidal law. The average angle of attack is 2.89, the amplitude of the angular excursion is 2.41 and the frequency of oscillation is of 50Hz. The Mach number at infinity is 0.6.

The computational domain is $\Omega = 30c \times 20c$, where $c$ is the chord, and the profile is positioned so that the computational domain extends for $10c$ upwards and downwards, $10c$ upwind and $20c$ downwind. The computational grid is $(4.8 \times 10^3)^2$. The simulation has been carried out starting from a uniform initial condition corresponding to the unperturbed flow. Time integration is pursued until the hysteresis cycle is periodic, i.e., after about two cycles of oscillation.

We present in Fig. 11.4 typical snapshots of the Mach field where the coalescence of the characteristics forms a transient shock on the suction side of the airfoil. The hysteresis cycle is shown in Fig. 11.5 where the computational results are contrasted to the experimental ones. The computational results are in good agreement with experimental data reported in AGARD R-702.

A collection of 65 snapshots of the flow primitive variables is taken over one period of oscillation once the flow is completely established. The size of the rectangle including the oscillating airfoil $\mathscr{R}_i$ is $1.15c \times 0.2c$, that of $\mathscr{R}_e$ is $2.5c \times 1.0c$.The

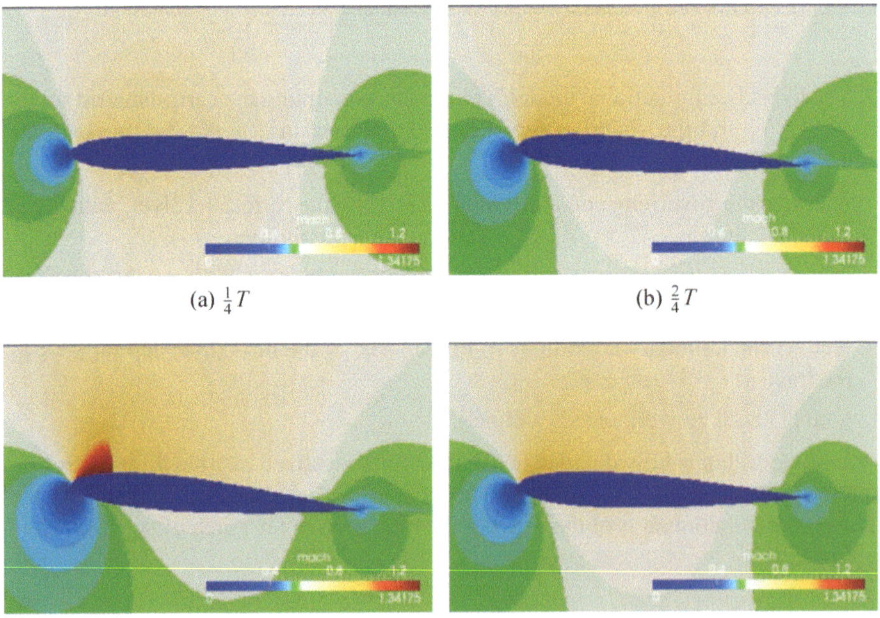

(a) $\frac{1}{4}T$ (b) $\frac{2}{4}T$

(c) $\frac{3}{4}T$ (d) $\frac{4}{4}T$

**Fig. 11.4.** Typical Mach number snapshots

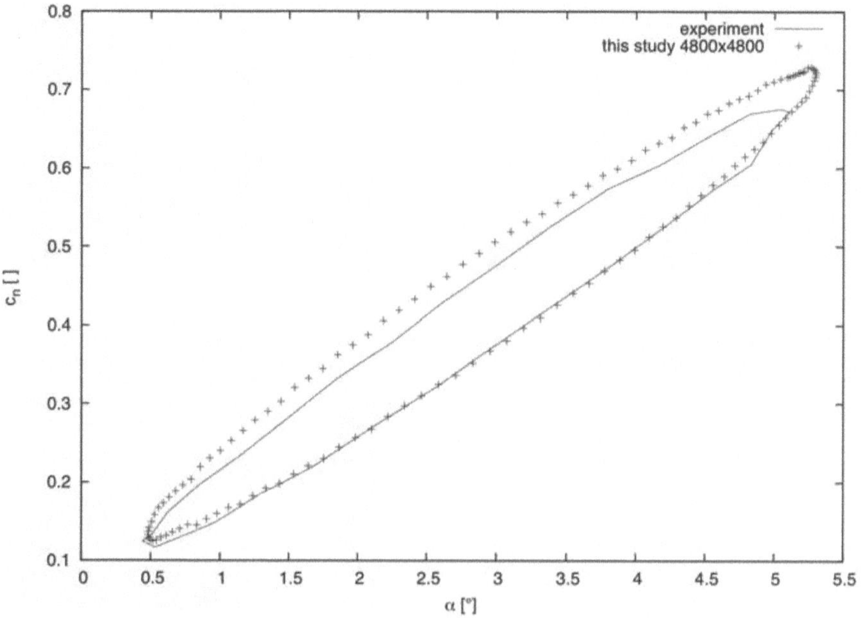


**Fig. 11.5.** Normal force coefficient vs. angle. Full cartesian simulation and experimental results from AGARD R702

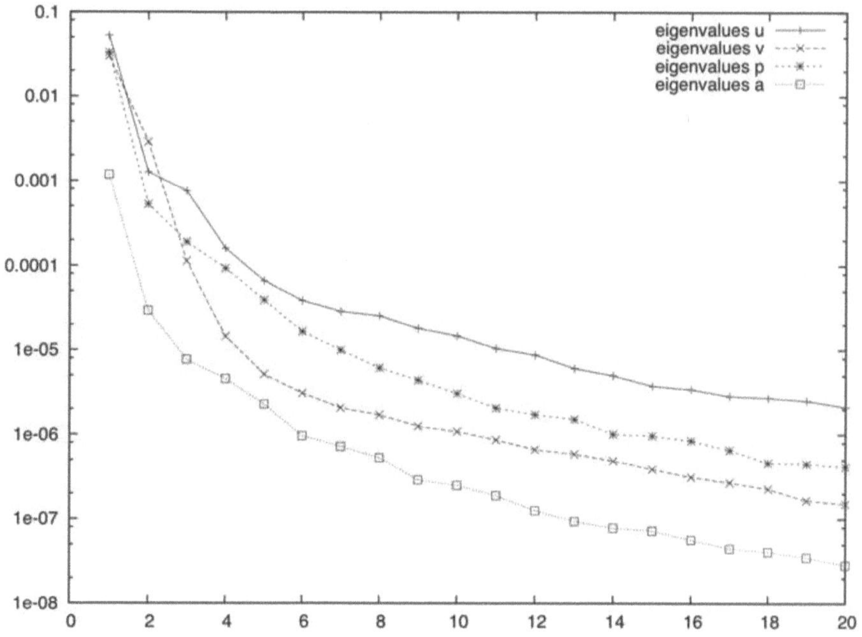


**Fig. 11.6.** Eigenvalues of the snapshot correlation matrix for horizontal velocity $u$, vertical velocity $v$, pressure $p$ and speed of sound $a$

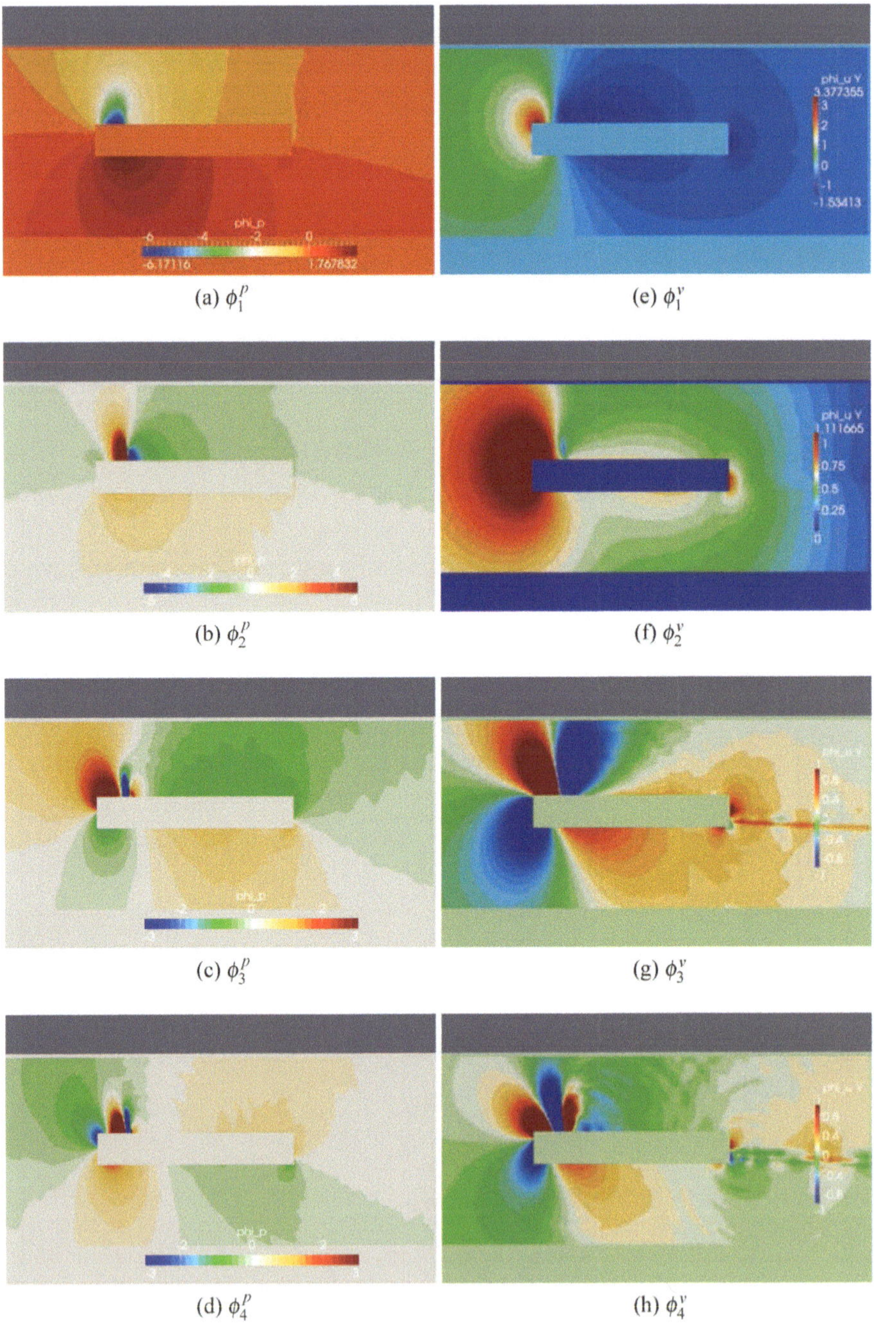


(a) $\phi_1^p$ (e) $\phi_1^v$

(b) $\phi_2^p$ (f) $\phi_2^v$

(c) $\phi_3^p$ (g) $\phi_3^v$

(d) $\phi_4^p$ (h) $\phi_4^v$

**Fig. 11.7.** First four POD modes. Left column pressure, right column vertical velocity

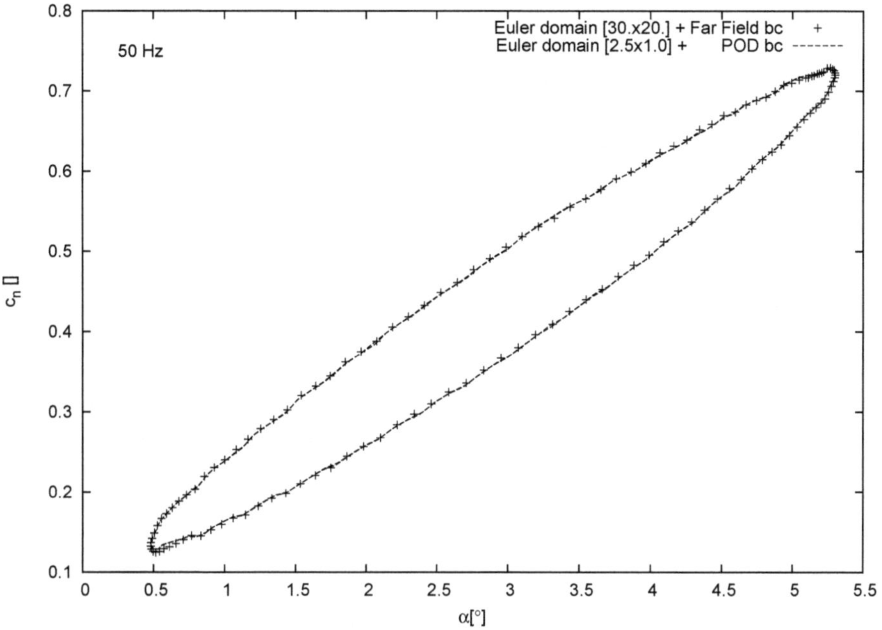


(a) 50*Hz*

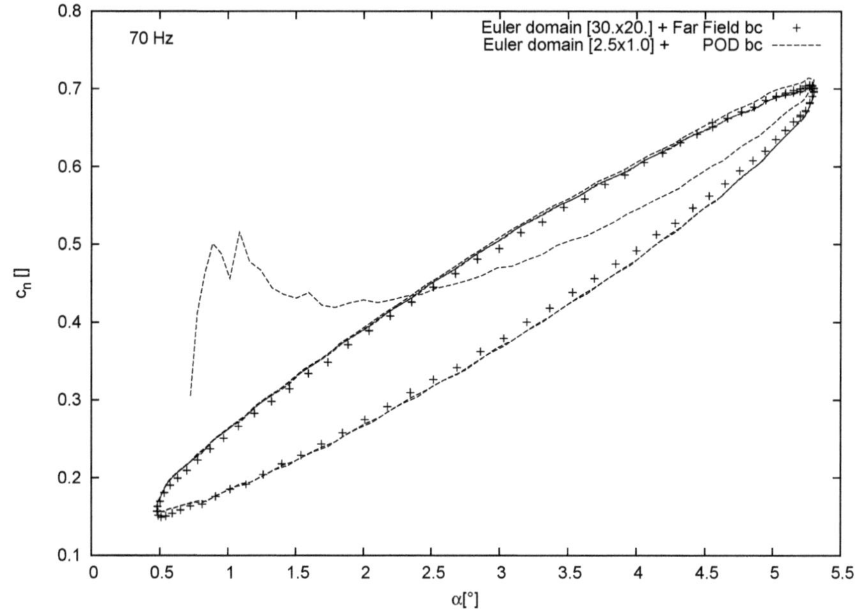


(b) 70*Hz*

**Fig. 11.8.** CT1 case. Normal force coefficient vs. angle. Full computation vs. hybrid ROM. POD modes are build from the 50Hz simulation only ratio between the grid points of the full computational domain and those of the hybrid ROM is approximately 260. This ratio corresponds to the CPU time reduction observed between the full computation and the hybrid ROM.

The eigenvalues of the snapshot correlation matrix are shown in Fig. 11.6. The first four eigenvalues account for about 99% of the database energy for each of the quantities considered. In Fig. 11.7 the first four POD modes for pressure and vertical velocity are shown. The third and fourth mode, whose energetic contribution is of less than 1% on average, show higher spatial frequencies.

In Fig. 11.8 we present the normal force coefficient of the actual hybrid simulation for the CT1 test case at 50Hz and at 70Hz. The 50Hz case corresponds to the snapshots used to build the POD modes. Therefore, this test case is designed to check to what extent the hybrid ROM is able to recover the original solution in the optimal situation. In Fig. 8a we show the comparison between the hysteresis curves obtained via the hybrid ROM and that relative to the full computation. The match is perfect. This means that the non-local boundary condition on $\partial\mathscr{R}_e$ (that corresponds to the trace projection operator) is indeed a very good approximation of the transmission conditions between $\partial\mathscr{R}_e$ and $\partial\Omega_c(t)$.

However, the most promising result is that for 70Hz shown in Fig. 8b. Here the hybrid ROM solution, with a boundary operator derived for the 50Hz case, is contrasted to the full simulation at 70Hz. The hybrid ROM starts from an arbitrary initial condition and after a short transient matches almost perfectly the full computation at 70Hz. This case represents a remarkable situation where the ROM leads to a reliable prediction for a case which was not previously included in the database used to build the POD modes.

In Fig. 11.9 the time history of the coefficients of the pressure modes are depicted. The coefficients pertinent to the Full Order Model are obtained by projecting the snapshots on the POD basis. The coefficients of the hybrid model are those obtained by the above method. An excellent match can be noticed for the first mode, both for 50Hz and 70Hz. For the higher modes still the comparision is very good but slight differences in amplitudes are present. Consequently the presented method is capable to determine the optimal coefficients also for cases which are not included in the database. The error in the force coefficent hysterisis may be decreased by using a more representative database.

### *11.3.2 Discussion*

The hybrid ROM implementation here described has limited impact on existing full CFD codes: it is easy to implement since it reduces to a non-local boundary condition. The only addition operation to perform is a projection of the interior domain iterative solution in the space spanned by the POD modes. The validation results that we present show that this method is accurate also for flow conditions that were not included in the database used to build the POD modes. This is due to the fact that the ROM takes care of flow features that are in principle weakly dependent on the specific geometry inside $\mathscr{R}_i$. Hence, a case not encompassed in the flow database is

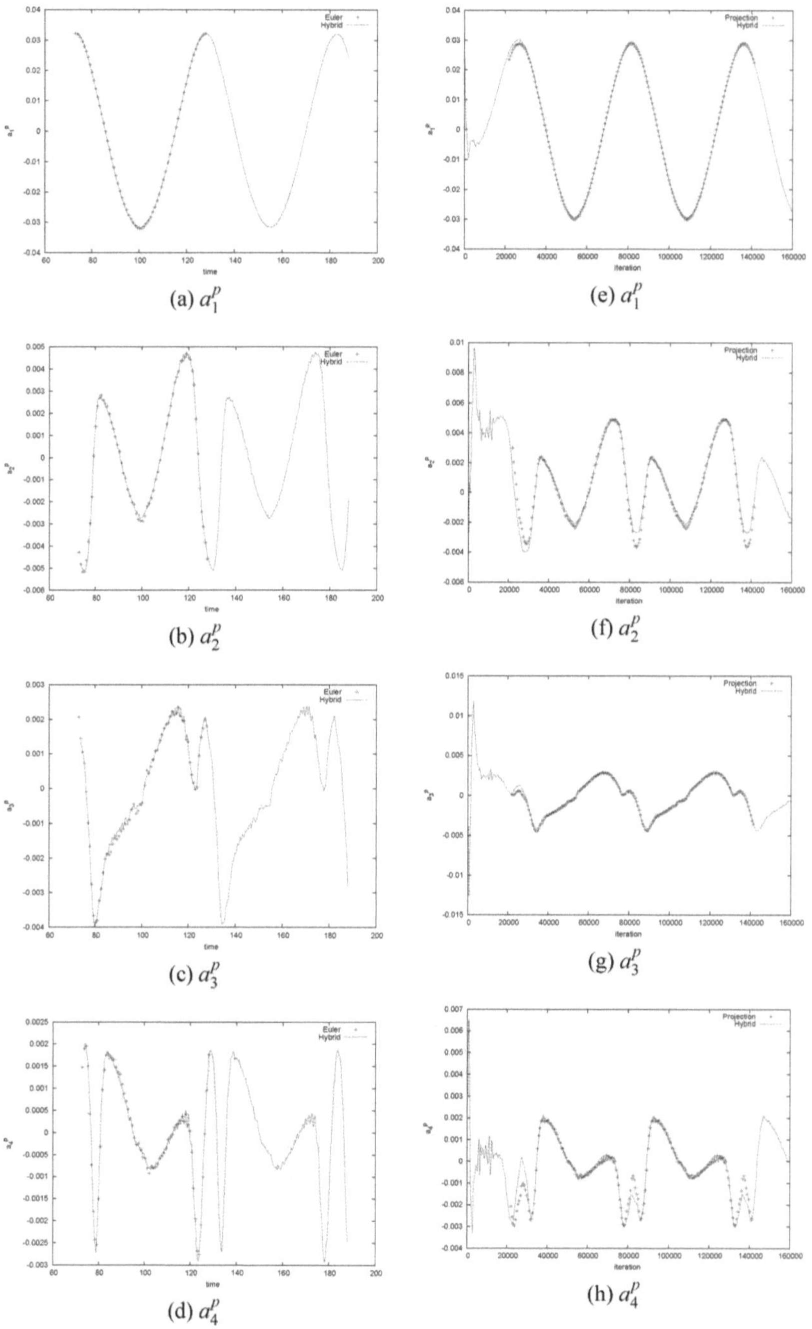


(a) $a_1^p$ (e) $a_1^p$

(b) $a_2^p$ (f) $a_2^p$

(c) $a_3^p$ (g) $a_3^p$

(d) $a_4^p$ (h) $a_4^p$

**Fig. 11.9.** Coefficients of the first four pressure POD modes. Comparison between full order and hybrid model for 50Hz (left) and 70Hz (right) likely to be better approximated by the reduced basis. The whole procedure can be seen as the computation of an empirical Green function of the far field.

## 11.4 ROM by Optimal Transport

Here we describe a non-linear interpolation of the snapshots so that the POD modes may more accurately represent solutions for points in the parameter space that were not included in the database from which they where derived. For a complete survey of this field, see [12, 13]. For an efficient method to numerically solve this problem without obstacles see [9] and references therein.

In order to fix ideas, we consider the case of an oscillating airfoil as in the CT1 test case, for given oscillation amplitude ($\alpha_m = 2.5$deg, $\alpha_0 = 4.$deg) but for several oscillation frequencies. For given phase of the oscillation, i.e. for given pitch of the airfoil our plan is to map the solution for $f = 30$Hz into that of $f = 70$Hz. Thanks to this mapping we can determine a non-linear estimate for the solutions at given pitch for $30Hz < f < 70$Hz.

### *11.4.1 Transport*

In Fig. 11.10 a conceptual description of transport is shown. Given a point $\xi \in \Omega_0$, where $\Omega_0 \subset \mathbb{R}^d$ is a reference configuration, transport at time $t$ is described by a mapping $X(\xi,t)$. The point $x = X(\xi,t)$ belongs to the actual physical configuration $\Omega \subset \mathbb{R}^d$. Let us consider a point $x$ in the actual physical configuration. The inverse mapping, denoted by $Y(x,t)$ (called otherwise backward characteristics), identifies the point in the reference configuration that has been transported by the direct map

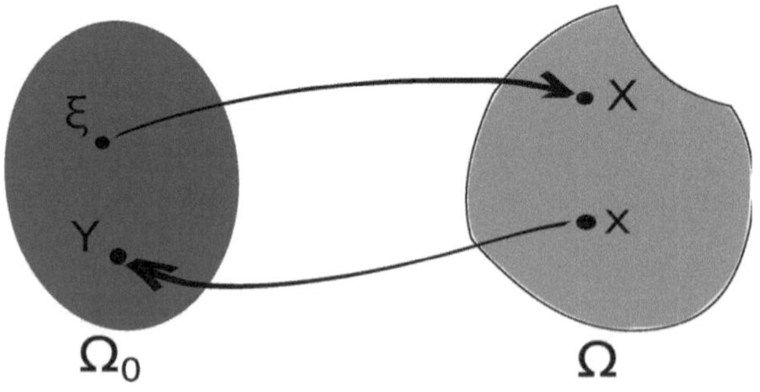


**Fig. 11.10.** Lagrangian description of transport: the reference configuration is $\Omega_0$, points $\xi \in \Omega_0$ are transported by the direct mapping in $X(\xi,t)$. Given the actual configuration $\Omega$, a point $x \in \Omega$ is sent back to its counterimage in the reference configuration by backward characteristics, i.e., the inverse mapping $Y(x,t)$

in $x$ at time $t$. The following relations hold:

$$
\begin{aligned}
&x = X(\xi,t), \;\; \xi = Y(x,t), \\
&Y = X^{-1}, \;\; [\nabla_\xi X][\nabla_x Y] = I,
\end{aligned}
\tag{11.1}
$$

where $[\nabla_\xi X]$ is the jacobian of the transformation $X(\xi,t)$ and $[\nabla_x Y]$ its inverse, i.e., the jacobian of the inverse mapping. Also, we have:

$$
\begin{aligned}
&\partial_t Y + \mathbf{v}\cdot\nabla_x Y = 0, \;\; Y(x,0) = x \\
&\mathbf{v}(x,t) = \partial_t X, \;\; X(\xi,0) = \xi,
\end{aligned}
\tag{11.2}
$$

where $\mathbf{v}$ is the velocity field.

Let us consider, as an example, the inviscid Burgers equation:

$$
\partial_t \mathbf{v} + \mathbf{v}\cdot\nabla_x \mathbf{v} = 0. \tag{11.3}
$$

This equation describes a pressure-less Euler flow. Since no force is acting on the medium, each component of the velocity field is purely advected. In lagrangian coordinates we have:

$$
\partial_t^2 X(\xi,t) = 0 \;\; \Longrightarrow X(\xi,t) = \xi + \mathbf{v}(\xi,0)t. \tag{11.4}
$$

The solution consists of particles moving on straight lines (no acceleration).

In order to determine the mapping, we define a suitable optimal transport problem. Let us associate a scalar density function $\rho(u) \geq 0$ to the solution $u(x,t)$ , in such a way that:

$$
{}_{\Omega}\rho(x,t)\;dx = 1, \; \forall t \in \mathbb{R}^+ \tag{11.5}
$$

so that the non-negative density is normalized to 1 for all times. The choice of the density function is for the moment arbitrary. If $u$ is a non-negative scalar and satisfies this normalization, it may be directly used as a density function.

Let $\rho_i$, $i = 1,2$ be the snapshots of the density function. The optimal transportation problem relative to this density pair is defined as:

$$
\begin{aligned}
&X^*(\rho_1,\rho_2) = Arg\inf_{\tilde{X}}\left\{ {}_{\Omega}\rho_1(\xi)|\tilde{X}(\xi)-\xi|^2\;d\xi \right\}, \text{ subject to} \\
&\rho_1(\xi) = \rho_2(\tilde{X}(\xi))\det(\nabla_\xi \tilde{X}).
\end{aligned}
\tag{11.6}
$$

The optimal mapping $X^*$ minimizes the cost of the $L^2$ transport (Monge) problem, among all the changes of coordinates $\tilde{X}(\xi)$ locally keeping constant mass between the densities 1 and 2. The solution to this problem exists and is unique and stipulates that the lagrangian velocity is the gradient of a (almost everywhere) convex potential $\psi(\xi)$.

In particular the same problem can be rewritten in the Eulerian frame of reference. The optimal conditions for the minimum are the familiar conservation law for the density and the previously introduced inviscid Burgers equation. The main difficulty of the problem is that this system is equipped with initial and final condition for the density but no initial condition for velocity. We therefore introduce an approximate Monge mapping as follows:

$$(\rho_1 - \rho_2) = \frac{1}{2}\nabla\cdot((\rho_1 + \rho_2)\nabla\psi) \tag{11.7}$$

so that $\nabla\psi = \mathbf{v}(x,0)$ and the inviscid Burgers equation 11.3 can be used to propagate in time the solution.

### *11.4.2 Results*

In order to illustrate the method, we have considered the pressure distribution at maximum pitch of the NACA0012 at Mach=0.6 corresponding to a set up similar to the CT1 test case. The densities $\rho_1(x)$ and $\rho_2(x)$ correspond to the pressure distributions. Given two sets of snapshots corresponding to two different frequencies of oscillation, we define $\rho_1(x)$ and $\rho_2(x)$ as the pressure corresponding to 30Hz and 70Hz, respectively. This can be done for each phase angle of the oscillation. The numerical scheme employed to determine the Monge approximate mapping ($\psi$) is a simple finite-difference second-order method. This initial mapping velocity ($\nabla\psi$) is used then as the initial condition for the transport problem. The initial pressure distribution $\rho_1(x)$ corresponds to $t = 0$ and $\rho_2(x)$ to a final time arbitrarily set to 1. The solution at any pseudo time $t$ between 0 and 1 corresponds to a non-linear interpolation of the solution at a frequency of oscillation of $30 + (70 - 30)t$. See Fig. 11.11. In this picture the actual solutions at 30Hz, 50Hz and 70Hz are shown in terms of pressure isolines. It should be remarked that the solution at 50Hz is not a linear interpolation of the solution at 30Hz and 70Hz, see Fig. 11.12. The pressure distribution at 50Hz, see Fig. 11.13, is found thanks to the non-linear interpolation. One-dimensional plots corresponding to a segment in a smooth region and in a region where the shock is present are shown. These results show that the non-linear interpolation method presented here can be used to determine overall reasonable estimates of intermediate snapshots of high-fidelity simulations not present in the database.

## 11.5 System Identification Using ROM in Tumor Growth Modeling

In this section ROMs are applied to system identification in tumor growth modeling. A complete description of the method is presented in [4].

In this work reduced order modeling is applied to the solution of an inverse problem as tool of solution regularization. In particular a set of semi-empirical eigenfunctions is built for each patient, exploiting the organ geometry retrieved from the first clinical exam. So the method is "patient specific". The eigenfunctions are then used in order to estimate parameters when new data are available from subsequent exams.

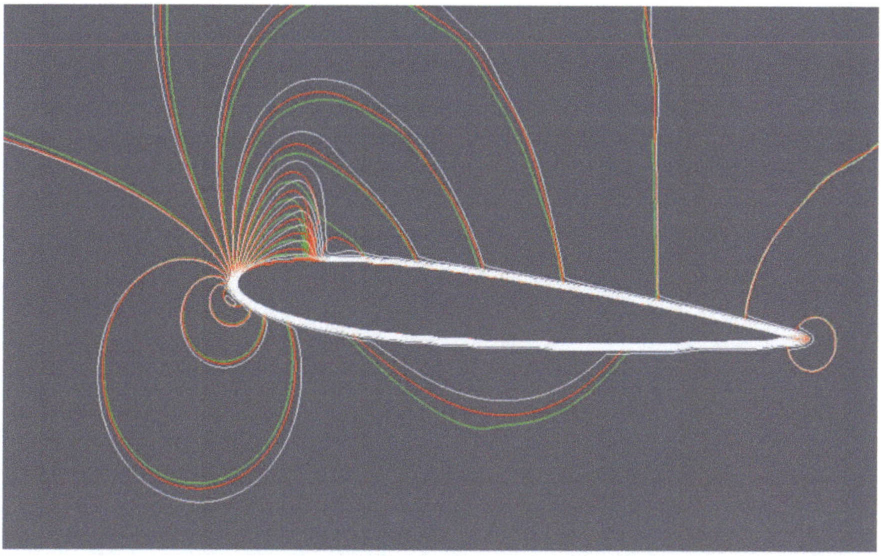

(a)

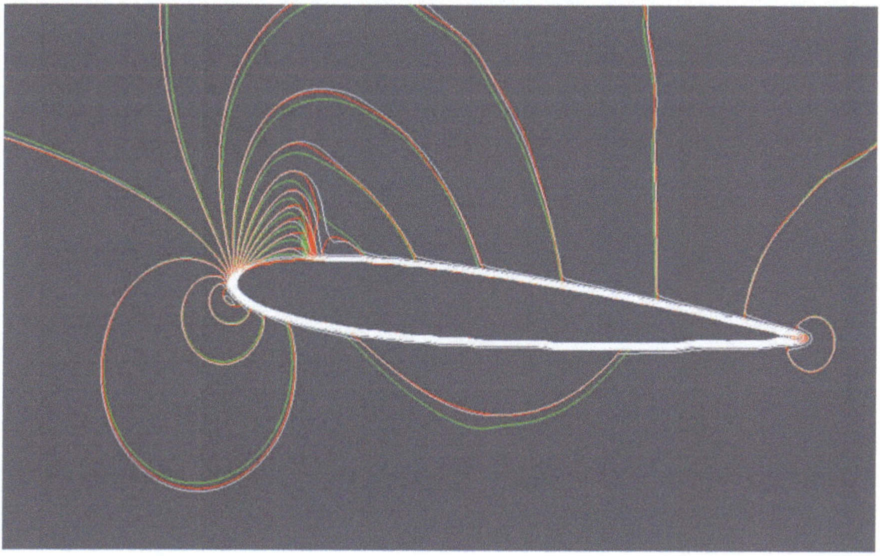

(b)

**Fig. 11.11.** (a) Iso-pressure lines of the solution at 30Hz (white), 50Hz (red), 70Hz (green) in the region of definition of POD; the white isolines correspond to the initial condition of the Monge problem. (b) Results of the Monge interpolation: estimated pressure snapshot at 50Hz. Estimated solution in white, actual solution in red. Green: actual solution at 70Hz

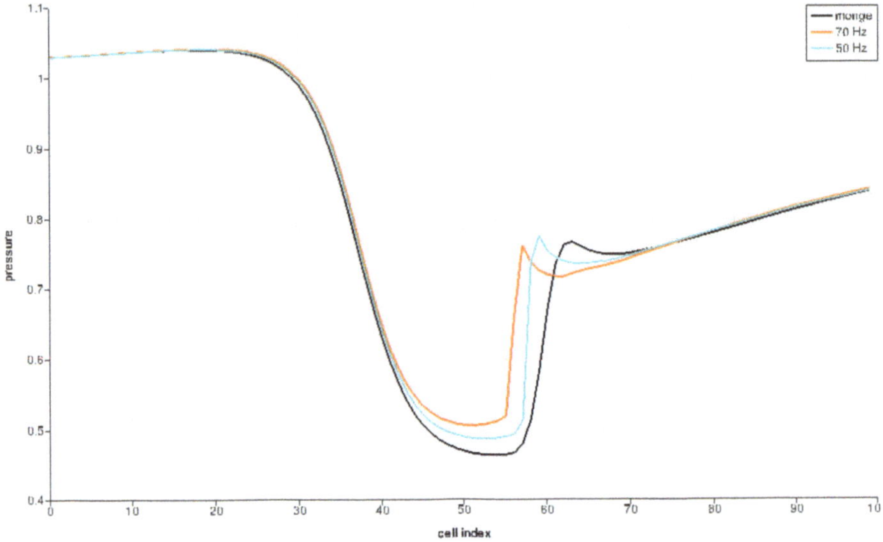


(a)

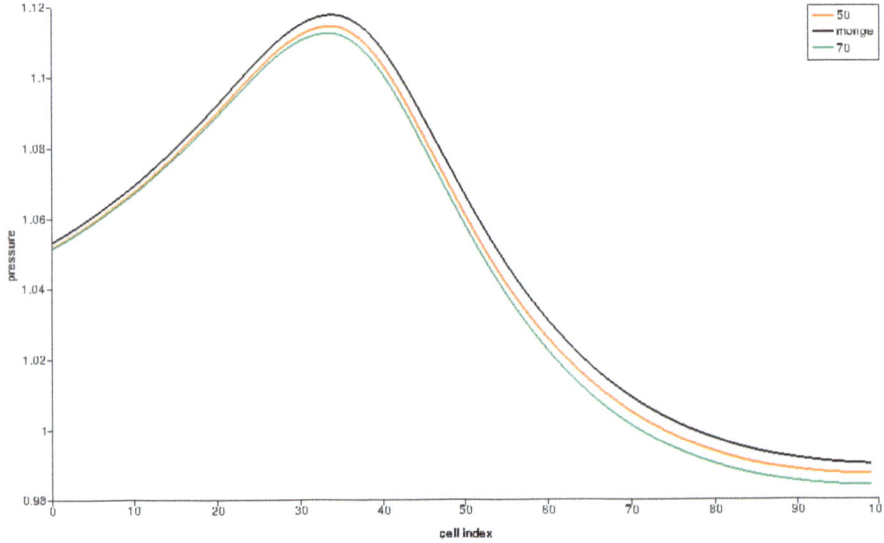


(b)

**Fig. 11.12.** Initial condition for the Monge problem (30Hz) and actual high-fidelity solutions (50Hz and 70Hz). (a) curves on a segment parallel to the abscissa where the pressure shows a shock wave; (b) solution on segment where the pressure is regular. The intermediate solution (50Hz) is not a linear interpolation of the initial condition (30Hz) and final condition (70Hz). "Monge" denotes here the initial condition of the Monge problem corresponding to the high-fidelity model at 30Hz

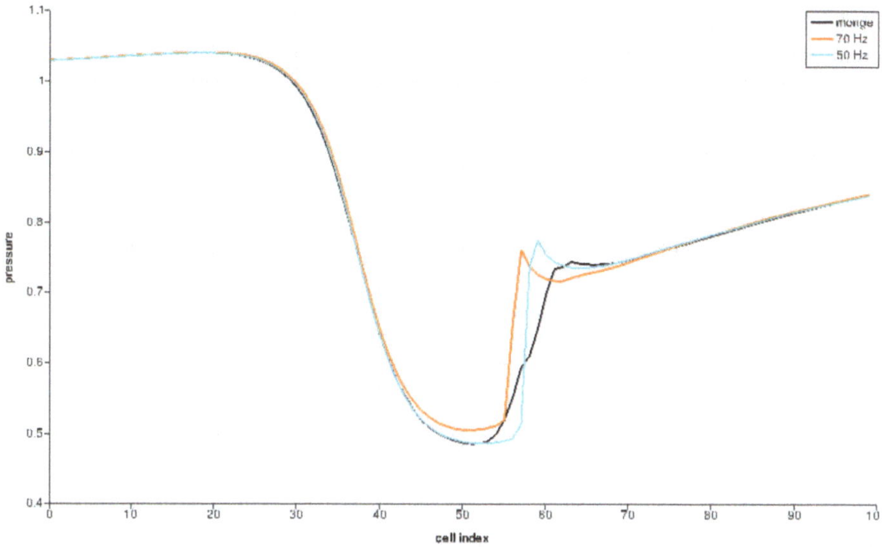


(a)

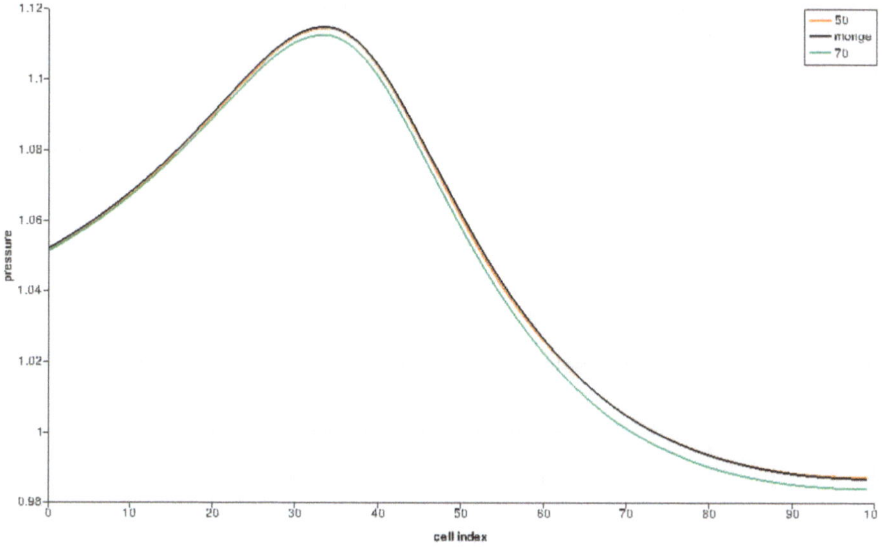


(b)

**Fig. 11.13.** Results of the Monge interpolation at 50Hz. The continuity equation and the inviscid Burgers equation are integrated starting from the initial conditions (see Fig.11.12). The pictures show the high-fidelity model results compared to those of the non-linear interpolation at 50Hz. (a) solution on a segment parallel to the abscissa where the pressure shows a shock wave; (b) solution on segment where the pressure is regular. These are typical results across the field

The macroscopic models for tumor growth are represented by a set of PDEs accounting for the phenomenological aspects of the pathology. For the present case, the system reduces to a set of non-linear parametric coupled PDEs that describes the evolution of a three-specie saturated reacting flow in a porous, isotropic, non-uniform medium.

The tumoral tissue is composed by two different phases, denoted by $P$ and $Q$. The density $P$ represents the number of dividing cells per unit volume, $Q$ is that of the necrotic cells. The healthy tissue is the phase denoted by $S$. Equations for $P$, $Q$ and $S$ read:

$$\frac{\partial P}{\partial t} + \nabla \cdot (\mathbf{v}P) = (2\gamma - 1)P, \tag{11.8}$$

$$\frac{\partial Q}{\partial t} + \nabla \cdot (\mathbf{v}Q) = (1 - \gamma)P, \tag{11.9}$$

$$\frac{\partial S}{\partial t} + \nabla \cdot (\mathbf{v}S) = 0. \tag{11.10}$$

where the velocity $\mathbf{v}$ models the tissue deformation and $\gamma$ (called the hypoxia threshold) is a scalar function of the nutrient concentration. If enough nutrients are available then $\gamma = 1$ and the tumor cells proliferate, otherwise they die. The healthy tissue evolves through an homogeneous conservation equation.

Assuming that $P + Q + S = 1$ in every point of the domain, a condition for the divergence of the velocity field is derived. This condition, coupled with a Darcy law, allows to describe the mechanics of the system:

$$\nabla \cdot \mathbf{v} = \gamma P, \tag{11.11}$$

$$\mathbf{v} = -k(P,Q)\nabla \Pi. \tag{11.12}$$

The scalar function $\Pi$ plays the role of a pressure (or potential), and $k$ is a permeability field, satisfying:

$$k = k_1 + (k_2 - k_1)(P + Q), \tag{11.13}$$

where $k_1$ represents the constant porosity of the healthy tissue and $k_2$ is the porosity of the tumor tissue.

The equation describing the nutrients has the following form:

$$-\nabla \cdot (D(P,Q)\nabla C) = -\alpha PC - \lambda C, \tag{11.14}$$

where $\alpha$ is the oxygen consumption rate for the proliferating cells, $\lambda$ is the oxygen consumption coefficient of healthy tissue and $D(P,Q)$ is the diffusivity. Boundary conditions and sources are set up according to the nature of the organs considered and will be detailed later on. The diffusivity may be written as:

$$D = D_{max} - K(P + Q). \tag{11.15}$$

The link between the nutrients concentration and the population dynamics is provided by:

$$\gamma = \frac{1 + \tanh(R(C - C_{hyp}))}{2}, \tag{11.16}$$

where $R$ is a coefficient and $C_{hyp}$ is called the hypoxia threshold. The resulting hypoxia function thus satisfies $0 \leq \gamma \leq 1$.

For this simple model the state variable set may be defined as $X = \{P, Q, C, \Pi\}$. The observable is defined to be $Y = P + Q$, as result from discussions with medical doctors about what is measured by CT scans in the case of lung metastases. One can not distinguish on images the cell species composing the tumor, but only the tumor mass. The control set consists in all the undetermined scalar parameters describing tissue properties (such as $k_1$,$k_2$,$D_{max}$, $K$), the tumor activities (nutrient consumptions $\alpha$, $\lambda$, and $C_{hyp}$), and the fields describing the initial non-observed conditions needed to integrate the system ($P(x,0)$).

### *11.5.1 Regularized Inverse Problem*

The observable evolution is governed by:

$$\dot{Y} + \nabla \cdot (Y\mathbf{v}) = \gamma(C)P. \tag{11.17}$$

the divergence of the velocity field obeys:

$$\nabla \cdot \mathbf{v} = \gamma(C)P - \frac{\int_\Omega \gamma P \, d\Omega}{\int_\Omega (1-Y) \, d\Omega}(1-Y), \tag{11.18}$$

where the expression relative to Neumann boundary condition for the pressure field was retained. In the case of Dirichlet boundary conditions the second term of the right hand side of this equation vanishes. The curl of the Darcy law reads:

$$k(Y)\nabla \wedge \mathbf{v} = \nabla k(Y) \wedge \mathbf{v}. \tag{11.19}$$

and the equation for the oxygen concentration field is written:

$$\nabla \cdot (D(Y)\nabla C) = \alpha PC + \lambda C. \tag{11.20}$$

The definition of the hypoxia function, $\gamma$, is unchanged.

The repeated index summation convention is used from now on. The non-observable variables are expressed as combination of POD modes:

$$\begin{aligned} P &= a_i^P \phi_i^P \quad i = 1,\ldots,N_P; \\ C &= a_i^C \phi_i^C \quad i = 1,\ldots,N_C; \\ \gamma P &= a_i^{\gamma P} \phi_i^{\gamma P} \quad i = 1,\ldots,N_{\gamma P}; \\ \mathbf{v} &= a_i^{v} \phi_i^{v} \quad i = 1,\ldots,N_v, \end{aligned} \tag{11.21}$$

where $a_i^{(\cdot)} = a_i^{(\cdot)}(t)$ are scalar functions of time, $\phi_i^{(\cdot)} = \phi_i^{(\cdot)}(\mathbf{x})$ are functions of spatial coordinates.

The dimension of the empirical functional space, *i.e.*, the number of POD modes used to reconstruct the solution, is chosen such that if additional POD modes are included, the reconstruction of a given field does not vary up to a certain error value that, in this work, was fixed at $10^{-4}$ in $L^2$ norm.

Substituting these expressions in the system Eqs. (11.17) and (11.20) we obtain:

$$\dot{Y} + a_i^{(v)} \nabla \cdot (Y \phi_i^{(v)}) = a_i^{(\gamma P)} \phi_i^{(\gamma P)}, \tag{11.22}$$

$$a_i^{(v)} \nabla \cdot \phi_i^{(v)} = a_i^{(\gamma P)} \phi_i^{(\gamma P)} - \frac{\int_\Omega a_i^{(\gamma P)} \phi_i^{(\gamma P)} \, d\Omega}{\int_\Omega 1 - Y \, d\Omega} (1 - Y), \tag{11.23}$$

$$a_i^{(v)} k(Y) \nabla \wedge \phi_{\mathrm{i}}^{(v)} = a_i^{v} \nabla k(Y) \wedge \phi_i^{(v)}, \tag{11.24}$$

$$a_i^{(C)} \nabla \cdot (D(Y) \nabla \phi_i^{(C)}) = \alpha a_j^{(P)} a_i^{(C)} \phi_j^{(P)} \phi_i^{(C)} + \lambda a_i^{(C)} \phi_i^{(C)}, \tag{11.25}$$

The hypoxia function $\gamma$, Equation (11.16), is multiplied by $P$, in such a way that the product $\gamma P$ is:

$$a_i^{(\gamma P)} \phi_i^{(\gamma P)} = a_j^{(P)} \phi_j^{(P)} \frac{1 + \tanh(R(a_i^{(C)} \phi_i^{(C)} - C_{hyp}))}{2}. \tag{11.26}$$

The system Eqs. (11.22–11.25) was finally solved by a least square approach under certain constraints that are introduced below. At a given time (say $t_0$), the snapshot $Y(t_0)$ and a subsequent snapshot $Y(t_1)$ are used to perform the computation of the time derivative. Let the residual of the l-th equation be $R_l$. We write $F = \sum_l R_l^2$ and

$$\left( a_i^{(\cdot)}(t_0), \pi_j \right) = \operatorname{argmin}(F) \tag{11.27}$$

where $a_i^{(\cdot)}$ are the expansion coefficients for the variables $P, C, \mathbf{v}, \gamma P$ and $\pi_j$ are the parameters to be identified.

The first constraint is linked to the fact that Eq. (11.25) is an homogeneous equation with respect to the coefficients $a_i^{(C)}$. If $C_{hyp} < 0$ the trivial solution is a solution for the whole system Eqs. (11.22) and (11.26). In order to prevent the identification of a system with unphysical solutions we get one scalar constraint from the boundary. In the case of Dirichlet boundary conditions $C = C_0$ on $\partial \Omega_C$ where $\Omega_C$ is a blood vessel domain, one scalar equation is obtained of the form:

$$\sum_i \left( \frac{\sum_j b_j^i}{\lambda_i^{1/2}} \right) a_i^{(C)}(t) = 1, \quad \forall t. \tag{11.28}$$

where $b_j$ and $\lambda_j$ are the eigenvalues and the eigenfunctions of the autocorrelation matrix used to build the modes for the variable C, respectively.

The second constraint to be imposed results from the observation that, since in the inverse problem the equation for the variable $P$ is not solved, the latter does not automatically satisfy: $0 \leq P \leq 1$ and therefore this is a constraint (fundamental for the population dynamics) to be imposed. To this end the residuals are penalized as follows:

$$\tilde{F} = F + c_1(\max\{a_i^{(P)} \phi_i^{(P)}\} - 1) + c_2(-\min\{a_i^{(P)} \phi_i^{(P)}\}) \tag{11.29}$$

where $c_1, c_2$ are positive constants, set in such a way that penalization does not affect the stability of the procedure (in the present work $(c_1, c_2) \in [1.0, 2.5]e-2$).

In order to decrease the computational cost of the procedure a third constraint is imposed to define a feasible set of solutions. The solution is sought so that the admissible values of the POD coefficients are sought in an interval $I_k$ that is obtained from $I_k^{db}$, the interval to which POD coefficients of the simulated solutions belong, by a stretching factor $1+\delta$ where $\delta$ is a suitable positive constant. In all the following simulations the value $\delta = 0.1$ was adopted.

The hypothesis that two subsequent snapshots are close in time, or, in other words, that the time between two snapshots is small if it is compared with the characteristic evolution time of the phenomenon, is very optimistic. In order to relax this hypothesis, instead of using first order finite differences, that is equivalent to perform a linear interpolation between the snapshots, a different kind of interpolation is used. However, an higher order finite difference scheme, equivalent to a polynomial interpolation, would require a large number of snapshots. As an alternative, still assuming that only two images are available, an additional hypothesis about the growth rate could be retained. Here, two cases are considered. In the case of exponential growth we write:

$$\dot{Y} \approx Aexp\{\zeta t\} + Bexp\{-\zeta t\} = f(\zeta), \tag{11.30}$$

where $A, B$ are chosen in such a way that the two available snapshots are interpolated. One parameter, $\zeta$, is free and enters the residual minimization process. The first equation of the system (11.17–11.20) becomes:

$$f(\zeta) + \nabla \cdot (a_i^{(v)} \phi_i^{(v)} Y) = a_i^{(\gamma P)} \phi_i^{(\gamma P)}. \tag{11.31}$$

In the case of a logistic-type growth we proceed in a similar way. We take

$$Y \approx AG(\omega, \sigma) + BG(-\omega, -\sigma) \tag{11.32}$$

where

$$G(\omega, \sigma) = \frac{\omega e^{\omega t}}{\omega - \sigma e^{\omega t}}. \tag{11.33}$$

As before $A$ and $B$ are adjusted such that the snapshots are interpolated. In this case, however, we are left with two free parameters ($\omega$ and $\sigma$) that are found within the residual minimization process. The inverse problem finally takes the form of a non-linear algebraic optimization problem, that is solved using a Newton trust region method.

### *11.5.2 Realistic Case Application: A Comparison with a Standard Sensitivity Approach*

In Fig. 11.14 four scans covering an evolution over 45 months are presented of some lung metastases of a primary tumor affecting the thyroid (Courtesy Institut Bergonié). Even though this patient is affected by several metastases, only the study of the one marked in Fig. 11.14(a) will be presented. It is a quasi-steady metastasis, which grows very slowly and thus need only to be monitored. The results obtained by means of a sensitivity technique are presented, when only the first two scans were used in

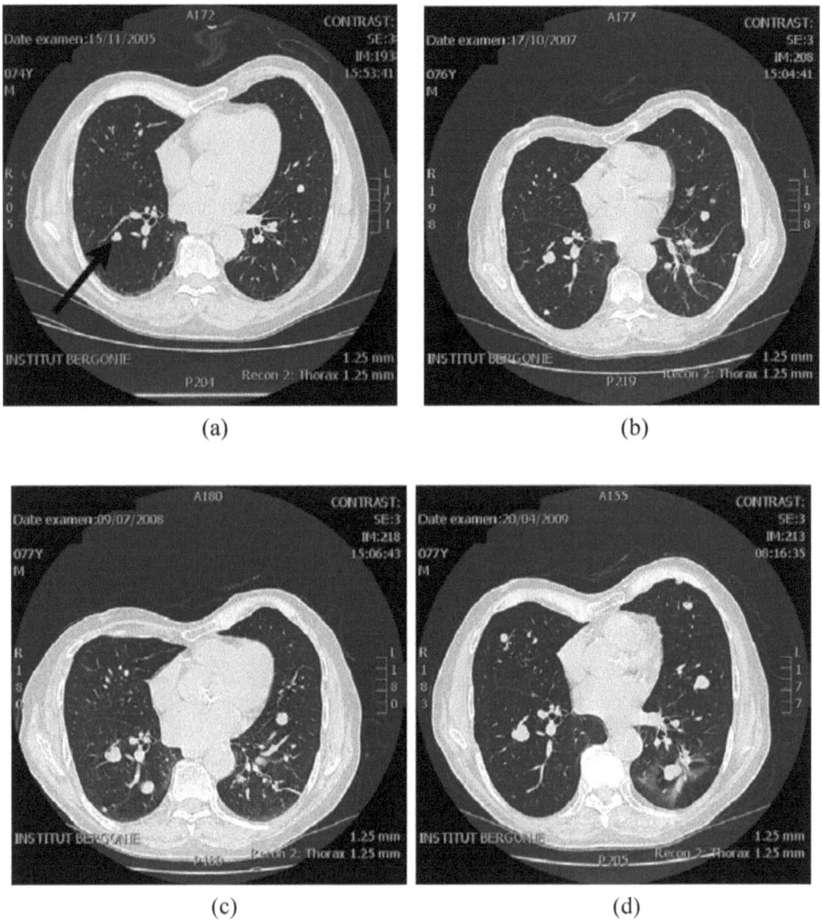


**Fig. 11.14.** Scans: (a) November 2005; (b) October 2007; (c) July 2008; d) April 2009

order to identify the system. This means that the first two images were used as data set to solve the inverse problem and find the set of control. Then, the direct simulation were performed covering the entire evolution and the result has been compared to the data of the subsequent exams.

The control set consists in the parameters and in the initial distribution for the proliferating cell density. In this particular test the initial density distribution for proliferating cells is taken:

$$P(x,0) = A\exp\left\{-\delta\Phi^2\right\}, \tag{11.34}$$

where $\Phi$ is the level set for the tumor, $A$ the amplitude and $\delta$ the steepness.

This system is solved at $t = 0$, taking the second image at $t = 0.3$. The time derivative is approximated by a logistic interpolation. In this particular case it is

**Table 11.3.** Data set and results for realistic case, fitted with the parameters identified by ROM: 6 volumes measures are taken from 2D scans, resolution 1.25*mm*

| | | | | | | |
|---|---|---|---|---|---|---|
| *Month* | 0 | 21.0 | 24.5 | 36.0 | 40.5 | 45.0 |
| *Area* | 4.2e-3 | 6.5e-3 | 8.1e-3 | 9.7e-3 | 1.03e-3 | 1.10e-3 |
| $\mathscr{E}_{Sens}(\%)$ | 0.0 | 1.8 | 2.47 | 2.02 | 1.94 | 1.36 |
| $\mathscr{E}_{ROM}(\%)$ | 0.0 | 1.9 | 2.50 | 2.80 | 8.67 | 6.12 |
| $\|Y-Im\|_{Sens}$ | 0.0 | 0.22 | 0.24 | 0.35 | 0.31 | 0.24 |
| $\|Y-Im\|_{ROM}$ | 0.0 | 0.23 | 0.26 | 0.38 | 0.36 | 0.32 |

equivalent to solve the reduced order model for the elliptic equations and to couple them with the residual approximation for the observable. The system is cheap from the computational stand point, its solution taking only few minutes on a standard laptop. The system was initialized with several initial conditions in order to check the stability and the presence of local minima.

The database used for the present case consists of 768 direct simulations, realized by sampling the parameters values appearing in the model as well as the parameters introduced to represent the initial distribution of proliferating cells (namely $A$ and $\delta$). A set of 20 time snapshots was retained from each of the simulations.

In Table 11.3 the errors are compared between the sensitivity approach (when two images are taken into account) and the reduced order model. The *ROM* performs quite well in terms of volume in the first part of the growth. For what concerns $L^2$ norms and in the second part of the growth sensitivity has substantially better results. The most relevant fact is that the two approaches show similar behavior in the very beginning (ROM is solved at $t = 0$). It is interesting that the reduced order model allows to get a correct solution on a time scale that is sufficiently large, *i.e.* on a scale comparable with the interval between two subsequent medical exams. In Fig. 11.15 the fitting curves are shown, confirming essentially what commented about the errors. Let us remark that the two methods starts with exactly the same trend, so that the Reduced Order Model approach results in an approximation of the Sensitivity one in $t = 0$. The Error contours for the third image (*i.e.* the first prediction) are shown for the two methods in Fig. 11.17. On the left, the result of the sensitivity is shown, the reduced order model is on the right. The differences between the two residuals are minimal, showing the ability of the reduced approach to mimic sensitivity.

### *11.5.3 A Fast Rate Tumor Growth*

In order to see if the method is robust enough to perform the identification in a very aggressive case, an exponential fast growth is studied. In Fig. 11.18 the evolution of a metastatic nodule is shown; the evolution takes about six months, the scans are taken at approximately constant rate. The problem is the following one: given the

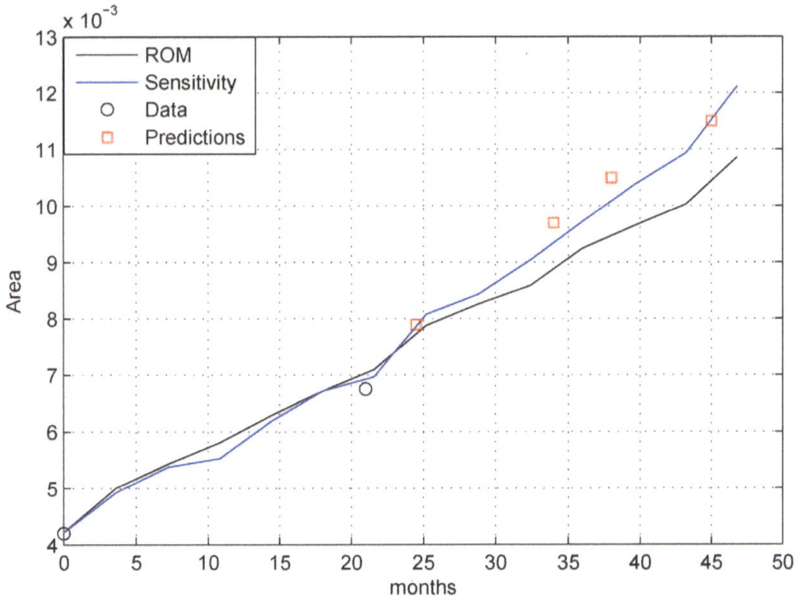


**Fig. 11.15.** Area as function of time, for the Reduced Order Model (black line) and for the Sensitivity approach (blue line)

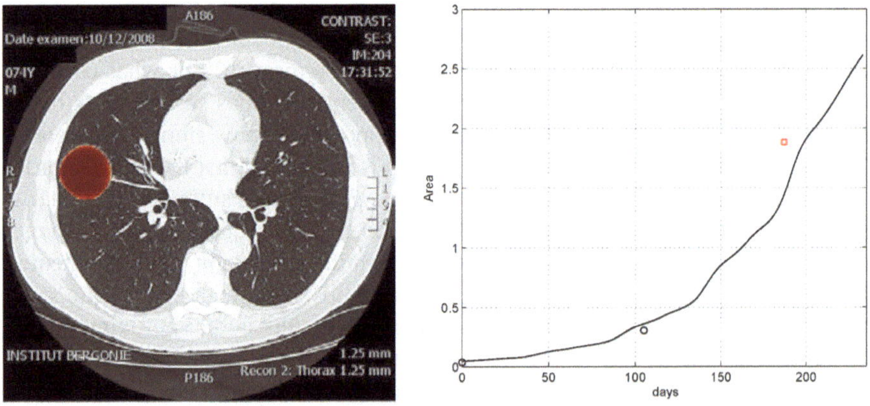


(a) (b)

**Fig. 11.16.** Results: (a) superposition of simulation and geometry; b) volume curve with respect to days

first two scans, we try to recover the third one, after having performed the parameters identification.

A database was build varying all the parameters in uniform intervals. The database consists in 128 simulations. For each one, 20 time frames are taken. The minimization takes about 20 minutes on one standard CPU. In Fig. 11.16(a) the superposition

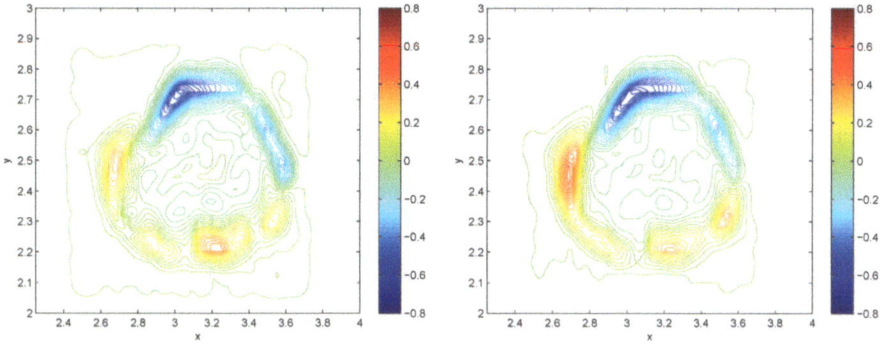


(a) (b)

**Fig. 11.17.** Zoom on the tumor: difference (signed absolute error) between the third scan and the solution when the identification is performed by (a) sensitivity; (b) ROM

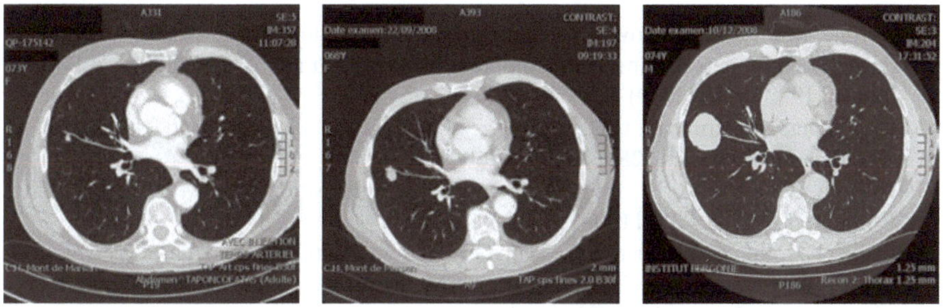

(a) (b) (c)

**Fig. 11.18.** Fast growing tumor: scan at (a) June 2008; (b) September 2008; (c) December 2008

of the simulation to the realistic geometry is shown, at the time corresponding to the third scan. The result is satisfactory, the volume not being too far from the measured one. The error is essentially a shape error. The model tends to regularize the shape, so that the simulated tumor is closer to a spheroid with respect to the real tumor. In order to prevent this error to arise two strategies are possible: the first one consists in modifying the model such that its dynamics is less regularizing and the second one consists in changing the control set.

In Fig. 11.16(b) the volume curve is plotted with respect to days. There is a certain error in volume at the time corresponding to the third scan, but, in terms of time, it is about 15 days on a time interval of 6 months. For such a growth, featured by a high rate and a large final volume, not enough mechanics have been accounted for. As a matter of fact, tumor expansion causes some compression in the tissues and the constraints imposed by the thorax are not negligible.

## 11.6 Conclusions

We have presented a set of methods where ROMs have been used to solve problems in applications. ROMs where not directly used for simulation, but instead as an auxiliary numerical expidient in conjunction with full model simulations or available data observations. Future investigations will need to improve model accuracy and robustness with respect to parameter variations, with the objective of accurate and robust predictive ROMs.

**Acknowledgements** This research is funded in part by the EU FP7 project FFAST, ACP8-GA-2009-233665. We thank Dr. Jean Palussière at the Institut Bergonié, Bordeaux, for selecting the patients and for fruitful modeling discussions.

## References

1. Barone, M.F., Kalashnikova, I., Segalman, D.J., Thornquist, H.K.: Stable Galerkin reduced order models for linearized compressible flow. Journal of Computational Physics **228**(6), 1932–1946 (2009)
2. Bergmann, M., Bruneau, C.H., Iollo, A.: Enablers for robust pod models. Journal of Computational Physics **228**, 516–538 (2009)
3. Buffoni, M.R., Telib, H., Iollo, A.: Domain decomposition by low-order modelling. Computers & Fluids **38**, 1160–1167 (2009)
4. Colin, T., Iollo, A., Lombardi, D., Saut, O.r: System identification in tumor growth modeling using semi-empirical eigenfunctions. Mathematical Models and Methods in Applied Sciences **22**(06), 1250003–1 (2012)
5. Galletti, B., Bruneau, C.H., Zannetti, L., Iollo, A.: Low-order modelling of laminar flow regimes past a confined square cylinder. J. Fluid Mech. **503**, 161–170 (2004)
6. Gorsse, Y., Iollo, A., Telib, H., Weynans, L.: A simple second order cartesian scheme for compressible Euler flows. Journal of Computational Physics **231**(23), 7780–7794 (2012)
7. Holmes, P., Lumley, J.L., Berkooz, G.: Turbulence, coherent structures, dynamical systems and symmetry. Cambridge University Press, Cambridge (1996)
8. Iollo, A, Lanteri, S., Désidéri, J.A.: Stability properties of POD–Galerkin approximations for the compressible Navier–Stokes equations. Theoretical and computational fluid dynamics **13**(6), 377–396 (2000)
9. Iollo, A., Lombardi, D.: A lagrangian scheme for the solution of the optimal mass transfer problem. Journal of Computational Physics **230**, 3430–3442 (2011)
10. Lombardi, E., Bergmann, M., Camarri, S., Iollo, A.: Low-order models: Optimal sampling and linearized control strategies. Journal Europeen des Systemes Automatises **45**(7–10), 575–593 (2011)
11. Sirovich, L.: Turbulence and the dynamics of coherent structures. Parts I, II and III. Quarterly of Applied Mathematics **XLV**, 561–590 (1987)
12. Villani, C.: Topics in optimal transportation. American Mathematical Society (2003)
13. Villani, C.: Optimal Transport, old and new. Springer-Verlag, Berlin Heidelberg (2009)
14. Weller, J., Lombardi, E., Iollo, A.: Robust model identification of actuated vortex wakes. Physica D **238**, 416–427 (2009)

# MS&A – Modeling, Simulation and Applications

**Series Editors:**



**Editor at Springer:**

Francesca Bonadei
francesca.bonadei@springer.com

1 L. Formaggia, A. Quarteroni, A. Veneziani (eds.)
Cardiovascular Mathematics
2009, XIV+522 pp, ISBN 978-88-470-1151-9

2. A. Quarteroni
Numerical Models for Differential Problems
2009, XVI+602 pp, ISBN 978-88-470-1070-3

3. M. Emmer, A. Quarteroni (eds.)
MATHKNOW
2009, XII+264 pp, ISBN 978-88-470-1121-2

4. A. Alonso Rodríguez, A. Valli
Eddy Current Approximation of Maxwell Equations
2010, XIV+348 pp, ISBN 978-88-470-1934-8

5. D. Ambrosi, A. Quarteroni, G. Rozza (eds.)
Modeling of Physiological Flows
2012, X+414 pp, ISBN 978-88-470-1934-8

6. W. Liu
Introduction to Modeling Biological Cellular Control Systems
2012, XII+268 pp, ISBN 978-88-470-2489-2

7. B. Maury
   The Respiratory System in Equations
   2013, XVIII+276 pp, ISBN 978-88-470-5213-0

8. A. Quarteroni
   Numerical Models for Differential Problems, 2nd Ed.
   2014, XX+656pp, ISBN 978-88-470-5521-6

9. A. Quarteroni, G. Rozza (eds.)
   Reduced Order Methods for modeling and computational reduction
   2014, X+332pp, ISBN 978-3-319-02089-1

For further information, please visit the following link:
http://www.springer.com/series/8377

FSC
www.fsc.org
MIX
Papier aus verantwortungsvollen Quellen
Paper from responsible sources
FSC® C105338

If you have any concerns about our products,
you can contact us on
**ProductSafety@springernature.com**

In case Publisher is established outside the EU,
the EU authorized representative is:
**Springer Nature Customer Service Center GmbH**
**Europaplatz 3, 69115 Heidelberg, Germany**

Printed by Libri Plureos GmbH
in Hamburg, Germany